Süßwasserflora von Mitteleuropa

Band 1 Chrysophyceae und Haptophyceae (

Band 2/1 Bacillariophyceae (Naviculaceae) (

Band 2/2 Bacillariophyceae (Epithemiaceae, Bacillariaceae, Surirellaceae) (1988, Nachdr. 1997)

Band 2/3 Bacillariophyceae (Centrales, Fragilariaceae, Eunotiaceae) (1991)

Band 2/4 Bacillariophyceae (Achnanthaceae. Kritische Ergänzungen zu Navicula (Lineolatae) und Gomphonema. Gesamtliteraturverzeichnis für Teil 1–4) (1991)

Band 3 Xanthophyceae 1. Teil (1978)

Band 4 Xanthophyceae 2. Teil (1980)

Band 5 Cryptophyceae und Raphidophyceae

Band 6 Dinophyceae (Dinoflagellida) (1990)

Band 7 Phaeophyceae und Rhodophyceae

Band 8 Euglenophyceae

Band 9 Chlorophyta I (Phytomonadina) (1983)

Band 10 Chlorophyta II (Tetrasporales, Chlorococcales, Gloeodendrales) (1988)

Band 11 Chlorophyta III (Chlorophyceae p. p.: Chlorellales, Protosiphonales)

Band 12 Chlorophyta IV (Chlorophyceae p. p.: Stichococcales, Microsporales; Codiolophyceae: Ulotrichales, Monostromatales)

Band 13 Chlorophyta V (Chlorophyceae p. p.: Chaetophorales, Trentepohliales, Chlorosphaerales)

Band 14 Chlorophyta VI (Oedogoniophyceae: Oedogoniales) (1985)

Band 15 Chlorophyta VII (Bryopsidophyceae: Cladophorales, Sphaeropleales)

Band 16 Chlorophyta VIII (Conjugatophyceae I: Zygnemales) (1984)

Band 17 Chlorophyta IX (Conjugatophyceae II: Zygnematales: Mesotaeniaceae; Desmidiales)

Band 18 Charales (Charophyceae) (1997)

Band 19 Cyanophyceae

Band 20 Schizomycetes (1982)

Band 21 Mycophyta (Phycomycetes, Fungi imperfecti, Lichenes etc.)

Band 22 Bryophyta

Band 23 Pterido- und Anthophyta 1. Teil: Lycopodiaceae bis Orchidaceae (1980)

Band 24 Pterido- und Anthophyta 2. Teil: Saururaceae bis Asteraceae (1981)

Addresses of the authors:
Dr. Kurt Krammer (Hindenburgstr. 26a, 40 667 Meerbusch, Germany)

Professor Dr. Horst Lange-Bertalot (Abtlg. für spezielle Botanik, Botanisches Institut, Siesmayerstr. 70, 60 054 Frankfurt am Main, Germany)

Addresses of the translators:
Nina Bate (Environment Protection Authority, 40 City Road, Southbank, Victoria 3006, Australia; GPO Box 4395QQ, Melbourne, Victoria 3001, Australia)

Andrew Podzorski (Hessenplatz 6, 60 487 Frankfurt am Main, Germany)

Jeanne Bukowska, Monika Michel, Dr. Jean Prygiel (Agence de l'Eau Artois-Picardie, 200 rue Marceline, Centre Tertiaire de l'Arsenal – B. P. 818, 59 508 Douai Cedex, France)

Addresses of the editors:
Professor Dr. Burkhard Büdel (Fachbereich Biologie, Abteilung Allgemeine Botanik, 13/274, 67 653 Kaiserslautern, Germany)

Dr. Georg Gärtner (Institut für Botanik, Sternwartestr. 15, 6020 Innsbruck, Austria)

Dr. Lothar Krienitz (Institut für Gewässerökologie und Binnenfischerei, Alte Fischerhütte, 16 775 Neuglobsow, Germany)

Dr. Gijsbert M. Lokhorst (Rijksherbarium Leiden/Hortus Botanicus, P. O.Box 9514, 2300 RA Leiden, Holland)

Die Deutsche Bibliothek – CIP-Einheitsaufnahme

Krammer, Kurt:
Bacillariophyceae / Kurt Krammer ; Horst Lange-Bertalot. – Heidelberg ; Berlin : Spektrum, Akad. Verl.
(Süßwasserflora von Mitteleuropa ; Bd. 2)
Part 5. English and French translation of the keys / Engl. transl. by Nina Bate ; Andrew Podzorski. French transl. by Jeanne Bukowska . . . – 2000
ISBN 3-8274-1030-4

Satz: Typomedia Satztechnik GmbH, Ostfildern
Druck und Verarbeitung: Franz Spiegel Buch GmbH, Ulm

Bacillariophyceae

Part 5: English and French translation of the keys

Kurt Krammer
Horst Lange-Bertalot

English translation by
Nina Bate (keys)
Andrew Podzorski (general part)

French translation by
Jeanne Bukowska · Monika Michel
Jean Prygiel (keys)

Spektrum Akademischer Verlag Heidelberg · Berlin

Preface of the authors

We have compelling reason to publish this volume 2/5 containing all determination keys translated into English and French versions.
Even though already planned it may take a while to realise a far-reaching novel concept of the diatom flora in English. The reasons for that are obvious:

- A professional description in English has revealed to be more complex than expected by the authors. Frankly, we have underestimated this problem.
- The rapid progress in the knowledge of taxonomy effects nomenclature and classification. The possibility of re-interpreting the microstructures of diatoms via scanning electron microscopy mainly results in a change of concepts.
- It may evoke didactic problems to transfer the abundance of knowledge in the flora to the reader and user in an overall comprehensive manner without being misleading.

The main problems are described in the following two sections: firstly, we will focus on the linguistic problem and how to solve it. Secondly, we will describe the taxonomic problem using a modified and supplemented version of the preface of the reprinted volume 2/1 from 1997.

Why did we write the volumes 2/1–4 dealing with diatoms in German and not in English?

Native speakers of excellent English may hardly imagine how difficult it is for a general biologist with another mother tongue to write in (good) English. In spite of having learnt English at school for several years and despite being more or less experienced in conversation with native speakers of English in the UK and elsewhere, our written English remains "poor English" – and is despite of these facts consuming a considerable amount of our restricted working time.
You are kind to accomodate this "poor English" when we correspond by letter and by e-mail or during our presentations at congresses. On the other hand, it is relatively simple for us to read even your more complicated or sophisticated phrases in scientific papers. Difficulties arise however, when we try to offer more than very simple phrases to express more complicated ideas in discussions. It is also deflating to receive scientific texts returned from kind reviewers that are covered with corrections and suggestions for linguistic improvements. Furthermore, we emphasize the astonishing fact that reviewers from different English speaking nations, and individuals from the same nation, give conspicuously different suggestions for linguistic improvements.
A specific example: A diatomist from Italy revealed the comments of a reviewer who criticised harshly the linguistic (not the scientific!) quality of his submitted paper. The British reviewer could not have known that this paper had been translated from Italian to English by a friend who was born, educated and graduated in natural sciences in Boston, Mass. USA. Consistent with this verdict it can be no surprise that another reviewer disparaged the linguistic efforts of a professional translator from Poland as largely unacceptable. Indeed, an internationally acceptable form of English seems rather difficult to achieve; in fact there appears to be no common standard that we can follow.
May we ask: is it really necessary to provide such "generous" help which, ultimately, generates stress for a poor non-native English speaking phycologist?

In this context we know that, as professionals, foreign people who have studied English at university should still give their papers to native speakers for corrections – if they want to rule out linguistic shortcomings. This, however, is quite a different matter. Our fields of interest are far from the standards of professionals of the English language and literature. We want no more than to transfer some biological news to our small subset of the scientific community, and being understandable should be sufficient.

A few decades ago it was common for native English speaking diatomologists to read and understand the basic diatom literature, even if it was written in French, German, sometimes also in Danish, Italian or Spanish. Comparatively, in the past, fiction was also traditionally read in these original languages by the "educated classes". These linguistic abilities are not actually extinct, but you find them progressively less common amongst readers. Contrast this with the emphasis amongst European and American business professionals to learn other languages. Nevertheless, an author of belletristic or novelistic literature will hardly ever write in a language other than his mother tongue. For other languages there are – indeed – translators.

Why did we not ask an English native speaker for help? Searching for help invariably creates delays leading up to publication. From experience, we know all too well that one can rarely find anyone who fulfils all the requirements. Other than an obvious ability to write good English, such a person would need:

- A command of the German language to understand what we mean.
- A knowledge of the concepts and the specific terminology of diatom research.
- Enough time and interest to translate our texts or at least to revise them thoroughly in addition to their own occupational activity.
- Readiness to give this help, at short notice, throughout the period of time necessary for publication.

To find such a colleague who is competent and ready to give help is – as we have learnt – very difficult. Perhaps we will find someone who is willing to volunteer such help in response to this lamentation.

As the scientific community of our discipline what can we learn from this? We authors, with mother tongues other than English, will have to accept sooner or later that only this language will be read by the majority of colleagues. This means: English will become the "Lingua franca" of diatom research. So far this has not actually happened. In the past Latin was the "Lingua franca" in nearly every discipline of science, literature, the fine arts and Christian religions. It is used even today, for example, as required by the rules of the International Code of Botanical Nomenclature (ICBN). Diagnoses of new taxa of recently discovered plants still have to be written in Latin to be validly published. In the near future this rule will probably be replaced by the alternative of doing such in Latin or English as already exists for fossil plants. It is likely in the next future that all this literature will be in English. It may take a little longer for our colleagues either side of the Atlantic to decide on which form this English should take.

In our view, the native speakers (which do not exist anymore in the case of Latin) ought to accept similar conditions used in the application of Latin. More than a simplified form of English should not be required by the rules to be valid. This is so for the Latin used by taxonomists which was, and still is, a simplified form, far removed from the lingual quality of classical Latin. The classical Latinist would criticise this "Kitchen-Latin", that is written and understood generally by scientists, in the same way the English native speaker

criticises the "poor English" of colleagues. In spite of this, the "poor Latin", analogous to good Latin, is written in a quasi unconstrained way and without a great fear of mistakes – not least by our colleagues in the USA and elsewhere – and is accepted generally in that way. A representative of the subject of classical Latin may lament this, but it is a fact with a long tradition.
In German speaking countries the recent colloquial German of millions of youth, and the written and spoken media for that matter, is far from being good or at even correct German. Nevertheless, we understand what they mean and have to tolerate it. Is it not like the situation in English speaking countries?
We in the scientific community should really agree upon recognising a comparable and simplified English with much more flexibility within the meaning of a modern Lingua franca. The considerable stress for a non-native speaker could be reduced and no serious problems for a native speaker would arise. This means that the latter should be ready to tolerate a simplified and perhaps a somewhat curious but understandable form of English, even in publications. Of course the reviewer should ensure that the "poor English" is unambiguous, but by accepting the simplified form the reviewer is free to focus more carefully on the science. Let us hope for such, as energy which is being consumed by the stress of linguistic efforts during translation can be used to promote the quality and quantity of scientific achievements.
The authors wish to cordially thank the expert colleagues, who translated text and keys from German to English, Nina Bate and Andrew Podzorski, as well as to French, Jeanne Bukowska, Monika Michel and Jean Prygiel.

Genera – Species – Names

14 years have passed since volume 2/1 of the "Süßwasserflora" was published – is it still up to date? The previous edition of Hustedt (1930) was still in use 50 years later, sought after in antiquarian bookshops and also reprinted. Over such a long period it inevitable became less actual but until the new Süßwasserflora 1986–1991 remained important as a flora for the identification of diatoms and also as a reference. Consequently this much more modern flora can be considered to be still sufficiently current. The reader will be able to find adequate answers to most taxonomic and ecological questions. However, it has to be remembered that the number of known taxa has increased considerably in the meantime. These are not simply names for different forms of appearance within cell cycles, but biological or at least statistically supported groups. These "Sippen" often have a special indicative value. Interested readers can find them in the following list of publications, which have been released since 1986. Since publication of the first edition, a lot more species are also known especially within the genera *Navicula* (s.str.), *Cymbella*, *Pinnularia* and *Gomphonema*. Regarding some new groupings from the older genera the number of known species has more than doubled. Complex and didactically difficult recording is still needed to put such a lot of species in a sensible, generally understandable and still clear order in a new revised flora for identification. This will delay the release date of the revision even further. It is comparatively easy to distinguish that taxa which in volumes 1–4 are still often combined to "Sippenkomplex"; they guarantee better clarity, however, with the concession of less differentiation and therefore less accuracy. A quite different problem are suggestions of new assignments to genera which can hardly be assessed meanwhile. Their value or advantage for the practice of identification requires critical judgement. Some seem to be convincing improvements. Others are more or less "neutral" and one can, provided sensible reasons are given, agree with them. Just as well,

with no disadvantage one can do without them. Another group can be seen as "pastime" of their authors because they tend to sort out single features rather than species to sensible genera. The trend of the last 10 years was leading to a large number of new genera, often monotypic "mini"-genera, which do not solve any problems. In some cases it is the historical random choice of nomenclatorial type species (typus generis), which then is followed by complex new ways of combinations. Some author's opinions is that genera – as with species – have to be by law of nature "forced" units. Consequently this opinion is inevitably leading to an increasing number of more and more and narrower and narrower defined genera sharing all features with their nomenclatoric type species (typus generis). The publications cited below include ongoing discussions of this question on which the clarity and the relative stability of the names depends to a large extent.

For the reader or user of the "Süßwasserflora" the species will be important in the first place rather than controversial discussions regarding the genera in which the species should have to be placed. In our opinion the (older) synonyms for the designation of a species are at least as precise as the "modern-fashionable" combinations of names. Nobody should feel forced to follow the new suggestions, at least not at this early stage. This applies even in cases in which a revision is generally sensible. Breaking off single groupings to new genera always gives grounds for new splittings with different quality. Basically a general concept was missing right from the beginning of this new development. The formal realisation of the splitting was started far too early.

The user of this diatom flora would be well advised to wait for the further development and – nomenclatorially and taxonomically speaking – for a hopefully not chaotic preliminary end of the discussion.

Limnologists, biologists and geographers can, in the meantime, still obtain adequate information from this flora.

Correct or allegedly incorrect generic classification – a problem not only for diatoms

The primary decision: one would like to establish a new species or a new infraspecific taxon. The rules of the ICBN require a binary combination. Therefore, the new species has to be assigned to a certain genus. If possible this should represent the "correct" and "final" name. Frequently, it appears to be ambiguous which of the established genus names should be the correct one. The name should rely on the latest version of taxonomic classification. However, the diatoms are far beyond a finalised version of genus classification. To date, we know two prominent species of Pteridophytes from the holarctic: beech fern and oak fern. Since Linné, their species names have been combined with 11 and 9, respectively, genus names. And who knows whether the currently used combinations are yet the final ones? Typically, the popular names have never changed.

A closer look to seed plants may be informative. About 25 (heterogeneous) genera have been sufficient to assign the nearly 2000 species among the cactuses. Subsequently, the genera have been increased to a number of 200. Today, only 92 are accepted. This leads to the conclusion that, in contrast to the species category, genera are weak units.

If, strictly speaking, a newly described species does not belong anymore to *Navicula*, for instance, this combination can not be considered any longer. A different binary term has to be found. This may include the use of an already

existing genus or the establishment of a new genus. Nevertheless, to some extent it is ambiguous if a specific decision turns out to be correct. In principle, it is possible to establish a new genus for each new species, in case none of the existing genera are suitable. We do not favour this procedure though.
The description of the distinct species is of primary importance. The potential name of the genus is of secondary importance and depends on the latest state of classification. However, the particular identification of the species has precedence over the genus.
The current trend is named "generomania" a term that is not necessarily exaggerating. We also have to follow this trend to some extent, in order not be considered as conservative. The search for new genera constantly brings about new name combinations for the same old taxons. However, we regret that there is little effort to search for new unknown species of diatoms, of which – according to the literature between – 200,000 and 10,000,000 still may exist (this numbers are unverifiable probabilities).

Introductory remarks to Volume 2/5

In contrast to vascular plants, the species composition of diatoms distributed all over the world show a high degree of similarity. Many diatom associations in South Chile or Australia differ only slightly from those in Europe, assuming the respective biotopes possess similar chemical and physical properties. This is the main reason why the volumes 2/1–2/5 on Bacillariophytes from "Freshwater Flora in Central Europe" are world-widely used by biologists, ecologists, hydrobiologists and geologists to identify diatoms. In addition, these volumes are the most comprehensive literature that is currently available to identify diatoms. Meanwhile, translations, particularly of the identification codes, circulate between diatom researchers of different countries. The present volume shall provide access to the translated (in English and French) identification codes for all interested researchers.
The content of volumes 2/1–2/4 does not result from the order described in "Classification of Bacillariophytes", vol. 2/1, p.79, 80, but does result solely from practical considerations when paying attention to the families: Foremost, we published those families that were investigated at the best. However, in the present volume we could follow the given order of the classification (vol. 2/1, p.79) when describing the codes in English and French. At first, we describe the Centrales and at second the Pennales with their two subsidiarities Araphidineae and Raphidineae. For a better overview it is generally noted in which volumes each family and genus are described. Furthermore, we always denounce the volume numbers when we refer to figures (e.g. vol. 2/1 Fig. 87:2). Therefore, the present volume represents also a detailed table of contents for the volumes 2/1–2/4.
Since the publication of "Bacillariophytes" comprehensive revisions have been made available for some genera, like, for instance, for the genera *Navicula* in a narrow sense (in volume 2/1 Naviculae lineolatae), *Pinnularia* and a bigger part of the genus *Cymbella* (see: following literature overview). The most important insights resulting from these adaptations are quoted in the footnotes for the genera *Cymbella* and *Pinnularia*. Others were interspersed in the codes for *Navicula*. Many other groups were also revised and supplemented (e.g. Achnanthaceae, Eunotiaceae, the genus *Aulacoseira*), though there were only few sites in the codes to add the supplements.
To make the user of the flora familiar with the terminology, the morphology of the frustule the biology of diatoms, the methods to prepare the frustules and the

taxonomical concept, the identification codes are supplemented with translated versions of the most important chapters from the general part of vol. 2/1. The mentioned supplements are not included in the French classification codes. The reader should consult the English classification codes. Therefore, we have marked those French codes with a (*) that have enclosed a footnote in the English codes.

Literature in addition to the literature list in vol. 2/5

For the reader with particular interest in the advances of taxonomy we recommend the publications listed below containing citations of the newer literature.

Fourtanier, E. & Kociolek, J. P. (1999): Catalogue of the diatom genera. Diatom Res. **14:** 1–190.

Hofmann, G. (1994): Aufwuchsdiatomeen in Seen und ihre Eignung als Indikatoren der Trophie. Bibl. Diatomologica 30:1–239.

Krammer, K. (1992a): *Pinnularia.* Eine Monographie der europäischen Taxa. Bibl. Diatomologica **9**: 1–353.

– (1992b): Die Gattung *Pinnularia* in Bayern, Hoppea **52:** 1–308.

– (1997a): Die cymbelloiden Diatomeen. Eine Monographie der weltweit bekannten Taxa Teil l. Allgemeines und *Encyonema* part. Bibl. Diatomologica **36**: 1–382.

– (1997b): Die cymbelloiden Diatomeen. Eine Monographie der weltweit bekannten Taxa Teil 2. *Encyonema* part., *Eucyonopsis* and *Cymbellopsis*. Bibl. Diatomologica **37**: 1–469.

– (2000) *Pinnularia.* In: Lange-Bertalot (ed.) Diatoms of Europe, Diatoms of European Inland waters and comparable habitats. **1:** 1–702.

Lange-Bertalot, H. (1993): 85 neue Taxa und über 100 weitere neu definierte Taxa ergänzend zur Süßwasserflora von Mitteleuropa. Vol. 2/1–4. Bibl. Diatomologica **27**: 1–454.

– (1995): *Gomphosphenia paradoxa* nov. spec. et nov. gen. und Vorschlag zur Lösung taxonomischer Probleme infolge eines veränderten Gattungskonzepts von *Gomphonema*. Nova Hedwigia **60**: 241–252.

– (1996): Rote Liste der limnischen Kieselalgen Deutschlands. Schr.-R. f. Vegetationskunde **28**: 633–677.

– (1996): *Kobayasia* gen. et spec. nov. Iconographia Diatomologica **4:** 277–287.

– (in preparation): *Navicula* s. str. and *Navicula* s. lato. In: Lange-Bertalot (ed.) Diatoms of Europe, Diatoms of European Inland waters and comparable habitats. **2:**

Lange-Bertalot, H. & Genkal, S. I. Diatoms from Siberia I. Iconographia Diatomologica **6:** 1–295.

Lange-Bertalot, H., Külbs, K., Lauser, T., Nörpel-Schempp, M. & Willmann, M. (1996): Dokumentation und Revision der von Georg Krasske beschriebenen Diatomeen-Taxa. Iconographia Diatomologica **3**: 1–358.

Lange-Bertalot, H. & Metzeltin, D. (1996): Oligotrophie-Indikatoren, 800 Taxa repräsentativ für drei diverse Seen-Typen, kalkreich – oligodystroph – schwach gepuffertes Weichwasser. Iconographia Diatomologica **2** :1–390.

Lange-Bertalot, H., Metzeltin, D. & Witkowski, A. (1996): Hippodonta gen. nov. Umschreibung und Begründung einer neuen Gattung der Naviculaceae. **4**: 247–266.

Lange-Bertalot, H. & Moser, G. (1994): Brachysira, Monographie der Gattung. Bibl. Diatomologica **29**: 1–212.

Metzeltin, D. & Lange-Bertalot, H. (1998): Tropical diatoms of South America I. Iconographia Diatomologica **5**: 1–695.

Metzeltin D., & Witkowski, A. (1996): Diatomeen der Bären-Insel. Iconographia Diatomologica **4**: 3–233.

Moser, G. (1999): Die Diatomeenflora von Neukaledonien, Systematik, Geobotanik, Ökologie, ein Fazit. Bibl. Diatomologica 43: 1–205.

Moser, G., Lange-Bertalot, H. & Metzeltin, H. (1998): Insel der Endemiten. Geobotanisches Phänomen Neukaledonien. Bibl. Diatomologica **38**: 1–464.

Moser, G., Steindorf, A. & Lange-Bertalot, H. (1995): Neukaledonien, Diatomeenflora einer Tropeninsel, Revision der Collection Maillard und Untersuchung neuen Materials. Bibl. Diatomologica **32:** 1–340.

Reichardt, E. (1994): Zur Diatomeenflora tuffabscheidender Quellen und Bäche im südlichen Frankenjura. Ber. Bayer. Bot. Ges. **64**: 119–133.

– (1995): Die Diatomeen in Ehrenbergs Material von Cayenne, Guyana Gallica (1843). Iconographia Diatomologica **l**: 1–107.

Reichardt, E. & Lange-Bertalot, H. (1991): Taxonomische Revision des Artenkomplexes um *Gomphonema angustum – G. intricatum – G. vibrio* und ähnliche Taxa. Nova Hedwigia **53**: 519–544.

Reichardt, E. (1999): Zur Revision der Gattung *Gomphonema.* Iconographia Diatomologica **8**: 1–203.

Round, F., Crafword, R. & Mann, D. G. (1990): The Diatoms. Biology & Morphology of the genera. Cambridge, Univ. Press, 747 p.

Witkowski, A. (1994): Recent and fossil Diatom flora of the Gulf of Gdansk, Southern Baltic Sea. Bibl. Diatomologica **28** :1–313.

Witkowski, A., Lange-Bertalot, H. & Metzeltin, D. (2000): Diatom flora of marine coasts I. Iconographia Diatomologica 7: 1–950.

Contents

I. General part

This is a partial translation of the general part in the vol 2/1 of this diatom flora. All fig.-numbers in the general part refer likewise to vol. 2/1.

1 Delimitation of the Bacillariophyceae

The Bacillariophyceae are characterised above all, by the following:
1. They are singled-celled, and are either free-living, bound by mucous to a substrate, form colonies bound by mucous threads, or form band-shaped colonies.
2. The cell-wall is in the form of a rounded or linear box, consisting of a box-bottom (hypotheca) and a box-lid (epitheca) of almost equal sizes.
3. Silicon dioxide (silicic acid) in varying degrees of hydration is the main cell-wall component. In addition to this, a thin organic membrane of diatotepin, an acidic polysaccharide, has been found on the cell-walls of almost all diatoms investigated.
4. The chloroplasts have a characteristic brown colour resulting from the brown xanthophyll fucoxanthin overlying the a and c chlorophylls.
5. Energy is stored in the form of the polysaccharide chrysolaminarin and drops of fat.
6. Ciliated cells, where the flagellum is oriented in the direction of motion, are found in the male gametes of the centric diatoms.
7. With few exceptions, they are diploid.
8. Most species have a raphe, an organelle, unique to the Bacillariophyceae, used in gliding over suitable substrates.

The smallest species found so far have a diameter of c. 2.5 µm and belong to the centric diatoms. The largest species have diameter resp. length of over 2 mm. Diatoms live not only in fresh- and salt-water, but also in soil and even in relatively dry situations, e.g. on rocks and walls, or under translucent quartz stones in deserts. In principal, they can live anywhere with sufficient moisture and light for photosynthesis. Planktonic species predominate in the oceans, seas and large freshwater lakes, where they form an important food-source for small shrimps and other life-forms. Smaller inland lakes also have planktonic species, but the majority live as epiphytes, epiliths or as epipelics, as in the littoral zones of the sea. Recent diatoms, as well as fossil and subfossil diatoms are of great importance. Subfossil diatoms occur on the beds of all large bodies of water in the form of whole or broken frustules of planktonic and benthic diatoms. Fossil diatoms, of either fresh- or saltwater origin, are found in deposits of Kieselgur or diatomite all over the world. They are not only an important raw material for industry, but are also stratigraphic indicators in geology.

Reproduction is predominantly asexual, through mitotic division. Sexual reproduction in the pennate genera occurs when two almost equally developed haploid gametes join to form a zygote, which then develops into an auxospore. In the centric genera, all investigated species have proved to be oogamous, with a large auxospore developing from the fertilised egg.

2 Terminology

The terminology used with diatoms is based on LM and EM investigations. As will be shown in the section on methods, many of the structures observed under LM (abbreviations will be explained in the glossary) do not represent actual morphological features. This is because objects smaller than 2 µm even with the best optics, cannot be resolved correctly, and are only "represented". Two neighbouring structures may thus be shown as being separate, although not details of the individual structures can be made out. Very many older terms should thus not be interpreted in a literal sense, but in connection with a particular definition. Here are two examples:

Raphe comes from ancient Greek and means "seam", a term normally implying the growing or bringing together of two similar, or nearly similar parts. The raphe however, as we know, is not a union, but a more or less complexly formed slit.

During sexual reproduction, two haploid gametes unite to form an "auxospore". The term has nothing to do with the sexless gametes (spores) of other lower plants. The term auxospore is however so established, that it will remain in modern literature.

In contrast, with the new terms introduced in this book, an attempt is made to reconcile meaning of the name with the definition.

It is still an open debate, as to whether or not traditional terms used in light-microscope diatom-diagnoses can be brought into line with the structure descriptions, as is the case in EM research. A notable example of this is seen in the transapical striae of pennate diatoms, which actually look like lines under the LM in most diatoms (fig. 23: 3). Under the SEM however, outside views of the valve reveal one or more rows of dot- or slit-shaped holes (foramina) are seen, whereas inside views show highly diverse structures of quite disparate morphology. According the level of focus in LM, the minimal depth of focus at high apertures will reveal either a light microscopic representation of the foramina, the underlying areolae, or perhaps only the border between the rib elements. The term transapical striae in LM thus encompasses so wide a range of actual morphological features, that in only a few cases does it describe the true underlying structure. The terms used in this volume will thus as far as possible be reconciled with the actual structures as seen under LM, and new terms introduced where deemed necessary.

An exhaustive terminology for the structural elements of the silicified cell-wall is found in Ross et al.(1979). The terms in the following glossary are drawn for the most part from the latter work, but with a sharper delimitation between the terms for LM and SEM. New terms were required to describe the fine structure of some pennate diatoms, and general terms were used in a somewhat broader sense.

2.1 Glossary

Abbreviations – The following abbreviations were used: str. = transapical striae, LM = light microscope, EM = electron microscope, REM (SEM) = scanning electron microscope, TEM = transmission electron microscope.

Aerophilic – regularly found out of water-bodies, e.g. occurring in the air-water interface, and (or) temporarily living in dry habitats (moss, water-doused rocks, moist earth, water-meadows, pluvial habitats).

Alveoli – lat. depression, furrow. Trough-shaped, transapical depressions on the inner side of the valve. Apically, they are delimited by the transapical ribs, and towards the valve edge by one or more rows of areolae. On the inside they are either completely open (figs 5: 1–8), or as in many *Pinnularia* species, in all species of *Caloneis* and *Gomphoneis*, partly occluded by an inner wall of alveoli (figs 7: 3–6). Ross *et al.* (1979) use the term alveolus only for the latter structure.

Annular ledge – a pseudoseptum opposite the sulcus furrow on the inner side of the valve probably stabilising the circular cross-section of the valve. It is either built solely from the sulcus furrow (*Aulacoseira alpigena, A. ambigua*), or is an annular ridge s. s. that may extend deep into the cell lumen (*A. distans*). Geometrically, it always has the shape of an annulus. Annular ledges are also found in species of other genera (e.g. *Cyclotella*), although never in association with a collar or sulcus furrow.

Anulus – a hyaline ring in the centre of the valve, often containing a few areolae in *Cyclotella, Cyclostephanos* and *Stephanodiscus.*

Apical – lat. apex = point. In pennate diatoms it refers to the cell poles.

Apical-axis – the longitudinal axis in pennate diatoms (figs 1: 1–5).

Apical furrows – furrows on both sides of the raphe and lying within the median costa.

Apical plane – a plane at right angles to the valve surface and running along the apical axis (fig. 1: 2).

Apical pore-fields – group of pores (see later) at a pole (e.g. *Gomphonema* fig. 11:6) or at both valve-ends (e.g. many araphid genera (figs 11: 1, 3) and *Cymbella* sensu stricto (figs 11: 5, 7, 8)). Those pores not closed by vela produce a secretion which solidifies into fine threads with which the cells are fixed to an appropriate substrate.

Areolae – chamber-forming perforations, rounded to angular in cross-section, in the valve wall. They are closed either on the outside, or inside, by a velum (figs 4: 5, 6; fig. 5: 1–8).

Areolar foramen (foramen) – Perforation in the outer side of an areola-type that is usually much smaller than average due to the bounding foramen lips. In many cases a thin velum is stretched across the inside of the areolar foramen, but this velum lies more commonly across the inner opening of the areola. The form of the areolar foramen and it's position is species specific, and thus an important taxonomic character. Only large foramina can be accurately studied in LM, and SEM examination is usually needed to see precise details. Smaller foramina appear only as "puncta" or "lineolae" (see later) under LM. During preparation sometimes not only the fine sieve-membranes, but also the foramina lips may be destroyed. Many SEM- and TEM-micrographs thus do not show the true structure of foramina, but corrosion artefacts.

Areolar rows – transapical rows of areolae.

Auxospore – cell (zygote) arising from the sexual fusion of two haploid gametes. The resulting cell grows to many time the size of the mother cell, and has a weakly silicified cell-wall with a structure that differs markedly from that of the vegetative cells (perizonium figs 24: 1–2).

Axial area – in pennate diatoms this is an areola-free zone on either side of the apical axis (fig. 1: 4).

Axial ribs – strongly developed apical ribs parallel to both raphe branches (fig. 13: 4c). They can be on the inside (*Frustulia rhomboides,* figs 4: 3–4) or on the outside (*Brachysira vitrea*, fig. 15: 6) of the valve. The axial ribs are not identical with the median costa (= sternum), which runs along the entire length of the apical axis.

Bands (segments) – single elements of the cell girdle. In the centric and araphid

diatoms they can occur in such numbers as to become the dominant character of the frustule.
Bilaterally-symmetrical – a term describing the symmetry in plants, animals or organs resp. organelles, in which two mirrors-images are formed when the structure is bisected. The opposite is radially symmetrical.
Bundled areola rows – several areolae rows lying close together by each alveolus (e.g. *Stephanodiscus*).
Calyptra – the candle-snuffer shaped valve in *Rhizosolenia*.
Canal raphe – a raphe whose fissure is in a canal-shaped hollow linked to the inner cell through a slit.
Canopy (conopeum) – a thin sheet of silica attached to the axial area. It covers a small, in extremes the entire, areolate portion of the valve (figs 17: 3–4).
Carinal dots – see fibulae
Carinoportulae – the connecting structures lying in the valve middle of some species of *Orthoseira* (Crawford 1981).
Cell wall, silicified – identical with frustule.
Central – the valve middle.
Central area – hyaline area in the valve middle. In some cases this is identical with the central nodule. Frequently there is no visible border between the central and the axial areas, and the two thus form a single hyaline zone.
Central fissure – superficial notch at the proximal ends of many median raphes. It's form is species specific and therefore an important taxonomic character (figs 15: 1–5).
Central nodule – a nodular thickening, of varying degree and extent, of the median costa within the central area (fig. 1: 5). In extreme cases the central nodule reaches the valve margin, and is then termed a stauros (fig. 14: 8).
Central pore – a rounded extension of the raphe-slit at the proximal raphe ends (fig. 15: 1–5). Central pores run normally perpenolicular to the valve surface, and seldom at an angle.
Central costa – see median costa.
Chamber – a general expression for a round or elongated hollow in the valve (generally an areola).
Chitinous setae – flexible, often very long, setae issuing usually from marginal or occasionally from central strutted processes.
Cingulum – part of the cell girdle associated with a single valve.
Coating membrane (diatotepum) – covering of organic material on the inner and outer sides of the silica cell wall (not a biomembrane like the plasmalemma).
Costate fibulae<D> – solid, transapical costae on the outer surface of the valve (Tryblionellae).
Collar – an annular sail around the margin of the valve surface in *Melosira nummuloides*.
Colliculate structures – structures having numerous, regularly spaced hollows and protrusions in the valve middle in some *Cyclotella* spp.
Collum – areola-free area at the distal ends of the mantel of *Aulacoseira* spp.
Complex raphe – an LM term for raphes where the "slot and key" alternate one or more times in the same valve half, where the direction of invagination of the raphe changes from valve-side to valve-side one, or more times in a single valve half (Fig. 13:7). Typical complex raphes are only found in a few of the large *Pinnularia* species.
Conspecific – identical with another taxon at the species level.
Convergent – striae that radiate towards the terminal nodules. In older literature the term divergent has also been used.

Copulae – mostly open elements of the cell girdle, set aside from the other girdle elements by their structure. Copulae often contain septae.
Corona – a structure formed from a ring of small connecting spines on the valve surface as in *Melosira nummuloides*.
Costae – longitudinal thickenings of the valve.
Costate fibulae – fibulae that have expanded to transapical walls (Paddock & Sims 1977; the primary costae of Schoeman & Archibald 1976).
Craticula – a strongly silicified rib-system within the frustule. According to Schmid (1975) it is a specialised form of the *valvae internae* (inner valve), resulting from increased osmotic pressure (*Navicula cuspidata*, fig. 24: 5).
Cribrum – evenly structured sieve membrane (hymen, velum) occasionally divided by silica connecting bars.
Cylindrical – centric diatom valves with a high mantle.
Diatotepum – part of the cell membrane (see later), consisting of acidic carbohydrate, that has a critical role in keeping the individual elements of the frustule (von Stosch 1981). The diatotepum is equivalent to the pectin-membrane of earlier literature.
Discus – valve surface in chain-forming centric diatoms, e. g. *Melosira, Aulacoseira*. The valve surface often has a marginal ring of connecting spines, occasionally with very small additional spines on the surface.
Discus-shaped – valves with a short mantel
Distal – distant from the middle, towards the end (opposite to proximal).
Dorsal – in diatoms that are asymmetrical along the apical axis, it is the side whose outer margin is more convex. The other side is the ventral side.
Dorsiventral – frustules in which dorsal and ventral sides can be distinguished.
Drum-shaped – valves of medium height in the pervalvar plane.
Electrolyte content – in the waterbodies considered here, the electrolyte content and conductivity mostly are well correlated, so that conductivity measurements reflect the electrolyte content well. The electrolyte content descriptions refer roughly to the following conductivities:

Very poor in electrolytes	– less than 50 µS/cm
Poor in electrolytes	– 50–100 µS/cm
Average electrolyte content	– 100–500 µS/cm
Rich in electrolytes	– over 500 µS/cm

Examples of electrolyte rich waters are, e.g. springs, the chalk-rich lakes of the Balkans (500–1000 µS/cm), and especially the salt-rich inland waters with sometimes over 1000 µS/cm.
Epicingulum – girdle component of the epitheca (see later).
Epitheca – epivalve + epicingulum. The larger valve with associated valve components that lies over the hypotheca, like a lid on top of a box. (fig. 1: 5).
Epivalve – the upper valve, lid-top of the epitheca. It consists of the valve face and that part of the valve, the mantle, that connects with the girdle (fig. 1:5).
Fasciculate striae – in centric species these are stria bundled together in sectors on the valve surface.
Filiform raphe – by small, but also sometimes by larger species, the raphe slit appears only as a fine line as a result of the resolution limits of LM.
Fascia – areola-free transapical bar at the valve middle. The valve surface over a stauros is always a fascia.
Fibulae (carinal dots) – support in the form of a silica strut, bridging the raphe-bearing keel on the inner side of the valve in many species with a canal raphe. The fibulae can end in one or more transapical striae, and be either solid, tubular or be flattened (fig. 14: 1).

Fenestra – external opening in the wings of *Surirella* (fig. 14: 6). The fenestrae are bordered by struts on their sides.
Fibular ribs – transapical ribs that bridge the raphe canal as fibulae on the inner side of the raphe keel. The fibular ribs often form transapical walls.
Fibulae wall knobs – knobs or button-like structures on the fibulae walls of almost all *Denticula* species studied. They lie either along the median line or on the raphe side.
Fimbrae – silica bridges between the valve and the valvocopula of the raphe-valve of *Cocconeis* species (Holmes et al. 1981). It is possible to distinguish between large second order bridges that appear to be "glued on", and small, regularly arranged primary bridges. Often, only the latter are present.
Folds – wave-like depressions as seen in transapical valve cross-section. At least apically they stretch along the entire length of the valve, on the outer side
Foramen, foramina – latin for opening, hole. The term will be used in the sense of it's original definition (Reimann 1960), and not be more restrictively re-defined. A foramen is thus taken to be an opening in the outer wall or chamber side-walls. The openings in the girdle are also foramina. When foramen is used without any qualifier, the areola foramen (see earlier) is implied. The other type of foramen (stigma foramen, pore foramen, girdle foramen) are defined separately. A valve surface with foramina is perforated, in contrast to the hyaline structural elements without foramina. Thin sieve-membranes can be stretched across the foramina.
Foramen border – structures delimiting the foramina,
Form – where "form" is meant in the sense of the formal taxonomic rank of "forma", the latin word forma will be specified. Form, or forms, as they are commonly used in the international taxonomic literature, are otherwise used as "neutral terms" (Mayer 1975). It is thus not formal taxonomic rank to which a particular population or taxon can belong to. Example: The "turris"-forms and "gautieri"-forms of certain Gomphonemas are regarded by us as not being formal taxonomic groups, but just as borderline variations. In addition to this, the word "form" is used in several places in it's normal literary way, as can be seen from the context.
Frustule – the complete silicified cell-wall, consisting of the epi- and hypotheka (fig. 1:5).
Fultoportula – hollow processes on the outside of the valve (Thalassiosiraceae), normally as a marginal ring.; tubuli (tubes) with 2–5 closely associated structures ("satellite pores") that penetrate the valve wall. They can be arranged in a marginal ring and/or otherwise arranged on the valve surface. Their organisation and number (including their presence and absence) are held as important taxonomic characters.
Geobotanical notes – specific localities are only mentioned in the diagnoses in exceptional circumstances, but occur in the photograph legends. Other geobotanical terms used are: boreal – cold-temperate, continental, north temperate. Temperate zone – the regions between the tropics and the polar circles whose average temperature in the warmest month exceeds +10 °C, and doesn't drop below –18 °C in the coldest month. In this area – the region covered by this flora (Europe and surrounded areas). Cosmopolitan – distributed world-wide in similar ecological biotopes. Nordic – in Eurasia, the region north of latitude 60 °N. Nordic-alpine – presence entirely, or almost entirely, restricted to the climatic conditions of the boreal, subarctic or arctic regions, as well as those of upper montane areas. Both hemispheres are included in this definition.
Group – a taxonomically undefined series of similar species.
Girdle – collective term for all structural elements between two valves.

Girdle-band – general term for all open and closed bands (segments) of the cell-girdle, i.e. valvocopulae, intercalary bands.
Häutungen ... – the formation of valves within a vegetative cell. This does not produce resting spores (von Stosch in Ettl et al. 1967).
Helictoglossa – a lip-shaped structure terminating the distal end of median- and canal-raphes on the inner surface of the valve. In many Naviculaceae there is only a faint trace of it on the terminal nodule (figs 16:3–8).
Hollow-fluted septae – the septae struts in the Epithemiaceae develop normally as grooves that fit onto the fibular wall like a rebate.
Hypotheca – the smaller of the two thecas of a frustule (fig. 1: 5, box-inner). See also epitheca.
Hyaline – description of unperforated parts of the valve, i.e. those parts lacking foramina (opposite: perforated).
Imbricatio, imbrication line – zig-zag line on one side of the valve girdle, e.g. in *Acanthoceros* or *Rhizosolenia*. The line is formed by the decussate arrangement of the open ends of the intercalary bands.
Infundibulum – characteristic structure on the valve surface *Surirella* section Fastuosae. According to Paddock (1985), it is comprised of a bowl and a stem. See also "pseudoinfundibulum".
Inner valves – valves that form within the normal frustule of vegetative cells. Inner valves are found in resting spores.
Intercostae – the areolate and alveolate area between the transapical costae.
Intercostal ribs – costae-like elements between the transapical costae. They form the transapical walls of the areolae. In many species these costae are in straight or curved apical lines (e.g. Naviculae lineolatae = *Navicula* in the present-day concept), giving the impression of longitudinal costae in LM.
Intermissio - the internal central fissure in many naviculoid diatoms. It is a species specific inward extension of the proximal ends of the raphe branches (figs 7: 3; 124: 9–11).
Interspace – the area between two fibulae (figs 12: 6, 7).
Initial cell (sporangial cell) – the first vegetative cell arising from the auxospore. It has a completely silicified cell wall, often completely different in valve structure and three dimensional form from normal vegetative cells. In pennate diatoms, the cell is more or less sausage-shaped, lacking a clear border between the valve face and mantle. Teratological forms are commoner than with normal cells. The division products of the initial cell are initial daughter cells, and are intermediate in form and structure between the initial cell and normal vegetative cells (fig. 23: 3).
Interstriae – hyaline lines between the areola rows.
Isolated puncta – see stigma, stigmoid.
Key and slot raphe – a complex form of the raphe slit in many Naviculaceae and species with a canal raphe, whereby both valve halves are intermeshed (Fig. 112: 2). Some raphe fissures have a double keys and slots (fig. 13: 7).
Key costa – a ridge running apically with the raphe key.
Köpfchen – the LM impression of a hollow-fluted septa in girdle view.
Labiate processes (rimoportulae) – a tube- or similarly shaped perforation in the cell wall. It's inner opening is formed into an elongated slit, often surrounded by lip-shaped structures (centric and araphid diatoms, fig 25: 6, 7).
Lateral area – a hyaline area running parallel to the axial area, and often merging with the central area near the central nodule (Naviculae lyratae = *Fallacia* Stickle & Mann and *Lyrella* Karajeva). In many *Amphora* species it is developed on one side only (fig. 13: 6).
Lateral raphe – in the median raphe of most species, two or more lines are visible in LM. These imitate the course of the inner- and outer-fissures, and

additionally in some cases, the raphe key and slot. In all of these cases the term lateral raphe is used in diagnoses (fig. 12: 5).

Linking spines – mostly a large number of spines serving to link single frustules into a chain (figs 20: 4–5).

Loculate areolae – areolae within a two-layered cell-wall (a sandwich system). Generally, a foramen in the inner- or outer-wall has a substantially smaller cross-section than the lumen of the individual areolae. On one side there is always a cribrum formed from thickened silica structures.

Longitudinal bands (longitudinal lines or furrows) – structures outside the median costa, running apically, and visible in LM. Morphologically, they are of varying structure (e.g. inner alveoli openings in *Pinnularia* costae in *Oestrupia*, longitudinal canals in *Neidium*.

Ligula – a toothlike protuberance on an open girdleband or intercalary band serving to intermesh the individual valve-elements.

Lineolae – in LM, the foramina, or the underlying areolae appear as lines in some species, as puncta in others. Lineolae is used to describe the former structure (fig. 12: 5).

Linking spine – structures connecting valves in various centric and pennate diatoms that form chains. In *Aulacoseira* they can be used to delimit species. Taxa with broad autecological tolerances in relation to electrolyte concentration (e.g. *Skeletonema subsalsum*) have very long spines in high salt concentrations, and almost rudimentary spines in low concentrations. It is often possible to distinguish between the spine base and the spine anchor (see elsewhere). The often hollow spine base is either the only element of the linking spine, or has a distal extension (spine anchor). Separating spines (see elsewhere) are specialised linking spines.

Longitudinal canal – a tubular canal either in the median costa as in *Diploneis* (fig. 17: 2), or marginal as in *Neidium* (fig. 17: 1).

Mantle border – imagined line separating the valve surface from the valve mantle. In many cases it is clearly defined by right- or obtuse-angled valve edges, by a hyaline area, or by a change in the form and organisation in the foramina. There are however many valves (e.g. initial cells), where such differentiation does not exist.

Marginal – near the mantle edge.

Median costa (central costa, sternum) – all costae systems that lie along the median in pennate diatoms. Along it's middle it is dissected by the raphe-fissure in the Monoraphidae and all Naviculaceae. The following are components of the median costa: central nodule, the costae systems running along both sides of the raphe fissure, the terminal nodules, and the valve mantle at the poles (fig. 6: 3).

Median raphe (central raphe) – raphe systems that lie in the median costa (Achnanthaceae, Naviculaceae), and that consist of two raphe branches separated by the central nodule. The raphe fissures end directly inside the cell (fig. 1: 6), in contrast to a canal raphe.

Morphotype – a poorly defined and/or form of questionable stability. Although not formally recognised, it refers to those individuals occurring close to the middle of the character distribution curves.

Nodules – thickenings in the cell wall that are closely associated with the raphe system, and occurring at the valve ends (only at the valve end in Surirellaceae) (fig. 1:5).

Ocellulimbus - the ocellus in various *Fragilaria* species. There is a continuous gradation to poorly delimited apical pore fields.

Ocellus - an area sharply delimited by specialised structures, and often raised.

Orbiculus - a special apical structure in some species foramina the subgenus *Achnanthes.*
Papillae - wart-like, round protrusions in the middle of the valve surface in some centric diatoms. Similarly formed depressions lie between the papillae.
Parallel – striae running at right-angles to the apical axis. The term is imprecise as radial and convergent striae are often parallel.
Passage pore – perforations linking neighbouring areolae, or areolae with the longitudinal canals.
Perizonium – a weak, ring-shaped silicification of the auxospore (figs 23: 1, 2).
Pervalvar axis – the axis through the centre of both thecas (figs 1: 1–5).
Pleural band (pleura) – the distal band of the cell girdle.
Plicate raphe fissure – a simply structured raphe fissure, V-shaped in cross section (fig. 12: 1).
Polar – identical with terminal.
Pores – circular wall perforations without vela.
Poroid areolae - areolae in a single-layered cell wall. Foramina of specialised shape and vela are absent from both the inner and outer surfaces of the valve.
Portula – in canal raphes an opening in the inner wall of a raphe canal to the cell interior, or the inner resp. outer opening of the alar canal. (Paddock & Sims 1977) (fig. 12: 6).
Proximal – lat. *proxime* = the nearest. Positioned nearer to the cell middle. In the Naviculaceae, nearer the central nodule.
Pseudoinfundibulum – a simple variant of the infundibulum in *Surirella* section Pinnatae having only the bowl P, the stem being absent.
Pseudonodulus - marginal or submarginal areolate structure. Under LM, *Actinocyclus normanii* it appears as a bright dot at the edge of the radiate areolae.
Pseudoseptum – short, expanded transverse walls, generally parallel to the valve-surface at the poles of some frustules (figs 3: 6–8).
Pseudosulcus - a more or less deep channel between the valves of the chain-forming centric diatoms. The form of the pseudosulcus is taken from the form of the valve surface. Flat valve surfaces have no, or only a flat pseudosulcus. Strongly convex valve surfaces form a deeply penetrating pseudosulcus between the valves. They can have great taxonomic significance, e.g. in *Aulacoseira.*
Puncta – LM portrayal of foramina, areolae and intercostal ribs. There is no correlation between the clarity of the puncta in LM, and the species size, or number of puncta resp. lineolae per stria. It is thus a good and constant character. Large puncta can appear faint under conditions of poor contrast. In diagnoses, "coarsely punctate" can mean a few faint puncta/striae.
Quincunx – a term taken from Roman coinage, meaning the arrangement of puncta as on five of a dice. Double-puncta are always present on the striae in alternate rows.
Radial – a striae pattern where they point away from the central nodule.
Raphe – slit-shaped opening in the valve surface, serving as an organelle for movement. All valves with raphes have two symmetrical raphe branches. In the Achnanthaceae and Naviculaceae, the raphe lies within the median costa. Species with a canal raphe however, have the raphe in the angle between the valve surface and the mantle, or raised on a specialised raphe-keel.
Raphe canal – in species with a canal raphe, the raphe ends in an apically running, tube shaped canal. This in turn connects with the inside of the cell via the alar canals of portulae (figs 12: 6–7).
Raphe costa – the costal structures accompanying the raphe slot on the outer and/or inner side of the valve. Outer raphe costae are found in *Navicula*

diluviana, and inner raphe costae in the Naviculae lineolatae. Particularly well developed raphe costae are called axial costae.

Raphe fissure – the arrangement of the outer and inner fissures (Fig. 12: 5, 13: 4) is often taxonomically very significant.

Raphe keel – apically running solid ridge rising above the valve surface, bounded distally by a median raphe or a raphe canal with canal raphe. A wing (see elsewhere) is a special development of the raphe keel.

Raphe key – see raphe slot.

Raphe ledge – a ledge on both sides of the raphe in *Amphora* species that covers a part of the axial area and the areolate valve (fig. 13: 6). The raphe ledge is a special form of the conopeum.

Raphe side – the valve side with the canal raphe in the Bacillariaceae and Epithemiaceae.

Raphe slit – see raphe.

Raphe slot – right-angled or keel-shaped bar which fits into the raphe key in key and slot raphes (fig. 12: 2).

Residuum – the particularly well-developed dorsal part of the valve mantle in many *Amphora* species. Normally it is separated from the valve surface by a hyaline area (Fig. 13: 6).

Reverse lateral raphe – a special form of the median raphe, where the proximal inner and outer fissures overlap (Fig. 12: 5).

Resting spore – a resting stage formed from a complex series of divisions in vegetative cells. It enables survival of unfavourable climatic conditions.

Rica – a modified cribrum (superfluous terminology).

Rimoportula (labiate process) – a tube penetrating the valve wall. On the outer valve surface is either only one foramen, or an elongated structure looking like a thorn in LM. The foramen or process in centric diatoms lies either near the normal marginal spines, or is displaced towards the mantel, resp. valve surface. On the inner side of the valve it is shaped into a lip-like structure.

Satellite pore – pores (usually 2–5) surrounding the fultoportulae.

Secondary stigma – these surround one or more of the main stigmata in pennate diatoms. In structure, they are the same, just smaller (Fig. 10: 4).

Section – this term is not used in it's strict taxonomic sense in this flora, but just for a group of similar species.

Separating spines – short or long, mostly pointed spines without an anchor that facilitate the separation of cell chains at particular positions (e.g. *Aulacoseira granulata*).

Septae, marginal –sail-like septae on the valvocupola or intercalary bands.

Septa seam – where septa extensions abut medially or submarginally. Either the seam is visible in SEM (*Denticula tenuis* has two, one on each side), or one or two holes are visible in LM. As these holes or abutments often occur in the same taxon, both are described as septa seams here.

Septum – in contrast to the pseudoseptum, the septum is not attached to the valve, but to the copulae, and is flat to undulating. The copulae are often open, so that the septum is most frequently found opposite the opening (Fig. 3: 3–5).

Setae – long, brush-shaped, silicified outgrowths on the outer side of the valve in some centric diatoms.

Shadow lines – strengthened radial striae in the marginal region of some *Cyclotella* species.

Spine, spinule – projections on the valve surface occurring either singly (e.g. *Gomphonema africanum*, fig. 17: 5) or in numbers (spinules, *Surirella bifrons* f. *punctata* fig. 17: 5). In some cases, connecting spines may join single cells to form chains (e. g. *Aulacoseira* and many araphid diatoms, some Naviculaceae, fig. 17: 7–8).

Spine-base – the base in connecting spines, on the distal end of which is a wide structure, the spine-anchor.

Spine anchor – structures at ends of the spine that link mechanically with the next cell in a chain. Their form is species-specific.

Sporangial form – see initial cell.

Stauros – a broad central nodule almost reaching the edge of the mantle, appearing as a fascia under LM.

Sternum – the "pseudoraphe" of araphid diatoms, i.e. an areola-free apically running stria (Round 1979).

Stigma – a canal-shaped perforation near the central area, not closed by a velum. Stigmata are morphologically unique in contrast to the normal areolae in that they possess their own alveolus with a specialised structure. They are neither homologous nor analogous with the stigmata of the flagellates.

Stigma foramen – outer opening of the stigma canal.

Stigmoid – simple form of the stigma where the foramen, canal and the alveolus are scarcely differentiated from the areolae of the valve surface ("isolated puncta" in the LM).

Stub – rudimentary form of a strut (Fig. 6: 8).

Sulcus – a constriction in the valve mantel just before it's distal ends in species of *Aulacoseira*. Three characteristic structures are found close to the "sulcus":
1. the usually weakly developed sulcus furrow on the outer side of mantel.
2. a pseudoseptum, the annular ridge (see earlier), is on the reverse side of the sulcus, on the inside of the valve.
3. the mostly very short, areola-free, distal part of the mantel (collum).

Sulcus furrow – a taxonomic character used in LM by focussing on the valve edge. It can be troughlike (*Aulacoseira subarctica*) to a sharp cut (*Aulacoseira subarctica*), and is more or less well-developed in a few taxa.

Suture – line of union between the valve and the girdle, or between individual girdle elements.

Symmetry – in taxonomic terms, the valves of many pennate diatoms are termed bilaterally-symmetrical. Morphologically however, the valves of almost all species are asymmetrical along the apical axis due to the Voigt fault, the raphe and various structural irregularities. Taxonomically, only the difference between dorsiventral- and this qualified bilateral-symmetry is of significance.

Tangential striae – in centric species these are areola rows that are not radially organised.

Teratological form – a valve-shape or structure that differs significantly from the norm, though still clearly identifiable with a particular species (Fig. 18: 7).

Terminal (polar) – at the valve ends in pennate diatoms.

Terminal area – hyaline area at the valve poles (e.g. *Navicula capitata*).

Terminal cap – strengthened part of the valve at the poles.

Terminal costa – a costa between the terminal nodules and the valve mantle at the poles. The terminal fissure runs in this costa on the outer side of the valve (Fig. 16: 3, 4). Two terminal costae are sometimes present (*Cymbella aspera*).

Terminal fissures – a shallow split on the outside of the valve in the distal raphe-ends in the Naviculaceae (fig. 16: 3–4). The terminal fissures are notches in the terminal area, and their shape and form are important taxonomic features.

Terminal nodule – a siliceous thickening at the distal ends of the raphe (fig. 1:5).

Theca – valve and girdle together (such a combined term does not exist in the Prasinophyceae).

Tignum – strutlike structures in the alveoli of some pennate diatoms (e.g. *Cymbella aspera*, fig. 7:1). In a rudimentary form they a short stumps (truncus, fig. 14: 6).

Transapical axis – the axis that runs parallel to the valve surface and a right angles to the apical axis (Fig. 1: 1–4):

Transapical plane – the plane in which the apical and pervalvar axes lie (Fig. 1: 1).

Transapical costae – costae between the median costa and the valve mantle (Fig. 6: 3). Under LM the striae are visible between these costae.

Transapical striae (str.) – depending on focus in LM, rows of foramina areolae or alveoli running between the transapical costae. They are described as being parallel when at ninety degrees to the apical axis, radial when angled away from the central nodule, and convergent when angled towards the central nodule (Fig. 1: 4).

Transapical partition – planar valve extension running transapically (*Diatoma*). In some species with a canal raphe (e.g. some *Epithemia*, Fig. 12: 6) the fibulae can develop as transapical partitions.

Transverse – see transapical.

Valve surface – the part of the valve surrounded by the mantle. In the sausage-shaped initial cells of many pennate diatoms, the valve surface and mantle have the same structure. See mantle border.

Valvar plane – plane parallel to the valve surface.

Valve mantle – the side walls of the valve. In many cases it has a different structure from the valve face. Sometimes it is unperforated but may have the same structure as the valve surface. The structure of mantle can be separated from the valve face by a hyaline area (Fig. 9: 5).

Valvocopula – the first element of the girdle directly connected with the valve mantle. If these elements are open, then two may be present.

Vegetative cells – cells resulting from mitotic division of other cells and auxospores. In most species, a change of form follows each division, and the recognition of this is important in avoiding mis-identification (Fig. 1: 3).

Velum – a structured or unstructured thin silica membrane stretched across the inside of the foramen, or that closes off the inside of an areola. Cleaning with strong acid usually destroys the fine velum.

Ventral – in dorsiventral species, the less convex side. See further under "symmetry".

Voigt fault – shorter striae or other irregularities on one side of the axial area that are symmetrical to the transapical axis (Fig 24: 3, 4; 25: 2).

Wing – a particularly strongly pronounced keel raphe that runs around the entire valve (*Surirella*, fig. 14: 6).

Wing canal – "alar" canals that link the raphe-canal of a canal-raphe with the cell interior at every interspace (see earlier). The inner and external openings of the canal-raphe are called portulae. If a long wing-canal is absent, then only the portula can be seen.

2.2 Appendix to Glossary

English	German	French	Latin
alar canal	Flügelkanal	canal alaire	canalis alaris
alveolus	Alveole	alvéole	alveolus
apical	apikal	apical	apicalis
apical axis	Apikalachse	axe apical	axis apicalis
apical furrows	Raphefurchen	sillon apicale	sulcus longitudinalis
apical plane	Apikalebene	plan apical	planities apicalis
apical pore field	Apikale Porenfelder	champs de pores apicaux	area porellis apicicalibus
areola	Areole	aréole	areola
areola foramen	Areolenforamen	foramen de l'aréole	foramen areolarum
auxospore	Auxospore	auxospore	auxospora
axial area	Axialarea	aire axiale	area axialis
axial costa	Axialrippe	côte axiale	costa axialis
band	Band	bande	taenia
branch of the raphe	Raphenast	branche du raphé	ramus raphis
canal raphe	Kanalraphe	raphé en canal	raphe canalis
canopy	Conopeum	conopeum	conopeum
central area	Zentralarea	aire centrale	area centralis
central nodule	Zentralknoten	nodule central	nodulus centralis
central pore	Zentralpore	pore central	porus centralis
central fissure	Zentralspalte	fissure centrale	fissura centralis
complex raphe	Komplexraphe	raphé complexe	raphe complexa
cingulum	Cingulum	cingulum	cingulum
coating membrane	Hüllmembran	membrane enveloppante	membrana tegens
connecting band	Gürtelband	bande connective	copula
costa	Rippe	côte	costa
craticula	Craticula	craticule	craticula

English	German	French	Latin
diatotepum	Diatotepum	diatotepum	diatotepum
epicingulum	Epicingulum	épicingulum	epicingulum
epitheca	Epitheka	épithèque	epitheca
epivalve	Epivalva	épivalve	epivalva
fascia	Fascia	fascia	fascia
fenestra	Fenster	fenêtre	fenestra
fibula	Fibula	fibule	fibula
filiform raphe	Fadenraphe	raphé filiforme	raphe filiformis
foramen	Foramen	foramen	foramen
foramen border	Foramenlippen	lèvre du foramen	labium foraminis
frustule	Frustel	frustule	frustulum
girdle	Gürtel	ceinture	cinctura
helictoglossa	Helictoglossa	hélictoglosse	helictoglossa
hypotheca	Hypotheca	hypothèque	hypotheca
hyaline area	Hyaaline Area	aire hyaline	area hyalina
hymen (velum)	Siebmembran, Velum	vélum	velum
hypocingulum	Hypocingulum	hypocingulum	hypocingulum
initial cell	Erstlingszelle	cellule initiale	cellula prima
intercalary band	Zwischenband	bande intercalaire	copula
intercostae	Intercostae	eintercôtes	intercostae
intermissio	Intermissio	intermissio	intermissio
internal valves	Innere Schalen	valves internes	valvae internae
interspace	Interspatium	espace interfibulaire	insterspatium
isolated punctum	isolierter Punkt	point isolé	punctum isolatum
keel punctum	Kielpunkt (Fibula)	point carénal	punctum carinae
key and slot raphe	Nut- und Federraphe	raphé rainure	raphe conjungens
key costa	Federrippe	côte à languette	costa clavis
labiate process	Lippenfortsatz	processus labié	rimoportula
lateral area	Lateralarea	aire latérale	area lateralis

English	German	French	Latin
lateral raphe	Lateralraphe	raphé lateral	raphe lateralis
ligula	Ligula	ligule	ligula
lineolae	Lineolae	lineolae (stries lignées)	lineolae
linking spine	Verbinungsdorn	épine de jonction	spina ligans
longitudinal bands	Längsbänder	bandes longitudinale	staenia longitudinalis
longitudinal canal	Längskanal	canal longitudinal	canalis longitudinalis
mantle border	Mantelrand	bord du manteau	margo limbi
median costa	Medianrippe	côte médiane	costa media
median raphe	Medianraphe	raphé médian	raphe media
nodule	Knoten	nodule	nodulus
passage pore	Verbindungsöffnung	pore interloculaire	pervium
perizonium	Perizonium	perizonium	perizonium
pervalvar axis	Pervalvarachse	axe pervalvaire	axis pervalvaris
pore	Porus	pore	porus
portula	Portula	portule	portula
pseudoseptum	Pseudoseptum	pseudo-septum	pseudoseptum
punctum	Punkt	point	punctum
raphe	Raphe	raphé	raphe
raphe canal	Raphenkanal	canal raphéen	canalis raphis
raphe costa	Raphenrippe	côte du raphé	costa raphis
raphe keel	Raphenkiel	caréne du raphé	carina raphis
raphe ledge	Raphenleiste	liteau du raphé	regula raphis
raphe slit	Raphenschlitz	fente du raphé	scissura raphis
raphe fissure	Raphenspalt	fissure du raphé	fissura raphis
residuum	Residuum	residuum	residuum
resting spore	Dauerspore	kyste (statospore)	endocysta
secondary stigma	Sekundärstigma	stigma secondaire	stigma secunda
septum	Septum	septum	septum
seta	Borste	soie	seta

English	German	French	Latin
spine	Dorn	épine	spina
spinule	Dörnchen	spinule	spinula
stauros	Stauros	stauros	stauros
stigma	Stigma	stigma	stigma
stigma foramen	Stigmaforamen	foramen du stigma	foramen stigmae
stigmoid	Stigmoid	point isolé	stigmoideum
struts	Bälkchen	contrefort	stignum
stub	Stumpf	tronçon	truncus
strutted process	Stützenfortsatz	processus renforcé	fultoportula
suture	Sutur	suture	sutura
terminal area	Terminalarea	aire terminale	area teminalis
terminal cap	Polkappe	calotte terminal	epilleus terminalis
terminal costa	Teminalrippe	côte terminale	costa terminalis
terminal nodule	Endknoten	nodule terminal	nodulus terminalis
terminal fissure	Endspalte	fissure terminale	fissura terminalis
theca	Theka	thèque	theca
transapical axis	Transapikalachse	axe transapical	axis transapicalis
transpical costa	Transapikalrippe	côte transapicale	costa transapicalis
transapical partition	Transapikalwand	cloison transapicale	paries transapicalis
transapical plane	Transapikalebene	plan transapical	planities transapicalis
transapical stria	Transapikalstreifen	strie transapicale	stria transapicalis
valve face	Schalenfäche	face valvaire	frons
valve mantle	Schalenmantel	manteau	limbus
valvar plane	Valvarebene	plan valvaire	planities valvaris
Voigt fault	Voigt-Diskordanz	défaut de Voigt	inordinatio Voigtii
wing	Flügel	aile	ala
valvocopula	valvocopula	valvocuplela	valvocopula
velum	velum	velum	velum

3 Material and structure of the cell wall

With few exceptions, the cell walls of diatoms consist of a silica layer coated with a considerably thinner layer of organic material. They are laid down externally as an alloplasmatic product of the cytoplasm. On the inner side of the valve, this membrane is in close contact with an euplasmatic primary membrane, the plasmalemma.

In diatom preparations and in Kieselgur, only or almost only, the silicified cell wall is present. In spite of SEM images suggesting a robust structure, the cell wall is relatively delicate. An example is given by Reimann et al. (1966) with *Navicula atomus* var. *permitis* (called *N. pelliculosa*), where the median costa acts to stiffen the entire valve, it was 0.25 µm thick and 0.08 µm wide. The perforated valve was 0.13 µm thick, decreasing to 0.08 µm and less at the margins. Measurements on some *Cymbella* species have shown that the thickness of the structural elements are not necessarily proportional to the size of the species. Thus, the transapical costae in a 18 µm long *Cymbella minuta* were 0.7 µm thick, while only 1.4 µm thick in a *Cymbella aspera* that was almost 200 µm long. In many species, high concentrations of silica form the central and terminal nodules, which may reach several µm in diameter.

The silicate acid occurs as an amorphous silica gel. The degree of hydration varies in a wide range and the SiO_2 content of the cell from 10% in *Phaeodactylum tricornutum* to 72% in *Navicula pelliculosa* (Torri & Volcani in Schmidt et al., 1981).

Detailed research has been carried out to the SiO_2 metabolism in diatoms. It is not only important for valve construction, but also for normal cell metabolism inclusive of DNA production. Because of this, diatoms have become a model for studies in Si-metabolism.

The individual structural elements in a diatom frustule are not formed homogeneously from a single siliceous mass, but are the result of a two phase process, as indicated by their fine structure. This becomes immediately obvious when valves become exposed to corrosive influences (prolonged exposure to water, acids during sample preparation, fossilisation processes). All corroded valves have a somewhat jagged appearance as a result of the dissolution of partially soluble compounds from the valve (see fig. 3: 1–2). Such corroded valves look a little like fired porcelain shards. During morphogenesis loose aggregates of silicate clumps form (Schmid & Schulz 1979), which reach a diameter of 30–40 nm in *Navicula permitis*, and up to 200 nm in *Thalassiosira excentrica*. Pickett-Heaps et al. (1979) showed that the central nodule of *Navicula permitis* is first formed from such silicate aggregates. Only later are the interstices filled with another silicate compound to form a smooth surface. Normal investigative techniques do not reveal this development process. Drum & Pankratz (1964) suggested the existence of "structured subunits", but were unsure as to whether or not their observations were an artefact arising from section with the microtome knife.

Diatoms have an organic cell-wall (coating membrane) in addition to the silicate cell-wall. This can be recovered through treatment with hydrofluoric acid, and then viewed as a negative of the silicate valve under TEM. First found by Liebisch (1929), this membrane stains intensively with ruthenium red. More recent studies by von Stosch (1981) showed that the coating membrane consists in reality of two membranes.

- A primary coating membrane on either the inner, or outer surface of many diatoms and enclosing the silicate elements. It probably consists of a coagulate of certain cell compounds.
- The actual coating membrane that is probably identical with the pectin layer described by Liebisch (1929). Lying on the inner side of the valve, this membrane is not of pectin, as originally assumed, but of acidic polysaccharides. In order to avoid the confusing description, diatopectin membrane, von Stosch (1977) suggested using the term diatotepum for this coating membrane. His opinion was that the diatotepum is functionally much more significant than the primary coating membrane, in that it has, e.g. an important factor in stabilising the individual elements of the frustule.

Amongst species with normally silicified cell-walls are species where the cell-wall is either only weakly silicified, or even has wall elements that are composed entirely of organic material. In *Cylindrotheca fusiformis*only the raphe system and adjoining areas, plus some girdle bands, are silicified (Reimann & Lewin 1964; Reimann et. al. 1965). The only transapically oriented silicate structures are the fibulae which link the raphe costae. All these silica-elements are held together by a cell-wall consisting of organic material. *Cylindrotheca gracilis* is similarly only partially silicified (Schmid 1966), as are a number of other marine species.
The fibrillar processes of many Thalassiosiraceae, which facilitate floating in this planktonic species (Walsby & Xypolita 1977) should also be mentioned at this point. These consist primarily of crystalline beta-chitin (Falk et al. 1966, Blackwell et al. 1967, Herth & Zugenmaier 1977). The ultrastructure of such fibrillae has been fully described and the morphology illustrated in *Cyclotella cryptica* (Herth & Zugenmaier 1977) and in a number of Thalassiosiraceae (McLoughlin 1965). In *Thalassiosira weissflogii* these chitin fibrils are exuded from over 80 marginal pores, and a number of central, less regularly positioned pores (strutted processes).

4 Morphology of the cell-wall

Even today, diatom diagnoses are normally based on LM investigations. Almost all species can be thus identified, and only a few so far require EM study to be correctly identified.
In principal, users of the systematic part of this book need not be concerned with gaining a detailed knowledge of the cell-wall morphology. However, a cursory overview of diatom morphology is important for those wishing to delve deeper into diatom systematics. An accurate interpretation of LM images, and insight into taxonomic allegiances, is only possible through a thorough knowledge of morphology.
The introduction of SEM 2–3 decades ago provided the foundation for modern micro-morphological studies. LM not only failed to give a clear image of the valve fine structure, but also misrepresented it, while TEM mostly could not penetrate the valve and one had to resort to sectioning for detailed information. Many LM and TEM studies thus led to theories on fine structure that are now surpassed by more modern techniques. In spite of this, LM studies have led to numerous concepts that are still valid today (e.g. Hustedt 1926 a, b; 1928 a, b; 1929 a, b; 1935 a, b). The following comments are restricted to the scope of this book, in particular to recent species of inland waters, and still further primarily

to the structure of pennate diatoms. This can be justified in that well over 90% of all species in inland waters are pennate diatoms.

Species diversity in European inland water is divided very unevenly in the major morphological groups. About 7% are centrics and the rest pennate diatoms divided as follows: araphid diatoms 8%, somewhat over 6% Eunotiaceae, almost 7% Achnanthaceae (Monoraphideae), a little over 50% Naviculaceae (sensu lato), and finally slightly less than 20% taxa with a canal raphe (excluding the canal-raphe bearing forms of the Naviculaceae).

These figures relate solely to the division of the species between the individual groups, while saying nothing about how common they are. Large ponds and lakes produce large numbers of diatoms, consisting in freshwaters mainly of araphid and centric diatoms. Even the epilithic diatom population in fast flowing waters, or growing epiphytically on plants in standing water, is often composed primarily of araphid diatoms.

In all but a few weakly silicified species (e.g. *Cylindrotheca gracilis*), diatom valves are self-supporting, i.e. in contrast to the cellulose cell-wall of many algae and higher plants, the valve geometry remains intact after turgor pressure is released (Krammer 1981b). This knowledge has formed the basis of diatom preparation techniques. Fossil diatoms retain their form over millennia, even under the pressure of overlying deposits. The latter indicates that the construction of diatom valves not only stabilises it's own form, but also is tolerant to stress from additional loads. For benthic diatoms in fresh and salt water, these loads come from waves, currents, falling water and particulate rain. Additional stress from with the cell is also catered for. Only with a well supported skeleton can gametes, often with an amorphous outline, through almost spherical auxospores to characteristically shaped vegetative cells, be accommodated. It is very probable that the static form of the vegetative cell would be unattainable without the supporting skeleton. As this silica cell-wall is not elastic, as is a cellulose cell-wall, it must be able to withstand considerable internal changes in turgor pressure resulting from osmotic changes in the surrounding medium.

Valve structure studies show that it is part of a multifunctional system. Metabolite exchange with the environment for example, must coexist with this static system. The openings necessary for this exchange are partly closed by sieve-membranes permeable to ions. Although weakening the cell-wall, various solutions have evolved which cater to both structural and ion-exchange requirements.

A third factor is found where valves consist of a brittle material, which although having a relatively high resistance to pressure, possesses a markedly lower tensile strength, i.e. is flexible and hard to shear (Krammer 1981 b).

Long raphes also have a significant effect on the stability of the valve. The interaction of all these factors results in valves well adapted to their environment.

To summarise, in contrast to the elastic systems of many other plants, resulting from crossed fibres, the silicified cell-walls of diatoms form a more or less stable outer skeleton. It is adapted to varying load-pressures. Our current state of knowledge allows us only to conjecture and pose questions as to the interactions between various forces and the valve elements. Only further research will reveal why diatoms structures have the form they do.

4.1 Form types and symmetry

Kirchner (1878) and Schütt (1896) divided the Bacillariophyceae into two groups, the Centrales and the Pennales. There are sufficient similarities in the basic structure of the two groups to allow them to be treated together. Figs 1: 1–5 are sketches showing the most important structural elements, axes and planes of pennate diatoms. The cell-wall of diatoms (frustule, girdle) is similar to a box consisting of two similar halves, the lower half of which is termed the hypotheca, and the upper half, the epitheca. Each of these is in turn composed of two parts, the valve plus the accompanying girdle. In some species the valves are relatively flat, but usually the edge is strongly curved, forming a narrow to broad valve mantle. The valve surface, mantle and attached girdle are collectively known as the epi- or hypotheca.

Three axes and planes (figs 1: 1–3) are of importance in describing diatom valves (O. Müller, 1895). Apical axis (A) links the two poles and lies parallel to the valve surface. In the centric diatoms the unlimited number of "apical axes" are termed radial axes. The morphological center of both valves is joined by the pervalvar axis (P). It is at right angles to both the valve surfaces and to the apical axis. Running parallel to the valve surface, but at right angles to the apical axis, is the transapical axis (T). With these three axes, three planes can be defined which have a significant importance in the description of the diatom's morphology. The apical- and transapical axes lie within the valvar plane (fig 1: 1); the apical- and pervalvar axes, the apical pane (fig. 1: 2); and the transapical- and pervalvar axes, the transapical plane (fig. 1: 3). Sections perpendicular to the pervalvar axis are cross-sections.

These terms of O. Müller are not meant to be used in a mathematical sense. In geometry there are no curved axes or domed planes. In diatom descriptions however, such modifications are most useful. The axes can be straight or curved lines, the planes, bent or twisted, and divide the frustule in equal or unequal halves (figs 2: 4–20). Where the planes are divided into two unequal halves, the corresponding axes terminate in two unequally developed poles, and are termed heteropolar. The main planes are not all planes of symmetry. Morphological planes and planes of symmetry are only the same in species with all three axes isopolar (i.e. both poles have the same morphological structure, e.g. *Cyclotella*, figs 2: 1–3 and *Pinnularia*, figs 2: 4–6). All other forms in figs 2: 7–21 are either heteropolar along one axis, or have at least one curved axis and thus a warped plane. In the genera *Amphora* (figs 2: 7–9) and *Cymbella* the apical and pervalvar axes are curved and the transapical axis straight. The valvar and transapical planes are planes in the geometric sense, while the transapical plane is bent. In *Achnanthes*, *Entomoneis* and *Rhopalodia* (figs 2: 13–18) some axes and planes are bent several times or even twisted. In the cymbiform *Amphora ovalis* (figs 2: 7–9) the transapical axis is heteropolar, the other axes isopolar, while in the club-shaped ("clavate") *Gomphonema* spp. (figs 2: 10–12), the apical axis is heteropolar and the other axes isopolar.

A wealth of terms describing valve geometry and symmetry, in addition to those listed here, can be found in the diatom literature. As no consistent usage for these terms has been found, they will not be detailed here.

"Symmetry" is another term that is rather loosely used in diatom valve descriptions. In all species with a median raphe, both halves are differently structured. The central and terminal fissures produce asymmetries, as do the stigmata, Voigt fault and other details of the fine structure. A structural difference exists between the epitheca and hypotheca in all diatoms so that even symmetry on the valvar plane is not assured. Such features are however, not taken into account in comments on symmetry.

The outline of the frustule elements in the transapical plane (middle row of plate 2) seldom shows the typical box-like form of the valve (figs 2: 2, 2: 17, 2:11). One such example is seen with *Stauroneis phoenicenteron* (fig. 6: 7). Cymbella species display different types of three dimensional overall valve-

structure, where the box in transapical section is round, hexagonal or quadratic (Krammer 1981a).
In addition to the valve face and mantle, some the valve of some species possess a third structural element, the pseudoseptum. This differs considerably in structure from a true septum (comparable to a section of the valve girdle), which are linked by intercalary bands. Pseudosepta are found, e.g. at the valve ends of many *Stauroneis*- and *Gomphonema* spp. and *Rhoicosphenia abbreviata*. This is formed by an inward extension of valve mantle, parallel to the valve surface, at one or both planes (figs 3: 6–8, 16: 7–8).

4.2 Morphological structures of the cell wall

The observation of diatom valves under LM reveals a multitude of forms, which at first glance cannot be placed into relatively few structural types. Deeper morphological investigation, using TEM or SEM, quickly reveals more similarities than essential differences. This is not particularly surprising, as the combination of functionality and physical material properties is only met by a limited number of valve structures. Once morphogenesis is complete, the frustule represents a construction that is subject to specifications that are completely identical with those of engineering. Thus in diatom valves we find three basic structures, just as are known from engineering and construction:
1. Perforated plates
2. Latticed ribs
3. One- and two-valved, alveolate rib-systems
The highest level of development, optimising reduction in material with static properties, is found in the latticed rib-structures of many marine centrics and also in some pennate diatoms. Structures of perforated plates, such as are found in e.g. many monoraphid diatoms, have a relatively high raw material requirement. The alveolate rib-systems lie somewhere between these two.
A perforated plate is the simplest solution to maintaining a sound structure and achieving a large area of exchange between the cell inside and the surrounding medium. One finds this structure, where a silica plate is perforated in various patterns by pore-shaped areolae, in many monoraphid diatoms. A characteristic perforated plate is found in raphe-valve of *Cocconeis pediculus* (fig. 4: 1). These circular to oval (in cross-section) areolae are closed by a velum on the inner side. This design is however only seldom found in it's primitive form. Gerloff & Rivera (1978) show, e.g. an araphid valve of *Cocconeis pediculus* from Jordan that is considerably more complex than this basic structure. On the valve's inner side, the simple perforated plate appears to be retained, but towards the outside, the simple pores are divided many times by complex lips. This rugged valve surface presents a large area for exchange between the cell inside and the surrounding medium. A number of Naviculaceae have valves that correspond to the perforated-plate structure type. *Frustulia* spp. have valves particularly characteristic of this structure. In *Frustulia rhomboides* (figs 4: 2–4) the areolae in the perforated plate are relatively regularly distributed. The material between the pores forms narrow transapical and apical costae. Towards the outside, the areolae become narrow foramina (fig. 4: 2), closed off only on the inside by inwardly convex, hemispherical sieve-membranes (fig. 4: 4). Although these sieve-membranes are relatively stable, they are often destroyed by acid preparation, thus emphasising the basic structure (fig. 4: 3). In *Navicula s. l.*, the perforated plate structure is found in several species, e.g. *N. cuspidata*. Again, the basic form of the perforated plate structure is seldom found. More commonly, *Navicula s. l.* has structures transitional to the alveolate rib-systems.

A hint of this is seen in *N. placenta* (fig. 4: 8). Many species of *Neidium*, e.g. *Neidium affine* (fig. 4: 7), display a transapical arrangement of alveoli. The structure is however closer to the perforated plate structure than to that of the alveolate rib-systems.

All perforated plate systems have two things in common. Firstly, that the intercostal ribs are strongly developed, often as strong as the transapical costae. Secondly, in all species so far studied, mostly strongly developed, inwardly convex, hemispherical vela exist on the valve's inner surface.

Latticed rib structures combines good stability with the minimum use of structural material. Many of the existing types of valves with such a structure are composed of thin bars and walls (Gerloff 1963). These valves are always double-walled, the walls being joined by thin membranes arranged in hexagons or squares, which in turn build the walls of the areolae. Many variations of this type of structure are present in marine planktonic, centric diatoms. Although this type of structure is very rare in pennate diatoms, good examples can be seen in the genera *Pleurosigma* and *Gyrosigma*. *Pleurosigma angulatum*, with its hexagonal areolae, has even become the most famous test object researched. However, even TEM studies of the valve have left many aspects of the exact structure open to active discussion (Kolbe 1954; Hustedt 1945, 1952). SEM studies show that the valves of both *Pleurosigma* and *Gyrosigma* have a similar structure and differ by two characters: in *Pleurosigma* the areolae are hexagonal, but quadratic in *Gyrosigma*. This gives rise to three inner and outer rib-systems in *Pleurosigma* and two in *Gyrosigma*. The valve structure in *Gyrosigma attenuatum* shows both fragments in figs 4: 5–6. The inner and outer walls of the valves are made of a double T-shaped supporting rib-system, which serves to further strengthen the lattice rib-system. The chamber walls (f) are not always fully developed, and often have connecting apertures (h). The chambers have large round foramina with flat vela (a) on the side of the chamber innermost to the valve. The outer longitudinal costae have lateral lips (d), bordered by the narrow, apically stretched outer foramina (e), close to the areolae.

One- and double-valved alveolate rib-systems are commonest in the pennate diatoms, and is found in the araphid diatoms as well as in many diatoms with median- and canal-raphes. Many centric diatoms also have this structure. The valves are based on a solid rib-system that is able to withstand pressure (Krammer 1981 b, 1982 a). Figs 6: 1 and 6: 3 show rib-systems that are visible not only during morphogenesis, but also during corrosive disintegration, in a centric and pennate diatom. In the centric *Cyclotella* (fig. 6: 1) the main element of the rib-system is the central swelling (d), and is to a certain extent analogous to the central area of the pennate diatoms. The radial costae correspond to the transapical striae of the pennate species. Intercostae lie between the transapical/radial striae, and the chamber-forming areolae-systems in their outer walls form an important route for metabolite exchange between the inner valve and the surrounding medium. On the inner side of the valve neighbouring transapical costae form shallow to deep grooves, the alveoli. This is normally open to the inner side of the valve, but in many cases is partially closed off by an inner wall. In marginal cases such as *Cyclotella comta* or many *Caloneis* species (fig. 7: 6), only a small marginal opening remains between the alveolus and the cell inside. The length and breadth of this opening in the alveolus wall is often a good taxonomic feature. An exception was found by Serieyssol (1981) who showed that in *Cyclotella andacensis* var. *bauzilensis* the alveoli opening ranges in length from one half to almost the full length of the transapical striae as seen in LM. The alveoli ranges from being completely closed to being fully open in a taxon.

The individual elements in the alveolate rib-systems are modified in different ways. Several examples of different modifications to the alveolate rib-system are shown on plate 5. In principal, two groups can be distinguished. In one, the foramina (6) in the outer wall of the alveoli are covered by a velum, while the alveoli are completely open towards the inside (figs 5: 1, 2). This is the normal structure found in e.g. *Cymbella*. The second group (fig. 5: 7,8), in contrast has the foramen (4) open to the outside and a velum stretched over the entire inner surface of the areola (6). This structure is very widespread as in e.g. *Navicula*, *Frustulia*, *Gyrosigma*, *Amphora*, and in many species with canal raphes. Even with in *Cymbella* one species, *Cymbella pusilla*, has this valve structure. Different variations on the structure with **velum on the outside** can be seen in figs. 5: 1 to 5: 6. Variously thickened intercostal ribs (12) lie between the normally very strong transapical costae (10) and tend to be twice the height of the former. Exceptions can however be found in for example, some species of the genus *Gomphonema*. Here the intercostal ribs are almost as high as the transapical costae (compare with *Cymbella aspera* fig. 7: 1), though the alveolation almost disappears. The transapical- and intercostal-ribs form the walls of the areolae, on whose outside the foramina lips (5) shape the foramina within which fine vela are stretched. These are usually very simply formed and are only slit-shaped (fig. 5:1). They can however also be jagged, as in *Cymbella mexicana* (figs 5: 4, 7: 2), and then varied foramina structures are produced (fig. 9: 6). Expansive foramina furnished with stable vela are only rarely produced on the outer surface of the areolae (*Cymbella helmckei*, figs 5: 6, 9: 4). Generally, there is a tendency for large foramina to be split by the formation of many lips, both protecting and reducing the area of thinner vela. The foramina lips are often very delicate, often merging directly into the vela (fig. 5: 2). The velum and the foramina lips are often absent in SEM studies due to acid treatment of the valves during preparation (fig. 8: 1). Foramina descriptions do not always take this into consideration. One or more correspondingly dimensioned rows of areolae are present per intercosta (figs 6: 5, 6).

Multiple rows of areolae per intercosta are diagnostic characters for some genera, e.g. *Pinnularia* and *Caloneis*. Some *Pinnularia* species have open alveoli and 4–12 rows of areolae on the outer side of the valve (Krammer 1982a). These areolae are organised in a honeycomb structure, with small, round tube-shaped chambers occluded by a fine velum on their outer side. In the Maiores, Complexae and some other groups, the alveoli are closed off on the inside by small to large opening that can span the entire length of the alveolus. In these cases the longitudinal bands, which are nothing more than the LM image of the alveolar openings, are missing under LM. figs 7:3 and 7: 4 show such a valve with closed alveoli. In the latter, the fairly thin inner wall of the alveoli, which is convex towards the alveoli, is clearly visible stretched between the transapical ribs. Small areolae m and their openings n (without foramen margins) are shown in fig. 7: 4. In *Pinnularia maior* the alveoli chambers have a diameter of about 100 nm, with a total height of 300 nm. The valves of *Caloneis* are similar to those of *Pinnularia* species with partially closed alveoli, though here the inner alveolus opening is very small and located close to the valve mantle (*Caloneis silicula*, fig. 7: 6). Here, the inner alveoli walls are membrane-thin, and the transapical costae thus clearly visible (fig. 7: 6).

Whereas multiple rows of areolae are generic characters (in combination with other features) of *Pinnularia* nd *Caloneis*, they are useful in delimiting species or small groups of species in *Gomphonema*, *Achnanthes* and *Gomphoneis*. In species with open alveoli, the same valve structure can be seen even in different genera. Thus, the valves of *Achnanthes lanceolata* (fig. 6: 5) and *Pinnularia gibba* (fig. 6: 6) are very similar in the organisation of the areolae.

In some centric diatoms with an alveolate rib structure e.g. *Cyclotella* (Helmcke & Krieger 1952, Round 1970, Lowe 1975, Serieyssol 1981), intercostae with two or more rows of areolae, often increasing in number towards the edge, are found between the robust radial costae. The radial costae are normally not only visible around the inner alveoli openings (e in fig. 6: 2), but also on the outer side of the valve around the closed alveoli (fig. 8: 5). In many species, independent of genus, the areolate alveoli are restricted to the valve surface, though in others they go well over onto the valve mantle. Such is the case with *Stauroneis phoenicenteron* (fig. 6: 7) where they run evenly from the valve surface to the valve mantle. Other groups in contrast, e.g. *Gomphoneis* species, have open alveoli on the valve surface which become closed towards an on the valve mantle (fig. 7: 5), and lead to several rows of fine areolae (m) radiating outwards. Various structures of as yet unknown function and of probably no structural significance can be found within the areolae and alveoli. Examples of these are seen in the delicate struts, strut-stumps (fig. 5: 1–3) and various papillae (figs 5: 5, 6: 8) that are in close association with outer chamber openings (see Mann 1981).

The second basic structure of valves with alveolate areolae is where a **large internal velum** is present (fig. 5: 7, 5: 8). The valves of the *Navicula* s.str. are typical examples of this. The areolae have foramina without velae to the valve's outer surface (4) and large velae, convex towards the valve interior. The areolae are often apically elongated (fig. 6: 4). There is an even transition from this structural type to the latticed rib structure, e.g. the valve of *Amphora ovalis* (figs 7: 7 and 7: 8). Narrow, but very high transapical costae are connected on the valve's inner side by irregularly arranged, rounded intercostal ribs (d). Transapically elongated velae (g), convex towards the valve's inner side, are stretched between these costae. A thin wall (r) furnished with narrow slits (v) limits the outer surface of the areolae. The alveolation on the valve's inner surface is very weakly developed.

This type of costae-system in it's various forms is not only also found in the araphid diatoms, but also in groups with a canal raphe, e.g. *Nitzschia*, *Epithemia*, *Surirella*. In some araphid diatoms the valve structure is made even more complex by transapically running cross-walls that strengthen the valve in addition to the transapical costae. In many groups with a canal raphe, the valves have cross-wall like fibulae (fig. 12: 6, 14: 4). These will be dealt with in detail in the discussion of the canal raphe.

In modern taxonomy the species-specific organisation and **form of the foramina** is a particularly important character. Under LM only their organisation can be used in diagnosis, while TEM or even better, SEM studies of uncorroded foramina may clarify difficult questions. The foramina can be transapically (fig. 8: 3) or apically (fig. 8: 2) elongated; can be in single or multiple rows (figs 8: 4–7); and can be level with the outer surface of the valve (fig. 8: 3), or can lie in the channels of the transapical costae (fig. 8: 5, 8: 8). SEM studies have also shown that the area of an individual foramen has little correlation to species size. Thus the foramina of the large *Cymbella aspera* (fig. 8: 2) appear as particularly delicate wall openings.

In general, the foramina of the most Naviculae are very simple structures. Only where special lip-formations appear as in *Cymbella mexicana* (fig. 9: 6) or in some *Gomphonema* species (fig. 9: 1), is a more complex form encountered. Particularly complex subdivisions of foramina are formed by the foramina lips as in the Achnanthaceae, Epithemiaceae and other groups with a canal raphe, and in particular in the centric diatoms. An example of a complicated structure is seen in *Epithemia intermedia* (fig. 9: 2).

The foramina of the areolae in many species are arranged on the valve surface and on part, to all of the valve mantle with a particular pattern. In some cases, e.g. *Caloneis silicula* (fig. 9: 3), foramina can be found directly over the ribs (arrow). These either end blind, or more commonly, are connected to the alveoli by slanting areolae. A hyaline area is often found at the mantle margin, e.g. in some species of *Neidium* (fig. 9: 5). Such longitudinal stripes as they are called in *Neidium*, are the outer walls of longitudinal canals (fig. 17: 1) lacking areolae. The form of the foramina can either be the same (fig. 9: 5) or different (fig. 9: 6) on the valve surface and valve mantle.

In general, the interior rib-systems as described above strengthen the valves. In many cases there is an additional external rib-system. The greatly dorsally expanded valve mantle of *Amphora aequalis* (fig. 13: 6, 18: 2) is strengthened in this way. The areolae are occluded internally or externally by a membrane thin enough to allow ion exchange. These thin silica structures are not load resistant and are usually integrated in strengthening structures. None of these membranes is built like a sieve. Apart from the fact that perforations are not required for metabolic exchange, the velae observed under SEM have a lower extinction quotient near fine tears at the tear-points (arrow) than in the sieve areas. This absorption or deflection of electrons implies that membranous silica structures exist even within the sieve pores. The sieve character of the velum exists only at the level of the silica molecule.

Fixing the thin vela is achieved in several ways. Internal vela are often large and thick. External membranes are, in contrast, often small and thin. They are either stretched over very narrow, elongated foramina, or the foramina are divided by multiple lips which then serve to strengthen the velum. These lip structures grade evenly in their thickness to that of the velum. Even within the velum, it's thickness varies and this strengthens it further. An overview of these structures, which can also be important diagnostic characters at the species level, is given in Lange-Bertalot & Simonsen (1978) and in Lange-Bertalot (1980) for *Nitzschia* species and in Mann (1981) for other groups.

An example of the fine structure of a velum with regularly spaced thinner and thicker regions is seen in *Nitzschia intermedia* (fig. 9: 7). The arrows point to tears in the velum. In very narrow sieve membranes additional thickening in the membrane in unnecessary, e.g. most *Cymbella* spp.. Such vela appear structureless under normal TEM magnifications and are often missed. Under the SEM they produce enough secondary electrons to allow detection (fig. 9: 4, arrow).

4.3 Wall perforations without a velum

Apart from areolae closed by a velum, which from their total area alone account for the lion's share of metabolic exchange between the cell and the environment, the silica cell wall has a number of otherwise structured perforations. In these, the plasmalemma is in direct contact with the surrounding medium. These structures are:

1. Stigmata and isolated punctae.
2. Apical pore fields and other mucilage pores.
3. Labiate pores (labiate processes) and similar structures.
4. Raphe slits.

4.3.1 Stigmata and stigmoids ("isolated puncta")

Structures close to the central nodule on the dorsal or ventral surfaces are particularly widespread in *Navicula*, *Cymbella*, *Gomphonema* and *Gompho-*

cymbella. Additionally, they structures can be found on the central nodule itself. These structures are areolae forming canals going either vertically or obliquely through the cell wall. Velae are nowhere to be found in association with simple stigmoids or stigmas. As with many other types of wall perforation, these structures possess a canal-like chamber, the stigma- or stigmoid areola; an outer opening, the stigma foramen; and an inner opening that always ends in an alveolus. This alveolus has either the same structure as other alveoli, and often is the end of the alveolus closest to the central nodule; or is differently formed and is either separated from or connected with the other alveoli. There is a continuous range of variation between these extremes. Even closely related species can range from having relatively simple structures similar to the stigmoids (e.g. *Cymbella aspera*), to having complex forms with jagged foramen areolae (e.g. *Cymbella lanceolata*). Relative simple stigmoids are particularly often found in species of *Navicula* (one or two punctae), *Gomphonema* (normally a single punctum, fig. 10: 7, though occasionally up to four), and *Cymbella* subg. *Encyonema*, and sometimes in *Cymbella* subg. *Cymbopleura*. In may *Encyonema* species the foramen of the stigmoids is the last foramen of the central transapical costa, and so differs in size from the other foramina that it is easily distinguished in both SEM and TEM. There are other examples where the differences are not marked. The stigmoids, after penetrating the cell wall at right angles, ends in the central dorsal alveolus. The latter is much narrower at it's proximal ends than elsewhere.

In many species of *Gomphonema* and *Navicula*, e.g. *N. latens* and *N. constans*, the stigmoids is set off from the last areola of the central transapical costa. These stigmoids do not end the proximal ends of the alveoli.

Stigmas vary from isolated punctae in four points:

1. The stigma foramina is not only a somewhat enlarged areolae foramina, it differs in form.
2. The stigma canal ends directly in the valve interior instead of the alveolus, and differs substantially in structure from normal alveolation.
3. In *Cymbella* the stigmas are mostly on the ventral side, penetrating the central nodules in all except a few species.
4. The stigma canals go through the cell wall at an oblique angle, as opposed those of the stigmoids which go through at right angles.

A relatively simple form of stigma is seen in *Cymbella aspera*. At the end of the middle 10–20 transapical costae, there are small, round, funnel-shaped stigma foramina. These connect with unusually differentiated, narrow, transapically elongated alveoli via somewhat oblique canals. These alveoli are usually much more complex, with glandlike, split margins as in *Cymbella cistula* (fig. 10:15), and are often set apart from the other alveoli. These structures are widespread in the genus *Cymbella sensu stricto* , where according to species, 1–10 stigmas can be present. Both in the species with several stigmas, e.g. *Cymbella schimanskii*, and in those with normally one stigma, their number can vary, and thus this character is of only of qualified use in taxonomy. In addition, structures transitional to the areolae are associated with the stigmas. Thus in *Cymbella affinis*, a double areola-like structure occurs between the large stigma close to the central nodule and the normal areolation. In *Cymbella schimanskii*, a number of secondary stigmas are found between each of the large, funnel-shaped main stigmas and where the foramina (left) merge into the normal areolation. The foramina of the stigmas differ from normal areolae not only by their structure, but also by their orientation. *Cymbella simonsenii* has thus apically elongated areola foramina (fig. 10: 6), and transapically elongated sigma foramina (arrow).

Stigmas differing from those just described lie roughly in the middle of the

central nodule. In *Cymbella mexicana* (fig. 10: 2–3)the stigma foramen lies in a pit between the two central pores of the raphe ends. The stigma canal goes through the central nodule at right angles and ends in a radially symmetrical alveolus with margins fluted like the petals of a flower.
The function of stigmas and isolated punctae is as yet unknown. Their constancy in many species and even between populations or during life-cycle shape changes is similarly uncertain. Thus, although in many cases they are used in species delimitation, in others, for that of subspecies delimitation, caution should be observed.

4.3.2 Apical pore fields and other mucilage producing structures

The fine pores of the apical pore fields can be differentiated only with difficulty in LM. Some of the larger pores of many centric diatoms and some araphid diatoms, e.g. *Diatoma*, *Tabellaria*, *Synedra* and *Eunotia* are however more easily observed in LM. Until recently, these pores were called mucous or jelly-pores as it was thought, particularly in the colonial species, that they supplied connective material or were involved in binding a cell to the substrate. Mucous is certainly produced by the pores of the apical pore-fields. As with silk-glands, the secreted liquid quickly hardens in contact with water and serves to anchor araphid diatoms to the substrate in fast-flowing waters, and to unite cells into complex colonies in e.g. *Cymbella* and *Gomphonema*.
The apical pore fields normally consist of many pore canals without membranes (fig. 11: 8) that either end directly in the cell interior, or, as in *Cymbella aspera* (fig. 11: 5), in an alveolate system of costae. Many araphid diatoms and *Cymbella* sensu stricto. have apical pore fields at both valve poles. Heteropolar diatoms, e.g. *Gomphonema* and *Rhoicosphenia*, only have pore fields at the acute pole (the "foot pole"). The foramina of the pore fields in the araphid diatoms differ only slightly, or not at all, from the foramina of the other areolae, except that they are more densely packed (figs 11: 1, 3). In the Naviculaceae however, the foramina of the pores of the apical pore fields often differ substantially from those of the other areolae. In *Rhoicosphenia abbreviata* the foramina of both pore fields are broad, transapically elongated slits in contrast to the narrow, apically elongated foramina of the areolae. Mann (1981) showed that the distribution of the apical pores in this species showed some parallels to the distribution of the areolae on the rest of the valve face. Species of *Gomphonema* in contrast have apical pore foramina that differ greatly not only in structure, but also in their distribution from the other foramina of the valve (fig. 11: 6). The areola foramina of *Gomphonema acuminatum* are delimited by lateral lip-like extensions in the shape of a horseshoe. Here the pore canals of the apical pores are round and specialized foramina lips are absent. Similar structures are seen in the apical pore fields of many *Cymbella* species. (figs 11: 7, 8). Some populations of *Cymbella lanceolata* however, have elongated outer openings (Krammer 1982c).
Structures other than the apical pore fields are certain to be involved with mucilage production. Most species of *Cymbella* subg. *Encyonema* live in mucilage tubes, though the organelles involved in their production are unknown. Stephens & Gibson (1979) found various mucilage secreting structures in the septate girdle bands of *Mastogloia* species. They are responsible for the production of the strong mucous threads which bind the cells onto the substrate.

4.3.3 Labiate- and strutted processes

In addition to the structures just described for mucous and jelly production, both centric and pennate diatoms have a number of other structures whose function is only partly known.

In the centric diatoms, the most common of these are two types of wall perforations: rimoportulae (labiate processes) and fultoportulae (strutted processes).They are called processes as they are usually not simply areola-like structures in the cell wall, but have tube-forming constructions on valve's inner side (labiate processes) or on the valve's outer side (strutted processes). The rimoportula (labiate processes, figs 25: 6, 7) has been found in almost all centric diatoms and many araphid diatoms. The name derives from the lip-shaped, elongated extrusions on the valve's. inner side. The slit in the lip can either be straight or angled, with the entire structure either sitting directly on the silica wall, or be raised on a tube-shaped structure. The tube-shaped perforation of the wall is not occluded by a membrane. Both the form and arrangement of the labiate processes varies in different genera and species, and in many cases can be used as a good taxonomic character. All labiate processes are arranged submarginally on the valve edge. In *Druridgea*, isolated groups each of c. 12 labiate processes are arrayed in a ring round the valve mantle (Hendey & Crawford 1977).

The second large group of wall perforations with external processes in the centric diatoms are the fultoportulae or strutted processes. These are found in some genera of the Thalassiosiraceae, generally organised as a marginal ring between the valve face and mantle. Some species (e.g. *Cyclotella comta*, Lowe 1975) have structures similar to the marginal fultoportulae in the center of the valve face. Fultoportulae can also be arranged in a double ring. Without exception these structures are hollow and can have tube-forming processes not only on the outside, but also on the inside surface of the valve. They can occur singly, or in groups. The tube-like perforations are almost always surrounded by three or four pores (Crawford 1981). We know much more about the function of the marginal fultoportulae than of the rimoportulae. Round (1970) showed that in *Stephanodiscus* species organic material was secreted from these canals; in *Thalassiosira* (Mc. Laughlin et al. 1965) and *Cyclotella cryptica* (Herth & Zugenmaier 1977), these secreted threads were found to be of chitin. With careful preparation, many of these threads are well preserved and can be observed with SEM. They have a relevance in the buoyancy of planktonic forms. In the Thalassiosiraceae, rimoportulae are present in addition to fultoportulae, which lead Crawford (1981) to conjecture that they secrete material to bind cells together in the chain forming species.

Labiate processes are also found in the araphid diatoms and the Eunotiaceae. In the former, their structure is the same as or similar to that in many centric diatoms (Hasle 1973) occur apically, or occasionally, in the center of the valve. In many species of the genus *Fragilaria* there is a strutted process at each, and sometimes just at one, valve pole. *Diatoma* has either one or two strutted processes (fig. 11: 3). A single strutted process is always in the middle of the valves of *Tabellaria* (fig. 11: 2). These labiate processes are relatively coarse structures and can easily be observed also in LM. They are more difficult to observe in *Eunotia*. where they probably lie in the mantle of the valve poles.

According to Hasle (1973), strutted processes have no association with colony formation or valve attachment to the substrate. The ability of labiate processes to secrete has been qualified by Crawford (1975) through his observations of labiate processes in *Melosira*.. TEM studies of ultrathin sections revealed concentrations of vesicles round the inner opening of these processes, suggesting that cell products may pass through their pores. The organisation of these processes itself, suggests that neighbouring cells could be united by secretions from them.

4.4 Raphes

Movement in diatoms is controlled by the raphe, an organelle which in it's more evolved forms, is very complex. It is closely integrated with the valve's static system and determines the structure of the rest of the valve. Three types of raphe can be distinguished in the Raphidinae:
1. *Eunotia* raphes
2. Median raphes
3. Canal raphes.
In all cases, the basic characteristic of the raphe is that it is a short or long, slit-like opening in the cell wall that disrupts the physical stability of the valve. The short, simple **raphe of** *Eunotia* has the smallest such effect. In most species the raphe runs only a short distance to the valve poles. The raphe fissure itself lies within the strongly built mantle, while the terminal fissures are in the thinner valve-face on corresponding costae (fig. 13: 1). In only a few species, e.g. *Eunotia flexuosa*, does a part of the raphe slit lie in the valve-face.
The **median raphe** with it's simple or complex raphe fissure, is the norm in the valves of the Naviculaceae and in a simple form, in the raphe valves of the Achnanthaceae. The raphe fissure opens directly into the cell or into an open furrow, and not into a longitudinal canal connected to the cell interior by openings or canals. With few exceptions, both raphe branches divide the valve into two sections which are held together by three nodes and at the polar regions of valve mantle. The diverse types of median costae result from this type of raphe construction.
Canal raphes, where the fissures, canals and portulae lie marginally and often on a specialised raphe keel, are the most complex. Most of such valves are characterised by the very varied development fibular structures. These can be so extended as to form separating walls that dissipate the tensions generated by this extreme form of raphe development. Canal raphes are found in the Epithemiaceae, Nitzschiaceae, Surirellaceae and in *Entomoneis* in the Naviculaceae (Schmid 1979, Paddock & Sims 1981).

4.4.1 Median raphes

Ever since O. Müller (1889) described the median raphe, an organelle unique in nature, it has provoked discussion about it's evolution and structure. Only through study of ultrathin sections under TEM and of broken valve fragments and SEM has it been possible to lay aside much of the speculative nature of these deliberations. Toman & Rosival (1948) showed in a TEM investigation that the raphe was a continuous fissure, and Reimann et al. (1965) that the raphe fissure is not closed by a membrane. Drum et al. (1966) demonstrated that the raphe fissure did not contain protoplasm (the streaming of which was previously thought to be the cause of movement), and that fibrils were responsible for locomotion.
The basic structure of the median raphe is of two longitudinal fissures lying more or less in the middle of the median costa, and separated by the median nodule. In transapically heteropolar genera, e.g. *Cymbella*, *Amphora* or *Gomphocymbella*, the median costa, and thus the raphe fissure, lie more or less ventrally displaced, adding to the asymmetry of the valve. At the distal end of

the raphe fissure there is a terminal nodule on the inner surface of the valve. This often has a lip-like process, the helictoglossa.

The simplest median raphe is seen in the raphe valve of many Achnanthaceae. A raphe valve fragment of *Cocconeis pediculus* where the central nodule (a) was disrupted during preparation is shown in fig. 13: 2. The walls of the raphe slit (b) transect the relatively thick pored surface at right angles. Robust, vertically oriented, round central and terminal pores delimit the raphe slit proximally and distally. Additionally, a terminal nodule with a weakly developed helictoglossa is located distally on the valve's inner surface. The typical median raphe with all it's individual elements is seen, with few exceptions, in the Naviculaceae. Functionally, all species with a median raphe have their central system of costae comprised of a collection of important static elements. These are, from the center of the valve towards the valve end: the central nodule, the costae on either side of the raphe-slit, the terminal nodules with their adjacent terminal costa, and the polar valve mantle.

The raphe slit differs in structure not only between species, but also along it's axial course (figs 12: 1–5). In the middle of each branch of the raphe it appears to us a plicate raphe-slit in the shape of a "V" on it's side (fig. 12:s1), or less frequently as a key and feather raphe (fig. 12: 2). Although there are intermediates between the two forms (fig. 12: 3), the key and feather raphe has qualitatively different structure form the simple, plicate raphe-slit. This is made clear by the function of the raphe slit. Firstly it is an important organelle for locomotion (dealt with later), and secondly it both splits and unites the both valve halves. The latter is achieved through hinge-like interlocking along wide stretches of the raphe branches. In contrast to this, the plicate raphe is fairly primitive, only being found where the general valve structure is stabilised by it's massive nature. Movements of both valve halves are restricted and loss of plasma through the raphe slit is prevented. This simple type of raphe-slit is found above all in small species like *Achnanthes minutissima* or *Navicula diluviana* (k in fig. 12: 3). In more advanced key and feather raphe, the tongue-shaped feather (fig. 13: 8) fits precisely into the raphe groove (figs 12: 2, 13: 3, 13: 5, 13: 7). This construction not only hinders pervalvar shearing, but also shearing movements of the valve along the raphe slit itself. Particularly good examples of this are seen in the complex raphes of some *Pinnularia* species, which have two feather and key elements in certain cross-sections (fig. 13: 7, arrow). Such a construction clamps both valves firmly together. Such a complex structure is normally only found in the central portion of the median raphe (fig. 12: 5) where, as in *Cymbella cymbiformis*, the outer raphe-slit is bowed strongly dorsally. Proximally and distally the raphe-slit is vertical in most species. In some *Cymbella* species there is proximal twist of the key and feather, giving what is called a reverse lateral raphe (d in fig. 12: 5). The course of the outer and inner slits as well as the edges of the key and feather are important taxonomic characters and are used in many diagnoses.

If no difference between the outer and inner slits can be seen in LM, the raphe is described as filiform. The presence of two or more parallel lines denotes a lateral raphe, and here the sections where raphe is lateral are often specified. Filiform and lateral raphes are not morphological terms, but simply a result of the limits of resolution under LM. Normal oil immersion objectives can differentiate between two lines that are 0.3–0.5 µm apart.

In the Naviculaceae there are in general five forms of the proximal raphe ends:

1. The raphe-slit widens continuously from the about a third to one quarter of the way towards the central nodule.

2. The raphe-slit runs unchanged to the central nodule and then makes a short turn to the left or right, or in both directions, forming a "T".
3. The raphe-slit widens suddenly at the central nodule to form a round or oval hole, the central pore.
4. As 3, but superficial slits continue from the pore in the same or reverse direction. These slits can have widely differing structures.
5. All the structures mentioned in 1–4 are absent and the raphe-slit lies in a variously formed groove-like central fissure.

The raphe structure varies very little and is thus a good character for lower taxonomic categories. Examples of such fissure-like structures are shown in figs 3: 1, 15: 1, 15: 3, and 15: 5.

The proximal raphe ends are most commonly delimited by differing forms of canal-like expansions, a well-tested structural form often used in engineering to dissipate tearing strain in brittle materials. The central pore can form a hole as in *Stauroneis phoenicenteron* (fig. 15:2), where it is additionally bound by the raphe costa. A modification to this round central pore is seen in the hammer-shaped central pore of *Frustulia rhomboides* (fig. 4: 2), and again in it's terminal pore (fig. 15: 8). Generally however, the proximal and distal ends of the raphe are seldom similarly developed. An example of this can be seen in the short rudimentary raphes of the "araphid" valve of *Rhoicosphenia abbreviata* in which pores terminate each raphe branch (Mann 1982). In valves with an especially strong valve mantle, specialised structures delimiting the raphe slit at the valve end are absent (figs 15: 6, 16: 8). A well developed terminal nodule (figs 16: 3–8) delimits the raphe slit. More or less longitudinal oriented terminal fissures on the valve's outer surface dissipate tension towards the valve mantle (figs 16: 1–2). These terminal fissures are notches on the terminal costa, and widen between the terminal nodule and the polar region of the valve mantle (b in figs 16: 3–4). Very large species such as *Cymbella aspera* can have two such terminal costae which however have only a single terminal fissures. In many species there is a further safety mechanism against tearing within the terminal fissures. It is either in the form of pore-shaped pit (fig. 16: 1, arrow), or a further nick in the terminal fissure itself (fig. 16: 2, arrow). A lip-like structure, the helictoglossa, delimits the raphe slit on the valve's inner side in many species (fig. 16: 3–8).

Raphe slits, as already intimated, are accompanied by various structures in different groups. The raphe ledge of many *Amphora* species (b in fig. 13: 6) and the conopeum of many *Navicula* species deserve a particular mention here.

The **raphe ledge** in many *Amphora* species is an important stabilising element of the valve, imparting reinforcement in the same way as an L- or T-beam, to the valve in addition to that given by the median costa.

The **conopeum** is a thin, membranous structure that is often destroyed during preparation. It may cover an area just in the region of the axial area (*Navicula bacillum*, fig. 17: 4), or almost all of the alveolate vale surface (*Navicula festiva*, fig. 17: 3). In both cases the edge of the conopeum is corroded, and it is thus incomplete. Structures of this type were first described by Kaczmarska (1979) and by Sims and Paddock (1979), though an indication of their presence in some *Mastogloia* species was given by Ricard (1975). Since then this type of structure has been found in many species, e.g. *Navicula monoculata*, *Navicula pygmaea*, *Navicula pseudomuralis* (Lange-Bertalot 1980). They are rarely observed under LM, where their presence is suggested, as in *Navicula festiva*, by an area of very poorly contrast where it covers the transapical striae in the valve center.

4.4.2 Canal raphes

While the median raphe lies in the valve surface along or slightly ventral to the apical axis, the canal raphe lies along the valve edge or has moved onto a keel (figs 12: 6–7). It thus strongly influences the valve stability, but in completely different way from the valve-splitting central raphe. There are two important features:

1. The valve-surface itself is generally relatively simply built, as tension does not have to be dissipated at a few, strongly developed nodules.
2. The weakened stability of the valve margins is corrected by pillar-like structures referred to as fibulae. These fibulae have quite different structures according to the position of the canal raphe and the valve construction. They vary according to species, genus and family from short supporting pillars (fig. 14: 1) to dividing walls stretching across the valve (figs 16: 6, 14: 4).

The canal raphe is the typical raphe construction of all species in the Epithemiaceae, Nitzschiaceae and Surirellaceae, although it is also found in some species of *Entomoneis* in the Naviculaceae (Schmid 1979, Paddock & Sims 1981).

The organisation of the canal raphe in *Epithemia* and *Surirella* is shown in figs 12: 6 and 12: 7. In contrast to the median raphe, where the raphe slit opens into the valve interior, the raphe slit (a) of the canal raphe is associated with an apically running raphe canal (i), which as in *Epithemia* (fig. 12: 6), runs into the valve edge or even the valve surface. In *Nitzschia* it is generally raised over the valve surface in a raphe keel, while in *Surirella* it is often in winglike extensions (fig. 14: 6) on all four corners of the frustule. Canal raphes lying in the valve surface, connection to the valve interior is via simple opening (k), the portulae. Those that lie in a raised keel or in wings (Paddock & Sims 1977), have a canal-shaped connection (alar canal), with a portula at each end, to the valve interior.

Plicate raphe-slit development is commonest in canal raphes, but the key and feather construction is also known. Nodules are normally small in comparison to the other features. The central nodule is most obvious in *Epithemia* species, in the Nitzschiaceae it is often rudimentary, and in many Surirellaceae it is shifted towards one valve pole. There are several different types of terminal nodules and helictoglossa.

4.5 Longitudinal canals

In LM, many species have "longitudinal lines" particularly around the margins, e.g. *Pinnularia*, *Caloneis* and *Neidium*. It was recognised very early (Cleve 1894) that these structures evolved several times. On investigating several species of *Neidium* Hustedt (1935) came to the conclusion that their longitudinal lines were silica costae, differing from other longitudinal lines only by their strength and greater separation. It was only the in-depth studies of many *Neidium* species by Sims and Paddock (1979) that revealed longitudinal canals (fig. 17: 1). These are normally furnished with pores on both the inner- and on the outer valve surfaces, thus enabling metabolite exchange with the surrounding medium. Additionally, these canals possess openings connecting them with the neighbouring areolae.

In contrast to *Neidium*, Hustedt (1935) correctly identified longitudinal lines near the central costae in *Diploneis* species as canal-like cavities, once again with internal and external pores. According to Sims and Paddock (1979) all longitudinal canals run the length of the valve without interruption. The two canals (fig. 17: 2) are neither connected near the central nodule, nor are they in contact with the raphe.

4.6 Surface structures

4.6.1 Spines, connecting spines and setae

A great variety of solid processes projecting from the silica valve are covered by these terms. They have widely differing forms and functions, reaching a peak of complexity of form in many marine diatoms (c.f. Schmid 1983). In the freshwater forms discussed here these structures are relatively simple, excepting the strutted processes of the Thalassiosiraceae. The latter have been treated elsewhere under the heading of secretary organs. Other members of this family, e.g. *Cyclotella chaetoceros* and *Stephanodiscus hantzschii* have very long setae. Exceptionally long setae are also found in the centric genera *Biddulphia*, *Rhizosolenia* and *Chaetoceros*.

Pennate diatom species in various genera have, in contrast, single to numerous spines and spine-like structures on the valve surface. An example of a polar spine can be seen is African *Gomphonema* species illustrated in fig. 17:5. The massive spine of *Surirella capronii* sitting on it's solid base close to the broad valve pole is noteworthy. *Surirella tenera* var. *nervosa* posses a similar structure. Apart from these large spines, small spines scattered over the entire valve surface (fig. 17: 6) can be found in several species, e.g. *Surirella turgida* and *Surirella linearis*.

The Naviculaceae also has some species with spinules scattered over the valve surface. An example here is given by *Anomoeoneis vitrea* where the entire surface is covered in fine spinules (fig. 15: 6).

Linking spines aid chain formation particularly in the genus *Aulacoseira* and in araphid diatoms, though they do occur occasionally in other genera (figs 17: 7–8). The term "linking spine" when referring to structures aiding chain formation in araphid diatoms should not be taken literally. These "spines" are not actually pointed, but spathulate (fig. 17: 7), and the valves are not joined mechanically, but probably by the spines adhering to the other valve. A pointer to the organic nature of the bond is given by the fact that such chains fall apart on cleaning with strong oxidants.

4.6.2 Other surface structures

In many species the valve surface is not smooth, but has a relief formed in various ways (figs 18: 1–5). This feature can be a significant taxonomic character. Transapical costa are often not only well developed internally, but also externally. This is clearly seen in many chain forming species (e.g. *Fragilaria*), and hinted at in *Eunotia* species (fig. 18: 1), some *Navicula*- (e.g. *Navicula pseudoscutiformis* fig. 8: 8), and in many *Nitzschia* species.

Assorted variations of furrow-like depressions on either side of the central nodule ("moon spots") are found in some *Pinnularia* and *Caloneis* species (fig. 18: 5).

Robust costal systems can strongly influence the external valve structure, e.g. the costae in the dorsal part of the valve mantle of *Amphora aequalis* (fig. 18: 2), or in many *Hantzschia* species. *Cymatopleura solea* has an undulate outer valve surface with a characteristic fine structure (fig. 18: 3). Many *Pinnularia* species have a characteristic rise- and furrow-structure which as shown in figs 18: 4, 5, forms a differential character at species level.

4.7 Teratological structures

Many greatly disparate deviations from normal valve development are covered by this term, and today their causes are to some extent known. A "typical" teratological development is characterised by a departure from the typical pattern of structures for a species. fig. 18: 7 shows on the left, a "normal" and on the right a teratological valve of *Cymbella (Encyonema) elginensis*. In this type of teratology, it is common that the organisation of the transapical striae changes, but the areola structure, and thus the number of punctae in 10 µm remains almost the same (fig. 18: 6). Raphe structures are seldom effected (Krammer 1979b). Not only can the structure change, but also the valve shape, and to such a degree that the form can hardly be associated with the normal taxon. As such variation not infrequently appears throughout a population, there is a danger that they may be treated as a new taxon. This has in fact happened in the past.
Teratological forms can be experimentally induced in different ways. They appear commonly in old cultures, whereby i. a. the valve length is subnormal, or as in e.g. *Pinnularia brebissonii* (Hostetter et. al. 1976), where the organisation of the raphe and transapical striae is the complete opposite of normal. Schmid (1980) and Schmid et al. (1981) artificially induced widely varying teratologies during valve morphogenesis of *Anomoeoneis sphaerophora* via e.g. treatment with colchicine. All stages through to the total destruction of valve and raphe structures were achieved. These observations on old cultures and the experimental results show that the environment could have a significant effect on valve structure.

4.8 Structural elements of the cell girdle

In the freshwater diatoms considered in this volume, the cell girdle is second in importance to characters of the valve as a source of taxonomic characters, with the exception of some centric and araphid diatoms. This results primarily from the fact that the methods of light microscopy are less suited to investigation of the cell girdle, and that valve views dominate in LM. The relative similarity of the cell girdle in the Naviculaceae makes it difficult to use as an unambiguous diagnostic group of characters. Finally, it is primarily the secondary characters of the girdle, e.g. septae, that are of diagnostic use.
Generally, there are taxa in which the valve dominates, e.g. the genera *Navicula*, *Cymbella*, *Nitzschia*, and *Surirella*; and others where the girdle is the dominating structure, e.g. many centric and araphid diatoms, *Eunotia*, and *Amphora*. Von Stosch (1974) can be consulted for further details. He distinguished three types of bands that make up the cingulum, i.e. the part of the girdle attached to the valve:
1. The valvocopula – an intermediary segment joining the cell girdle to the valve mantle.
2. Copulae and intercalary bands follow on from the valvocopula and have a structure different from the distally positioned elements.
3. Pleura and pleural bands terminate the cingulum or valve girdle distally.
This degree of differentiation is however, only possible in a few valve girdles. In most cases the individual bands or segments hardly differ and are termed girdle bands. Many araphid diatoms and members of the Naviculaceae have septae in association with one or more of the girdle elements (figs 3: 3–5), then referred to as intercalary bands. These are all closed rings, with the same outline as the valve. Additionally, most species possess open bands which often appear under

LM as hairpin- like structures (fig. 10: 2). As a rule these segments are so arranged that the open sides of neighbouring segments are on opposite sides of the girdle. Several open bands can, however, be stacked in parallel above each other. The proximal and distal margins of these bands often have different structures, with one side smooth, and the other with a row of perforations that are mostly visible in LM, at least in larger forms (fig. 19: 3). A tonguelike projection that fits into a recess in the previous band or the polar valve mantle is often present (Krammer 1981a). In centric diatoms this projection is called the ligula, and in the pennate diatoms there is at least a functionally similar structure (figs 19: 2, 7). The number of bands varies strongly between genera. Two to three bands are usually present in the Naviculaceae, though there are exceptions with numerous girdle bands, e.g. many *Amphora* species (figs 19: 5–6). The latter is also true for many genera of araphid diatoms (figs 19: 1, 2, 4), *Eunotia* and *Epithemia* (fig. 19: 7) to name but a few examples.

Only in relatively few cases do the structures of the valve and bands lead to a stable mechanical union. It is thus not surprising that oxidation of the organic substances of the cell walls leads to dissembly of the individual elements. An organic connection between the valve and girdle bands can be assumed, with the perforations in the girdle bands allowing, so to say, doweling with organic substances. Von Stosch (1981) was of the view that the diatotepum has a prominent importance in holding together the valve and girdle bands. Siliceous structures binding the valve and valvocopula together have been described by Gerloff & Rivera (1979) in *Cocconeis pediculus*. The valve and girdle of this species can only be separated after destruction of these silica structures. Similar structures can be seen in *Cocconeis placentula* (fig. 25: 3). In many families of pennate diatoms species can be found with intercalary band walls running more or less parallel to the valve surface. These septae penetrate the cell to varying depths. They can be narrow bands at the periphery, or almost span the entire frustule and be divided into chambers. They can be relatively flat as in many *Epithemia* species, bent as in *Tabellaria* (fig. 3: 3), or undulate as in the marine *Grammatophora* (fig. 3: 5). In many genera they are the main diagnostic character, e.g. *Mastogloia* (fig. 3: 4). During frustule morphogenesis, they grow inwards from the intercalary bands forming a suture, often clearly visible in LM, where the elements join.

4.9 Inner valves

Individuals with a second valve within the frustule occur sporadically both in wild populations and in culture. Von Stosch (1967) showed that two different phenomena, division and resting spore formation, which appear similar at first glance, are involved here. Fundamental differences in their formation, morphology and physiology exist between them. Both processes are the result of exogenously induced metabolic disturbances and thus suboptimal conditions for growth, e.g. in old cultures.

Morphologically they differ in that by division, the form of the inner valve hardly differs from the normal valves of the vegetative cell (e.g. *Eunotia serpentina* in Hustedt 1930–1959, fig. 28), while the resting spores differ both in their three dimensional form and in their fine structure from vegetative cells (e.g. *Eunotia "soleirolii"*, von Stosch & Fecher 1979). The principal difference is in the function of the two inner valve containing forms: divisions do not serve as resting spores and continue division immediately favourable conditions occur, and there is no complex germination procedure. In contrast, resting spores can remain inactive for weeks or months after favorable conditions return.

4.9.1 Inner valve formation (Häutungen)

The processes involved in this type of inner valve formation have be dealt with in depth by von Stosch (1967) and Geitler (1980). According to Geitler, inner valve formation in *Hantzschia amphioxys* and *Achnanthes coarctata* can occur in several ways. Firstly two daughter cells can be made in such a way that one, or two, daughter cells result. Alternatively, a mitosis can occur, but both daughter nuclei remain within the same protoplast. Inner valve formation is thus associated with an unequal cell division. In contrast to resting spores, these cells do not accumulate reserves. If they are transferred to fresh medium, division resumes (Geitler 1980) and at first two daughter cells are produced, each with an epivalvar inner valve, that then separate as normal. The inner valves are identical in structure to the outer valves, possess a normal raphe, and cells with a earlier inner valve can move normally.

4.9.2 Resting spores

Resting spores, also called "double valves, inner septae, inner thecae, or double thecae", serve to tide over unfavourable conditions. They differ from vegetative cells in five ways (von Stosch & Fecher 1979):
1. They are a result of a specific sequence of development.
2. They have a characteristic morphology.
3. The cells are rich in reserves, especially oils.
4. They are unable to germinate without having been exposed to a cold period of a certain length.
5. The can survive dormant for up to three years.

Von Stosch & Fecher (l.c.) studied the events leading up to spore formation and subsequent spore germination in a *Eunotia* species. Protoplast differentiation in a pervalvar direction starts the process. The plastids concentrate in the hypotheca. A mitosis with an unequal cytokinesis follows, leading to two physiologically and morphologically different daughter cells. The larger of the two survives and forms a robust, strongly valve convex in the direction of the smaller valve. Now the plastids close to this new valve are displaced, and the same process is repeated on the other side of the resting spore. In girdle view, the newly formed resting spore is bordered by the two colorless, narrow cells on both sides. In contrast to vegetative cells, the cell wall of the resulting resting spore has no raphe, and could be mistaken for a *Fragilaria* when examined in isolation. Rows of areolae are, however, present on the inner side of the valve, though they appear to be closed on the valve's outer surface. Girdle bands can also be present in resting spores, though in smaller numbers than in vegetative cells.

Germination is almost as complex a process as spore formation. After an obligatory resting period, two unequal divisions signal the onset of germination.

Meridion circulare builds resting spores in a similar was to *Eunotia pectinalis* varieties (Geitler 1971). A well known example of resting spore production is *Navicula cuspidata*, though a complexity is added by the presence of a "craticula". This craticula (fig. 24: 5) is supposed to appear in response to raised osmotic pressure. Schmid (1975) suggested that it served as an inner skeletal element balancing the raised osmotic pressure, and a seal for the raphe and areolae. The actual inner valve forms within these craticulae (each valve has one). In it's three dimensional form and fine structure, this inner valve is

completely different from *Navicula cuspidata* and was thus described by Peragallo (1893) as *Navicula cuspidata* var. *heribaudii.*
Resting spores are also known from many species of centric diatoms. In many cases a cell will first form a particularly thick valve. Once again an unequal division occurs resulting in one small and one large cell. The former degenerates, while the latter deposits a second thick layer of silica over the plasmalemma. According to von Stosch and Fecher (1979) there is a physiological difference between these and the resting spores of pennate diatoms. In culture, they germinate within a few days under the appropriate conditions.
An equal division however occurs in very many centric diatoms and neither cell degenerates. Acytokinetic mitosis occurs (von Stosch & Drebes 1964).
A review of resting spores and their physiology, together with copious references is found in Hargraves & French in Fryxell (1983).

5 Reproduction

5.1 Asexual reproduction and changing the outline

Reproduction in diatoms is by mitotic division, interrupted when a particular stage is reached by a cell bisexual reproduction, i.e. auxospore formation.
An exact knowledge of the reproductive sequence has a special importance in taxonomy, as fundamental changes in taxonomic characters can occur during the process. These changes additionally may be influenced by ecological factors. This is the reason why studies and identification of diatoms should never be made on single individuals, but on populations. For the same reason even the discovery of a few specimens of a particularly distinct taxon, should not be followed by a new species description. It is precisely such "species" or "varieties" that end up as synonyms of older, adequately described taxa. In diagnoses, such taxa can be recognised by very narrow ranges for measurements, or by the comment "very uncommon". There are of course exceptions, and these will detailed later.
The rigid siliceous shell of most diatoms determines special rules for division over generations. One obvious feature is the reduction in cell size, the mechanism of which was recognized and described by MacDonald (1869), Pfitzer (1871) and Tomaschek (1873). Division is in the valvar plane along the pervalvar axis, as determined by the physical restraints of the frustule. The cell contents swell, forcing the hypo- and epitheca apart though not separating them. Occasionally the girdle is widened by the production of additional intercalary bands, making the cell considerably wider than an undivided vegetative cell. After mitotic division, which is generally concurrent with chloroplast division, the daughter nuclei and chloroplasts migrate in opposite directions along the pervalvar axis. Cell wall deposition for the daughter cells after protoplast division. These new cell walls are always laid down within the epi- and hypotheca of the mother cell, so that the former's valves form the hypotheca of the daughter cells. Lastly, the elements of the girdle are complimented.
This short description of division shows that two cell sizes are produced by each division:
1. One cell the same size as the parent cell – that with the epitheca of the parent cell as epitheca.

2. One cell about a double valve thickness smaller than the parent cell – that with the hypotheca of the parent cell a epitheca.
In this way the average size of a vegetatively reproducing population will decrease with time. This feature of aging in diatoms makes them unique amongst unicellular organisms.
Volume changes with size, and in the case of *Melosira varians* (von Stosch 1965) it can be reduced to 1/20 of the maximum cell content. The shape of the cell can change during this process to such an extent that a large specimen appears to have little in common with a small one. A detailed study of morphological changes during the life-cycle of pennate diatoms was made by Geitler (1932 i. a.). The conclusions were as follows:
1. The apical axis length decreases not only in absolute, but also in relative terms, more quickly than the transapical axis, i.e. small cells are relatively broader than large cells.
2. The same is true for the valve mantle and girdle, so that in girdle view smaller cells are wider than large cells.
3. The valve outline of smaller cells is "simpler" and more rounded than that of lager cells.
4. The fine structure of valves (costae, punctae, lineolae) change little in size in comparison to the apical axis. Their number per unit area remains, within narrow limits, constant.
5. The thickness of valves, girdle bands and intercalary bands decreases with each division.
Geitler (l.c.) noted that particularly elongated valves are most effected by the above named rule 1 (the length-breadth ratio change), while in more rounded cells, e.g. *Cocconeis placentula*, only detailed measurements show any changes. Rule 3 has a strong effect on the development of the valve middle and ends. In a number of genera, valves are swollen in the middle, a feature that, according to Geitler (l.c.) can be traced back to auxospore production. This tends to be indistinct or absent from smaller cells. Capitate or elongated cell ends also tend to become less marked as cell size reduces, and concavities and ledges evened out.
The above rules are not without exceptions. An example here is outline of *Achnanthes hungarica* which becomes more complex with progressive divisions (Geitler 1980). It should also be mentioned that cell physiology changes with reducing cell size. Achnanthes species slowly loose the ability to produce mucous stalks. Pennate and centric diatoms become sexual at a particular cell size, but if the cell size decreases further, the ability to become sexual is lost (Geitler 1932, von Stosch 1954).
Division frequency varies greatly between individual species. In cultures of *Navicula seminulum* it was less than 24 hours.
Not all diatoms change their form during vegetative division. Some can evade this process by various techniques. Diatoms with particularly thin cell walls often have girdle bands possessing limited elasticity. According to Hustedt (1969), it is almost impossible to detect size differences in a *Fragilaria* chain of 2000 cells, which results from a calculated eleven division cycles. Membrane stretching and other similar mechanisms (von Stosch 1955) can also have a similar effect.
From time to time "secondary growth" is discussed. This concept does not however include the cell-size maintaining strategies mentioned above (Hustedt 1930, Geitler 1932). The only way a species can regain it's maximal cell size is via auxospore formation. Experimentally, von Stosch (1965) was able to produce vegetative cell enlargement by osmotically enlarging protoplasts fully or

partially removed from the parent cell. New valves were subsequently produced by the protoplasts.
If small cells find no suitable mate for auxospore production while they are becoming sexual, they divide further and finally die. In taxonomic literature this type of cell is often referred to as a "Kümmerform".

5.2 Sexual reproduction – auxospore formation

Taxonomically, the post-auxospore cells are of particular significance, in that they allow the maximal cell size to be restored. The small end-products of vegetative division are called the mother-cells; the cells arising from the allogamous union, the daughter cells.
Ever since the studies of Karsten (1898, 1900) and the classic work of Klebahn (1896) on auxospore production in *Rhopalodia gibba*, important observations on sexual reproduction in diatoms have been made, in particular by Geitler and von Stosch. As only a brief insight into the relationships that occur in auxospore formation can be given here, detailed information should be sought in the reviews of Geitler (1932, 1958, 1969, 1979), von Stosch (1950, 1951, 1958a, 1958b, 1964), Fritsch (1965), Drebes in Werner (1977) and Ettl (1980).
Those who want to find auxospores in their samples should fix them well, so that the cohesion between parent- and daughter-cells is not disrupted, and that the cytological details remain visible. Spring and autumn are the best seasons for finding auxospores.
All diatoms having sexual reproduction are diploid. Meiosis in the diploid parent cells of the pennate diatoms produces non-flagellate haploid iso- or anisogametes. The latter differ only in the ways in which they move during copulation. The conjugation of two gametes results in a zygote which grows subsequently into an auxospore. Some exceptions to this pattern can be found however. A prerequisite for sexual reproduction is that a particular cell size is reached during vegetative division. Cells can only become sexually active within a narrow cell-size range. If a sexual partner is not found, and division continues, then the cells will die.
Allogamy and isogamy in the auxospore formation of *Epithemia zebra* var. *saxonica* (Sunda form) (fig. 22) was described by Geitler (1932, 1977). When the gamete parent cells have reached 32–66 µm (in contrast to an initial cell length of 150–180 µm), two position their ventral sides together, even although a large size difference exists (fig. 22: 1). Mucilage, in particular from the cell poles, binds the cells together. The protoplast separates from the cell wall, vacuoles disappear, and the protoplast contracts apically and transapically while remaining unchanged along the pervalvar axis (figs 22: 1–2). At the same time, a jelly is produced within the cell which differs from the "bonding jelly" i.a. in it's refractive properties. Cytokinesis now occurs in connection with a reduction division (fig. 22: 3), and is followed by the second meiosis. The latter can also occur in the undivided protoplasts. Finally, the two new protoplasts rotate by about 45 (figs 22: 4–5), though this can be 90 in other species. Two gametes survive from those produced by each parent cell in the two division phases. During this process, the epi- and hypotheca part until an slitlike opening appears. Each gamete pair then forms a connecting conjugation tube from conjugation jelly (fig. 22: 6–12). The nuclei first migrate into the tube, and then the rest of the protoplasts and chloroplasts. Nuclear fusion occurs in the tube middle, and the resulting zygote lays down a protecting wall. Each zygote produces an auxospore oriented in the direction of conjugation and at right angles to the apical axis of the parent cells (fig. 22: 13). Through the production

of large amounts of mucous, the auxospore forces the valves of the parent cells apart, eventually surpassing them in size. When both valves are seen simultaneously, valve views are presented.

Different species modify this pattern in different ways. In gametogenesis two normal nuclei usually result from the first meiotic nuclear division. One variation on this is where the nuclei become pyknotic or are reabsorbed (Geitler 1932, 1952). After the completion of meiosis I, chloroplast division followed. One of the two nuclei produced during meiosis II may degenerate and either be reabsorbed by the protoplast, or remain in a pycnotic state within the gametes (Geitler 1951, 1952) also gave examples where both nuclei remain functionally intact within the gametes.

The sequence of events in the isogamous conjugation of *Epithemia zebra* is also found (Geitler 1932) in *Denticula*, *Rhopalodia* and *Amphora*. In *Epithemia zebra* the apical axes of the parent cell and auxospore are at right angles (fig. 22: 13). An exception to this rule was given by Mann (1982), where in *Rhoicosphenia abbreviata* in spite of isogamous conjugation, the valves of the parent and daughter cells lie parallel. Sessile and motile gametes are however, more commonly produced (functional anisogamy). Once again, the apical axes of parent- and daughter cells lie parallel, an arrangement very common in the Naviculaceae.

The fact that the auxospore formation method is strongly dependant on how motile a species is, is supported by studies so far done on araphid species. Here, either two auxospores, as in *Fragilaria fasciculata* (syn. *Synedra affinis*), or one auxospore is formed from a single parent cell as a result of vegetative division. In addition, there are many species in which auxospore formation follows a simpler path. Thus, two gamete mother cells can each produce a gamete, each of which in turn produce a single auxospore (some *Cocconeis* species, *Eunotia formica* i. a.). Finally, a parent cell can produce an auxospore through automixis or autogamy, i.e. the conjugation of two gametes of one parent cell, or two nuclei of a single parent cell (Geitler 1935, 1952).

In contrast to the pennate diatoms with their morphologically identical gametes, centric diatoms are oogamous and produce sperm and egg cells. This process was first described by von Stosch (1950) in *Melosira varians*. Later, von Stosch and Drebes (1964) observed the entire cycle of auxospore production in vivo in the marine *Stephanopyxis turris*. The cells of this species start producing gametes when they are only 2/5 to 1/5 of their maximum size. A number of flagellate haploid sperm are produced after three sequential mitoses followed by a complete meiosis (Manton & von Stosch 1966). In older literature the sperm were referred to as "microspores" in reference to their relative size. The female cell undergoes meiosis in which one each of the daughter cells degenerates so that at the time of fertilisation, the egg has only one functional nucleus. The egg-cell then elongates, pulling thecas apart so that the plasma can expand slightly outwards in the middle. It is at this point that the sperm can penetrate the egg. Fusion of the male and female nuclei then occurs, and the zygote covers itself with a polysaccharide membrane covered in silica scales. This membrane expands with the zygote as it grows to form a large auxospore. Turgor pressure within the auxospore reduces parallel with it's first mitosis, and the plasma separates from the zygote membrane on one side by plasmolysis. It is on this side that the first silica valve is formed. One nucleus of the mitotic division degenerates. After a further mitosis, the second valve is formed.

Auxospore production in other centric diatoms follows a similar pattern, with variations in the number of egg-cells or sperm per gamete mother cell. In *Melosira varians*, sperm tend to be produced in narrower chains, oogonia in broader (von Stosch 1950). The antheridia hold four sperm after meiosis.

Present day diatom systematics is almost entirely based on valve structure. Geitler however pointed out that there are small populations that distinguish themselves from other similar populations by their behaviour at division. Thus *Cymbella cesatii* var. *paradoxa* is differs from *Cymbella cesatii* var. *cesatii* partly in it's measurements, but most in the details of it's sexual reproduction (Geitler 1975). In *Cymbella cesatii* var. *cesatii* mating is either ventral-ventral, dorsal-ventral or dorsal-dorsal all with equal probability, while var. paradoxa dorsal-dorsal mating is the rule.
The weakly silicified cell wall of the auxospore, the **perizoneum**, was recognised in and described from LM studies last century. In his definitive investigations on auxospore production in *Rhopalodia gibba*, Klebahn (1896) showed that fully formed auxospores are enveloped in numerous ring-shaped, wall-elements. More recent detailed research on the perizonium exists for *Rhabdonema* (von Stosch 1962) and *Rhoicosphenia* (Mann 1982).
According to Mann, an isogamous conjugation in *Rhoicosphenia* produces two zygotes, each of which grows into an auxospore. During auxospore development the perizonium is successively built from about 35 ring-shaped transverse and 5 apically running longitudinal bands (fig. 23: 1). Progressively more transverse bands appear as the auxospore elongates. Mann (1982) recognised three types of band. The band lying around the center of the cell (P) is a broad closed ring with frayed out margins, and circling ridge in it's middle (primary band). Immediately adjoining this band are two incomplete rings (S and T), whose open sides lie on the ventral side (fig. 23: 2). The 2–3 secondary bands (S) on either side of the primary band differ substantially from the adjoining c. 15 tertiary bands.
The five longitudinal bands (LP, LS) lie under the transverse bands on the ventral side of the auxospore, leaving the dorsal side free. Two secondary longitudinal bands (LS) lie on either side of a broad central band (primary longitudinal band: LP). The transverse bands are organised in a similar way to a bipolar cingulum in that each element overlaps it's distal neighbour. The perizonium does not cover the ends of the auxospore, and it here that the initial cells formed within the perizonium first appear. Initial cell formation within the perizonium is shown in fig. 23: 2. First, the epivalve is furnished with three girdle bands laid down under the longitudinal-band free, dorsal side of the perizonium. Hypovalve formation then occurs under the longitudinal bands of the perizonium.
Perizonium formation follows a similar pattern in *Rhabdonema* (von Stosch 1962), *Cymbella* and *Pinnularia* (Mann 1982). In *Cocconeis pediculus* the process is a little different.

6 Euplasmatic structures

Many of the intracellular organelles of diatoms are similar to those of higher plants and other algal groups. The standard works of Robards (1974) or Metzner (1981) can be consulted for details. The golgi body is most similar to that of higher plants, with it's dictyosomes, chondriom, endoplasmic reticulum and ribosomes. Class specific characteristics are however found in the plastids. Their morphological structure and position can, in some cases, be species specific. Closely associated with the chloroplasts are pyrenoids. Again, form and position can be significant at the species level.
In LM one can also observe the nucleus, nucleoli and a very well-developed

vacuole. This vacuole is normally so large that the cytoplasm exists as a thin wall covering filling all the alveoli and areola, but not the raphe slit. Diatom cells that have been freed of their silica with fluoric acid have a negative image of the valve on their surface (Schmid 1980). The large vacuole is bisected by a bridge of cytoplasm in most pennate diatoms (thoroughly described by e.g. Lauterborn 1896). Strands of cytoplasm from the bisecting bridge and peripheral cytoplasm often traverse the vacuole.

The nucleus is normally situated in the middle of the cytoplasmic bridge, and is enveloped in a double membrane during interphase. Each of these membranes is c. 7.5 nm thick and are separated by a 10–40 nm wide perinuclear gap. A number of pores up to 100 nm in diameter have been found in the nuclear membrane (Robards 1974). The outer membrane is furnished with endoplasmic reticulum. Little is known about the exact form of the DNA containing structures in the interphase caryoplasma.

The nucleolus contains RNA, small quantities of DNA, and protein. It is supposed that one nucleolus is present for each haploid chromosome set in the nucleus. A number of perinuclear dictyosomes surround the nucleus, and may play a function in valve formation. Blunn & Evans (1981) found that in *Achnanthes subsessilis*, the perinuclear golgi apparatus produces mucous releasing vesicles. This mucous is rich in polysaccharides and is thought to be secreted through the raphe. The dictyosomes (golgi apparatus) were first described by Pfitzer (1871) in *Pinnularia* as "Doppelblättchen" and "Doppelstäbchen". They are 0.5–2.5 µm in diameter and occur in pairs. Drum & Pankratz (1964) and Stoermer (1965) determined that they have the same structure as in higher plants. Refer to the literature cited above for further details on the finestructure of the dictyosomes, with their layers of membrane-surrounded chambers. Today, the prime function of dictyosomes has been shown to be material synthesis.

While cell nucleus is the information center of the cell, the **mitochondria** are the powerhouses of the cell. In diatoms they normally lie peripherally, between the chromatophore and the plasmalemma, although mitochondrial branches can enter into the alveoli and areolae. Caloneis amphisbaena has a mitochondrion or branch thereof in every alveolus (Walker, Sicko-Goad & Stoermer 1979). Mitochondria are highly complex and contain both RNA and DNA, though differently organised from that in the nucleus.

Many organelles are bound by a membrane. It's function is important and the structure is the same in most details to that of the elementary membrane described by Robertson (1964). The nuclear membrane has already been mentioned. Other important membranes are the alloplasmic cell membrane (plasmalemma), the membranes of the vacuole (tonoplast), that of the SDV-vesicle (silicalemma). The plasmalemma with it's 10 nm thick wall, is a very thick membrane acting as a permeability barrier for resorption and secretion. The tonoplast is much thinner and functionally very different from the plasmalemma. In contrast to the latter, it remains semipermeable after cell death. Enclosed in it's tonoplast, the **vacuole** acts as the garbage can for metabolic wastes and as a reserve for storage products of the cell as in other plants. These are additionally important in osmotic regulation.

In studies on living diatom cells **storage products** are particularly noticeable, particularly the strongly refractive oil droplets in the cytoplasm. The most important storage product in diatoms is, however, the polysaccharide chrysolaminarin, which is store in solution in special vacuoles. These vacuoles are also strongly refractive and can easily be mistaken for oil droplets. Spherical bodies embedded in the cytoplasm and having no relation to other cell organelles, were described by Hustedt (1930b) and Geitler (1932, 1977) in the Epithemiaceae.

According to Drum & Pankratz (1965), they are surrounded by a membrane. Both Drum & Pankratz and Geitler suggest that these spherical bodies are symbiotic, prokaryotic organisms.

Of all organelles, the **chloroplasts** and their accompanying **pyrenoides** have been most used in the past in diatom taxonomy. Mereschkowski (1903) split the diatoms into groups based on chloroplast structure. More recently, Cox (1979, 1981), pointed out the importance of chloroplasts in the taxonomy of naviculoid diatoms. Additionally, Geitler reported on the taxonomic significance of chloroplasts in various studies. Generally, chloroplasts in species of *Cymbella* subg. *Encyonema* lie on the ventral side of the valve, while in most other species with dorsally turned terminal fissures, they lie dorsally. Pyrenoid structure and position may sometimes also be used to separate groups. In LM, chloroplasts may appear plate-, ribbon- or grain-shaped (Heinzerling 1908). Two are present in many pennate diatoms. They lie on either side of the girdle bands and stretch from on terminal nodule to the other. I n some cases the chloroplasts are united by a bridging element in the middle. In *Cymbella* one of the chloroplasts lying on the dorsal or ventral side of the girdle, possesses rag-like processes towards the ventral or dorsal side. More complex arrangements, where the chloroplasts are often divided, are found in many genera. In many genera, chloroplasts of differing form may be present. *Pleurosigma* has both ribbon-shaped and grain-like chloroplasts. In species where the cell is divided into chambers by septae, the chloroplasts often produce extensions into the individual chambers.

Chloroplast ultrastructure is sketched in figs 21: 1–2. As in other algae, they are isolated from the cytoplasm by a double membrane, and differentiated from the plastids of higher plants in the details of their lamellar structure and pigments. The organisation of the thylakoids is class-specific character. Diatom lamellae are comprised of three thylokoids, whereby the outermost lamellae (girdle lamellae) are saucer-shaped and enclose all other lamellae (fig. 21: 2). The enclosed thylakoid lamellae without exception, lie along the longitudinal axis of the plastid. A fold of endoplasmic reticulum (figs 21: 1–2 chloroplast ER) that is connected to the double membrane of the nucleus, encloses the chloroplast in addition to the chloroplast membrane.

Living diatoms are characterised by their yellow-brown colour. This results from the presence of different xanthophylls, e.g. fucoxanthin, diatoxanthin, diadinoxanthin and various carotinoids. The various pigments can be present in different combinations, so that a cell may appear yellow, olive-green, olive-brown or brown. Pinnularia viridis thus earned it's specific name from the green colour of it's chloroplasts.

A naked **pyrenoid**, penetrated by a few simple lamellae, is closely associated with the chloroplast (fig. 21: 1). The lamellae are always comprised of only two thylakoids. The number of lamellae (1–3) is constant within a species (Drum & Pankratz 1964). Most of the authors mentioned studied pyrenoids surrounded by a membrane, except for the membraneless pyrenoid of *Gomphonema parvulum*. There still is no consensus of opinion as to the function of the pyrenoid, but it is thought to be involved in metabolic processes. In contrast to other algal groups, pyrenoids have so far, no significance in diatom systematics.

7 Valve morphogenesis

The origins of the diatom cell wall were thoroughly studied over the last decades. Si-metabolism received special attention. As only an overview of the resulting finding can be presented here, the reader is referred to the reviews of Schmid et al. (1981) and Volcani (1981). Morphologically, there is a major difference between the formation of a hypotheca resulting from vegetative division and the creation of a completely new valve from an auxospore. In both cases the process of silicification is identical. Valve formation follows a strict pattern that has been described by Reimann et. al. (1965, 1966) for vegetative division. After a cell has divided, a continuous, flat bag comprised of an elementary membrane, the silicalemma forms under the plasmalemma of the newly formed cell surface. It is important for the polymerisation of silicic acid required for valve formation. Within this vesicle the SiO2material for areas close to the raphe or median costa is first secreted, then that for the rest of the valve (fig. 21: 1). Silica deposition thus does not occur evenly over the entire valve surface, but discretely, following a predetermined pattern. The plasmalemma is left lying on the surface of the valve during morphogenesis of most species investigated. The inner layer of the silicalemma then takes over the function of the plasmalemma in the new valve. Finally, the plasmalemma produces the organic layer of the valve.

Pleura formation follows a similar pattern, though each band has it's own silicalemma and silicification proceeds from the mantle edge outwards. Once again, the organic layer (diatotepum) of the valve is formed last.

The silica deposition vesicles (SDVs) surrounding the silicalemma contain stored silicates in the form of small nodules, 30–40 nm in diameter (Schmid 1979). During morphogenesis in *Navicula permitis* (as *N. pelliculosa*), these nodules can be clearly seen around the central nodule (Chiappino & Volcani 1977). In the second stage of silicification, the internodular spaces are filled until a homogenous wall is produced. The silica produced during this second phase appears to be soluble in water and acid, as the nodular structure is clearly visible in corroded valves (fig. 3: 1). Nodule size differs between species.

Silicification in raphe-bearing valves begins around the central nodule irrespective of the raphe position, and at the median costa in rapheless valve. In forms with a median raphe, the median costae lying on either side of the raphe are formed outwards from the central nodule (Geitler 1932, Reimann 1960, Schmid 1976 etc.). The median costae grow along the apical axis towards the poles, showing first a "growth zone" of nodule deposition, and then a "reinforcement zone" where the nodules are consolidated as described above. One half of the median costa often (*Navicula permitis*) grows faster than the other, so that the two parts meet at a predetermined position between the central and terminal nodules. This is the position of the Voigt-discordance (fig. 23: 5) as described by Voigt (1943, 1956). Depending on the position of the canal raphe, silicification begins towards the valve margins on either side of the raphe slit. Going outwards from the costae system of the raphe, first the transapical costae or other transapically running structures, and then the apical costae or the intercostal costae are laid down. In the frustules of the Surirellaceae the central nodule is shifted towards the pole (the head-pole in heteropolar forms), and this is where silicification begins (Schmidt 1979 a, b).

Silicification in centric diatoms generally begins in the centre of the new hypovalve and progresses centrifugally towards the margins. Centripetal development has been described from Cyclotella and Melosira, though further investigation is required here.

Geitler (1963) hypothesised that the formation of each new valve is associated with mitosis, whereby one of the nuclei produced by this acytokinetic process degenerates. This hypothesis has since been verified in many studies, and has been observed not only in vegetative division, but also during auxospore formation (von Stosch & Kowallik 1969). Von Stosch & Fecher (1979) and Geitler (1980) showed that this is also true for the formation of inner valves.

The previously mentioned **Voigt discordance** was described by Voigt (1956) as a regular "structural irregularity" occurring particularly in the Naviculaceae, and lying ... to the transapical axis, directly on the axial area (figs 24: 3; 25: 2). It is most easily seen in species with a broad axial area and regularly arranged transapical striae. According to Voigt, the relative distance of these structural irregularities from the central nodule is species specific. Species with terminal fissures bent in the same direction have the Voigt discordance on the valve side pointed to by the terminal fissures. Detailed morphological investigations have shown that the Voigt-discordance is often found in it's typical form. Many exceptions exist in which there is asymmetry due to only one discordance being visible on one valve, or where structural irregularities occur at the valve margin or even in the region of the mantle (figs 24: 1, 2, 4; fig. 25: 1). Future studies should be directed at providing a morphogenetic explanation of these observations.

Valve formation in auxospores differs substantially from that in vegetatively dividing cells. Firstly, the haploid gametes and diploid zygotes are naked, having no silicified cell-wall. Silica scales can be found in fully grown auxospores of centric diatoms. In pennate diatoms a belt of weakly silicified tyre-like bands completely enclose the cell (the perizoneum). Actinkinetic mitoses within the perizoneum produce the epi- and hypovalves of the new vegetative generation. Their form and structure are quite different from that of normal cells (fig. 23: 2). These initial cells are characterised by one valve having no a girdle, and an enlarged valve mantle; and the other valve having a normal girdle with girdle bands. The initial cells are also known as sporangial cells. Up to five divisions are necessary for the characteristic form and structure of a species to be reached. The products of each division can be quite different. In the Naviculaceae the initial cell and the first division products differ from normal cells in their shape, valve-pole structure, and in various irregularities, reminiscent of teratologies, in the fine structure. Their valves are often undulate in the middle. The number of punctae or lineolae in 10 μm is virtually identical with that of normal cells however, and can often be used to align this type of cell with those of populations under study. Figure 23: 2 shows the organisation of perizoneum and the newly formed cell of *Rhoicosphenia abbeviata*as described by the in-depth investigations of Mann (1982).

Although much has been learned about cell-wall morphogenesis in the past years, an understanding of the mechanics of the processes involved is still in it's early stages. Thus we know little about which structures have their development controlled genetically, and which are influenced in other ways, e.g. lines of tension.

8 Movement

Few problems have aroused as much interest with diatoms in the past as how they move. All the details differ from those of all other plants and animals. Their ability to move was what spurred Ehrenberg to put forward thesis that diatoms were animals. His main argument was however, that diatoms had a mouth in the region of the central nodule, as evidenced by the aggregation of colored particles at this location when diatoms were placed in an appropriate suspension.

Although Ehrenberg's animal hypothesis was given up in the second half of the nineteenth century, it was only with the advent of better optics at the end of that century that new observations lead to further hypotheses. Only diatoms possessing a raphe were found to be able to move. O. Müller (1889, 1894) and Lauterborn (1894) delved into raphe construction and discussed at length the mechanics of movement. Müller's hypothesis was that a jelly filament like an endless band moved along the inner and outer raphe slits and through the central pore and hypothetical pores at the raphe ends. This idea so caught people's imagination that it is still encountered today, even although disproved by Lauterbach. In the second half of this century new methods have been used to investigate movement in diatoms, and this is reflected by the numerous publications on the subject. Reviews or particularly important results can be found in Drum & Hopkins (1966), Harper & Harper (1967), Gorden & Drum (1970), Harper in Werner (1977), Edgar (1979, 1982) and Edgar & Pickett-Heaps (1982). Accounts of the earlier studies can be found in Fritsch (1935), Drews & Nultsch (1962) and Lewin & Guillard (1963).

In principle, two problems must be addressed with respect to diatom locomotion:

1. A physiological one in which chemical energy is converted into mechanical energy.

2. A mechanical-fluid dynamic one in which mechanical energy is transferred into locomotion.

The first problem has be mostly resolved in the last few years through investigations into the ultrastructure of the cell plasma. Thus, Drum (1966), Drum & Hopkins (1966) and later other authors found e.g. bundles of microfibrils in the form of contractile elements, similar to the fibrils of muscles, in the cytoplasm under the raphe slit of *Caloneis amphisbaena*. microfibrils are cell structures found everywhere where contraction and motility are involved (cf. Sitte in Metzner 1981). In striated muscle the microfibrils contain two proteins that are involved in contraction: actin and myosin. They play a decisive function in the conversion of chemical energy into mechanical energy. Edgar & Picket-Heaps (1982) showed that NBD-phallididin could be isolated from the plasmalemma-associated bundles of contractile, microfibrils under the raphe of *Navicula cuspidata*. The authors were of the opinion that this indicated the actin-like nature of the microfibrils. The bundles of fibrils found by Drum & Hopkins (1966) and Nutsch (1974) are probably associated with the translocation of vesicles (the "crystalline bodies" of the authors cited). The energy related processes and their control are probably also steered by them.

The second problem, that of the translation of mechanical energy into locomotion, has attracted a number of hypotheses, none of which have yet been verified (cf. Drum & Hopkins 1966, Harper in Werner 1977, Edgar 1982, Edgar & Pickett-Heaps 1982). Some events are however recognized by most authors as a prerequisite to locomotion. Linear flow, in the opposite direction to that of diatom movement, in one or more raphe-systems, and the contact of the raphe-systems with the substrate are two such prerequisites. Diatoms reach speeds of over 20 µm.s-1with their locomotion system. The performance of the diatom locomotion system varies markedly between raphe-systems and species. Harper & Werner (1977) quoted speeds for e.g. *Amphora ovalis* of 4.5 µm/s, *Nitzschia linearis* of 24 µm/s, while Williams (1965) recorded a speed of 240 µm/s in one *Nitzschia* species.

9 Methods

9.1 Preservation of field samples

The preservation method used for collected diatom material depends on how it will be used. If only the silicified valves are to be examined, then one only has to add chemicals that prevent rotting. A few crystals of thymol per tube, a little chinosol or most commonly, formaldehyde are the normal preserving agents. Enough drops of the latter to bring the concentration up to a 1–2% solution is all that is needed. Kolbe (1948) however pointed out that very week formalin solutions could damage the fine structure. It is known that formalin breaks down into alcohol and formic acid. Reimann (1960) demonstrated that even in water with an extremely low formalin concentration, silicic acid was released from the silica valves of diatoms. On the other hand, the quantity of SiO_2 dissolved is so small that after 21 months in formalin solutions of up to 10% concentration, no change in fine structure is visible under an electron microscope. To eliminate the risks of erosion, the formalin solution can be buffered to c. pH 7.5. Hexamethylentetramine is often used for this purpose.

In bottom samples collected for ecological study, it is important to know if the species found are subfossil or recent. In this case the cell contents must be fixed. For studies on the chloroplasts, small quantities of material can be covered with a 10% formalin solution. The material must then be taken into 96% ethanol, where it can be permanently stored. Material for cytological studies an be fixed in a 96% alcohol and glacial acetic acid mixture (5:1). This mixture does not keep. Small amounts of material are quickly plunged into this solution for fixation. 96% alcohol on it's own can also be used.

Living and dead cells can be easily differentiated if acetic carmine is used after preservation in formalin. According to Stosch & Reimann (1970), even species with a weakly to barely silicified cell wall can be detected in this way.

9.2 Separation from inorganic substances

An initial examination of the raw material should always be made, as the preparation method is dependent on material composition. If delicate, lightly silicified, or almost unsilicified forms are present in the material as revealed by aceto-carmine staining, the act of fractionating the material can lead to valve disruption, and it is thus best avoided. Should raphe-bearing pennate diatoms predominate in a fresh sample, then the living cells can themselves be induced to leave the substrate. The entire sample should be placed on a saucer and a fine cloth sieve placed on top and the whole moistened. After one night, most diatoms will have wandered into the cloth and it can be carefully removed and processed. Eaton & Moss (1966) stated that normally about 90% of all living mobile cells can be collected this way.

Extracting diatoms from preserved material is mostly done with sieves and sedimentation. The former separates the cells from coarser material and filamentous algae, and the latter the diatoms from sand. Plastic sieves of various sizes available in any supermarket are suitable for the first task. Sedimentation is done in 250 ml beakers. Where larger diatoms are present, the sedimentation time must be short, just a couple of minutes, otherwise the cells will settle with the sand and thus be lost. Pennate diatoms ranging in size from 30–60 µm settle at c. 1cm in 10 minutes. Sand and coarse mud settle at ca. 1 cm in 1–2 minutes, and diatoms smaller that 10 µm in length/diameter take up to 2 hours for 1 cm. The latter value is the same for the finest mud fractions, and thus a clean diatom extract is not always possible. Solutions with selected specific gravity's (e.g. Thoulet'sche solution) can be used for a better separation, but the technique is very involved and seldom comes into question.

9.3 Decalcification and separation from organic substrates

Excepting material from calcium-poor waters (e.g. the dystrophic water of northern Europe and high moors), it is almost always necessary to dissolve calcium traces in dilute hydrochloric acid and then to rinse the sample. This is particularly important if further processing with sulphuric acid is needed, as otherwise a calcium sulphate-diatom precipitate will form. Depending on the sample type as described in the previous section, the decalcification is done either before or after fractionating. In the former, plant fragments, overgrown stones etc. are placed in a beaker with dilute HCl and boiled for a while. Not only are the calcium deposits dissolved this way, but mucous stalks and other mucous structures are reduced to sugars, thus releasing the diatoms from the substrate. After letting the sample cool, the coarse remains are sieved off and the filtrate rinsed until it is circumneutral.

9.4 Removing organic remains

Most structures in the diatom cell wall are so fine that optimum conditions for LM examination must be achieved. Thus, the organic components of the cell must normally be removed. Many methods have been developed to do this, and they all have their own advantages and disadvantages. Which is chosen depends largely on how the sample is to be studied. Preparation techniques can be slit into four basic groups:

1. Heating
2. Treatment with acid together with oxidative materials
3. Hydrogen peroxide treatment
4. Enzyme treatment.

Heating has the great advantage that the frustules remain intact, and that the process is relatively simple. Together with enzyme treatment, these methods allow chain formation in centric and pennate diatoms to be studied. Difficult taxonomic problems with genera having differing valve structures (*Achnanthes*, *Cocconeis*) may be solved by these methods. This method lends itself to samples with little material or *is* containing very delicate samples. Samples must, however be fairly clean. Heating is also used when the girdle is to be examined with LM or SEM, as the frustule disintegrates to valves and individual girdle element in methods 2 and 3 above.

Well cleaned raw material is taken from distilled water and spread on a number of cover slips. After drying, the coverslips should be placed on mica sheets, resting on the metal grid of a normal laboratory tripod. Flaming is with a bunsen burner adjusted to maximise oxidation by heating, but without distorting or melting the coverslip, and never so high that the silica elements glow. Artefacts are produced if the latter occurs. The cell contents first carbonise, then oxidise to CO_2. When the sample is pure white, the process is complete. Thermostatically controlled electric heating plates or ovens can of course be used instead of a bunsen burner. When using a heating plate, the mica sheet or a polished steel sheet supporting the coverslips can be placed directly on the heater. A ceramic tile can be placed over the coverslips to raise the temperature. The ashed remains of the organic material are often troublesome. They can however easily be removed by picking up the coverslips, after cooling, in forceps and gently rinsing them in distilled water. There is no danger of loosing material as the flaming process bonds the material to the coverslip. This method is most often used material freshly collected on an excursion is the be immediately examined. All that has to be done is to take a couple of drops of material that has been well shaken in water from the same location, spread them on coverslips, let them dry, flame them and rinse well with distilled water. Rinsing will not only remove any remaining ash, but also soluble salts, so that a useable slide can be made. In calcium rich samples, a few drops of HCl can be added to the first lot of distilled water used for rinsing, and this in turn removed by rinsing in pure distilled water.

Acid oxidation is still the most commonly used method of preparing diatoms. It effectively removes all organic parts of a cell, including the diatotepin covering membrane, but has the disadvantage that the silica structures of the cell wall are more likely to be damaged. The acids dissolve one of the solid phases of silicic acid out of the cell wall, so that under higher SEM magnifications, the cell wall appears more or less jagged structure (Fig. 3: 1). Often, the fine sieve membranes, foramina lips and similarly delicate structures are damaged. Acid treated diatom valves used for SEM or TEM studies must thus always be viewed with the knowledge that some fine structures will appear as artefacts, or even be entirely missing. In LM studies such damage is of little significance. Foramina enlargement through corrosion may even enhance contrast. The advantage of most methods using acid oxidation is that samples containing much organic material can be cleaned without problem. Additionally, many methods lead to the separation of both valves, thus optimising LM studies.

All acid oxidation methods work either with strongly hygroscopic, concentrated acids, or more commonly, with strong oxidising agents in acid solution. Their use is strongly dependent on the available technical resources. In the absence of an effective fume cupboard for example, all methods employing boiling acids can only be done with the very greatest caution. On environmental grounds they should be entirely avoided. Hendey (1964) using glass components, constructed an apparatus that absorbed the toxic, corrosive gases produced when diatoms were boiled in acid.

Boiling methods: Boiling in concentrated sulphuric acid is commonest. Here, diatom material is sedimented in an Erlenmeyer flask or beaker and the remaining liquid decanted. Concentrated sulphuric acid is then poured over the sample to a depth of 5–10 mm. Technically, sulphuric acid is quite enough to effect cleaning. After boiling for about 20 minutes (with a boiling rod and covered with a watch glass) a spatula full of potassium nitrate is added to the hot acid, or in the case of very impure material, somewhat more. Momentary colourlessness occurs, and after cooling the diatoms form a pure white precipitate. This can be freed of acid by rinsing as described above.

Cleaner material can be prepared by boiling in nitric acid. Clearing with potassium nitrate is not required. Organic material can also be digested by boiling in chromo-sulphuric acid 70% perchloric acid until the brown coloration that appears at first, disappears. According to von Stosch & Reimann (1970), samples boiled in perchloric acid show no visible signs of damage.

Cold methods: Even in the "cold methods" diatoms are heated in acid, though they are never boiled. The addition of concentrated sulphuric acid to a diatom suspension leads to a strongly exothermic reaction. The "cold method" using sulphuric acid and potassium permanganate (Hustedt 1969). This method can be used anywhere and usually gives satisfactory results. A disadvantage however is that the frustules remain mostly intact. Concentrated sulphuric acid is poured over clean material containing as little water as possible. The acid is mixed with the material by twirling the reaction vessel. A small quantity of concentrated potassium permanganate is then added, and again mixed in by twirling the reaction vessel. This mixing action should be repeated several times at intervals. After about 15 minutes concentrated oxalic acid is added until the sample is completely decolorised. The sample must then be thoroughly rinsed. Good results are achieved where small amounts of relatively clean material is to be processed. If the sample contains more organic remains, or if a large quantity is to be cleaned, or if one must be certain that all organic parts of the cell are oxidised, then the following method should be used. Samples strongly acidified with sulphuric acid are placed on a boiling- or warming plate adjusted to c. 100 °C. A little concentrated potassium permanganate solution is then added. After five minutes the sample should be inspected, and where the solution is becoming colourless (indicated that oxidation has exhausted the permanganate), a further quantity of potassium permanganate should be added. It is often necessary to add more permanganate several times, before discoloration no longer occurs. Oxalic acid is then added to decolorise the remaining red or brown potassium permanganate, and to halt oxidation.

Plankton material containing little contamination can be cold-cleaned using 50% nitric acid. At temperatures of 50 °C, 12 hours are required for cleaning. Again, the reaction vessel should be twirled from time to time.

Oxidation of organic material by **hydrogen peroxide** is not so effective as with acids, but is much gentler. It is best used with samples that require little cleaning, such as plankton, and where corrosion should be limited, as in TEM and SEM studies. Van der Werff (1955) let diatom material stand for 15 minutes to 24 hours in 30% H_2O_2 before adding potassium permanganate dropwise, or boiled the material in 30% H_2O_2 as potassium permanganate was added. If too much permanganate was added, then a few drops of HCl were added. Potassium dichromate can be used in place of potassium permanganate.

A particularly gentle cleaning method was described by Swift (1967). Diatom material was placed in a 60 ml quartz tube or dish with 30% H_2O_2 and then irradiated with an air-cooled UV lamp. All organic material should be oxidised after about two hours. Swift (l.c.) used a 30 cm long, 1200 W mercury lamp, while Schmid (1976) used a 280 W "Astralux" light for the same length of time.

A combination of H_2O_2 and hydrochloric acid is usually recommended for fossil material. Schrader (1971) modified the procedure of Kanaya (1959), boiling a material for 15 minutes in concentrated H_2O_2' then adding concentrated HCl to the hot mixture, and finally boiling the sample for a further 15 minutes.

Enzyme treatment is primarily used where TEM and SEM investigations are involved, as anything which would effect the ultrastructure of the silica elements should be involved. As the polysaccharide enclosing membrane remains untouched by this method, information about organisation of the individual elements of the frustule and about chain structure can be obtained. If one is to work with only partly silicified species (e.g. *Cylindrotheca gracilis*), then this is the only method that can satisfactorily be used.

Von Stosch (1953) and Reimann (1960) introduced pancreatin to the techniques opf microscopy, as an enzyme to destroy organic components. Reimann equated the results with those achieved by sulfuric acid-potassium nitrate treatment. Diatoms should first be rinsed in distilled water, then treated with acetone and aqueous phenol to remove fats, rinsed with distilled water again, and finally digested in a 2% solution of phosphate buffered pancreatin (pancreatinum absolutum, E. Merck, Darmstadt) at 40 °C for 24 hours (von Stosch, 1981). At 38 C the dissolution of the organic cell contents takes a few hours. The sample is finally rinsed once more in distilled water. If all organic components are to be removed, then the sample must then be treated with a saturated, aqueous solution of chloral hydrate or with "Eau de Javelle" (Merck, 13% active chlorine).

9.5 Preparation of fossil material

If the material is Kieselgur, then it can be treated as a cleaned recent sample. Rocky samples including other sedimentary material is more difficult to work with. Such samples must first be reduced to a paste. Hustedt (1924) poured a hot, concentrated solution of sodium sulphate over such samples. Crystallisation during the cooling process disrupts the hard material. Repeated treatment of this type will achieve the desired result.

Directions for further processing are given by Schrader (1969). Calcium is first removed by boiling the material in HCL. After rinsing, it is boiled in 0.5% aqueous NaOH, then rinsed again and boiled for 20 minutes in concentrated sulphuric acid. Potassium nitrate is then added until the solution clears. After rinsing again, it is boiled in the 0.5% aqueous NaOH until the precipitate has swelled to such an extent that it is completely broken down by a last acid treatment (quickly boiling in sulphuric acid). Rinsing until the acid is leached out ends the procedure.

Hustedt (1969) stated, that often just boiling in a 10% perhydrite solution is sufficient.

9.6 Preparation of slides

Most of the ultrastuctural details of diatoms lie at the limits of resolution of light. In addition to this, all normal mounting media used in cytology have a refractive index similar to that of diatom valves. Slides with diatoms mounted in these media are too low in contrast for a satisfactory investigation. Diatoms are thus either enclosed in air, which gives a very high contrast, or in a medium with a much higher refractive index than the diatom valves. In both cases, a cleaned diatoms suspension in distilled water (if glycerine was present, it must be rinsed out) must be diluted until it appears only slightly cloudy. A single drop of this suspension is then placed on a well-cleaned coverslip. The latter are in practice stored in distilled water to which has been added a little surfactant (e.g. Agepon). After cleaning the coverslips with a cotton cloth, the thin remaining film of surfactant ensures that the diatom suspension spreads evenly. Normally, round or quadratic, 12–18 mm coverslips are used, whereby 15 mm circular coverslips have gained popularity over 18 mm ones as the diatom suspension spreads more easily on the latter. After the diatom suspension has spread, the coverslips must not be touched until dry, as vibration can easily cause the diatoms to clump, thus spoiling the slide.

High refractive index mounting media: Until a few decades ago, "Styrax" was most commonly used as a diatom mountant. Although it's refractive index of 1.62 is mostly sufficient for the investigation of diatom valves, it has now been almost entirely replaced by media of a higher refractive index. Three media are generally used today: "Hyrax", r.i. 1.71 (Hanna 1930); "Naphrax", r.i. 1.69 (Flemming 1954); amd "Pleurax", r.i. 1.73 (Hanna 1949)(refractive indices after

Meller 1985). Hyrax (supplier: Custom Research & Development Inc., 8500 Mt. Vernon Rd., Auburn, California 95603, USA) is a neutral, synthetic resin soluble in xylene or toluene. Without the solvent, it has an r. i. of 1.71 and melting point of 120–130 °C. It's slight yellow colour is not noticeable in thin films. Commercial Hyrax is dissolved in 20% toluene and forms an easily dosable fluid. To use this mountant, a hot-plate capable of melting it is required. An inverted, temperature-controlled clothes iron has proved a useful alternative. Diatom-coated coverslips are first placed on the heated surface to drive off the last moisture. After cooling, a small quantity of fluid Hyrax is dropped onto the coverslip with a pipette or glass rod, and the coverslip placed back on the hot-plate until all the solvent has evaporated. Once cool, the coverslip is then placed on the middle of a glass slide, which is then warmed until the mountant melts and spreads evenly. The method described for Naphrax also gives good results with Hyrax .

Naphrax can either be purchased (Northern Biol. Supplies, 31 Cheltenham Ave., Ipswich 1PI 4LN, UK), or can be made (see Fleming 1943, 1954), if a chemical laboratory is available. It is usually dissolved in toluene. To make a permanent slide a drop of Naphrax is placed in the middle of a slide. To prevent the diatoms being disturbed on the coverslip, it is held, diatoms uppermost, over a spirit flame or small bunsen burner, or best of all, placed on a hot plate for a short time to bond the diatoms to the coverslip. This coverslip is then placed on the Naphrax drop on the slide, and the slide warmed until the Naphrax "boils" for 3 seconds. Finally, the slide is quickly cooled by placing it on a metal sheet. Glass slides used in this method should be cleaned with, and stored in, absolute alcohol, being dried on a warm-plate before use.

Pleurax cannot be purchased, but can be relatively easily made (see von Stosch 1974, Meller 1985) in a chemical laboratory. According to the measurements of Meller however, Pleurax made using von Stosch's techniques had an r. i. of 1.73, instead of 1.8 as is usually cited in the literature. Better results can be obtained (Meller 1985) if a naphthalene/naphthalene sulphide resin mixture is used. Refractive indices of 1.78 to 1.81 can be obtained this way. This mountant can be used in the same way as Naphrax, although there is no data on the effect of aging. No changes were noted (Meller in litt.) after six months. A prerequisite however is the use of pure chemicals and careful preparation.

10 **Taxonomic concept of this flora** (version expressed on the XIV International Botanical Congress Berlin 1987)

Conventional species creation has led us into a dilemma of vast proliferation of taxa. Minor problems at the beginning, in the last century; almost insoluble problems now. This is mainly because differences which have formerly been regarded as sufficient for species differentiation become successively smaller and smaller and finally disappear as criteria suitable for identification. Any taxonomic limitation on the species level must become arbitrary in such continuous ranges of forms, unless the naming of local populations or clones or individuals is desired.

Diatom taxa which we usually consider as species are abstractions from presumable real species. Indeed each and all are based on one single or a small number of individuals while as biosystematic unities they are subject to

extensive subjective assessment. This is the source of conflicts about indistinct, inadequate species concepts in various classes of unicellular organisms. In theory we ought to differentiate between three categories:
1. Real species in the sense of population biology.
2. Taxonomic species in form of critically reflected morphologic entities characterised by different stages of the cell cycle, and by autecologic and synecologic data.
3 Inadequate is a primitive creation of typological species based on a single or very few specimens without further information. The rules of the ICBN permit such "subminimal" taxa if correctly published. However, they often reveal as "pseudotaxa".
In practice, however, we seldom discriminate between these categories. What actually occurs in each biotope is the presence of individuals. By far the greater majority of these are clonally derived individuals, which arise from successive vegetative reproduction by binary division. Individuals are much less frequently produced from a single parent by autogamy or apomixis (for instance parthenogenesis), or, as is usually the case in higher plants, directly from a biparental zygote. Together all such offspring, which are actually or potentially inter-fertile, build communities sharing the same gene pool. They should be reproductively isolated from other similar populations. The problem with the diatoms is that the population and the species boundaries remain obscure because regular monitoring of their sexual behaviour in nature on a world-wide basis is not feasible. Crossing and breeding experiments are bound to a necessarily elaborate, often unsuccessful methodology. Nevertheless only this can provide a fundamental starting point for understanding species in terms of population biology and evolutionary theory. The resulting entity is the so-called biological species.
This is the first of two fundamentally different research approaches which are recognised in the systematic treatment of organisms and for diatoms in particular.
The second is the inductive method: as far as diatoms are concerned this is traditionally based on light-microscopically comparisons of valve outlines and structure, since other classificatory methods, e.g. with the aid of protoplasmic features, have been shown to be extremely inadequate in many cases or (at least for the present) impracticable as far as all occurring taxa are in question. This is particularly the case when paleobotanical aspects are involved. The systematic units delimited in this way must be considered morphospecies, in the sense of Ernest Mayr (Mayr 1975). The most important methodological criteria on which they are defined are similarity and difference. The question is, whether the two approaches lead us to equivalent entities or not. Algologists firmly believe, or vaguely hope, that ultimately all visible features of the diatoms themselves will reveal where the species boundaries lie. Certainly, morphologic features are rarely isolation-relevant. However, if important as constructive elements, they demonstrate the result of biological adaptation. Thus, they are no less relevant than other secondary, which means indirect, criteria of a biospecies, for instance non-sexual behaviour or physiology. According to Mayr, on the basis of a number and specificity of morphological differences, one should be able to find an "indicator" of reproductive isolation; that means features which allow the deduction of such to be traced. That is a minor problem in higher organised organisms. Whether or not this is possible in certain unicellular organisms, where such significant features can hardly be detected, is still under discussion. Normally they can be found more ore less clearly between genera and taxonomical species groups. The question arises as to whether Mayr's advice, as such, is helpful for diatom taxonomists with

respect to the species level. We suggest that the chance of finding a taxonomic species in good accordance with a real species is high. Thus, if many individuals of all developmental stages of their peculiar life cycle and many populations from different localities are the basis of a taxon. The taxon then is to be understood as a well founded hypothetical species, which might be confirmed or refuted. However, the probability is low if a taxon is based on a very few specimens or even tends to zero if only one specimen is the random sample in discussion.

It is however interesting that successful crossing experiments can produce contradictory results compared to morphological "character taxonomy". Thus, Geitler (1973) discovered barriers to cross-fertilisation between microspecies within one morphospecies. This indicates approaching twin-species, which contain sexually isolated but morphologically identical units. Conversely, polymorphism can occur in plant species: similar to the example of the garden cabbage varieties or the domestic dog with its interbreeding races. This, projected upon the conventional diatom morphospecies, would undoubtedly lead to their being awarded the status of species.

Another, opposite example in the animal kingdom is the brown trout. Formerly isolated races with different morphological characteristics could be induced by fishery interests to interbreed towards an undifferentiated race-mixture. On the other hand, among the brown seaweed *Ectocarpus siliculosus*, populations from European and American coasts are not inter-fertile, though morphologically completely identical.

Various opinions are held by biologists with a background in recognition theory as to whether a strict separation between the taxonomic and population biological species definition is absolutely necessary, or whether the taxonomic species concept with the addition of biological information does not ultimately also lead to recognition of the true species. Even the biological species concept has its very weak points. In particular its application to groups of organisms which lack strict biparental reproduction is not without problems, and must be modified or superseded by using other concepts, in particular the syngameon theory of the Dutch botanist Lotsy, published in 1916 and 1925, almost forgotten later on and discussed again by Grant (1976) with reference to higher plants. The ability to cross is the criterion for a syngameon and not the existence or absence of isolation mechanisms, in contrast with the biospecies.

Autogamy and apomixis have been detected in various diatom taxa. But very little is known about morphological or other biological consequences which, however, might be important biosystematically with regard to reduced panmixis. Nothing is known of phenomena such as polyploid complexes or hybridism. All these phenomena are well known in higher plants and also in various cryptogamic groups. Conspicuous consequences arise to the biospecies concept, since complexes of uniparental hybrid clones, microspecies and semispecies accompany the original sexual species. This is obvious in angiosperms such as *Hieracium*, *Crepis*, *Rubus*, *Citrus*. They exhibit high genetic uniformity compensated for by rich modification capacities, in particular as pioneer plants from the ecological point of view.

Though there was no evidence until recently, we must be able to presuppose these capacities also for diatoms. In fact, to all appearances they seem to occur in the form of indefinable clusters around certain generic subgroups such as many "weak" taxa of lanceolate Nitzschiae.

Irrespectively thereof, what is needed, if we have to carry out a study which is based entirely on the taxonomic classificatory principle? We should know that it is provisional, as a currently practicable classification, aware of its inherent shortcomings and restricted biological evidence. Unfortunately this classificatory principle has fallen into disrepute not only on theoretical grounds, but also because of the conceptless, extremely uncritical practice of many of its practitioners. One polemical species definition is: "A species is what the author understands by a species". That is not so very far removed from the current situation of contemporary species creation. It is not refutable and thus not scientific. In fact the principle of authority, which is a psychologically influenced aspect, plays an important role in the acceptance or rejection of opinions. Better known authorities often succeed with comparatively weak arguments, if such are even given in support of their decisions.

The Rules of the International Code of Botanical Nomenclature (ICBN) can provide little assistance in deciding here. In particular they are not appropriate to direct how to discern between species. However, at least the priority principle offers a guideline. Already established taxa have priority – they are the reference point, a basis for later new descriptions. Each author should clearly present the features on which a new taxon differs from already established ones. It is essential that each diagnosis contains a differential diagnosis, something which has been largely unconsidered in practice. Not that thereby the problem of species definition in diatoms would be solved since even then description and the evaluation of characteristics remain overburdened with other subjective "judgements". But it would be possible to see to what extent a younger taxon should be "taken seriously", independent of the authority behind it.

When comparisons with apparently similar established older taxa are partially or entirely omitted, evaluation of the new taxon can be appropriately orientated. This is also the case where the differentiating characteristics given are exclusively those recognised as falling within the normal variability for populations of that genus.

How did this dilemma of the vast proliferation of taxa occur ?

Numerous new taxa have been shown to be synonyms of older taxa because the authors had absolutely no knowledge of the latter.

The generally known "image" of similar older taxa is often unrelated to the type material, false or uncertain. Sometimes it is simply determined from illustrations, often second-hand.

It has been shown to be a great disadvantage, and of little sense to ignore the actual spectrum of variation by overtly choosing only quite specific (namely type-specimen-like) individual forms for diagnosis and illustration. Thus, other forms of the life-cycle or minor variations provided opportunities for the description of "new species".

The drawings of older taxa are often so inadequate or subjectively drawn that the intended form cannot be recognised with certainty again. Apart from this the sparse diagnostic information is often even less helpful because it may fit to numerous other taxa.

The probability that overlaps will occur in the characteristics used for the definition of taxa increases steadily. The network of taxa forms an increasingly narrower reticulate pattern. In particular in genera with few distinguishing characteristics there are always too few recognisable character combinations for new taxa.

How should conclusions be reached so that the typologically moulded diatom systematics can at least begin to satisfy the existing desire for order pragmatically? For a series of scientific disciplines, for instance applied hydrobiology, ecology, geology, information as to whether the recurrent appearance of a particular form really represents a definable species (in terms of population biology) or not, is no essential.

Nevertheless such practical functions cannot be fulfilled when unlimited possibilities for the creation of new taxa can be so excessively and uncritically

exploited. One result is already apparent – the majority of hydrobiologists or ecologists are simply resigned to ignoring new taxa.
Therefore our suggestion: The taxonomist in practice will state continua and discontinua, he will describe as new what is apparently different and will synonymize what is supposedly identical. He will have to substantiate and to justify his decisions. His procedures will only be correct in terms of scientific theory, if his hypotheses are formulated in such a manner that they can be falsified (i.e. disproved) by new findings and – if so – can be replaced by new hypotheses.

II. Keys from the Bacillariophyceae-vols. 2/1–2/4

IIA English keys

Classification of the Bacillariophyceae

(According to Simonsen (1979), with additional comments by Simonsen (pers. comm.). Only the families and genera found in inland waters and estuaries are listed).

A. Order **Centrales**

I. Suborder **Coscinodiscineae**
1. Family **Thalassiosiraceae** Lebour 1930, emend. Hasle 1973
Aulacosira Thwaites (?)
Cyclotella Kützing
Skeletonema Greville
Stephanodiscus Ehrenberg
Thalassiosira Cleve
2. Family **Melosiraceae** Kützing 1844
Melosira Agardh s.str.
3. Family **Coscinodiscaceae** Kützing 1844
Coscinodiscus Ehrenberg
4. Family **Hemidiscaceae** Hendey 1937, emend. Simonsen 1975
Actinocyclus Ehrenberg

II. Suborder **Rhizosoleniineae**
1. Family **Rhizosoleniaceae** Petit 1888
Rhizosolenia Brightwell
2. Family **Biddulphiaceae** Kützing 1844
Acanthoceros Honigmann
Attheya West
Biddulphia Gray
Terpsinoe Ehrenberg
3. Family **Chaetoceraceae** H. L.Smith 1872
Chaetoceros Ehrenberg

B. Order **Pennales**

I. Suborder **Araphidineae**
1. Family **Fragilariaceae** Hustedt 1930
Asterionella Hassall
(*Centronella* Voigt)
(*Ceratoneis* Grunow)

Diatoma Bory
Fragilaria Lyngbye
Meridion Agardh
(*Opephora* Petit)
Synedra Ehrenberg
Tabellaria Ehrenberg
Tetracyclus Ralfs

II. Suborder **Raphidineae**
1. Family **Eunotiaceae** Kützing 1844
Actinella Lewis
Eunotia Ehrenberg
Peronia Brébisson & Arnott
(*Semiorbis* Patrick)
2. Family **Achnanthaceae** Kützing 1844
Achnanthes Bory
Cocconeis Ehrenberg
3. Family **Naviculaceae** Kützing 1844
Amphipleura Kützing
Amphora Ehrenberg
Anomoeoneis Pfitzer
Berkeleya Greville
Caloneis Cleve
Cymbella Agardh
Diatomella Greville
Didymosphenia M. Schmidt
Diploneis Ehrenberg
Entomoneis (*Amphiprora*) Ehrenberg
Frustulia Rabenhorst
Gomphocymbella O. Müller
Gomphonema Ehrenberg
Gyrosigma Hassall
Mastogloia Thwaites
Navicula Bory
Neidium Pfitzer
Oestrupia Heiden
Pinnularia Ehrenberg
Pleurosigma W.Smith
Rhoicosphenia Grunow
Scoliopleura Grunow
Stauroneis Ehrenberg
4. Family **Epithemiaceae** Grunow 1860
Epithemia Brébisson
Rhopalodia O.Müller
5. Family **Bacillariaceae** Ehrenberg 1840
Bacillaria Gmelin
Cylindrotheca Rabenhorst
Cymbellonitzschia Hustedt
Denticula Kützing
Gomphonitzschia Grunow
Hantzschia Grunow
Nitzschia Hassall
Simonsenia Lange-Bertalot
6. Family **Surirellaceae** Kützing 1844

Campylodiscus Ehrenberg
Cymatopleura W.Smith
Stenopterobia Brébisson
Surirella Turpin

According to the classification of Simonsen (1979), as well as most other classification systems which are not discussed here, the Bacillariophyceae are divided into two Orders on the basis of symmetry and the arrangement of the rib systems and areolae. In the first, the **Centrales**, the valves are either circular, with a shape which is derived from a circle or show a symmetrical centre in the middle of the valve. Areolar and rib systems are in most cases radial or quasiradial. There are, however, variations on this basic pattern, which may reach tri- or bi-polar form. For example, the Biddulphiineae can be bi-polar, with many species which may show symmetry which can be mistaken for that of the pennate diatoms. This also applies for example in many species of the Cymatosiraceae, a family newly described by Hasle, von Stosch and Syvertsen (1983). This family has been placed in the Centrales because of the presence of flagellate male gametes and other characteristics which are only found in this Order. Marine species of the Centrales which differ from the circular form are not considered in the following keys.
In the second Order, the **Pennales**, the valves are always elongated and the structures are oriented around a median axis. As the name implies, the rib system follows a feather-like pattern. There is a more or less pronounced median rib system – a central keel with a number of ribs extending from it on both sides.
The two Orders can easily be differentiated by the type and number of principal axes on the valve face:
1a Valves with radial axis (vol. 2/3 Fig 2.1R) . . **A. Order Centrales** (p. 59)
1b Valves with two main axes, a longer apical axis (vol. 2/3 Fig 1.1A) and a shorter transapical axis (vol. 2/3 Fig 1.1T) . . **B. Order Pennales** (p. 67)
Most of the species which belong to the Centrales are marine planktonic forms, whereas the Pennales are found mainly in the littoral region of marine and inland waters. Although the classification presented here begins with the Centrales, the Family Naviculaceae (Pennales) is presented in the first volume.

A. Order Centrales (vol. 2/3)

Key to the genera of the Centrales, described in this flora
1a Cells with numerous open intercalary bands and imbrication lines . . **2**
1b Cells not as above **3**
2a Cells with one long seta at each end . . . .
13. ***Rhizosolenia*** (vol. 2/3 p. 84) (p. 67)
2b Cells with two long setae at each end . . . .
. . . . **11.** ***Acanthoceras*** (one species, vol. 2/3 p. 83)
3a Cells with high valve mantles and usually with well developed girdles, usually forming long compact chains **4**
3b Cells discus to drum-shaped, valve mantle shorter, girdle not usually very conspicuous, cells single or in short chains **8**
4a Cells very weakly silicified, usually somewhat deformed by acid cleaning .
. . . . **10.** ***Skeletonema*** (vol. 2/3 p. 81) (p. 66)
4b Cells more strongly silicified **5**
5a Discus with a distinctly delineated centre, structure different to the marginal part of the valve **6**

5b Discus structure not as above 7
6a Discus distinctly areolate . . . . . . **2. *Orthoseira*** (vol. 2/3 p. 12) (p. 61)
6b Only rows of connecting ribs distinguishable on the discus . . . . **3. *Ellerbeckia*** (vol. 2/3 p. 17) (p. 61)
7a Areolae on the mantle only barely distinguishable with the light microscope . . . . **1. *Melosira*** (vol. 2/3 p. 7) (p. 60)
7b Areolae on the mantle clearly visible with the light microscope (see also *M. lineata* morphotype *juergensii* . .)**4. *Aulacoseira*** (vol. 2/3 p. 19) (p. 61)
8a Cells with elliptical valves and long spines, bent sideways, at the apices of the valve . . . . **12. *Chaetoceros*** (one species, vol. 2/3 p. 84)
8b Valve form not as above 9
9a Valves with two large ocelli . **14. *Pleurosira*** (one species, vol. 2/3 p. 86)
9b Valves without ocelli 10
10a Areolae arranged in structural sectors 11
10b Areolae not arranged in structural sectors 13
11a Valves with loculate areolae arranged in fascicles (structural sections) · In smaller cells the fascicles are quite indistinct. Valves without radial ribs or changes in levels of the valve surface 12
11b Valves with distinct radial ribs . . . .
9. *Stephanocostis* (one species, vol. 2/3 p. 80)
12a Pseudonodulus visible at the valve margin with careful focussing, without strutted processes in the central area of the valve face . . . . **15. *Actinocyclus*** (one species, vol. 2/3 p. 88)
12b Pseudonodulus absent, one to several strutted processes present in the central area of the valve . . . . . . **8. *Thalassiosira*** (vol. 2/3 p. 77) (p. 66)
13a Areolae rows/striae increasing from the centre to the valve edge. . . . 14
13b Areolae rows only at the valve edge, the centre with a different structure .
5. *Cyclotella* (vol. 2/3 p. 40) (p. 63)
14a Interstriae forked marginally . . **6. *Cyclostephanos*** (vol. 2/3 p. 61) (p. 65)
14b Interstriae not forked . . . . . . **7. *Stephanodiscus*** (vol. 2/3 p. 65) (p. 65)

1. *Melosira* Agardh 1827 nom. cons. (vol. 2/3 p. 7)

Typus generis: *Melosira nummuloides* (Dillwyn) Agardh

Key to species:

1a Valves with a ring-like keel on the margin of the discus (vol. 2/3 Fig. 8: 1–8) . . . . **4. *M. nummuloides***
1b Valves without ring-like keel on the discus 2
2a Discus with a single large puncta in the middle, many cells with inner valves (vol.2/3 Fig. 9: 1–13) . . . . **5. *M. dickiei***
2b Discus structure not as above 3
3a Mantle and girdle with hardly any structure visible with the light microscope 4
3b Mantle and girdle with distinct structures visible with the light microscope 5
4a Cells rectangular in girdle view (vol. 2/3 Fig. 4: 1–8) . . . . **1. *M. varians***
4b Cells elongated octagonal in girdle view (vol. 2/3 Fig. 6: 1–5) . . . . **2. *M. moniliformis* var. *octagona***
5a Structure of the discus, mantle and girdle relatively coarse (Fig. 5: 1–7) . . . . **2. *M. moniliformis***
5b Frustule comparatively finely punctate (vol. 2/3 Fig. 7: 1–9) . **3. *M. lineata***

2. *Orthoseira* Thwaites 1849 (vol. 2/3 p. 12)

Typus generis: *Melosira americana* Kützing 1844

Key to species:

1a Disci without carinoportulae in the middle 2
1b Disci with distinct carinoportulae 3
2a Mantle almost absent (vol. 2/3 Fig. 13: 1–8) . . . . . ***5. *Melosira arentii***
2b Mantle relatively high (vol. 2/3 Fig. 6: 6–8) . . . . . **6. *Melosira undulata***
3a 4–6 large, shining flecks visible on the discus margin by focussing up and down (vol. 2/3 Fig. 11: 6–9) . . . . **3. *O. dendrophila***
3b Without shining flecks 4
4a Girdle bands with rows of distinct puncta at the margin, carinoportulae very small (vol. 2/3 Fig. 12: 8–12) . . . . **4. *O. circularis***
4b Girdle bands irregularly punctate 5
5a Connecting spines short (vol. 2/3 Fig. 10: 1–11; 11: 1–4) . **1. *O. roeseana***
5b Connecting spines long (vol. 2/3 Fig. 12: 1–7) . . . . . **2. *O. dentroteres***

3. *Ellerbeckia* Crawford 1988 (vol. 2/3 p. 17)

Typus generis: *Melosira arenaria* Moore ex Ralfs 1843
Only 1 species with two varieties, *Ellerbeckia arenaria* (Moore) Crawford.

4. *Aulacoseira* Thwaites 184 (vol. 2/3 p. 19)

Typus generis: *Melosira crenulata* Kützing 1844

Key to species:

1a No areolae on the mantle or only one transverse row on the mantle edge and the mantle end 2
1b Two or more areolae on the pervalvar striae 3
2a Mantle without areolae (vol. 2/3 Fig. 33: 12–17) . . . . . **14. *A. perglabra***
2b One distal row and one proximal row of areolae on the mantle (vol. 2/3 Fig. 36: 1, 2) . . . . **15. *A. lirata* var. *biseriata***
3a 2–3 areolae on the pervalvar striae 4
3b Always more than 3 areolae on the pervalvar striae 11
4a Discus with only one marginal ring of small areolae (vol. 2/3 Fig. 32: 1–9) . . . . **12. *A. tethera***
4b Arrangement of areolae on the discus not as above 5
5a Arrangement of areolae very variable, in addition to disci with 2–3 marginal rings of areolae there are disci with areolae over the whole surface, but usually the centre field is free of fully developed areolae (vol. 2/3 Fig. 23: 12–17). . . . . **14. *A. perglabra***
5b Always with areolae on the entire face of the discus 6
6a All valves in sample with 2–3 areolae on the pervalvar striae . . . . **9. *A. distans* var. *tenella***
6b Most valves in a sample with more than 3 areolae on the pervalvar striae, in addition to valves with 2–3 areolae on the pervalvar striae 7
7a Discus with relatively irregular arrangement of areolae, areolae moderately large (vol. 2/3 Fig. 29: 1–22 . . . . **9. *A. distans***
7b Large areolae forming regular structures on the discus 8
8a Collar short (vol. 2/3 Fig. 30: 2–7) . . . . **9. *A. distans* var. *nivalis***

---

* Krammer (2000) established the new genus Brevisira with this species.

8b Collar long (vol. 2/3 Fig. 30: 1–9) . . . . **13.** *A. pfaffiana*
9a Puncta (areolae) on the mantle very large, pearl-like, usually less than 10 puncta/10µm on the pervalvar striae or puncta very irregularly arranged on the striae **10**
9b Arrangement of areolae on the mantle not as above **11**
10a Mantle height always less than the diameter in all cells in a sample (vol. 2/3 Fig. 34: 1–12) . . . . **15.** *A. lirata*
10b Mantle height greater than the diameter in most cells in a sample (vol. 2/3 Fig. 37: 1–10) . . . . **18.** *A.crassipunctata*
11a Pervalvar striae parallel to the pervalvar axis, only very slightly diagonal or only weakly diagonal to the pervalvar axis on small cells in a chain . . **12**
11b Pervalvar striae of all cells distinctly diagonal to the pervalvar axis, with the exception of the separation cells **19**
12a At least some of the puncta on the pervalvar striae of the mantle more or less oblong **13**
12b All puncta on the mantle rounded or square (with the light microscope) **15**
13a Fewer than 15 puncta/10 µm on the pervalvar striae (vol. 2/3 Fig. 26: 1–9; 27: 1–12) . . . . **7.** *A. crenulata*
13b More than 18 puncta/10 µm on the pervalvar striae **14**
14a Disci coarsely areolate, usually no areolae in the middle region or areolae faintly outlined (vol. 2/3 Fig. 35: 1–13) . . . . **16.** *A. lacustris*
14b The entire discus face finely areolate (vol. 2/3 Fig. 36: 3–18) . **17.** *A. tenuis*
15a Chains with separation cells with differing arrangement of areolae, and with long spines (vol. 2/3 Fig. 20: 1–9) . . . . **2.** *A. muzzanensis*
15b End cells without a different arrangement of areolae **16**
16a Pervalvar striae very finely areolate, more than 20 striae/10µm (vol. 2/3 Fig. 31: 16, 17) . . . . **11.** *A. laevissima*
16b 16 or fewer pervalvar striae/10µm **17**
17a Valves almost always longer than broad (vol. 2/3 Fig. 22: 1–12) . . . . **4.** *A. islandica*
17b Valves always broader than high **18**
18a Collar relatively short (vol. 2/3 Fig. 29: 1–23 . . . . **9.** *A. distans*
18b Collar always as long as the rest of the valve mantle (vol. 2/3 Fig. 23: 1–10) . . . . **13.** *A. pfaffiana*
19a Cells with separation cells which have parallel pervalvar rows of larger areolae and some particularly long linking spines (vol. 2/3 Fig. 18: 1–12; 19: 1–9) . . . . **1.** *A. granulata*
19b No such separation cells present **20**
20a Linking spines very large **21**
20b Linking spines smaller **22**
21a Walls relatively thin, ringleiste (circular ledge) narrow (vol. 2/3 Fig. 24: 1, 3–6; 25: 1–15) . . . . **6.** *A. italica*
21b Walls thick, ringleiste (circular ledge) broad (Fig. 28: 1–11) . **8.** *A. valida*
22a Discus face finely punctate, linking spines (visible with EM) with bifid anchors (Fig. 21: 1–16) . . . . **3.** *A. ambigua*
22b Discus face with at most a marginal ring of puncta, linking spines (visible with EM) not as above **23**
23a Linking spines pointed, without anchor, emerging from two pervalvar ribs (vol. 2/3 Fig. 23: 1–11) . . . . **5.** *A. subarctica*
23b Linking spines short, broad with branched anchor, each rising from one pervalvar rib (vol. 2/3 Fig. 2: 6; 31: 1–15; 32: 10–16) . . . **10.** *A. alpigena*

5. *Cyclotella* (Kützing) Brébisson 1838 nom. cons. (vol. 2/3 p. 40)

Typus generis: *Cyclotella tecta* Håkansson & Ross 1984 (= *Cyclotella distinguenda* Hustedt 1927)

Key to species:

1a Valve edge without shadow lines **2**
1b Valve edge with shadow lines **26**
2a Centre of the valve with indistinct structures or no structures at all (see also **22a**) **3**
2b Centre of the valve distinctly structured **12**
3a Centre of the valve tangentially undulating **4**
3b Centre of the valve not as above **9**
4a Centre of the valve oval **5**
4b Centre of the valve not as above **6**
5a Without strutted processes in the centre (vol. 2/3 Fig. 46: 6–8; ?9–11) . . . . **8. *C. iris***
5b Numerous strutted processes in the centre (vol. 2/3 Fig. 46: ?9–11, 12, 13) . . . . **9. *C. michiganiana***
6a Centre with scattered indentations and raised areas (colliculate structure) **7**
6b Centre field structure not as above (vol. 2/3 Fig. 44: 1–10, see description p. 44) . . . . **3. *C. meneghiniana***
(vol. 2/3 Fig. 44: 11a,b, see description p. 45) . . . . **4. *C. gamma***
7a Diameter usually greater than 20 µm (vol. 2/3 Fig. 45: 1–8) . **5. *C. striata***
7b Diameter less than 20 µm **8**
8a Margin finely structured (vol. 2/3 Fig. 46: 1a, 1b) . . . . . . **6. *C. caspia***
8b Margin coarsely structured (vol. 2/3 Fig. 46: 2–5) . . **7. *C. hakanssoniae***
9a Relatively large species, greater than 25 µm (vol. 2/3 Fig. 43: 12–14) . . . . **2. *C. plitvicensis***
9b Relatively small species, less than 20 µm **10**
10a Cells forming colonies (vol. 2/3 Fig. 52: 3) . . . . **22. *C. delicatula***
10b Usually found as individuals **11**
11a Undulating radial structures arranged in sectors in the centre of the valve (vol. 2/3 Fig. 52: ?1–2, 4–6, ?7–9) . . . . **23. *C. comensis***
11b Centre of the valve relatively smooth (vol. 2/3 Fig. 51: ?7, 10–14) . . . . **20. *C. cyclopuncta***
12a Centre with star-shaped structure **13**
12b Centre structure not as above **16**
13a Small forms, less than 10 µm **14**
13b Larger forms, greater than 10 µm **15**
14a Cells forming colonies (vol. 2/3 Fig. 49: 11) . . . . . . **17. *C. glomerata***
14b Cells found as individuals (vol. 2/3 Fig. 49: 5–8) . **16. *C. pseudostelligera***
15a The elements of the star shaped structure in the centre without structure (vol. 2/3 Fig. 49: 1a-4, ?9) . . . . **15. *C. stelligera***
15b The elements of the star shaped structure in the centre areolate (vol. 2/3 Fig. 64: 1–8) . . . . **35. *C. rossii***
16a Small forms, between 3 and 10 µm **17**
16b Larger forms, greater than 25 µm **19**
17a Central field of the valves with circular papillae **18**
17b Papillae in the central field of the valves not circular (vol. 2/3 Fig. 50: 12) . . . . **19. *C. trichonidea***
18a Central field flat, cells circular (vol. 2/3 Fig. 50: 1–11, 13, 14; 51: 1–5) . . . . **18. *C. ocellata***
18b Central field undulating (vol. 2/3 Fig. 48: 4a-7) . . . . . **14. *C. tripartita***

19a Central field radially striate 20
19b Central field not as above 22
20a Cells found as individuals (vol. 2/3 Fig. 47: 1) . . . . **10.** ***C. elgeri***
20b Cells found in colonies 21
21a Colonies more chain-like (vol. 2/3 Fig. 56: 1a-2; 64: 9–11) . . . . **32.** ***C. planctonica***
21b Colonies more irregular (vol. 2/3 Fig. 64: 12–14) . . . . . **33.** ***C. socialis***
22a No structure in the central field 23
22b Distinct structure in the central field 24
23a Central field oval (vol. 2/3 Fig. 65: 1–3) . . . . **36.** ***C. schumannii***
23b Central field round (vol. 2/3 Fig. 43: 1–11; Fig. 51: 6–8, 16, 18) . . . . **1.** ***C. distinguenda***
24a Central field without crater-like structures 25
24b Central field with more or less well-ordered structure, appearing punctate (vol. 2/3 Fig. 65: 5, 6) . . . . **37.** ***C. krammeri***
25a Large central area with scattered indentations and raised areas (colliculate structure) (vol. 2/3 Fig. 47: 2) . . . . **11.** ***C. areolata***
25b Relatively small central zone, irregularly punctate (vol. 2/3 Fig. 47: 3a-4) . . . . **12.** ***C. fottii***
26a Centre of the valve with radial structure 27
26b Valve centre not as above 32
27a Marginal area with finer structure 28
27b Marginal area with coarser structure 30
28a Valves strongly radially undulating (vol. 2/3 Fig. 54: 3a-4b; 55: 1, ?2–7b; 56: ?3a-5; 57: ?4, 5; 61: ?1–5b) . . . . **24.** ***C. bodanica*** **var.** ***lemanica***
28b Valves not radially undulating 29
29a Central area with very fine structure, puncta on the radial striae run over into the marginal area (vol. 2/3 Fig. 53: 1–6; 54: 1–2; 57: 1–3) . . . . **24.** ***C. bodanica*** **var.** ***bodanica***
29b Central area very loosely structured, no transition of the middle puncta on the radial striae to the marginal area (vol. 2/3 Fig. 58: 1–4b) . . . . **24.** ***C. bodanica*** **var.** ***affinis***
30a Cells form colonies of a few to many cells (vol. 2/3 Fig. 60: 7–10, see description page 59) . . . . **30.** ***C. praetermissa***
(vol. 2/3 Fig. 59: ?1–3; 60: ?1–5, see description page 58) . . . . **31.** ***C. quadrijuncta***
30b Cells found as individuals 31
31a Valves very delicately structured (vol. 2/3 Fig. 63: 1–7) **34.** ***C. glabriuscula***
31b Valves with distinct structure (vol. 2/3 Fig. 62: 1–6, 9–12) . **29.** ***C. comta***
32a Small forms, less than 7 µm 33
32b Large forms, greater than 7 µm 34
33a Marginal structure visible, large central area (vol. 2/3 Fig. 51: 19–21) . . . . **21.** ***C. atomus***
33b Structure of entire valve distinguishable, small central area (vol. 2/3 Fig. 65: 8a-10) . . . . **38.** ***C. wuethrichiana***
34a Stellate structure in the central area, formed of large areolae (vol. 2/3 Fig. 47: 5, 6; 48: 1a-3) . . . . **13.** ***C. antiqua***
34b Structure of the central area not as above 35
35a Valves elliptical (vol. 2/3 Fig. 61: 6a-10) . . . . **28.** ***C. austriaca***
35b Valves round 36
36a Central zone radially punctate (vol. 2/3 Fig. 59: 4) . . . . **25.** ***C. styriaca***
36b Central zone tangentially undulating 37

37a Marginal structure finer (vol. 2/3 Fig. 59: 6) . . . . **26.** ***C. stylorum***
37b Marginal structure more coarse (vol. 2/3 Fig. 59: 5) . . **27.** ***C. baikalensis***

6. *Cyclostephanos* Round in Theriot et al. 1987 (vol. 2/3, p. 61)

Typus generis: *Stephanodiscus* (*bellus* A. Schmidt var.?) *novae zeelandiae*, Cleve 1881

Key to species:

1a Valves arched 2
1b Valves flat 5
2a Radial striae, radial ribs on the outer margin of the valve face (vol. 2/3 Fig. 67: 8a-9b) . . . . **6.** ***C. dubius***
2b Valves structure not as above 3
3a Radial striae coarsely areolate and interstriae bifurcate (vol. 2/3 Fig. 66: 1a-3) . . . . **1.** ***C. novaezeelandiae***
3b Radial striae finely areolate, interstriae bifurcate 4
4a Valves greater than 15 µm (vol. 2/3 Fig. 66: 4a,b; 67: 1a-2) **2.** ***C. damasii***
4b Valves less than 15 µm (vol. 2/3 Fig. 67: 5) . . . . . . **4.** ***C. costatilimbus***
5a Valves with interstriae not bifurcate (vol. 2/3 Fig. 67: 6a,b) . . . . **5.** ***C. tholiformis***
5b Valves with interstriae bifurcate (vol. 2/3 Fig. 67: 3, 4) . **3.** ***C. invisitatus***

7. *Stephanodiscus* Ehrenberg 1846 (vol. 2/3 p. 65)

Typus generis: *Stephanodiscus niagarae* Ehrenberg 1846 (designated by Boyer 1927)

Key to species:

1a Cells considerably longer than broad, usually forming long chains (vol. 2/3 Fig. 74: 10–11) . . . . **11.** ***S. binderanus***
1b Cells considerably shorter than broad 2
2a Structure of the valve edge irregular 3
2b Structure of the valve edge regular 7
3a Diameter of the largest valves in the growth cycle always greater than 60 µm (vol. 2/3 Fig. 68: 1–3, 5; 69: 1a, 1b; 70: 1) . . . . . . **1.** ***S. niagarae***
3b Diameter of the largest valves in the growth cycle always less than 60 µm 4
4a Diameter less than 20 µm(vol. 2/3 Fig. 71: 6; 72: 1–2b) **4.** ***S. agassizensis***
4b Diameter greater than 20 µm 5
5a Position of the strutted processes on both valves of one frustule different (either in the central area or at the edge) (vol. 2/3 Fig. 73: 4a-5b) . . . . **7.** ***S. galileensis***
5b Position of the strutted processes not as above 6
6a One or more strutted processes in the central area of the valve (vol. 2/3 Fig. 68: 4a,b; 69: 4, 5: 71: 1–2b) . . . . **2.** ***S. rotula***
6b No strutted processes in the central area of the valve (vol. 2/3 Fig. 69: 3; 70: 3; 71: 3a-5b) . . . . **3.** ***S. neoastraea***
7a Diameter less than 12 µm 8
7b Diameter greater than 12 µm 13
8a Valves flat 9
8b Valves with concentrical undulations 10
9a With strutted processes on the valve faces (vol. 2/3 Fig. 74: 1–4) . . . . **9.** ***S. parvus***

9b Without strutted processes on the valve face (vol. 2/3 Fig. 75: 4–11; 76: 1–3 . . . . **13. *S. hantzschii***
10a Structure on the valve face fine **11**
10b Structure on the valve face coarser **12**
11a Structure of the rows of areolae only clear at the margin (vol. 2/3 Fig. 74: 8a-9) . . . . **10. *S. vestibulis***
11b Structure of the rows of areolae distinct over all (vol. 2/3 Fig. 74: 5–7) . . . . **8. *S. minutulus***
12a Recent (vol. 2/3 Fig. 75: 1a,b?; 2a-3b) . . . . **12. *S. medius***
12b Fossil (?recent) (vol. 2/3 Fig. 76: 7a,b) . . . . **15. *S. oregonicus***
13a Valves flat (vol. 2/3 Fig. 75: 4–11; 76: 1–3) . . . . **3. *S. hantzschii***
13b Valves with concentric undulations **14**
14a Recent **16**
14b Fossil **15**
15a 4 bundled rows of areolae at the valve margin (vol. 2/3 Fig. 73: 1a-2) . . . . **6. *S. aegyptiacus***
15b 2–3 bundled rows of areolae at the valve margin (vol. 2/3 Fig. 76: 7a,b) **15. *S. oregonicus***
16a 4–6 bundled rows of areolae at the valve margin (vol. 2/3 Fig. 76: 4a-6) . . . . **14. *S. transsylvanicus***
16b At most 3 bundled rows of areolae at the valve margin **17**
17a Structure of the margin regular (vol. 2/3 Fig. 72: 3a-4) . . . **5. *S. alpinus***
17b Structure of the margin slightly irregular (vol. 2/3 Fig. 75: 1a,b?; 2a-3b) . . . . **12. *S. medius***

8. *Thalassiosira* Cleve 1873 (emend. Hasle) (vol. 2/3 p. 77)

Typus generis: *Thalassiosira nordenskioeldii* Cleve 1873

Key to species:
1a Valves with rows of areolae arranged in fascicles **2**
1b Valves with rows of areolae arranged in more or less radial rows . . . **4**
2a Valve diameter greater than 25 µm **3**
2b Valve diameter less than 25 µm (vol. 2/3 Fig. 77: 5a, b) . . **2. *T. virsurgis***
3a Valves with only one central strutted process (vol. 2/3 Fig. 77: 1, 2) . . . . **1. *T. nordenskioeldii***
3b Valves with several central strutted processes (vol. 2/3 Fig. 77: 6a, b) . . . . **3. *T. baltica***
4a Valves with one ring of central strutted processes (vol. 2/3 Fig. 77: 3, 4) . . . . **4. *T. weissflogii***
4b Valves with one central strutted process
(vol. 2/3 Fig. 78: 1–3) see description p.79 . . . . **5. *T. proschkinae***
(vol. 2/3 Fig. 60: 6a, b) see description page 80 . . . . . **6. *T. pseudonana***

10. *Skeletonema* Greville 1865 (vol. 2/3 p. 81)

Typus generis: *Skeletonema barbadense* Greville 1865

Key to species
1a Pseudosulcus absent in closely spaced cells (vol. 2/3 Fig. 84: 5–10; 85: 1–3) . . . . **1. *Skeletonema subsalsum***
1b Pseudosulcus distinct in closely spaced cells (vol. 2/3 Fig. 85: 4–8) . . . . **2. *Skeletonema potamos***

13. *Rhizosolenia* Ehrenberg 1843; emend Brightwell 1858 (vol. 2/3 p. 84)

Typus generis: *Rhizosolenia americana* Ehrenberg 1843

Key to species:

1a Bristle more or less at the centre of the calyptra, sutures of the intercalary bands and the imbrication lines usually not very clear with the light microscope (vol. 2/3 Fig. 86: 1–4) . . . . **1.** ***R. longiseta***

1b Bristle on the side of the calyptra, sutures of the intercalary bands and the imbrication lines distinct in wet mounts (vol. 2/3 Fig. 86: 5–8) . . . . **2.** ***R. eriensis***

B. Order Pennales (vol. 2/3)

Valves without raphes . . . . **I. Suborder Araphideae (p. 67)**
Valves with raphes . . . . **II. Suborder Raphidineae (p. 74)**

I. Suborder Araphidineae

In this flora only one Family, Fragilariaceae.

Family Fragilariaceae Hustedt 1930 (vol. 2/3 p. 90)

Family type genus: *Fragilaria* Lyngbye

Key to the genera of the Fragilariaceae:

1a Intercalary bands of the cells with septa **2**
1b Intercalary bands of the cells without septa **3**
2a Valves with clearly visible transverse walls . . . . **1.** ***Tetracyclus*** (vol. 2/3 p. 90) (p. 67)
2b Valves without transverse walls . . **5.** ***Tabellaria*** (vol. 2/3 p. 104) (p. 69)
3a Valves always with distinct transverse walls **4**
3b Valves without partition walls (compare also with vol. 2/3 Fig. 118: 1–10) **5**
4a Valves heteropolar . . . . **3.** ***Meridion*** (one species, vol. 2/3 p. 101)
4b Valves isopolar . . . . **2.** ***Diatoma*** (vol. 2/3 p. 93) (p. 68)
5a Cells almost always heteropolar, mostly star-shaped colonies, rarely forming zig-zag colonies . . . . **4.** ***Asterionella*** (vol. 2/3 p. 102) (p. 68)
5b Not with the above combination of characters **6**
6a Valves with partly closed alveoli (EM necessary to confirm this character), brackish water species (see vol. 2/3 Fig. 136: 8, 9) . . . . **6.** ***Synedra*** (one species, vol. 2/3 p. 111)
6b Not with the above combination of characters **7**
7a All cells in a population heteropolar, brackish water species, fine structure of the areolae as shown in vol. 2/3 Fig. 118: 20 . . . . **8.** ***Opephora*** (one species, vol. 2/3 p. 165)
7b Not with the above combination of characteristics . . . . **7.** ***Fragilaria*** (vol. 2/3 p. 113) (p. 69)

1. *Tetracyclus* Ralfs 1843 (vol. 2/3 p. 90)

Typus generis: *Tetracyclus lacustris* Ralfs 1843

Key to species:

1a Valves shorter than 30 µm, outline linear-elliptical to circular (vol. 2/3 Fig. 89: 8–20) . . . . **3. *T. rupestris***
1b Valves longer than 30 µm, transapically widened in the centre, cross-shaped, rarely rhomboid **2**
2a Valve with a inflated central region, resulting in broad, cross shaped, rarely rhomboid valves, inflated region never emarginate (vol. 2/3 Fig. 87: 1–8; 88: 1–8; 89: 1–6) . . . . **1. *T. glans***
2b Central inflated region distinct and emarginate (with a convex notch) (vol. 2/3 Fig. 89: 8–20) . . . . **2. *T. emarginatus***

2. *Diatoma* Bory 1824 nom. cons. (vol. 2/3 p. 93)

Typus generis: *Diatoma vulgaris* Bory de Saint-Vincent 1824

Key to species:

1a Usually more than 6 dividing walls/10 µm, valves commonly form zig-zag bands **2**
1b Usually fewer than 5 dividing walls/10 µm, valves most often form compact bands **5**
2a Valves with labiate processes at both ends, transapical striae very delicate, not distinguishable with the light microscope. Outline spindle-shaped or linear, without constricted ends (vol. 2/3 Fig. 92: 6; 96: 11–21) . . . . **4. *D. moniliformis***
2b Valves with a labiate process only at one end **3**
3a Valve width 5 µm or less, valves long and narrow (vol. 2/3 Fig. 96: 1–9, 10) . . . . **3. *D. tenuis***
3b Valves more robust, width greater than 6 µm **4**
4a Valves linear with distinctly separate, capitate ends (vol. 2/3 Fig. 92: 5; 95: 8–14) . . . . **2. *D. ehrenbergii***
4b Valves mostly broad, linear or elliptical-lanceolate, if linear, then usually without set off ends (vol. 2/3 Fig. 91: 2, 3; 93: 1–12; 94: 1–13; 95: 1–7; 97: 3–5) . . . . **1. *D. vulgaris***
5a Valves linear with distinctly set off, usually capitate ends (vol. 2/3 Fig. 102: 4–10) . . . . **7. *D. anceps***
5b Ends not capitate, at most somewhat set off **6**
6a Valves very robust, mostly longer than 40 µm, striae distinct, 18–22/10 µm (vol. 2/3 Fig. 97: 6–10; 98: 1–6) . . . . **5. *D. hyemalis***
6b Valves more delicate, mostly shorter than 40 µm, striae very delicate, 22–35/10 µm (vol. 2/3 Fig. 91: 1; 92: 1–4; 98: 7; 99: 1–12) . **6. *D. mesodon***

4. *Asterionella* Hassall 1850 (vol. 2/3 p. 102)

Typus generis: *Asterionella formosa* Hassall 1850

Key to species:

1a Girdle view with triangular broadened ends (vol. 2/3 Fig. 103: 1–9; 104: 9, 10) . . . . **1. *A. formosa***
1b Girdle view only slightly broadened at the foot pole (vol. 2/3 Fig. 104: 1–8) . . . . **2. *A. ralfsii***

5. *Tabellaria* Ehrenberg 1840 (vol. 2/3 p. 104)

Typus generis: *Tabellaria trinodis* Ehrenberg 1840 (*T. fenestrata*) (Lyngbye) Kützing 1844

Key to species:

1a Valve concave in the middle or approximately elliptical (vol. 2/3 Fig. 105: 9–16) . . . . **5. *T. binalis***

1b Valve convex in the middle with ends inflated **2**

2a Edge of the valve without spinules (visible by focussing), (isolated) intercalary bands open at the ends (always?) and without rudimentary septa; combined with the following characteristics: central axial area narrow in the centre, linear, ends distinctly capitate, rimoportula distinct, quite close to the middle puncta in the inflated central area, 4 intercalary bands with closely spaced septa in cells ready to divide (vol. 2/3 Fig. 105: 1–4) . . . . **1. *T. fenestrata***

2b Edge of the valve with delicate to coarse spinules, not with the above combination of characteristics **3**

3a Spinules relatively large, longer than 0.5 µm (EM), edge of the valves parallel between inflated areas, always 4 intercalary bands with septa, rimoportula in the axial area, but distinctly displaced towards the margin of the inflated region (vol. 2/3 Fig. 105: 5–8) . . . . . **2. *T. quadriseptata***

3b Differing in one or more of the above characters, particularly with a greater number of septate intercalary bands **4**

4a Valves 10–16 µm wide in the centre, always with one rimoportula near the ends, which can be seen as regular short transapical lines or distinct puncta by focussing with the light microscope (vol. 2/3 Fig. 107: 1–6) . . . . **4. *T. ventricosa***

4b Valves less than 10 µm wide, only a single rimoportula near the middle of the valve (vol. 2/3 Fig. 106: 1–13) . . . . **3. *T. flocculosa-complex***

7. *Fragilaria* Lyngbye 1819 (vol. 2/3 p. 113)

Typus generis: *Fragilaria pectinalis* (O.F. Müller) Lyngbye 1819 (?*Fragilaria capucina* Desmazières 1825)

Since 1991 (publication of the first edition of vol. 2/3) numerous new species, new combinations (nov. comb.), changes of rank (nov. stat.) and of specific names (nov. nom.) have been published in several volumes of Bibliotheca Diatomologica and Iconographia Diatomologica.

Key to species:

1a Valves tri-polar, star-shaped **59**

1b Valves bi-polar **2**

2a Valves strongly to weakly heteropolar **55**

2b Valves isopolar **3**

3a Apical axis weakly sickle- or banana-shaped or valves greater than 100 µm long and sharply widened in the centre and irregularly curved to bent . **53**

3b Apical axis usually straight **4**

4a Frustules pelagic, forming star-shaped aggregates **51**

4b Not with the above combination of characteristics **5**

5a Valves similar to *Diatoma tenuis*, with irregularly arranged transapical costae, some of which are barely visible, and some stronger in appearance (vol. 2/3 Fig. 118: 1–7) . . . . **25. *F. incognita*** in part

5b Not as above 7
6a Cell aggregate 'comb-shaped', the individual frustules touching in girdle view either only in the centre or at the ends as well (vol. 2/3 Plate 116) . 50
6b Not as above 7
7a In girdle view striae running far over the edge of the mantle, as a result there is little or no hyaline area visible (vol. 2/3 Fig. 128: 1–10) . . . . 20. ***F. nitzschoides***
7b Not as above 8
8a Valves with very short striae along edge, axial area therefore expanded to a large central area (tribes with narrow somewhat needle-shaped valves are not considered here) 42
8b Not with the above combination of characteristics 9
9a Central area symmetrical and abruptly set off from the axial area, encircled with a distended ring 41
9b Central area not as above or absent 10
10a Valves large (length 120–500 µm) with wedge-shaped to capitate broadened ends (vol. 2/3 Fig. 123: 1–3) . . . . 30. ***F. dilatata***
10b Not with the above combination of characteristics or less distinct from forms of *F. ulna* and *F. biceps* 11
11a Valves needle-shaped, gradually narrowing from the centre to the ends . 12
11b Valves approximately elliptical, widely lanceolate or convex, concave or linear in the centre and not narrowed until the ends 13
12a Valves on average in the population at least 20 times longer than wide . . 38
12b If valves on average in the population shorter in relation to width, see also the following taxa: 6. *F. delicatissima* in part, 5. *F. tenera* in part, 4. *F. famelica* in part, 1. *F. capucina* species complex in part, 26. *F. ulna* species complex in part
13a Valves more or less concave in the middle 14
13b Not as above 17
14a Largest valves with width greater than 8 µm, if the valve is more undulating, then the middle indentation is the most strongly pronounced (vol. 2/3 Fig. 128: 11–14; 129: 1, 6) compare with *F. lata* (vol. 2/3 Fig. 129: 5) and individual forms in the species complex around *F. ulna* . **21. *F. constricta***
14b Valve width usually less than 8 µm 15
15a 2.5–3 µm wide valves, sides with long shallow concavity (vol. 2/3 Fig. 111: 25–28) . . . . **23. *F. alpestris***
15b Valves broader and/or concave only in the central area 16
16a Valves with two undulations and coarsely punctate (vol. 2/3 Fig. 130: 19, 20) . . . . **43. *F. robusta***
16b If the above combination of characteristics is not present, compare with the following taxa: 1.5. *F. capucina* var. *mesolepta*, 12. *F. parasitica* var. *subconstricta*, 8. *F. crotonensis*, 1. *F. capucina* species complex in part (several groups), 2. *F. bidens* in part, 4. *F. famelica* in part, 34. *F. construens* in part, 40. *F. oldenburgiana* in part, 22. *F. lata* in part, 15. *F. virescens* in part, 26. *F. ulna* species complex in part
17a (13) Valves more or less distended in the centre. 18
17b Valves linear, elliptical, lanceolate or usually not distended 24
18a Valves narrowly linear (3–3.5 µm wide) and slightly distended in the centre (vol. 2/3 Fig. 134: 26–31) . . . . 40. ***F. oldenburgiana***
18b Valves shorter and broader with more strongly enlarged area 19
19a Striae coarse, fewer than 12/10 µm and not conspicuously punctate (vol. 2/3 Fig. 133: 33–42) . . . . **37. *F. leptostauron*** in part
19b Striae finer, usually more than 12/10 µm 20
20a Ends with somewhat rhomboid/lanceolate outline, more elongate and

sharply rounded, (with EM) without small linking spines (vol. 2/3 Fig. 130: 1–8) . . . . **12. *F. parasitica*** in part

20b Not with the above combination of characteristics **21**

21a Axial area not differentiated or barely so . . . . **22. *F. lata*** in part

21b Axial area more or less distinctly differentiated **22**

22a Striae appearing coarsely punctate (vol. 2/3 Fig. 130: 25–30) . . . . **42. *F. pseudoconstruens*** in part

22b Striae at most delicately punctate **23**

23a Valves ends usually conspicuously elongate, rostrate(vol. 2/3 Fig. 116: 8–10) . . . . **11. *F. heidenii*** in part

23b Ends less elongate and more widely rounded (vol. 2/3 Fig. 132: 1–32) . . . . **34. *F. construens*** in part

24a (17) Valve edge more undulating (vol. 2/3 Fig. 132: 23–27), very rarely also *F. pinnata*. . . . . **34. *F. construens*** in part

24b Not as above **25**

25a Axial area very narrow to barely visible **26**

25b Axial area narrower to wider, but always still clearly visible **28**

26a Valves broader than 5 µm (vol. 2/3 Fig. 126: 1–10) . . . . **15. *F. virescens***

26b Valves to 5 µm wide **27**

27a Striae evenly spaced, 18–21/10 µm (vol. 2/3 Fig. 126: 11–20) . **17. *F. exigua***

27b Striae 13–17/10 µm, often unevenly spaced (vol. 2/3 Fig. 118: 11–16); in cases where striae are more dense compare with *F. incognita* without *Diatoma*-like, band shaped ribs (vol. 2/3 Fig. 118: 1–6) . **24. *F. bicapitata***

28a (25) Central area more or less distinctly unilateral, valve length usually less than 50 µm (vol. 2/3 Plate 108, 109); in addition other tribes or individual examples of these complexes . . . . ***F. capucina* var. *vaucheriae* and var. *perminuta***

28b Not with the above combination of characteristics **29**

29a Striae coarse, approximately 5–12/10 µm **30**

29b Striae usually more than 12/10 µm **32**

30a Polyhalobic oceanic forms (vol. 2/3 Fig. 136: 12, 13) . . **33. *F. investiens***

30b Tribes in freshwater and restricted in brackish water **31**

31a Length of the population usually limited to much less than 40 µm, striae not punctate, at most appearing lineolate (vol. 2/3 Fig. 133: 1–42) compare also coarsely striate tribes of *F. capucina* complexes **36. *F. pinnata*** in part

31b Length usually much greater than 40 µm and/or breadth around 5 µm or greater, compare also however coarsely striate tribes of *F. capucina* and coarsely punctate *F. minuscula* **61**

32a (29) Striae appear more or less distinctly punctate. **33**

32b Puncta of striae difficult to distinguish or not visible **36**

33a Valves linear to linear-lanceolate, usually with at least a suggestion of a set off central area in the axial area (vol. 2/3 Fig. 111: 4–17) . . . . **4. *F. famelica*** in part

33b Valves rounded to linear-elliptical, central area not set off from the axial area **34**

34a Valves mostly linear-elliptical with broad to flat rounded ends and very wide axial area (vol. 2/3 Fig. 129: 10–13); compare with also *Delphineis karstenii* (vol. 2/3 Fig. 129: 16, 17) . . . . **44. *F. zeilleri*** in part

34b Valves more strongly convex or rounded **35**

35a Valves broadly elliptical to approximately rounded (vol. 2/3 Fig. 130: 31–42); compare with also forms shown in vol. 2/3 Fig. 130: 21–23 . . . . **35. *F. elliptica*** in part

35b Valves linear-elliptical with blunt to moderately widely rounded ends (vol. 2/3 Fig. 132: 17–22) . . . . **34. *F. construens*** in part (*subsalina*-tribe)

36a (32) Valves without central area set off from the axial area **37**
36b Central area more or less distinctly clearly set off, Plate 108–112 compare also *F. famelica* in part . . . . . . **1. F. capucina**-species complex in part
37a Valves (except the smallest) strictly linear with cuneate narrowed ends (vol. 2/3 Fig. 127: 1–5A) . . . . **16. *F. neoproducta***
37b Valve outline variable, rarely strictly linear (vol. 2/3 Fig. 132: 1–32); compare with *F. oldenburgiana*, *F. famelica*, *F. capucina* tribe complex, *F. investiens* . . . . **34. *F. construens***-tribe complex in part
38a (12) Largest valve width 1.5–2 µm, striae barely distinguishable, more than 22/10 µm (vol. 2/3 Fig. 115: 15, 16) . . . . **7. *F. nanana***
38b Largest valve width at least 2 µm and/or striae clearly visible, up to 22/10 µm **39**
39a If the axial area is very wide in relation to the smallest width of 2–3 µm (for example vol. 2/3 Fig. 115: 8, 9), compare with the following taxa: 5. *F. tenera* in part, 6. *F. delicatissima* in part, 13 *F. capucina* var. *amphicephala*, species complex surrounding 32. *F. fasciculata* in part.
39b Axial area narrower **40**
40a If the length is usually distinctly less than 100 µm, compare with the following taxa: 5. *F. tenera* in part, 1. *F. capucina* species complex in part, 4. *F. famelica* in part
40b Length usually up to 100 µm or greater **61**
41a (9) Striae coarsely punctate (vol. 2/3 Fig. 136: 1–7) . . . **31. *F. pulchella***
41b Striae indistinct or not punctate (vol. 2/3 Fig. 111: 18–22) . **2. *F. bidens*** . . . . **1. *F. capucina* species complex** in part
42a (8) Valves rhomboid to broadly lanceolate or with distended enlarged area elongated close to the poles **43**
42b Valves linear, elliptical, narrowly lanceolate **45**
43a Striae very coarse, maximum of 12/10 µm, nevertheless never punctate, at most appearing lineolate (vol. 2/3 Plate 133) . . . . **36. *F. pinnata*** in part . . . . **37. *F. leptostauron*** in part
43b Not with the above combination of characteristics **44**
44a Ends narrowly drawn out and sharply rounded (vol. 2/3 Fig. 130: 1–8 . . . . **12. *F. parasitica***
44b Ends more broadly elongated and bluntly rounded (vol. 2/3 Fig. 130: 9–17); see also *F. pseudoconstruens* (vol. 2/3 Fig. 130: 24–30) and *F. construens* in part . . . . **41. *F. brevistriata***
45a (42) Valves linear to linear-elliptical **46**
45b Valve edge strongly curved **47**
46a Striae distinctly punctate, central area at most moderately wide (vol. 2/3 Fig. 130: 31–42), see also *F. zeilleri* (vol. 2/3 Fig. 129: 10–15), *F. construens* in part as well as *Delphineis karstenii* (vol. 2/3 Fig. 129: 16, 17) . . . . **35. *F. elliptica*** in part
46b Striae not punctate in appearance (vol. 2/3 Fig. 134: 1–8); compare with some examples of *F. investiens* . . . . **38. *F. lapponica*** . . . . **36.2. *F. pinnata*** var. ***intercedens***
47a (45) Valve length rarely greater than 25 µm, striae usually distinguishable as 1–3 puncta **48**
47b Valve length rarely less than 25 µm, striae not punctate, by EM without linking spines (vol. 2/3 Fig. 135: 1–18), compare with *F. investiens* (vol. 2/3 Fig. 136 : 12, 13) . . . . **32. *F. fasciculata*** in part
48a Striae around the edge, usually not distinguishable as two or three puncta . . . . **41. *F. brevistriata***
48b Striae usually distinguishable as two to three puncta with the light microscope **49**

49a Valves linear-elliptical (vol. 2/3 Fig. 129: 14, 15, 130: 17) . . . . **44.** ***F. zeilleri*** var. ***elliptica***
49b Valves rounded (vol. 2/3 Fig. 130: 21–23) . . . . **(?)43.** ***F. robusta***
. . . . **(?)42.** ***F. pseudoconstruens***
. . . . **(?)41.** ***F. brevistriata*** var. ***elliptica***
50a (6) Valves relatively short and greater than 5 µm wide in the middle (vol. 2/3 Fig. 116: 8–10) . . . . **11.** ***F. heidenii***
50b Valves elongated and less than 5 µm wide in the middle (vol. 2/3 Fig. 117: 1–4) . . . . **8.** ***F. crotonensis***
51a (4) Length of the valve up to 100 µm or much more, particularly "*Synedra acus* var. *ostenfeldii*" Krieger . . . . **26. Species complex around *F. ulna***
51b Length usually less than 5 µm **52**
52a Valves narrowly lanceolate with sharply rounded ends (vol. 2/3 Fig. 111: 23, 24) . . . . **3.** ***F. utermoehlii***
52b Valves elliptical to linear with bluntly rounded ends, frequently slightly distended in the middle (vol. 2/3 Fig. 134: 21–25) . . . **9.** ***F. berolinensis***
53a (3) Ventral side of the valve more or less distended in the region of the central area (vol. 2/3 Fig. 117: 8–14) . . . . **14.** ***F. arcus*** in part
53b Ventral side of the valve without such a locally narrowed and limited enlarged area **54**
54a Valves greater than 100 µm long, unevenly bowed with central enlarged area on both sides of the central area (vol. 2/3 Fig. 116: 6, 7); compare with auxospores and initial cells of other taxa . . . . **9."*F. montana*"**
54b Characteristic not present (vol. 2/3 Fig. 117: 15–16), individuals *in situ* found on zooplankton . . . . **13.** ***F. cyclopum***
55a (2) Striae around 20/10 µm or more **56**
55b Striae fewer than 16/10 µm **57**
56a Tribes in electrolyte-poor freshwaters (vol. 2/3 Fig. 126: 11–20) . . . . **17.** ***F. exigua*** in part
56b Tribes in coastal brackish water (vol. 2/3 Fig. 127: 9–15) **18.** ***F. subsalina***
57a (55) Striae 13–16/10 µm, finely punctate, axial area extremely narrow (vol. 2/3 Fig. 127: 16–21) . . . . **19.** ***F. schulzii***
57b Striae usually more widely spaced, not punctate, but lineolate or hyaline, axial area less narrow (tribe corresponding to the genus *Opephora* sensu auct. nonnull.) **58**
58a Tribes in coastal brackish water, spines visible in diagonal positions on the striae and not on the transapical costae (vol. 2/3 Fig. 134: 9–20 or 32, 33) . . . . ***Op. pacifica*** **(vol.2/3p.166)**
. . . . ***Op. olsenii*** **(vol.2/3p.166)**
58b Tribes found predominantly in freshwater and spines absent or, if spines are present, they are usually on the transapical costae (vol. 2/3 Fig. 133: 12–17 or 28–30), *Opephora* ***martyi*** sensu auct. nonnull., see also *F. berolinensis* (vol. 2/3 Fig. 134: 22.23) . . . . **36.4.** ***F. pinnata*** var. ***subsolitaris***
. . . . **37.3.** ***F. leptostauron*** var.***martyi***
59a (1) Elongate "arms", approximately 20–40 µm, width approximately 2 µm, striae 20 or more/10 µm (vol. 2/3 Fig. 117: 1, 2) . . . . . **10.** ***F. reicheltii***
59b Not as above **60**
60a Pole capitate, striae more than 13/10 µm (vol. 2/3 Fig. 117: 4–7A) . . . . **34.5.** ***F. construens F. exigua***
60b Striae clearly more widely spaced and pole not capitate (vol. 2/3 Fig. 117: 3) . . . . **36.3.** ***F. pinnata*** **var.** ***trigona***
61a (31, 40) Individual puncta cannot be distinguished on the striae, with EM areolae regularly arranged in double rows (vol. 2/3 Fig. 120: 1–5), compare

with *F. goulardii* (vol. 2/3 Fig. 123: 4) as well as a number of other tribes in the complex around *F. ulna* . . . . **29. *F. lanceolata***

61b Areolae arranged in single rows **62**

62a Pole widely rounded or more or less capitate, often somewhat spoon-shaped, central area usually not developed, length of the valve on average greater than 250 µm, width 5–10 µm, frustules may be aggregated into chains and may have small spines on the valve edge (vol. 2/3 Fig. 121: 1–5), presumably this is a heterogeneous species complex . . . . . **28. *F. biceps***

62b Not with the above combination of characteristics **63**

63a Frustules usually aggregated into chains and with small spines at the valve margin, valves 7–10 µm wide and on average in every population under 200 µm long (vol. 2/3 Fig. 121: 6–8) . . . . **27. *F. ungeriana***

63b Not with the above combination of characteristics, characteristics in variable combinations, particularly in the relationship of length to width, the form of the pole and the central area (vol. 2/3 Plates 119, 122) . . . . **26. *F. ulna* species complex**

II. Suborder Raphidineae:

The six families of this suborder which are found in inland waters are distinguished by the following characteristics:

Key to the Families:

1a Raphe present on only one valve from each frustule . . . . **2. Achnanthaceae** (vol. 2/4) (p. 82)

1b Both valves of a frustule with a raphe **2**

2a Very short branches of the raphe at the end of the valves, usually only curved into the valve face for a very short distance . . . . **1.Eunotiaceae** (vol. 2/3) (p. 74)

2b Raphe branches usually take up the whole length of the valve **3**

3a Valves with a median raphe . . . . . . **3. Naviculaceae** (vol. 2/1) (p. 93)

3b Valves with canal raphe **4**

4a Valves without keel . . . . **5. Epithemiaceae** (vol. 2/2) (p. 182)

4b Valves with keels **5**

5a Canal raphe lying in the valve face . **4. Bacillariaceae** (vol. 2/2) (p. 163)

5b Canal raphe running around the valve on the mantle edge . . . . **6. Surirellaceae** (vol. 2/2) (p. 184)

1. Family Eunotiaceae Kützing 1844 (vol. 2/3)

Family type genus: *Eunotia* Ehrenberg 1837

Key to genera of the Eunotiaceae:

1a Valves straight, wedge-shaped . **3. *Peronia*** (one species, vol. 2/3 p. 229)

1b Valves usually not wedge-shaped and straight, particularly somewhat curved around the apical axis and therefore dorsiventral **2**

2a Within each population all the valves usually heteropolar . . . . **2. *Actinella*** (one species, vol. 2/3 p. 229)

2b Within each population either all valves are isopolar or only some are heteropolar . . . . **1. *Eunotia*** (vol. 2/3 p. 169) (p. 74)

1. *Eunotia* Ehrenberg 1837 (vol. 2/3 p. 169)

Typus generis: *Eunotia arcus* Ehrenberg 1837

Since 1991 (publication of the first edition of vol. 2/3) numerous new species, new combinations (nov. comb.), changes of rank (nov. stat.) and of specific names (nov. nom.) have been published. See "Additions and corrections" in the second edition of vol 2/3 and in several volumes of Bibliotheca Diatomologica and Iconographia Diatomologica.

Key to the species groups:

1a Raphe ends extend in a narrow curve a short distance towards the middle of the valve on the valve surface (vol. 2/3 Fig. 137–140) . . . . **key group C (p. 81)**

1b Not as above 2

2a Valve ends elongate and nose-like (rostrate), distal raphe ends can only be seen by focussing through the valve, or if valve ends are not nose-like, then raphes are not recognisable at all 3

2b Distal raphe ends end in a shorter or longer curve on the valve surface (vol. 2/3 Plates 141–160), valve ends not rostrate . . . . **key group A** (p. 75)

3a Neither the raphe nor the terminal nodule distinguishable with the light microscope, valve very strongly curved to almost semicircular (vol. 2/3 Fig. 166: 8–11), see also *E. eruca* (vol. 2/3 Fig. 166: 6, 7) . . . . **53.** ***E. hemicyclus*** (vol. 2/3 p. 227)

3b Raphes exclusively within the valve mantle (only the terminal fissures may reach slightly into the valve surface), in valvar view they are therefore not distinguishable on the surface, only by focussing through the valve. Terminal nodules are more or less displaced from the poles – valve ends are therefore rostrate. If it is difficult to decide, the species is also in key group A (vol. 2/3, Plates 161–164 and part of 166) . . . . **key group B** (p. 80)

Key group A

Key to species subgroups:

1a Larger valves in length and/or width, at least some of the valves within the population longer than 30 µm and/or broader than 6 µm 2

1b Smaller valves, maximum length usually less than 30 µm and width less than 6 µm 4

2a Dorsal edge with two or more pronounced humps **subgroup Aa** (p. 75)

2b Dorsal edge a simple curve or flat undulation, or with a low more or less pointed hump or with a convex shoulder just before the end of the valve 3

3a Valves with variable distended area on the dorsal or ventral edge . . . . **subgroup** Ab (p. 76)

3b Dorsal edge with a simple curve . . . . **subgroup Ac** (p. 77)

4a (1) Dorsal edge always with a conspicuous shoulder before the end (vol. 2/3 Fig. 150: 8, 9), compare with *E. auriculata* (vol. 2/3 Fig. 160: 14) . . . . **41.** ***E. bactriana*** (vol. 2/3 p. 218)

4b Dorsal edge a simple curve or with one or more distended areas . . . 5

5a Dorsal edge with at least one conspicuous distended area or undulation . . . . **subgroup Ad** (p. 78)

5b Dorsal edge with a simple curve . . . . **subgroup Ae** (p. 79)

Subgroup Aa

1a Valves with two humps or undulations; examples with only sharply rounded shoulders at the ends (vol. 2/3 Fig. 150: 8, 9) . . **41.** ***E.bactriana*** examples with ventral edge with two humps (vol. 2/3 Fig. 160: 4, 5) . . . . **42.** ***E.gibbosa***

1b Valves with three or more humps (in exceptional cases more than 20), if valves with a rather flat undulating ventral edge (vol. 2/3 Fig. 141: 1–7), compare with *E. siberica* (vol. 2/3 Fig. 141: 8–10) . . . . **10.** ***E.pectinalis***

2a Valves in a population always with three humps (vol. 2/3 Fig. 146: 6–9) . . . . **44.** ***E. triodon***
2b Valves with more than three humps (vol. 2/3 Fig. 146 : 1–5), compare with *E. muelleri* (vol. 2/3 Fig. 146: 10, 11) . . . . **43.** ***E. serra***
3a (1) Valves with humps particularly strongly arched **4**
3b Humps more weakly arched **5**
4a Shoulders of the humps very steep, with an angle of approximately 90° (vol. 2/3 Fig. 149: 3–6), compare with *E. papilio* (vol. 2/3 Fig. 160: 9) as well as other taxa from tropical and sub-tropical regions . . . . **6.7.** ***E. praeruptia-papilio*** **complex**
4b If the shoulders are less steep and sinus between humps not as low (vol. 2/3 Plates 149, 150), see other taxa in the key group around *E. praerupta* as well as additional taxa from tropical and subtropical regions
5a (3) Dorsal edge markedly constricted near the ends, ends with a curved rostrate appearance (vol. 2/3 Fig. 150: 1–7) **6.6** ***E. praerupta*** var. ***bigibba***
5b Ends bluntly rounded, or flat, truncated **6**
6a Dorsal edge distinctly constricted at the end **7**
6b Dorsal edge not constricted or only weakly **8**
7a Ends rounded (vol. 2/3 Fig. 149: 8–19), see also *E. circumborealis* (vol. 2/3 Fig. 143: 16–23) . . . . **7.** ***E. diodon***
7b Ends truncated (vol. 2/3 Fig. 148: 9, 11, 12) . . . . **6.4.** ***E.praerupta bidens*** **complex**
8a (6) Ends narrowly rounded, occasionally with a denticle on the dorsal edge . . . . **7.** ***E. diodon***
8b Ends truncate or broadly rounded **9**
9a Valves relatively narrow in relation to their length, approximately 6–14 times longer than wide **11**
9b Valves broader, compact **10**
10a Ends flat, truncate (vol. 2/3 Fig. 148: 9, 11, 12) . . . . **6.4.** ***E. praerupta bidens*** **complex**
10b Ends broadly rounded (vol. 2/3 Fig. 158: 4–6) compare with *E. circumborealis* (vol. 2/3 Fig. 143: 16–23) and *E. implicata* (vol. 2/3 Fig. 143: 1–9A) as well as *E. zygodon* (vol. 2/3 Fig. 159: 8, 9) . **30.2** ***E. monodon*** **var.** ***bidens***
11a (9) Dorsal edge less convex, more linear, more than 12 striae /10 µm (vol. 2/3 Fig. 159: 1) . . . . **45.** ***E. ruzickae***
11b Dorsal edge more strongly convex, less linear, fewer than 12 striae /10 µm (vol. 2/3 Fig. 147: 18) . . . . **5.** ***E. arcus bidens*** **complex**

Subgroup Ab

1a Ventral edge with one distended area in the middle **2**
1b Ventral edge straight, weakly curved or flat undulations without a central distended area **8**
2a Dorsal edge constricted in the middle **3**
2b Not as above **4**
3a Ends conspicuously broadened, "paddle-shaped" (vol. 2/3 Fig. 141: 8–10) . . . . **12.** ***E. siberica***
3b Ends not as above (vol. 2/3 Fig. 141: 1–7) . . . . **10.** ***E. pectinalis***
4a Dorsal edge with a hump or undulation in the centre **5**
4b Not as above **6**
5a Ends somewhat cuneate, raphe distally running for quite a distance around the pole (vol. 2/3 Fig. 152: 8–12A) . . . . **29.** ***E. formica***
5b Ends broad to sharply rounded, raphe short, dorsal edge occasionally with multiple flat undulations (vol. 2/3 Fig. 141: 1–7) . . . . . **10.** ***E. pectinalis***
6a (4) Ends somewhat cuneate(vol. 2/3 Fig. 152: 8–12A) . . . **29.** ***E. formica***

6b Ends more broadly rounded 7
7a Valves large, width greater than 10 μm(vol. 2/3 Fig. 158: 1–3) . . . . **30.** ***E. monodon***
7b Valves more delicate, particularly narrower (vol. 2/3 Fig. 141: 1–7) . . . . **10.** ***E. pectinalis***
8a (1) Raphe ends only slightly turned into the valve face (vol. 2/3 Fig. 161: 1–7) . . . . **49.** ***E. sudetica***
8b Raphe ends more widely turned into the valve face 9
9a Ends approximately cuneate(vol. 2/3 Fig. 152: 8–12A) . . **29.** ***E. formica***
9b Ends bluntly rounded to broadly truncate 10
10a Breadth 4 μm, striae 16–20/10 μm, dorsal edge bowed in the middle (vol. 2/3 Fig. 156: 35–40) . . . . **38.** ***E. silvahercynia***
10b Not with the above combination of characteristics 11
11a Striae close together, more than 12/10 μm, delicate appearance (vol. 2/3 Fig. 159: 1) . . . . **45.** ***E. ruzickae***
11b Striae usually more than 12/10 μm, distinctly punctate (vol. 2/3 Fig. 141: 1–7) . . . . **10.** ***E. pectinalis***

Subgroup Ac

1a Valve width around 10 μm or more 17
1b Valve width less than 10 μm 2
2a Valves very strongly curved, particularly on their ventral edge, in extreme cases nearly semi-circular (vol. 2/3 Fig. 157: 1–12) . . . . . **33.** ***E. elegans***
. . . . **34.** ***E. arculus***
2b Valves less extremely curved 3
3a Ventral edge more or less strongly curved 4
3b Ventral edge weakly curved to straight 7
4a Valves narrow, at most 4 μm, terminal raphe ends somewhat curved back, but this can be difficult to distinguish (vol. 2/3 Plates 137–140), compare with *E. subarcuatoides*, as well as *E. steineckei*, *E. exigua*, *E. tenella*, *E. paludosa* (vol. 2/3 Plates 153–155), *E. arculus* (vol. 2/3 Fig. 157: 4–12) . . . . **1.** ***E. bilunaris***
. . . . **2.** ***E. naegelii***
4b Valves broader than 4 μm 5
5a Dorsal and ventral edges parallel up to the ends (vol. 2/3 Fig. 152: 1–3) . . . . **28.** ***E. parallela*** **var.** ***angusta***
5b Dorsal edge with ends more or less constricted 6
6a Ends broadly rounded (vol. 2/3 Fig. 157: 13–18) see also *E. rostellata* (vol. 2/3 159: 4, 5) . . . . **35.** ***E. septentrionalis***
6b Ends flat, truncated (vol. 2/3 Plate 147) . . . . **5.** ***E. arcus***
7a (3) Ventral edge relatively weakly, but still distinctly, curved 8
7b Ventral edge straight or barely curved 14
8a Valves with relatively delicate structure 9
8b Structure more coarse, striae distinctly punctate 13
9a Ends broadly rounded and with conspicuous lip-shaped terminal nodes or terminal areas (vol. 2/3 Fig. 151: 11–13) . . . . **32.** ***E. lapponica***
9b Not with the above combination of characteristics 10
10a Valves sharply narrowed to rostrate ends, dorsal edge with small spines or, in specimens found in high moors, without small spines (vol. 2/3 Fig. 157: 19–28) . . . . **25.** ***E. denticulata***
10b Not with the above combination of characteristics 11
11a More than 16 striae /10 μm, ends rostrate and curved dorsally (vol. 2/3 Fig. 154: 31–34) . . . . **19.** ***E. nymanniana***
11b Not with the above combination of characteristics 12

12a Dorsal edge with ends constricted, pole therefore more or less separate (vol. 2/3 Plate 147) . . . . **5.** ***E. arcus***
12b Pole not separate (vol. 2/3 Fig. 142: 7–15), see also *E. intermedia* (vol. 2/3 Fig. 143: 10), *E. tenella* (vol. 2/3 Fig. 154: 23–30) and *E. bilunaris* (vol. 2/3 Plates 137, 138) . . . . **14.** ***E. minor***
13a (8) Terminal raphe ends reaching close to the dorsal edge (vol. 2/3 Fig. 151: 1–10A) . . . . **27.** ***E. glacialis***
13b Terminal raphe ends shorter (vol. 2/3 Fig. 142: 1–6) . . . **11.** ***E. soleirolii***
14a (7) Ends rounded **15**
14b Ends truncate **16**
15a Valves as shown on vol. 2/3 Plate 137 . . . . **1.** ***E. bilunaris***
15b Valves not as above, striae more widely spaced (vol. 2/3 Fig. 142: 1–6), see also *E. pectinalis* (vol. 2/3 Plate 141) . . . . **11.** ***E. soleirolii***
16a (14) Width/length ratio between 3 and 7 (vol. 2/3 Plate 148) . . . . **6.** ***E. praerupta***
16b Width/length ratio between 6 and 15 (vol. 2/3 Plate 147) . . . **5.** ***E. arcus***
17a (1) Valve width at most 10 µm **18**
17b Valve width greater than 10 µm **21**
18a Ends with conspicuously large terminal nodule (area), striae delicate, around 20/10 µm (vol. 2/3 Fig. 151: 11–13) . . . . **32.** ***E. lapponica***
18b Not with the above combination of characteristics **19**
19a Dorsal and ventral edges parallel up to the ends (vol. 2/3 Fig. 10: 14–18) . . . . **28.** ***E. parallela***
19b Dorsal edge with ends constricted or valve gradually narrowing distally **20**
20a Ends bluntly rounded or inflated (vol. 2/3 Fig. 151: 1–10A) **7.** ***E. glacialis***
20b If the ends are broadly rounded to flat, truncated see *E. praerupta* and *E. arcus* (vol. 2/3 Plates 147, 148)
21a (17) Valve width 22–28 µm (vol. 2/3 Fig. 158: 7, 8) . . . . . **31.** ***E. clevei***
21b Valve width less than 18 µm **22**
22a Dorsal edge parallel up to the ends (vol. 2/3 Fig. 152: 1–7) **28.** ***E. parallela***
22b Dorsal edge more or less strongly constricted, forms with variable ends (vol. 2/3 Plate 148) . . . . **6.** ***E. praerupta***

Subgroup Ad

1a Dorsal edge weakly undulate, the indentation in the middle of the valve shallow (vol. 2/3 Fig. 156: 35–40) . . . . **38.** ***E. silvahercynia***
1b Not as above **2**
2a Dorsal edge with a sharply distended area in the middle **3**
2b Dorsal edge with two or more rounded humps **4**
3a Ventral edge undulating (vol. 2/3 Fig. 156: 27–34) . **24.** ***E. microcephala***
3b Ventral edge straight (vol. 2/3 Fig. 155: 22–37) . . . . **23.2.** ***E. paludosa*** **var.** ***trinacria***
4a (2) Dorsal edge with more than two humps **5**
4b Dorsal edge with two humps **6**
5a Humps evenly spaced, flatly rounded (vol. 2/3 Fig. 156: 1–22), see also *tridentula* forms of *E. exigua* (vol. 2/3 Fig. 153: 21–27) . **39.** ***E. muscicola***
5b Humps unevenly distributed, block-like appearance (vol. 2/3 Fig. 156: 23–26) . . . . **40.** ***E. cristagalli***
6a (4) Ends rostrate and bent dorsally, or capitate
(vol. 2/3 Fig. 153: 19, 20, 24) . . . . **17.** ***E. exigua*** **var.** ***bidens***
(vol. 2/3 Fig. 154: 18–22) . . . . **22.** ***E. rhynchocephala*** **var.** ***satelles***
6b Ends either broadly capitate or bent back and rostrate **7**
7a Ends bluntly rounded (vol. 2/3 Fig. 143: 1–9A) . . . . . **15.** ***E. implicata***

7b Ends broad to flatly rounded or obliquely truncated, striae mostly broadly spaced like those in *E. implicata* (vol. 2/3 Fig. 143: 16–23) . . . . **16. *E. circumborealis***

Subgroup Ae

1a Raphe ends at best only slightly turned into the valve face above the ventral edge (vol. 2/3 Fig. 164: 12–20), see also *E. siolii* (vol. 2/3 Fig. 165: 1–10) and additional species in tropical and subtropical regions . . . . **48. *E. rhomboidea***

1b Raphe ends as a rule easily distinguishable with high magnification . . **2**

2a Valves narrow, approximately 1–3 µm **3**

2b Valves usually wider than 3 µm **5**

3a Striae widely spaced, fewer than 15/10 µm (vol. 2/3 Fig. 150: 10–24), see also *E. tenella* (vol. 2/3 Fig. 154: 23–30) . . . . **26. *E. fallax***

3b Striae more closely spaced, more than 14/10 µm **4**

4a Ventral edge strongly curved (vol. 2/3 Plates 137, 138), see also so-called *falcata* and *subarcuata* forms and *E. subarcuatoides* as well as several tribes from the *E. exigua* species complex (vol. 2/3 Plates 153, 154) . . . . **1.2 *E. bilunaris* var. *mucophila***

4b Ventral edge only slightly curved or straight, valve width variable in different varieties, also with short pointed central hump (vol. 2/3 Plate 155), see also the taxa under couplet 4a, with which they are easily confused . . . . **23. *E. paludosa***

5a (2) Valves 3 to around 5 µm wide **6**

5b Valves greater than 5 µm wide **23**

6a Valves approximately bean-shaped, dorsal edge not at all constricted towards the ends (vol. 2/3 Fig. 143: 10–15) . . . . **37. *E. intermedia***

6b Not with the above combination of characteristics **7**

7a Ventral edge strongly curved (vol. 2/3 Fig. 157: 4–12), see also *E. elegans* (vol. 2/3 Fig. 157: 1–3) . . . . **34. *E. arculus***

7b Ventral edge moderately, weakly or not at all curved **8**

8a If the dorsal edge is less arched and ends more or less rostrate, see also the species complex around *E. exigua* (vol. 2/3 Plates 155, 156) and *E. arculus* (vol. 2/3 Fig. 157: 4–12)

8b Not with the above combination of characteristics **9**

9a Ventral edge straight or very weakly curved **21**

9b Ventral edge distinctly concave **10**

10a Dorsal edge strongly vaulted **14**

10b Dorsal edge moderately strongly vaulted **11**

11a Ends rostrate to capitate (vol. 2/3 Plates 153, 154, 138) . . **17. *E. exigua*** . . . . **20. *E. meisteri*** . . . . **36. *E. subarcuatoides***

11b Ends not set off or barely so, bluntly to broadly rounded. **12**

12a Dorsal edge strongly constricted at the ends (vol. 2/3 Plates 153, 154), see also *E. septentrionalis* (vol. 2/3 Fig. 157: 13–18) and *E. denticulata* (vol. 2/3 Fig. 157: 19–28) . . . . **17. *E. exigua*** . . . . **21. *E. tenella***

12b Dorsal edge weakly or not at all constricted **13**

13a Dorsal edge still noticeably constricted, as a result the ends appear somewhat set off (vol. 2/3 Fig. 142: 7–15), see also the variant of *E. implicata* with one hump (vol. 2/3 Fig. 143: 1–9) . . . . . . **14. *E. minor***

13b Dorsal edge not constricted towards the ends, raphe somewhat downturned, however these characteristics can be difficult to distinguish (vol. 2/3 Plates 137, 138) . . . . **1. *E. bilunaris falcata*** and ***subarcuata*** **forms**

14a (10) Dorsal edge sloping gradually from the middle to the end, in such a way that the valve width continually decreases **15**
14b Dorsal and ventral edge approximately parallel in the middle of the valve, constricted just before the ends **19**
15a Dorsal edge slightly undulating **16**
15b Dorsal edge not undulating **17**
16a Dorsal edge distally very strongly constricted, as a result the ends appear stalked and capitate (vol. 2/3 Fig. 154: 11–22) . . **22.** ***E. rhynchocephala***
16b Ends not as above, broadly rounded (vol. 2/3 Fig. 156: 1–22) . . . . **39.** ***E. muscicola***
17a (15) Terminal nodule (area) conspicuously pronounced, dorsal edge with or without denticles (vol. 2/3 Fig. 157: 19–28) . . . . . **25.** ***E. denticulata***
17b Terminal nodule (area) not conspicuous **18**
18a Striae narrowly spaced, around 20 or more/10 μm (vol. 2/3 Plate 153) . . . . **17.** ***E. exigua***
18b Striae widely spaced (vol. 2/3 Fig. 154: 23–30) . . . . **21.** ***E. tenella***
19a (14) Terminal nodule (area) noticeably pronounced (vol. 2/3 Fig. 157: 19–28) . . . . **25.** ***E. denticulata***
19b Terminal nodule (area) inconspicuous, very small **20**
20a Capitate ends appearing pedunculate, dorsal edge at most delicately undulating (vol. 2/3 Fig. 154: 11–22) . . . . **22.** ***E. rhynchocephala***
20b Not with the above combination of characteristics (vol. 2/3 Fig. 157: 13–18), see also *E. minor* (vol. 2/3 Fig. 142: 7–15) . **35.** ***E. septentrionalis***
21a (9) Dorsal edge strongly curved, rarely with two undulations (vol. 2/3 Fig. 154: 1–10), see also *E. rhynchocephala* (vol. 2/3 Fig. 154: 11–22) . . . . **20.** ***E. meisteri***
21b Dorsal edge less strongly curved or approximately parallel to ventral edge **22**
22a Dorsal edge only weakly constricted towards the end (vol. 2/3 Fig. 154: 23–30) . . . . **21.** ***E. tenella***
22b Dorsal edge strongly constricted, also with ventral edge distinctly constricted (vol. 2/3 Fig. 154: 11–22) . . . . **22.** ***E. rhynchocephala***
23a (5) Valve ends broadly capitate and/or broadly rounded to flat, truncated **24**
23b Valve ends not as above . . . . **14.** ***E. minor***
24a Length/breadth ratio 3–7 (vol. 2/3 Fig. 148: 4–8) . . . . **6.3.** ***E. praerupta-curta*** **complex**
24b Length/breadth ratio 6–15 (vol. 2/3 Plate 147) . . . . **5.** ***E. arcus***

Eunotia key group B

Key to species:
1a Dorsal edge with 2 or more conspicuous humps **10**
1b Dorsal edge without multiple humps, at most with weak undulations . **2**
2a Valve width 5–8 μm **3**
2b Valve width 2–6 μm **6**
3a Dorsal edge with a more or less distinct shoulder sloping towards the ends of the valve (vol. 2/3 Fig. 161: 1–4) compare with *E. veneris*, *E. pirla*, *E. carolina* (vol. 2/3 Plate 163) as well as *E. convexa* and *E. siolii* (vol. 2/3 Plate 165) . . . . **49.** ***E. sudetica***
3b Dorsal edge sloping gradually to the ends, without shoulders **4**
4a Valves with strongly convex dorsal edge, often bean-shaped (vol. 2/3 Fig.

164: 1–10), see also *E. sudetica* (vol. 2/3 Fig 161: 5–7) and *E. incisa* (vol. 2/3 Plates 161, 163) . . . . **50. *E. faba***

4b Valves more elongated, sometimes wavy (serpentine) in appearance . . **5**

5a Raphe not visible in valve view with focus in high position (vol. 2/3 Fig. 164: 1–10) . . . . **50. *E. faba***

5b Raphe ends very distal, poorly visible with focus in high position (vol. 2/3 Fig. 152: 1–7) . . . . **28. *E. parallela***

6a (2) Valve width approximately 4–6 µm **7**

6b Valve width approximately 2–4 µm **8**

7a Terminal nodule clearly set back from the pole, as a result the ends appear clearly “shark-like” (vol. 2/3 Fig. 161: 12–19, 163: 1–7) (that is like the mouth of the shark, underslung below the ‘nose’) . . . . . . **46. *E. incisa***

7b Terminal nodule closer to the pole, resulting in only slightly nose-like ends (vol. 2/3 Fig. 164: 11–20, see also *E. intermedia* (vol. 2/3 Fig. 143: 10–15) . . . . **48. *E. rhomboidea***

8a (6) Raphe ends not clearly distinguishable with focus in high position (vol. 2/3 Fig. 164: 11–20), see also *E. siolii* (vol. 2/3 Fig. 165: 1–9) . . . . **48. *E. rhomboidea***

8b Raphe ends visible as weakly curved with focus in high position . . . **12**

9a Striae widely spaced, around 13/10 µm, dorsal edge constricted near the ends (vol. 2/3 Fig. 154: 23–30), see also *E. fallax* (vol. 2/3 Fig. 150: 10–24) as well as *E. siolii* (vol. 2/3 Fig. 165: 1–9) . . . . **21. *E. tenella***

9b Striae narrower, around 20/10 µm, dorsal edge not constricted (vol. 2/3 Plate 155), see also the species complex around *E. exigua* (vol. 2/3 Plates 153, 154) . . . . **23. *E. paludosa***

10a (1) Dorsal edge with (4)5–10 humps (vol. 2/3 Fig. 166: 1–4) . . . . **52. *E. hexaglyphis***

10b Dorsal edge with 2 humps **11**

11a Valve width at the end at most 1/4 of the maximum width, terminal nodule clearly separate from the end (vol. 2/3 Fig. 161: 21–25) . **51. *E. bidentula***

11b If the ends of the valves are less wide, 1/2 to 1/3 of maximum width, terminal nodule is closer to the ends, see *E. diodon* and the group around *E. praerupta* (vol. 2/3 Plates 149, 150)

Eunotia key group C

Key tos pecies:

1a Valves large, width usually 5–10 µm, length usually much greater than 50 µm (vol. 2/3 Fig. 140: 7–18) **2**

1b Valves on the whole more delicate, width usually less than 5 µm (vol. 2/3 Plates 137, 138, 140: 1–6) **3**

2a Valves broader in the middle than distally (vol. 2/3 Fig. 140: 7) . . . . **4. *E. pseudopectinalis***

2b Valves the same width or ends more or less widened (vol. 2/3 Fig. 140: 8–18) . . . . **3. *E. flexuosa***

3a (1) Valves much longer than wide, width 2–4 µm **4**

3b Valves 4 µm to over 5 µm wide (vol. 2/3 Plate 137 in part) . **1. *E. bilunaris***

4a Valve ends not narrowed (vol. 2/3 Fig. 140: 8–18) . . . . **3. *E. flexuosa* species complex**

4b Valve ends gradually narrowed . . . .

5a Retrograde raphe ends at best distinguishable as short curves (vol. 2/3 Plates 137–139), see also *E. subarcuatoides* . . . . **1. *E. bilunaris***

5b Retrograde raphe ends distinguishable by focussing, valves usually less curved than *E. bilunaris* (vol. 2/3 Fig. 140: 1–6) . . . . . . **2. *E. naegelii***

2. Family Achnanthaceae Kützing 1844 (Vol. 2/4)

Family type: *Achnanthes longipes* Agardh 1824 syn. *Conferva armillaris* O.F. Müller 1783

Key to the genera of the Achnanthaceae:

1a Raphe valve with valvocopulae distinctly visible in valve view (with the light microscope), valvocopulae combined by fimbriae with the valve face . . . . **2. *Cocconeis*** (vol. 2/4 p. 83) (p. 92)

1b Not as above . . . . **1. *Achnanthes*** (vol. 2/4 p. 1) (p.82)

1. *Achnanthes* Bory 1822 (vol. 2/4 p. 1)

Typus generis: *Achnanthes adnata* Bory 1822

Achnanthes (sensu lato) is currently being split into many newly defined genera. Users of this flora must wait to see where the discussions will lead. It is the species, with their biological characteristics, that are important, not the names of the genera, which may differ according to the opinions of different taxonomists (see also the preface of this vol. 2/5).

Key to the subgenera

1a Valve faces and valve mantle coarsely punctate, there is a single punctate band in the broad girdle view; raphe-less valve usually with a more or less eccentric, narrow axial area (vol. 2/4 Plates 1 and 2) . . . . **1. *Achnanthes*** (vol. 2/4 p. 2) (p. 82)

1b Not with the above combination of characteristics; if the valve face is coarsely punctate, there is at most a single ring of puncta around the valve on the valve mantle, the single girdle band appears hyaline. The median axial area of the raphe-less valve is often elliptically to lanceolately broadened. . . . . **2. *Achnanthidium*** (vol. 2/4 p. 6) (p. 83)

1. Subgenus *Achnanthes* Bory 1822 (vol. 2/4 p. 2)

Typus subgeneris: *Achnanthes adnata* Bory 1822

Key to the species in the subgenus *Achnanthes*

1a Striae formed by 2 (rarely 3) rows of puncta (vol. 2/4 Fig. 1: 1) . . . . . **4. *A. longipes***

1b Striae formed of a single row of puncta **2**

2a Valves inflated in the middle and at the ends (Vol. 2/4 Fig. 2: 9, 10) . . . . **6. *A. inflata* var. *inflata***

2b Valve outline not as above **3**

3a Valves with more or less elongated ends, and more or less concave in the middle (vol. 2/4 Fig. 2: 1–8) . . . . **5. *A. coarctata***

3b Not with the above combination of characteristics **4**

4a Valve pole wedge-shaped, usually 7–8 striae/10 µm (vol. 2/4 Fig. 1: 2, 3) . . . . **1. *A. brevipes***

4b Valve pole broadly rounded, not distinctly wedge-shaped (cuneate), more than 8 striae/10 µm **5**

5a Axial area of the raphe-less valve bordered by a fleck-like structure (orbiculus) at both valve poles (vol. 2/4 Fig. 1: 11–15) . . . **3. *A. parvula***

5b Raphe-less valves without orbiculus **6**
6a Valves rhomboid-lanceolate, never concave in the middle (vol. 2/4 Fig. 2: 11, 12) . . . . **6. *A. elata* syn *A. inflata* var. *elata***
6b Valve outline otherwise, populations, at least in part, with valves weakly constricted in the region of the central area . . . .
7a Striae 13–15/10 µm (vol. 2/4 Fig. 1: 9, 10) . . . . ***A. islandica*** (no description given)
7b Striae more widely spaced, 9–12/10 µm (vol. 2/4 Fig. 1: 4–8) . . . . **2. *A. subsessilis* syn. *A. brevipes* var. *intermedia***

2. *Achnanthes* subgenus *Achnanthidium* (Kützing 1844) (vol. 2/4 p. 6) Hustedt 1933 non sensu Hustedt

Typus subgeneris: *Achnanthidium microcephalum* Kützing 1844 (designated by Reimer in Patrick & Reimer 1966)

Key to the species (species groups) in the subgenus *Achnanthidium:*
1a Raphe-less, or rarely raphe-less and raphe valve, with a "horse-shoe" shaped mark (vol. 2/4 Plates 41–48) . . . . **key group H** (p. 91)
1b Raphe-less valves without "horse-shoe" shaped mark **2**
2a Raphe-less and raphe valves show considerable difference in the density and the angle of the striae (for example vol. 2/4 Fig. 16: 1–21) . . . . . . **key group A** (p. 84)
2b Raphe-less and raphe valve show less difference, however the central area can be of different size or the striae of one valve can be interrupted by a hyaline area, or they can differ only in the density or only in the angle of the striae **3**
3a Striae appearing coarse and widely spaced, less than 20/10µm, nevertheless they are not distinguishable as individual rows of puncta (by EM there can be two or more rows of areolae, or the alveoli are overall not subdivided by areolae) . . . . **key group B** (p. 84)
3b Striae more widely spaced or distinguishable as single rows of puncta **4**
4a Striae around 40/10 µm, barely distinguishable with the light microscope, valves with beak-like, very elongated ends (vol. 2/4 Fig. 22: 31) . . . . **75. *A. gracillima*** (vol. 2/4 p. 54)
4b Not with the above combination of characteristics **5**
5a Striae on the valve face interrupted by a narrow hyaline area on both sides of the median, often the interruption appears only on the raphe-less valve . . . . **key group C** (p. 86)
5b Striae not interrupted or interrupted only at the edge of the valve mantle **6**
6a Raphe more or less diagonal, sigmoid or approximately straight and only bent at the distal ends, and then in opposite directions . . . . **key group D** (p. 87)
6b Raphe with completely straight ends or ends bent to the same side, in any case not distinguishable as bent to opposite sides **7**
7a Outline of the valves elliptical or elliptical-lanceolate or approximately rhomboid with broadly rounded ends. Ends not elongated or only slightly so . . . . **key group E** (p. 87)
7b Outline not as above, at least in the range within a population most individuals are distinctly more than twice as long as broad (outline tending to linear) or their ends are distinctly elongated **8**
8a Valve outline (on average in the population) linear, linear-elliptical or linear-lanceolate, valve breadth less than 5 µm (also narrow, markedly lengthened forms as shown in vol. 2/4 Plates 32–37) **key group F** (p. 89)
8b Not with the above combination of characteristics, valves broader or

strongly elliptical to lanceolate, often narrowing to capitate or rostrate ends . . . . **key group G** (p. 90)

Achnanthes (Subgenus Achnanthidium) key group A
(species with considerable structural difference between raphe and raphe-less valves)

Key to species:

1a Examples less than 10 µm long, raphe-less valve without striae (vol. 2/4 Fig. 15: 39–44) . . . . **26.** ***A. kuelbsii***
1b Both valves with structures **2**
2a Raphe-less valve with a distinct isolated punctum in the central node, striae more dense on the raphe-less valve, very difficult to differentiate (vol. 2/4 Fig. 29: 1–9) **3**
2b Not with the above combination of characteristics **4**
3a Striae on the raphe-less valve more than 40/10 µm (vol. 2/4 Fig. 29: 7–9) . . . . **69.** ***A. bremeyeri***
3b Striae on the raphe-less valve 30–36/10 µm (vol. 2/4 Fig. 29: 1–6) . . . . **68.** ***A. bahusiensis***
4a (2) Valves with more or less elongated, narrowed wedge-shaped (cuneate) or generally rostrate ends **7**
4b Ends broadly rounded **5**
5a Central area of the raphe-less valve forming a transverse band **6**
5b Central area less pronounced (vol. 2/4 Fig. 18: 20, 21) . . . . ***A. reversa*** (no description given)
6a Raphe-less valve with larger central area, as a result the striae are along the edge of the valve (vol. 2/4 Fig. 16: 15–21) . . . . **1.** ***A. lutheri***
6b Raphe-less valve with smaller central area (vol. 2/4 Fig. 16: 1–14) . . . . **30.** ***A. oblongella***
7a (4) Striae on the raphe valve distinctly punctate (vol. 2/4 Fig. 21: 1–17) . . . . **45.** ***A. clevei***
7b Characteristics not as above or very difficult to distinguish **8**
8a Raphe valve with fascia (transverse band) cross-shaped (vol. 2/4 Fig. 23: 1–27) . . . . **49.** ***A. exigua*** in part
8b Characteristics not as above **9**
9a Raphe-less valve with distinctly punctate striae, raphe valve without pronounced central area (vol. 2/4 Fig. 21: 18–27) . . . **46.** ***A. laterostrata***
9b Not with the above combination of characteristics (vol. 2/4 Fig. 23: 39, 40) . . . . **76.** ***A. dispar***

Achnanthes (Subgenus Achnanthidium) key group B
(raphe and non-raphe valves with coarse striae)

Key to species:

1a Valve outline on average in the population tending to a linear shape, width less that 5 µm (vol. 2/4 Fig. 30: 1–19, Fig. 36: 38–41, 37: 1–8) **25**
1b Not with the above combination of characteristics **2**
2a Valve ends broadly rounded, not elongated **3**
2b Valve ends more or less elongated **12**
3a Raphe valve with cross-shaped central area reaching out to the edge of the valve, somewhat asymmetrically widened (vol. 2/4 Fig. 19: 1–15) . . . . **41.** ***A. hungarica*** in part

3b Not as above **4**
4a Raphe-less valve with smaller central area **8**
4b Raphe-less valve with larger, particularly broader, central area, as a result the striae are placed along the edge **5**
5a Valve rhomboid-elliptical to rhomboid-lanceolate, usually shorter than 14 µm (vol. 2/4 Fig. 17: 22–34) . . . . **34. *A. montana***
5b Valve larger and/or elliptical without a tendency to rhomboid **6**
6a Raphe valve with narrower, more linear axial area **7**
6b Raphe valve with widened lanceolate axial area (vol. 2/4 Fig. 17: 14–21) . . . . **33. *A. rupestris***
7a Raphe valve with narrow to very narrow central area between the middle striae, usually reaching out to the edge (vol. 2/4 Fig. 16: 22–33) . . . . **28. *A. conspicua***
7b Central area broader, usually not reaching out to the edge (vol. 2/4 Fig. 17: 35–42) . . . . **35. *A. rupestoides***
8a (4) Raphe valve without a pronounced central area, marine forms (vol. 2/4 Fig. 40: 23–26) . . . . **90. *A. polaris***
8b Not with the above combination of characteristics **9**
9a Raphe valve with narrow to very narrow central area between the middle striae, usually reaching out to the edge (vol. 2/4 Fig. 16: 22–33) . . . . **28. *A. conspicua***
9b Not with the above combination of characteristics **10**
10a Raphe valve with larger central area approximately round (vol. 2/4 Plates 39, 40) . . . . **92. *A. delicatula*** in part
10b Central area smaller, widened across the valve by 1–2 shortened striae **11**
11a Tribes in electrolyte rich fresh and brackish water, striae on the raphe valve curved, strongly radial (vol. 2/4 Plates 39, 40) . **92. *A. delicatula*** in part
11b Tribes in electrolyte poor fresh water, striae on the raphe valve weakly radial (vol. 2/4 Fig. 18: 1–13)
outline elliptical-lanceolate . . . . **36. *A. distincta***
outline strongly elliptical . . . . **37. *A. stewartii***
12a (2) Distal raphe ends curved in opposite directions, end bluntly elongated (vol. 2/4 Fig. 18: 14–17) . . . . **38. *A. holstii***
12b Distal raphe ends curved in the same direction or straight **13**
13a Ends elongated into a bill shape **14**
13b Ends bluntly wedge-shaped or sharply rounded or bluntly elongated . **23**
14a Striae on the raphe-less valve face broken by a hyaline area running lengthwise down the valve on both sides (vol. 2/4 Fig. 22: 5–12) see also some tribes of *A. ploenensis* var. *gessneri* (vol. 2/4 Fig. 22: 25–28) . . . . **47. *A. kolbei***
14b Not as above **15**
15a Valves large, 30 to greater than 40 µm long, marine forms, striae on the raphe-less valve with a double row of puncta visible with the light microscope (vol. 2/4 Fig. 40: 21, 22) . . . . **91. *A. linkei***
15b Not with the above combination of characteristics **16**
16a Striae on the raphe valve extremely strongly radial in the centre (vol. 2/4 Fig. 23: 39, 40) . . . . **76. *A. dispar***
16b Not as above **17**
17a Striae also strongly radial on the raphe-less valve (vol. 2/4 Fig. 15: 45–48) . . . . **29. *A. lacunarum***
17b Striae on the rapheless valve approximately parallel to weakly radial . **18**
18a Valves only 12 µm long or smaller **19**
18b Valves larger **21**

19a Valves elliptical to linear-elliptical with abruptly set off, broadly elongated ends (vol. 2/4 Fig. 38: 13–24) . . . . **63.** ***A. daui***
19b Valves with narrow to bluntly rostrate elongated ends **20**
20a Ends narrow rostrate (vol. 2/4 Fig. 26: 31–40) . . . **60.** ***A. lemmermanni***
20b Ends short and broadly elongated (vol. 2/4 Fig. 38: 1–12) . **62.** ***A. grana***
21a (18) Valves with slightly convex edges, tending to be linear-elliptical, alveoli (seen with EM) undivided (vol. 2/4 Fig. 22: 13–30) . . . . **48.** ***A. ploenensis***
21b Valve edge strongly convex, alveoli (by EM) with 2 or more rows of puncta **22**
22a Striae (with EM) usually made up of double rows of puncta (vol. 2/4 Fig. 26: 47–49) . . . . . . **A. species "from Bonn"** (no description given)
22b Striae usually made up of more than two rows of puncta. . . . . **92.** ***A. deliculata*** in part
23a (13) Valve always less than 12 µm long, tribe from electrolyte poor fresh water (vol. 2/4 Fig. 38: 1–12) . . . . **62.** ***A. grana***
23b Valves in population on average considerably longer than 12 µm . . . **24**
24a Ends bluntly rounded or rounded wedge-shaped, outline elliptical to rhomboid-lanceolate (vol. 2/4 Plates 39, 40) . . **92.** ***A. deliculata*** in part
24b Ends subcapitate, gradually broadly elongated, valve outline tending linear-elliptical-lanceolate (vol. 2/4 Fig. 40: 14–20) . . . . . **93.** ***A. pericava***
25a (1) Central area never widened to the valve edge, valve strictly linear and not longer than 15 µm (vol. 2/4 Fig. 37: 1–8) . . . . **86.** ***A. nodosa***
25b Not with the above combination of characteristics **26**
26a Valve strictly linear, only up to 3 µm wide, less broad in the middle, central area usually reaching both valve edges (vol. 2/4 Fig. 36: 38–41) . . . . **85.** ***A. kriegeri***
26b Valve linear-lanceolate, on average in the population always broader than 3 µm, central area not reaching to valve edge, or doing so only on one side (vol. 2/4 Fig. 30: 1–19) . . . . **73.** ***A. thermalis*** in part

Achnanthes (Achnanthidium) key group C
(striae interrupted by a hyaline area)

Key to species:
1a Valves elliptical with broadly rounded ends (vol. 2/4 Fig. 28: 9–20) . . . . **66.** ***A. suchlandtii***
1b Valves with elongated ends **2**
2a Fewer than 15 striae /10 µm (vol. 2/4 Fig. 22: 5–12), see also individual forms of *A. ploenensis* var. *gessneri* (vol. 2/4 Fig. 22: 21–28) **47.** ***A. kolbei***
2b More than 15 striae /10 µm **3**
3a Middle part of the valve usually linear-elliptical; marine tribes, in brackish water or electrolyte rich freshwater (vol. 2/4 Fig. 26: 7–23) . **59.** ***A. amoena***
3b Middle portion of the valve elliptical, usually with more strongly convex edges, tribes in electrolyte poor freshwater (vol. 2/4 Fig. 26: 24–29) . . . . ***A. nitidiformis*** (no description given)

Achnanthes (Subgenus Achnanthidium) key group D

(raphe diagonal, sigmoid or with distal ends curved in opposite directions)

Key to species:

1a Raphe conspicuously diagonal or sigmoid 2
1b Raphe mainly straight, only curved in opposite directions at the ends . 5
2a Valve face strongly arched in 4 areas, so that parts of the structure of the striae are out of focus when viewed with the light microscope in one focal plane (vol. 2/4 Fig. 9: 1–10) . . . . 7. *A. flexella*
2b Not with the above combination of characteristics 3
3a Central area of the raphe-less valve developed on one side, brackish water forms (vol. 2/4 Fig. 29: 16) . **40.** *A.* ***pseudobliqua*** (no description given)
3b Not with the above combination of characteristics 4
4a Valves elliptical with broadly rounded ends (vol. 2/4 Fig. 12: 1–9) . . . . **11.** *A.* ***bioretii*** in part
4b Valves broadly lanceolate with narrowed wedge-shaped (cuneate) ends (vol. 2/4 Fig. 18: 18, 19) . . . . **39.** *A.* ***obliqua***
5a (1) Striae coarse, less than 15/10 µm 6
5b Striae more closely spaced 7
6a Valves with distinctly elongated ends (vol. 2/4 Fig. 18: 14–17) . . . . . . **38.** *A.* ***holstii***
6b Ends not elongated or only slightly elongated (vol. 2/4 Fig. 19: 1–8) . . . . **36.** *A.* ***distincta***
7a (5) Valves approximately linear with broadly rounded or cuneate narrowed ends; central area of the raphe valve broadened on both sides, forming a fascia (transverse band) reaching to the edge of the valve (vol. 2/4 Fig. 19: 1–15) . . . . **41.** *A.* ***hungarica***
7b Not with the above combination of characateristics 8
8a Central area of raphe-valve broadened, reaching to the edge on both sides (vol. 2/4 Fig. 23: 1–27) . . . . **49.** *A.* ***exigua*** in part
8b Central area asymmetrical, one side more strongly developed (vol. 2/4 Fig. 9: 14–22, Fig. 10: 1–11) . . . . 9. *A. laevis*

Achnanthes (Achnanthidium) key group E

(Outline elliptical or elliptical-lanceolate or rhomboid, always with broadly rounded ends)

Key to species:

1a Valves constricted in the middle (concave) (vol. 2/4 Fig. 10: 28–33) . . . . **16.** *A.* ***didyma***
1b Not as above. 2
2a Striae appear conspicuously punctate, without the use of special lighting effects 3
2b Striae not conspicuously punctate 7
3a Central area of the raphe-less or of both valves pronounced on one side 4
3b Central area approximately symmetrical 5
4a Valves broadly elliptical, found in electrolyte poor freshwater (vol. 2/4 Fig. 14: 11–18), compare with *A. lapidosa* (vol. 2/4 Fig. 27: 1–14) . . . . **22.** *A.* ***hintzii***
4b Valves more lanceolate, with more strongly narrowed ends, living in brackish water (vol. 2/4 Fig. 29: 17–22) . . . . **70.** *A.* ***punctulata*** . . . . **71.** *A.* ***pseudopunctulata***

5a (3) Valves elliptical with broadly rounded ends, found in electrolyte poor freshwater **6**
5b Valves with narrowed, cuneate, bluntly rounded ends, found in electrolyte rich to brackish water (vol. 2/4 Fig. 28: 1–8) . . . . **67. *A. subsalsa***
6a Central area of the raphe and raphe-less valve more or less the same, raphe more or less diagonal, at least in some individuals in a population (vol. 2/4 Fig. 12: 1–9) . . . . **11. *A. bioretii*** in part
6b Not with the above combination of characteristics (vol. 2/4 Fig. 10: 12–27) . . . . **10. *A. helvetica*** in part
7a (3) Valves rhomboid-lanceolate in outline in at least in part of the population **8**
7b Valves without the tendency to a rhomboid outline **10**
8a Valves very small, shorter than 10 µm, striae not distinguishable without special lighting techniques (vol. 2/4 Fig. 11: 14–17) . . . **77. *A. carissima***
8b Not with the above combination of characteristics **9**
9a Raphe-less valve with larger central area, as a result the striae are around the edge of the valve (vol. 2/4 Fig. 17: 22–34), see also *A. rupestris* (vol. 2/4 Fig. 17: 14–21) . . . . **34. *A. montana*** in part
9b Raphe-less valve with central area pronounced on one side (vol. 2/4 Fig. 14: 19–26) · If the combination of characteristics is not present and the frustule in girdle view is conspicuously spoon-shaped, see *A. altaica* part. (vol. 2/4 Fig. 20: 24–32), *A. rossii* (vol. 2/4 Fig. 20: 1–12), *A. rechtensis* part. (vol. 2/4 Fig. 20: 13–23) . . . . **23. *A. semiaperta*** in part
10a (7) Raphe-less valve conspicuously without structure, although radial striae can be resolved on the raphe valve at least with oblique lighting (vol. 2/4 Fig. 15: 39–44) . . . . **26. *A. kuelbsii*** in part
10b Not as above **11**
11a Raphe-less valve with much larger central area, as a result the striae are around the edge, also distally there is no narrower separate axial area . **12**
11b Central area smaller, at least distally there is a separate axial area, but the striae in this region are not shortened, or only slightly so **14**
12a Striae coarse, less than 24/10 µm (vol. 2/4 Fig. 15: 29–38), compare with *A. conspicua* part. (vol. 2/4 Fig. 16: 22–33) . . . . **27. *A. holsatica*** in part
12b Striae more dense **13**
13a Frustules slightly bent in girdle view, valves usually longer than 10 µm (vol. 2/4 Fig. 13: 1–20), by EM no rib-like raised areas between the areolae rows, compare also *A. levanderi* (vol. 2/4 Fig. 15: 8–18), *A. daonensis* (vol. 2/4 Fig. 12: 10–20) . . . . **19. *A. marginulata***
13b Not with the above combination of characteristics (vol. 2/4 Fig. 11: 9–13) . . . . **15. *A. scotica***
14a (11) Valves comparatively very small, usually shorter than 7 µm (vol. 2/4 Fig. 11: 1–8) . . . . **4. *A. curtissima***
14b Valves larger on average in the population **15**
15a Valves strictly elliptical, without a tendency to linear-elliptical **16**
15b At least the largest examples are tending to linear-elliptical **20**
16a Striae very dense, around 30/10 µm (vol. 2/4 Fig. 14: 1–10) . . . . **21. *A. subatomoides***
16b Striae more widely spaced, around 25/10 µm **17**
17a Raphes bent to opposite sides at the poles, distinguishable by careful focussing at least in the largest examples within the population (vol. 2/4 Fig. 10: 12–27) . . . . **10. *A. helvetica*** in part
17b Raphe ends always straight **18**
18a Central area forming a transverse band across the valve. Band formed by

the reduction in length of one or more striae on both sides (vol. 2/4 Fig. 18: 22, 23) . . . . *A. johncarteri* (no description given)

18b Central area of the raphe-valve small, only one or at most two striae are shortened **19**

19a Only one single striae is shortened on one side (vol. 2/4 Fig. 15: 1, 2) . . . . **17.** ***A. lacus-vulcani***

19b Populations have examples with one or two shortened striae on both sides (vol. 2/4 Fig. 15: 8–18), compare also *A. saccula* with smaller valve width around 4 µm (vol. 2/4 Fig. 15: 19–28) . . . . **18.** ***A. levanderi***

20a (15) Valve width usually around or under 4 µm **21**

20b Valve width usually greater than 4 µm **22**

21a Distal striae considerably more dense than proximal striae (vol. 2/4 Fig. 34: 25–34), compare also shorter examples of the broad-valved *A. minutissima* complex . . . . **83.** ***A. kranzii*** in part

21b Distal striae only slightly more dense than proximal striae (vol. 2/4 Fig. 15: 19–28), compare also shorter examples of *A. petersenii* (vol. 2/4 Fig. 37: 28–39), *A. pusilla* (vol. 2/4 Fig. 37: 9–18), *A. nodosa* (vol. 2/4 Fig. 37: 1–8), *A. rosenstockii* (vol. 2/4 Fig. 36: 32–37) . . . . **25.** ***A. saccula***

22a (20) Raphes bent to opposite sides at the poles, distinguishable with careful focussing at least in the largest examples within the population (vol. 2/4 Fig. 10: 12–27) . . . . **10.** ***A. helvetica*** in part

22b Distal raphe ends always straight **23**

23a Striae more widely spaced, less than 24/10 µm (vol. 2/4 Fig. 17: 1–13) . . . . **32.** ***A. kryophila***

23b Striae more dense, more than 26/10 µm **24**

24a On the raphe-less valve the transverse band of the central area is distinctly set off from the axial area (vol. 2/4 Fig. 12: 21–31) . . . . **13.** ***A. chlidanos***

24b On the raphe-less valve the axial area is widened in a lanceolate fashion forming a central area which is indistinctly set off (vol. 2/4 Fig. 12: 10–20) . . . . **12.** ***A. daonensis***

Achnanthes (Achnanthidium) key group F

(*Achnanthes minutissima* species complex and similar species)

Key to species:

1a Central area of the raphe valve rhomboid or rhomboid-elliptical, proximal raphe ends widely spaced (vol. 2/4 Fig. 33: 23–31) . . . . . . **79.** ***A. exilis***

1b Not with the above combination of characteristics **2**

2a If the valve width on average in the population is over 5 µm, see under key group E

2b Valve width on average in a population distinctly less than 5 µm . . . **3**

3a Valve outline variable, raphes distally bent to one side **4**

3b Raphe conspicuously straight distally (in doubtful cases this can only be determined by EM or by comparison with the figures of following taxa) **7**

4a Striae very dense proximally, more than 30/10 µm, axial area of the raphe-less valve wide, lanceolate, central area often one-sided, reaching to the edge (vol. 2/4 Fig. 28: 21–31), recently this taxon has been placed in the genus *Nupela* . . . . **65.** ***A. silvahercynia***

4b Not with the above combination of characteristics **5**

5a Proximal striae 25–30/10 µm, distal striae 35–40/10 µm, central area of the raphe valve forming a broad fascia, found in electrolyte poor water (vol. 2/4 Fig. 34: 25–34) . . . . **83.** ***A. kranzii***

5b Not with the above combination of characteristics **6**

6a Axial area of the raphe-less valve broad, lanceolate, central area usually reaching to the edge on one side (vol. 2/4 Fig. 30: 1–19) **73. *A. thermalis***
6b Not with the above combination of characteristics . . . . **81. *A. biasolettiana*** (species complex)
7a (3) Valves linear, more than 3 µm wide, valve ends only slightly narrowed, distal striae not more closely spaced than the proximal striae, or only slightly so **8**
7b Not with the above combination of characteristics **10**
8a Striae widely spaced, slightly more than 20/10 µm (vol. 2/4 Fig. 37: 9–18), compare also *A. nodosa* (vol. 2/4 Fig. 37: 1–8) and *A. kriegeri* (vol. 2/4 Fig. 36: 38–41) . . . . **87. *A. pusilla***
8b More than 24 striae /10 µm **9**
9a Around 30 striae /10 µm (vol. 2/4 Fig. 37: 28–39) . . . . **88. *A. petersenii***
9b Around 26 striae /10 µm (vol. 2/4 Fig. 37: 19–39), later described as *A. linearioides* . . . . **89. "*A. linearis*" sensu auct. nonnull.**
10a Frustules spoon-shaped in girdle view (not bent) and forming band-like aggregates (vol. 2/4 Fig. 34: 23, 24 . . . . **80. *A. catenata*** in part
10b Not with the above combination of characteristics (vol. 2/4 Plates 32–35) . . . . **78. *A. minutissima*** (species complex)

Achnanthes (Achnanthidium) key group G

Key to species:

1a Valves with ends elongated, rostrate or capitate **7**
1b Valves may have very weakly elongated ends **2**
2a Valves always less than 10 µm long, striae not distinguishable without special lighting effects (vol. 2/4 Fig. 11: 14–17) . . **7. *A. carissima*** in part
2b Without the above combination of characteristics **3**
3a Striae punctate **4**
3b Striae not distinguishable as punctate or puncta visible only with special lighting **5**
4a Central area small, more strongly pronounced on one half of the raphe-less valve (vol. 2/4 Fig. 27: 1–14), compare also *A. punctulata* and *A. pseudopunctulata* (vol. 2/4 Fig. 29: 17–22) . . . . **64. *A. lapidosa*** in part
4b Central area lanceolate, large (vol. 2/4 Fig. 28: 1–8) . **67. *A. subsalsa*** in part
5a (3) Valves usually linear, central area narrow, cross shaped, transverse . **6**
5b Valves elliptical, central area broader, not cross-shaped (vol. 2/4 Fig. 20: 24–32) . . . . **42. *A. altaica*** in part
6a Valves more than 6 µm wide, striae around 20/10 µm (vol. 2/4 Fig. 19: 1–15) . . . . **41. *A. hungarica*** in part
6b Valves around 4 µm wide, striae around 30/10 µm (vol. 2/4 Fig. 23: 28–32) . . . . **51. *A. subexigua*** in part
7a (1) Striae widely spaced, around 20/10 µm. Central and axial area not differentiated from each other (vol. 2/4 Fig. 22: 13–30), compare also *A.* aff. *lemmermannii* species from Bonn (vol. 2/4 Fig. 26: 47–49) . . . . **48. *A. ploenensis*** in part
7b Not with the above combination of characteristics **8**
8a Valves elliptical to linear-elliptical with capitate elongated ends, striae difficult to resolve, around 30/10 µm or more **9**
8b Not as above **13**
9a Valves more broadly elliptical, greater than 14 µm long, striae barely distinguishable with specialised lighting, more than 40/10 µm **10**
9b Not with the above combination of characteristics **11**

10a Raphe-less valve with punctate structure in the central nodule (vol. 2/4 Fig. 25: 7–12) . . . . *A. impexa* (no description given)
10b Raphe-less valve without puncta (vol. 2/4 Fig. 25: 1–6) recently found to belong to the genus *Nupela* . . . . **55. *A. impexiformis***
11a (9) Frustules conspicuously spoon-shaped in girdle view, forming band-like aggregates (vol. 2/4 Fig. 34: 23, 24) . . . . . . **80. *A. catenata*** in part
11b Not with the above combination of characteristics **12**
12a Valves less than 10 µm long (vol. 2/4 Fig. 24: 23–27) . . . **53. *A. stolida***
12b Valves greater than 10 µm long (vol. 2/4 Fig. 24: 1–7), compare also capitate forms of the *A. minutissima* species complex (vol. 2/4 Plates 32–35) . . . . **54. *A. pseudoswazi***
13a (8) Middle portion of the valve broadly elliptical to approximately square with broadly capitate elongated ends, central area of the raphe valve small, central area of the raphe-less valve very large (vol. 2/4 Fig. 26: 1–6) . . . . **58. *A. submarina***
13b Not with the above combination of characteristics **14**
14a Central area of both valves very small to absent, axial area narrowly linear to lanceolate (vol. 2/4 Fig. 25: 13–20), compare also *A. amoena* (vol. 2/4 Fig. 26: 7–23) . . . . **56. *A. ingratiformis***
14b Not as above **15**
15a Valves with three undulations formed by broadly capitate rounded ends and a similarly distended middle part of the valve **16**
15b Not as above **17**
16a Valves around 20 µm long or greater (vol. 2/4 Fig. 19: 17–22) . . . . **57. *A. trinodis***
16b Valves less than 14 µm long (vol. 2/4 Fig. 36: 32–37) . . . . **84. *A. rosenstockii*** in part
17a (15) Valves longer than 20 µm with broader central area and delicately punctate striae, around 30/10 µm (vol. 2/4 Fig. 31: 1–10) **74. *A. imperfecta***
17b Not with the above combination of characteristics **18**
18a Striae of both valves similar, around 20/10 µm, central area of the raphe valve forming a very narrow transverse band (vol. 2/4 Fig. 23: 33–38) . . . . **50. *A. ziegleri***
18b Not with the above combination of characteristics **19**
19a Central area of the raphe valve forming a transverse band reaching to the edge of the valve **20**
19b Not as above **21**
20a Striae approximately the same density on both valves, around 30/10 µm (vol. 2/4 Fig. 23: 28–32) . . . . **51. *A. subexigua*** in part
20b Striae on the raphe-less valve more widely spaced, fewer than 26/10 µm (vol. 2/4 Fig. 23: 1–27) . . . . **49. *A. exigua*** in part
21a (19) Striae 28–33/10 µm, raphes weakly bent distally, mainly found in standing waters (vol. 2/4 Fig. 20: 1–12) . . . . **43. *A. rossii*** in part
21b Striae on average somewhat more widely spaced. Raphe more or less straight distally, mainly found in flowing waters (vol. 2/4 Fig. 20: 13–23) . . . . **44. *A. rechtensis*** in part

Achnanthes (Subgenus Achnanthidium) key group H
(raphe-less valve with horse-shoe shaped mark)

Key to species:
1a Horse-shoe shaped mark on both the raphe-less and raphe valves (vol. 2/4 Fig. 47: 1–6) . . . . **99. *A. calcar***

1b Horse-shoe shaped mark usually only on the raphe-less valve **2**
2a Striae on the raphe-less valve coarse, in comparison the striae on the raphe valve are much closer together **3**
2b Striae of both valves of approximately the same distance apart **5**
3a Valves with rostrate ends **4**
3b Ends at most weakly elongated (vol. 2/4 Fig. 48: 1–16) . . . . **97. *A. oestrupii*** in part
4a Valves shorter than 20 µm (vol. 2/4 Fig. 48: 19–26) . . . **98. *A. peragalli***
4b Valves longer (vol. 2/4 Fig. 48: 17, 18) . . . **97. *A. oestrupii* var. *pungens***
5a (2) Striae around 20/10 µm or more, horse-shoe shaped mark relatively weakly marked (vol. 2/4 Fig. 14: 27–34) . . . . **24. *A. lauenburgiana***
Other species which do not belong to the *A. lanceolata* species complex in the widest sense : see under *A. hungarica*, *A. laevis* var. *diluviana*, *A. delicatula* species complex, *A. fonticolanceolata*, *A. semiaperta*, *A. thermalis.*
5b Without the above combination of characteristics **6**
6a Valves linear and abruptly slightly distended in the middle (vol. 2/4 Fig. 45: 1–11) . . . . **95. *A. fragilarioides***
6b Not as above **7**
7a Valves rather broadly elliptical with broadly rounded ends (not rhomboid), horse-shoe shaped mark with a second arc-shaped outline, striae very strongly radial at the ends (vol. 2/4 Fig. 46: 7–17) **96. *A. joursacense***
7b Without the above combination of characteristics, horse-shoe shaped mark with single or double outline, compare with diverse 'exotic' species which are not found in central Europe (vol. 2/4 Plates 44–46) . . . . **94. *A. lanceolata*** species complex

2. *Cocconeis* Ehrenberg 1838 (vol 2/4, p. 83)

Typus generis: *Cocconeis scutellum* Ehrenberg 1838 (designated by Boyer 1927)

Key to species
1a Central area of the raphe-less valves lanceolate, small, delicate forms . **2**
1b Central area of the raphe-less valve linear, mostly robust forms **5**
2a Raphe-less valves coarsely punctate, puncta mostly rounded **3**
2b Raphe-less valves more finely punctate **4**
3a Raphe-less valves with 6–9 striae/10 µm, raphe valves with approximately 22 striae/10 µm (vol. 2/4 Fig. 56: 1–13) . . . . **3. *C. disculus***
3b Raphe-less valves with 10–12 striae/10 µm, raphe valve with more than 30 striae/10 µm (vol. 2/4 Fig. 56: 14–17) . . . . **6. *C. pseudothumensis***
4a Delicate, round or transverse-elliptical puncta or lines on the striae of the raphe-less valves, usually greater than 20 striae /10 µm (vol. 2/4 Fig. 57: 8–31) . . . . **5.*C. neothumensis***
4b Raphe-less valves with more coarse puncta or lines, striae 11–14/10 µm (vol. 2/4 Fig. 55: 1–4; 56: 18–32) . . . . **4. *C. neodiminuta***
5a Raphe-less valves with a marginal ring of double rows of areolae (vol. 2/4 Fig. 50: 4, 6; 57: 5–7; 58: 1–13) . . . . **7. *C. scutellum***
5b Striae of the raphe-less valves consisting of simple puncta or lines up to the edge **6**
6a Frustules with strongly arched valves (vol. 2/4 Fig. 55: 5–8; 57: 1–4) . . . . **2. *C. pediculus***
6b Valves relatively flat (vol. 2/4 Fig. 49: 1–3; 50: 1, 2, 5; 51: 1–9; 52: 1–13; 53: 1–19; 54: 1–11) . . . . **1. *C. placentula*** (p. 93)

Key to the varieties of *Cocconeis placentula* Ehrenberg 1838

1a Axial area of the raphe and raphe-less valves diagonal to the apical axis (vol. 2/4 Fig. 51: 6–9) . . . . **1.5. var. *klinoraphis***
1b Axial area not diagonal **2**
2a Valves for the most part longer than 40µm **3**
2b Valves for the most part shorter than 40µm **6**
3a Striae on the raphe-less valves finely punctate, with 18–22 puncta/10µm (vol. 2/4 Fig. 51: 1–5) . . . . **1.1. var. *placentula***
3b Striae on the raphe-less valves more coarsely punctate **4**
4a Raphe-less valve with more than 16 striae/10µm with transapically lengthened fine areolae (vol. 2/4 Fig. 52: 1–13) . . . . **1.6. var. *lineata***
4b Raphe-less valves with fewer than 15 striae/10µm, areolae appearing as coarse or fine round puncta **5**
5a Raphe-less valve with coarse puncta, 7–8/10µm . . . **1.2. var. *intermedia***
5b Raphe-less valve with finer puncta, 10–12/10µm . . . . . **1.3. var. *rouxii***
6a Areolae of the raphe-less valve very large and transapically broadened, with moderate magnification seen as a shining inner portion with a dark border. Areolae more or less regularly arranged in longitudinal rows (vol. 2/4 Fig. 54: 3–11) . . . . **1.8. var. *pseudolineata***
6b Puncta finer **7**
7a Raphe-less valves with more than 23 striae/10µm, finely punctate . . . **8**
7b Raphe-less valves with fewer than 22 striae/10µm, more coarsely punctate **9**
8a Raphe-less valve with 30–38 striae/10µm . . . . . . **1.4. var. *tenuistriata***
8b Raphe-less valve with 24–26 striae/10µm (vol. 2/4 Fig. 51: 1–5) . . . . **1.1. var. *placentula***
9a Each striae with 3–5 broad solid areolae, apically arranged in regular longitudinal rows (vol. 2/4 Fig. 53: 1–19) . . . . **1.7. var. *euglypta***
9b Each striae with numerous areolae, apically in a zig-zag arrangement (vol. 2/4 Fig. 52: 1–13) . . . . **1.6. var. *lineata***

3. Family Naviculaceae Kützing 1844 (sensu Simonsen 1979; vol. 2/1)

Typus generis: *Navicula* Bory.

Key to the genera of the Naviculaceae:

1a Apical and/or transapical axis heteropolar **2**
1b Apical and transapical axis isopolar **7**
2a Apical axis isopolar, transapical axis heteropolar, valves dorsiventral . **3**
2b Apical axis heteropolar, transapical axis isopolar **4**
3a Valve mantle a little broader on its dorsal side than it is on the ventral side, residuum and intercalary bands absent . . . . **11. *Cymbella*** (vol. 2/1 p. 300) (p. 129)
3b Valve mantle usually much wider on the dorsal side than on the ventral side, a residuum and often many intercalary bands give the frustule a wide elliptical appearance . . . . **12. *Amphora*** (vol. 2/1 p. 342) (p. 138)
4a (2) Only one valve with completely developed raphe, the second valve with only short rudimentary raphes at the poles . . . . **16. *Rhoicosphenia*** (vol. 2/1 p. 381, see number 13, *Gomphonema*) (p. 139)
4b Both valves with fully formed raphes **5**

5a Alveoli closed at least in the mantle region; a longitudinal line apparent in valve or girdle view . . . .
14. ***Gomphoneis*** (vol. 2/1 p. 379, see number 13, *Gomphonema*) (p. 139)
5b Alveoli open in the valve and mantle region **6**
6a Both polar fissures of similar form, clearly bent backwards at an angle .
15. ***Didymosphenia*** (vol. 2/1 p. 380, see number 13, *Gomphonema*) (p. 139)
6b Both polar fissures as a rule of different form . . . . **13.** ***Gomphonema*** (vol. 2/1 p. 352) (p. 139)
7a (1) Cells with distinct intercalary bands and septa **8**
7b Cells without intercalary bands and septa **9**
8a Intercalary bands with chambered edges . . . . **19.** ***Mastogloia*** (vol. 2/1 p. 432) (p. 162)
8b Intercalary bands with extensive septa, without chambered edges . . . . **20.** ***Diatomella*** (vol. 2/1 p. 436) (p. 162)
9a (7) Valves or raphe S-shaped **10**
9b Valves or raphe not S-shaped **13**
10a Raphe at an angle on the valve and S-shaped, but valve not S-shaped . **11**
10b Valves and raphe S-shaped **12**
11a Raphe not lying on a keel . **7.** ***Scoliopleura*** (one species, vol. 2/1 p. 282)
11b Raphe lying on a keel . . . . . . **22.** ***Entomoneis*** (vol. 2/1 p. 438) (p. 163)
12a (10) Puncta (areolae) arranged in apical and transapical lines . . . . **10.** ***Gyrosigma*** (vol. 2/1 p. 295) (p. 129)
12b Puncta (areolae) arranged in primary transapical and secondary diagonal pattern . . . . **9.** ***Pleurosigma*** (vol. 2/1 p. 294) (p. 128)
13a (9) Valves with distinct raised longitudinal ribs . . . . **21.** ***Oestrupia*** (vol. 2/1 p. 436) (p. 162)
13b Raised longitudinal ribs absent from valve face **14**
14a Puncta on the striae not visible under the light microscope **15**
14b Puncta on the striae visible under the light microscope if optical conditions suitable **16**
15a Alveoli always closed, at least in the central region of the valve, with one to two small openings marginally or in the mantle region . . . . **17.** ***Caloneis*** (vol. 2/1 p. 382) (p. 143)
15b The alveoli open or closed with more or less wide inner openings in the valve region . . . . **18.** ***Pinnularia*** (vol. 2/1 p. 397) (p. 144)
16a (14) Central nodule with horns or cross -shaped **17**
16b Central nodule round, oval or elongate **18**
17a Central nodules with horn-like extensions . . . . **8.** ***Diploneis*** (vol. 2/1 p. 283) (p. 127)
17b Central nodule cross-shaped . . . **2.** ***Stauroneis*** (vol. 2/1 p. 236) (p. 122)
18a (16) Raphe bifurcate at the poles, valves with one or more submarginal longitudinal lines . . . . **6.** ***Neidium*** (vol. 2/1 p. 265) (p. 125)
18b Raphe not bifurcate at the poles, longitudinal lines absent **19**
19a Raphe accompanied by strong inner axial ribs **20**
19b Raphes not accompanied by inner axial ribs **21**
20a Valves spindle shaped, central nodule elongated, therefore the proximal raphe ends distant from one another . . . .
5. ***Amphipleura*** (vol. 2/1 p. 262) (p. 125)
20b Valves lanceolate, central nodule less elongated . . . . **4.** ***Frustulia*** (vol. 2/1 p. 258) (p. 124)
21a (19) Striae consist of a few short lineolae running transapically . . . . **3.** ***Anomoneis*** (vol. 2/1 p. 251) (p. 124)
21b Valves structure not as above . . . . **1.** ***Navicula*** (vol. 2/1 p. 84) (p. 95)

1. *Navicula* Bory de St. Vincent 1822 (vol. 2/1, p. 84)

Typus generis: *Navicula tripunctata* (O.F. Müller) Bory 1824

Key to the species groups (sections) of *Navicula*:
(The divisions are based partly on characteristics which may well be based on natural relationships and partly on practical criteria, in critical cases a species can be found in several places in the key.)

1a Valves have lateral areas in addition to central and axial areas, which usually appear to be united with the central area in an H-shaped figure (vol. 2/1 Fig. 65: 1–13), or, in consistently smaller valves (length <20µm), lateral areas which run on both sides between axial area and valve area appear more or less clearly as a hyaline longitudinal line, crossed by transapical striae (vol. 2/4 Fig. 66: 1–34) **2**
1b Valves without such a structure **3**
2a Valves with central and lateral area clearly connected in an H-shape (apart from *N. pygmaea*, this group has only a few species from saline and brackish waters) . . . . . **key group I "Lyratae"** (vol. 2/1 p.170) (p. 110)
2b Lateral areas with longitudinal ribs or similar structures which run apically indicated only by lines or between the axial and lateral area only one row of points each, so that a lyre shape is not obvious . . . .
key group J, spp. group around *N. monoculata* (vol. 2/1 p.173) (p. 110)
3a (1) Two or more distal striae interrupted by a circular or semi-circular hyaline area, in smaller specimens this structure can be reduced so much that only one or several pairs of exaggerated points appear directly at the axial area . . . . . . **key group K "Annulatae"** (vol. 2/1 p.178) (p. 111)
3b Not with the above combination of characters **4**
4a Striae comparatively very coarse (approximately 9/10µm), however without recognisable fine structure. Only represented here by one species (vol. 2/1 Fig. 82: 7, 8) . . . . **"Laevistriatae"**
. . . . **247. *N. elegans*** (vol. 2/1 p.236)
4b Not with the above combination of characters **5**
5a Striae form a pattern of three crossing systems of points. Only represented here by one species (vol. 2/1 Fig. 82: 5–6) . . . . **"Decussatae"**
. . . . **245. *N. placenta*** (vol. 2/1 p.235)
5b Not as above **6**
6a Central pores of the raphe with filamentous extensions bent laterally, as are the terminal fissures. Only represented here by one species (vol. 2/1 Fig. 65: 14, 15) . . . . **"Fistulatae"**
. . . . **46. *N. gibbula*** (vol. 2/1 p.235)
6b Not with the above combination of characters (however, compare with *N. mutica* in scanning EM, Fig. 53: 8) **7**
7a Raphe branches have inner and outer fissures which cross each other near the middle, striae are segmented into more or less transapically elongated coarse points. Only represented here by two species (vol. 2/1 Fig. 82: 1–4), in *N. tuscula* the striae form a double row of points at the edge of the valve (vol. 2/1 Fig. 81: 1–7) . . . .
. . . . **Species group of *N. tuscula* and *N. pseudotuscula*** (vol. 2/1 p.234)
7b Combination of characters does not apply **8**
8a Striae consist of lineolae (line-like foramina). In some forms, in particular small valved ones, they are only visible under the light microscope with the help of oblique illumination or darkfield, or striae are generally parallel **9**
8b Striae consist of more or less clear points or the finer structure can't be differentiated **12**

9a Striae more or less radial (often converging at the ends) . . . . **key group A "Lineolatae"** (vol. 2/1 p.88) (p. 97)
9b Striae and transapical ribs parallel or only slightly radial, in larger forms a more or less recognisable system of lines crossing at right angles is formed by regular alignment of the lineolae or rows of points with longitudinal ribs **10**
10a Valves spindle shaped with elongated cross-shaped central nodules, here only represented by two species (vol. 2/1 Fig. 52: 4–6) . . . . **"Fusiformes" [Haslea]** (vol. 2/1 p.132) (p. 104)
10b Central nodes not cross shaped, central area small, roundish to lanceolate **11**
11a Valve length usually greater than 20 µm . . . . **key group B "Orthostichae"** (vol. 2/1 p.124) (p. 102)
11b Valve length usually less than 20µm **key group C** (vol. 2/1 p.129) (p. 103)
12a (8) *Striae consisting of more or less recognisable line of points which flank the median rib, these however are not in combination with longitudinal sutures; or with obviously contrasting median costae, or axial area . **13**
12b Striae cannot be resolved as a line of points, or with other character combinations than 12a **17**
13a Central nodes form an elongate cross, with more or less eccentric stigma . . **key group E, species group around *N. mutica*** (vol. 2/1 p.147) (p. 106)
13b Without the above combination of characteristics **14**
14a Wide elliptical forms with broadly rounded ends which are not (or only slightly) extended (vol. 2/1 Fig. 59: 1–19; 60: 1, 2) . . . . **key group F, species group around *N. pseudoscutiformis*** (vol. 2/1 p.158) (p. 107)
14b Not as above **15**
15a Frustules with obvious intercalary bands visible in girdle view . . . . **key group G "Microstigmaticae"** (vol. 2/1 p.160) (p. 108)
15b Frustules without obvious intercalary bands **16**
16a Valves with relatively coarse radial striae similar to the Lineolatae, however with relatively dense, more or less point-like, foramina (usually >24/10 µm) instead of lineolae, transapical ribs therefore appear wider than the striae. Average length of valves in a population greater than 20µm or, if slightly less, wide and lanceolate. Often one, less often several, stigmata in the region of the central nodule . . . . **key group D, species group around *N. elginensis*** (vol. 2/1 p.134) (p. 104)
16b Striae consisting of widely separated coarse points (usually <24/10 µm, at least in the middle) · The remaining species of the "Punctatae" have very variable valve forms, largely marine or brackish water forms or living under conditions of wide osmotic fluctuations . . . . **key group H, species group around *N. pusilla*, *N. semen* and *N. humerosa*** (vol. 2/1 p. 165) (p. 109)

---

* Excluded here are the majority of small valved forms with striae which can be seen with the electron microscope to be formed of more or less point like foramina. These cannot be clearly seen without special optics. It is a problem that in the sections "Minusculae" and "Bacillares" sensu Hustedt (1961), individual species or even individuals (for example the initial cells) that have recognisably punctate striae are not considered at this point in the key. On the other hand other small valved forms from the *N. mutica* species group are traditionally identified as "Punctatae". Other forms without clear puncta which previously belonged to Lineolatae have to be considered on the basis of their similar appearance and other structural similarities.

17a (12) Valves usually narrow, linear to lanceolate in outline with more or less rostrate to capitate ends. Raphe usually with sharply angular terminal fissures. Striae strongly radial in the central region, converging abruptly near the ends, usually very dense (>30/10 μm to indistinguishable under the light microscope, compare with vol. 2/1 Fig. 79: 1–28) . . . . **key group L, species group around *N. bryophila* and *N. subtilissima*** (vol. 2/1 p.180)**(p. 111)**

17b Not with the above combination of characteristics **18**

18a *The median rib is clearly raised by flanking longitudinal grooves which are more or less strongly marked . . . . **key group M "Bacillares"** (vol. 2/1 p.184) (p. 112) (presumably homogeneous species group around *N. bacillum*, but often not clearly identifiable under the light microscope)

18b Longitudinal grooves not recognisable. Small to (rarely) medium forms with variable fine structure, without the character combinations of the other key groups . . . . **key group N (heterogeneous remainders) "Minusculae"** (vol. 2/1 p.205) (p. 115)

Navicula key group A – Navicula Lineolatae ("Subgenus Navicula") (vol. 2/1, p. 88)

Species group with the major characteristics of the typus generis: *Navicula tripunctata*. See additional species of this group in vol. 2/4, plates 59–73.

Key to species:

1a Valves, unlike the characteristic form of the "Lineolatae", with conspicuously coarse broad striae, in relation to the smaller valve length (almost always shorter than 30 μm). Pole usually without striae (see vol. 2/1 Fig. 16: 1–11 as well as vol. 2/1 Fig. 15: 8–11) **71**

1b Not with the above combination of characteristics **2**

2a Pole also not striate, central area large, striae 10–13/10 μm, more strongly radial (vol. 2/1 Fig. 34: 12, 13) . . . . **29. *N. globulifera***
If central area is small, striae about 20/10 μm, very weakly radial (vol. 2/1 Fig. 44: 12, 13) . . . . **63. *N. riparia***

2b Pole not conspicuously free of striae **3**

3a Striae markedly bent at the ends, relatively large valved forms (vol. 2/1 Fig. 41: 2) . . . . **55. *N. oblonga***

3b Striae not conspicuously bent **4**

4a Central pore of the branch of the raphe or the proximal ends of the raphe branch with the central pores conspicuously bent to one side, as a rule the central nodule is also displaced with one side appearing larger (see vol. 2/1 Fig. 37: 1–8 as well as Fig. 38: 1–15). Not considered here are forms with

---

* Species with a canopy which extends further over the axial area and which is distinguishable under the light microscope as two longitudinal lines (for example compare Fig. 17: 3 with 17: 4 as well as 83: 4–7) cannot be satisfactorily incorporated in a light microscope key. In such cases fine structural similarities (relationships) which are really systematically more important have to be ignored in favour of practical advantages. The separation of the species on the basis of these characters, which are generally very sound, cannot be certain with the light microscope, especially for small valved forms, so that some species which belong here will appear again in the remaining key groups.

raphe branch which is turned inward in the region of the central nodule from a more or less more marginal position in the middle of the median rib (see vol. 2/1 Fig. 39: 12, 13) **68**

4b Raphe ends with the central pores in the region of the central nodule lying in the middle, in other cases however not in the particular combination mentioned above **5**

5a Striae (usually) continuously radial to the ends **6**

5b Striae (usually) convergent at the ends **15**

6a Central area reaching to the valve edge, striae appearing coarse in relation to smaller valve length, 7–10/10 µm (vol. 2/1 Fig. 42: 13–15) . . . . **59.** ***N. costulata*** (compare with 60. *N. subcostulata* and 57. *N. lesmonensis*)

6b Central area not extended to the edge **7**

7a Striae at the central nodule considerably more widely spaced than the others; large valved forms (see vol. 2/1 Fig. 26: 1) . . . . . . **40.** ***N. hasta***

7b Striae in the middle not widely spaced or less extremely so **8**

8a Central pores markedly close to each other. Forms found in salt, brackish and fresh water with very high electrolyte concentrations **9**

8b Without the above combination of characteristics **10**

9a Conspicuously narrow lanceolate valves, lineolae barely discernible (vol. 2/1 Fig. 39: 6–11) . . . . **49.** ***N. duerrenbergiana***

9b Valves more broadly lanceolate or with linear to slightly concave edges, lineolae distinctly differentiated (vol. 2/1 Fig. 39: 4–5) **48.** ***N. pavillardii***

10a (8) Raphe branches with the central pores in the central nodules bent to one side (vol. 2/1 Fig. 38: 1–4) . . . . **42.** ***N. schroeteri***

10b Central pores more or less straight **11**

11a Middle striae alternately shorter and longer, ends broadly rounded (vol. 2/1 Fig. 40: 1, 2), compare with *N. splendicula* (vol. 2/1 Fig. 33: 1–3), with individual examples also with continuously radial striae **38**

11b Middle striae can be shortened, but not alternating with longer striae, ends not broadly rounded **12**

12a Ends slightly to distinctly rostrate **14**

12b Ends gradually narrowed, rounded points, not rostrate **13**

13a Valve width 6–9 µm, central area somewhat more strongly dilated transapically and not conspicuously asymmetrical (vol. 2/1 Fig. 36: 8) . . . . **38.** ***N. pseudolanceolata***

13b Valve width 9–12 µm, central area more or less dilated and lanceolate, usually markedly asymmetrical (vol. 2/1 Fig. 36: 10–12) . . . . **39.** ***N. concentrica***

14a Central area considerably large and rounded (vol. 2/1 Fig. 35: 1–4), compare with the type specimen *N. expecta* (vol. 2/1 Fig. 31: 15) · If the central area is broadened across the valve and lineolae very distinct (around 20/10 µm), then similar forms with slightly elongate ends, such as *N. pseudolanceolata* shown in vol. 2/1 Fig. 36: 9, should be considered . . . . **30.** ***N. trivialis***

14b Central area narrowly lanceolate (vol. 2/1 Fig. 36: 4–7) **37.** ***N. praeterita***

15a (5) Central area conspicuously wider transapically than in apical direction, almost always more than half the valve width, the transverse rectangular arrangement is characteristic of the ground form (all species with this variable central area are repeated at couplet number 30) **16**

15b Central area rounded, rhomboid, lanceolate or only weakly delineated . **30**

16a Valves more strongly arched, with the result that the striae at the edges appear conspicuously bent, marine and brackish water forms **29**

16b Not with the above combination of characteristics **78**

17a Valve length always less than 20 µm, without elongated ends, striae 14–18/10 µm (vol. 2/1 Fig. 35: 14–20) . . . . **35. *N. perminuta***
17b Not with the above combination of characteristics **18**
18a Valves always smaller than 30 µm, central area narrow, striae conspicuously coarse in relation to small valve size, less than 12/10 µm (vol. 2/1 Fig. 42: 16–17 compare with *N. costulata* Fig. 42: 13–15, *N. lesmonensis* Fig. 41: 8–11) . . . . **60. *N. subcostulata***
18b Not with the above combination of characteristics **19**
19a Large and moderately large forms, inner and outer raphe fissure running parallel almost over the total length to the strongly marked central pores, striae and lineolae conspicuously coarse, respectively 5–12/10 µm and 18–25/10 µm **26**
19b Not with the above combination of characteristics **20**
20a Lineolae usually greater than 30/10 µm, difficult to distinguish **23**
20b Lineolae about 25/10 µm, still relatively easily distinguished **21**
21a Valves with elongated rostrate ends
striae moderately radial (vol. 2/1 Fig. 31: 6, 7) **16. *N. subrhynchocephala***
striae more strongly radial (vol. 2/1 Fig. 31: 15) . . . . . . **18. *N. expecta***
21b Valves ends not elongated or very slightly so **22**
22a Striae strongly radial (vol. 2/1 Fig. 27: 12–17) . . . . **4. *N. cari***
22b Striae weakly radial (vol. 2/1 Fig. 28: 17–19) . . . . **8. *N. libonensis***
23a (20) Striae weakly radial **25**
23b Striae more strongly radial **24**
24a Valves widely lanceolate with more or less sharply rounded ends (vol. 2/1 Fig. 32: 16, 17) . . . . **22. *N. menisculus* var. *upsaliensis***
24b Valves narrow lanceolate with broadly rounded ends (vol. 2/1 Fig. 27: 12–17) . . **4. *N. cari*** compare with *N. tenelloides* with smaller valve size and less enlargement of the central area (vol. 2/1 Fig. 38: 16–21) as well as *N. angusta* with very variable form of the central area (vol. 2/1 Fig. 28: 1–5)
25a (23) Valves linear, length greater than 30 µm (vol. 2/1 Fig. 27: 1–3 compare with *N. recens* Fig. 27: 7–11 as well as *N. margalithii* Fig. 27: 4–6) . . . . **1. *N. tripunctata***
25b Valves lanceolate, length less than 30 µm (vol. 2/1 Fig 32: 1–4) . . . . . . **19. *N. veneta***
26a (19) Valve ends not elongated or only slightly elongated, but always bluntly rounded **28**
26b Valves with either sharply rounded and/or elongated rostrate ends . . **27**
27a Valve ends not elongated or only slightly so (vol. 2/1 Fig. 30: 2–4) . . . . **13. *N. meniscus***
27b Valve ends usually more elongated(vol. 2/1 Fig. 30: 5–8, Fig. 31: 1, 2) . . . . **14. *N. rhynchocephala***
28a (26) Striae 8–9/10 µm, lineolae 18–25/10 µm (vol. 2/1 Fig. 30: 1) . . . . **12. *N. peregrina***
28b Striae 8–9/10 µm, lineolae about 25/10 µm (vol. 2/1 Fig. 31: 3–5) . . . . **15. *N. slesvicensis***
29a (16) Central area usually asymmetrical, valve ends very variable (vol. 2/1 Fig. 39: 3) . . . . **47. *N. cancellata***
29b Central area symmetrical, valve ends cuneate, sharply rounded (vol. 2/1 Fig. 39: 2) . . . . **46. *N. arenaria* var. *rostellata***
30a (15) Central area small or not clearly set off from the axial area, with, in general, a lanceolate form **55**
30b Central area larger, form variable, rounded or rhomboid **31**
31a Striae at the central nodule more or less regularly alternately shorter and longer **32**

31b Striae around the central nodule shortened in some other way **41**
32a Ends elongated, rostrate **36**
32b Ends blunt, seldom sharply rounded, however not conspicuously elongated **33**
33a Ends blunt to broadly rounded **34**
33b Ends sharply rounded (vol. 2/1 Fig. 32: 17; 34: 5) **37**
34a Central area large, rounded, striae 13–16/10 µm (vol. 2/1 Fig. 39: 12, 13) . . . . **50.** ***N. bottnica***
34b Central area of a different form, striae usually considerably less than 13/10; in *N. digitoradiata* var. *minima* with more dense striae the central area is never large and round **35**
35a Lineolae coarse, 20–25/10 µm **38**
35b Lineolae indistinct, more than 30/10 µm (vol. 2/1 Fig. 34: 1–9) . . . . **27.** ***N. digitoradiata***
36a (32) Central area large, rounded (vol. 2/1 Fig. 35: 5–8), compare with *N. digitoradiata* var. *rostrata* sensu Hustedt (vol. 2/1 Fig. 39: 6) . . . . **31.** ***N. salinarum***
36b Central area small, with an irregular border (vol. 2/1 Fig. 32: 12–15) . . . . **21.** ***N. capitatoradiata***
37a (33) Central area transversely widened (vol. 2/1 Fig. 32: 17) . . . . **22.** ***N. menisculus*** **var.** ***upsaliensis*** part.
37b Central area smaller, with an irregular border, variable (see vol. 2/1 Fig. 34: 5 as well as individual forms on vol. 2/3 Plate 33) · Forms of what is presumably a species spectrum of *N. digitoradiata*, *N. stankovicii*, *N. cryptotenella* as well as *N.* species 1 and 2.
38a (11, 35) Raphe branches approximately straight, either with strongly marked central pore and very coarse striae (7–9/10 µm) or striae densely spaced (12–16/10 µm) and with weakly elongated ends **40**
38b Raphe branch curved or with a conspicuous intersection of the inner and outer raphe fissure close to the central nodule **39**
39a Lineolae very coarse, around 20/10 µm (vol. 2/1 Fig. 40: 3–7) . . . . **52.** ***N. striolata***
39b Lineolae less coarse, around 24/10 µm, altogether somewhat smaller valves with less bluntly round ends (vol. 2/1 Fig. 40: 8, 9) . . **53.** ***N. oppugnata***
40a (38) Striae 7–9/10 µm (vol. 2/1 Fig. 40: 1, 2) . . . . . . **51.** ***N. reinhardtii***
40b Striae 12–16/10 µm (vol. 2/1 Fig. 36: 1–3) . . . . **36.** ***N. splendicula***
41a (31) Central area large, one or both sides reaching at least approximately half the width of the valve **42**
41b Central area small or with an irregular border, formed by individual lengthened or widely spaced striae **54**
42a (31) Lineolae distinctly marked, up to approximately 25/10 µm **42**
42b Lineolae less distinctly marked, more than 25/10 µm **45**
43a Valves large (50–140 µm long), raphe somewhat filamentous, central nodule distinctly larger on one side, central area large, round to rhomboid, oligosaprobic freshwater forms (vol. 2/1 Fig. 41: 1) . . . **54.** ***N. vulpina***
43b Not with the above combination of characteristics **44**
44a Moderate-sized forms with elongated rostrate ends, striae more than 12/10 µm (vol. 2/1 Fig. 31: 6, 7) . . . . **16.** ***N. subrhynchocephala***
44b Larger forms with coarser striae, less than 12/10 µm (see also *N. expecta*, vol. 2/1 Fig. 31: 15) **26–28**
45a (42) Valves strongly arched, because of this the striae appear distinctly bent at the edges, marine and brackish water forms **29**
45b Not with the above combination of characteristics **46**
46a Striae only weakly to moderately radial **47**

46b Striae distinctly radial **51**
47a Valves linear (vol. 2/1 Fig. 27: 1–3) . . . . **1.** ***N. tripunctata***
47b Valves lanceolate to linear-lanceolate (but also compare special forms of *N. halophila*, vol. 2/1 Fig. 44: 18) **48**
48a Valves small, narrower than 6.5 µm **49**
48b Valves larger, particularly broader **50**
49a Valves more or less elliptical (vol. 2/1 Fig. 28: 13–15) . . . . **7.** ***N. cincta*** small "questionable forms"
49b Valves lanceolate (vol. 2/1 Fig. 32: 1–4) . . . . **19.** ***N. veneta***
50a (48) Striae 9–10/10 µm, lineolae around 30/10 µm (vol. 2/1 Fig. 27: 4–6) . . . . **2.** ***N. margalithii***
50b Striae 11–14, lineolae around 35/10 µm (vol. 2/1 Fig. 27: 7–11) **3.** ***N. recens***
51a (46) Valves (other than the smallest in a population) distinctly linear (vol. 2/1 Fig. 28: 1–5) . . . . **5.** ***N. angusta***
51b Valves lanceolate to approximately linear-lanceolate **52**
52a Central area with a regular border, more or less circular to rhomboid . **53**
52b Central area rhomboid (vol. 2/1 Fig. 29: 1–4) . . . . **9.** ***N. radiosa***
53a Valve width less than 8 µm (vol. 2/1 Fig. 31: 8–14) **17.** ***N. cryptocephala***
53b Valve width greater than 8 µm (vol. 2/1 Fig. 29: 5–7) . **10.** ***N. lanceolata***
54a (41) Valves elliptical to linear-lanceolate (vol. 2/1 Fig. 28: 8–15), see also *N. eidrigiana* (vol. 2/1 Fig. 28: 6, 7) . . . . **7.** ***N. cincta***
54b If valves conspicuously lanceolate see *N. cryptocephala* (vol. 2/1 Fig. 31: 8–14), *N. cryptotenella* (vol. 2/1 Fig. 33: 9–11), *N. menisculus* (vol. 2/1 Fig. 32: 16–25), *N. heimansii* (vol. 2/1 Fig. 29: 8–11), *N. phyllepta* (vol. 2/1 Fig. 32: 9–11)
55a (30) Lineolae appearing distinctly separate, 20–25/10 µm **56**
55b Lineolae more dense **59**
56a Striae more delicate, at least 16/10 µm (vol. 2/1 Fig. 41: 3, 4) . . . . **56.** ***N. gottlandica***
56b Striae coarser, maximum 14/10 µm **57**
57a Valve ends more or less bluntly rounded (vol. 2/1 Fig. 28: 6, 7) . . . . **6.** ***N. eidrigiana***
57b Valve ends more or less elongated, rostrate, slightly capitate or very sharply rounded **58**
58a Salt and brackish water forms, ends sharply rounded (vol. 2/1 Fig. 34: 10, 11) . . . . **28.** ***N. flanatica***
58b Forms from oligosaprobic fresh water, ends mostly rostrate to slightly capitate (vol. 2/1 Fig. 36: 4–7) . . . . **37.** ***N. praeterita***
59a (55) Striae approximately parallel to weakly radial **60**
59b Striae, at least in the middle, distinctly radial **62**
60a Exclusively very small forms with a maximum width of 4 µm, central pores close to each other **61**
60b Larger forms, central pores more or less separate from each other. species found in electrolyte-rich water or under intermittent osmotic pressure (vol. 2/1 Fig. 43: 1–11) . . . . **62.** ***N. halophila***
species found in very electrolyte poor bogs, poles non striate but may be difficult to distinguish (vol. 2/1 Fig. 44: 12, 13) . . . . . . **63.** ***N. riparia***
61a Without outline of central area, striae 17–20/10 µm (vol. 2/1 Fig. 35: 9, 10) . . . . **32.** ***N. salinicola***
61b Central area weakly outlined, striae 13–16/10 µm (vol. 2/1 Fig. 35: 21–24) . . . . **34.** ***N. incertata***
62a (59) Very small forms, maximum length 12 µm, with a weakly arched widened middle and conspicuously broadly rounded ends; membership in the "Lineolatae" is doubtful (vol. 2/1 Fig. 35: 11–13) . **33.** ***N. bremensis***

62b Valves longer **63**
63a Valve width to a maximum of 4 μm (vol. 2/1 Fig. 38: 16–20) . . . . **45.** ***N. tenelloides***
63b Valve width greater than 5 μm **64**
64a Median rib appearing conspicuously enlarged, central pores close to each other. Salt and brackish water forms, rarely also in very electrolyte rich freshwater (vol. 2/1 Fig. 32: 5–11) . . . . **20.** ***N. phyllepta***
64b Not with the above combination of characteristics **65**
65a Valves broadly lanceolate, striae 8–12/10 μm (vol. 2/1 Fig. 32: 16–25) . . . . **22.** ***N. menisculus***
65b Valves more narrowly lanceolate, striae more than 12/10 μm **66**
66a Raphe curved in the median area only in the central region of the valve, the remainder not able to be resolved because it runs very close to the edge of the very narrow axial area, only found in electrolyte poor water (vol. 2/1 Fig. 29: 8–11 including *N. heimansioides*) . . . . **11.*****N. heimansii***
66b Not with the above combination of characteristics **67**
67a Lineolae approximately 30/10 μm, typical forms linear-lanceolate (vol. 2/1 Fig. 33: 1–4) . . . . **23.** ***N. stankovicii***
67b Lineolae more dense, barely distinguishable, always lanceolate forms (vol. 2/1 Fig. 33: 9–11, 13–17) . . . . **24.** ***N. cryptotenella***
(vol. 2/1 Fig. 33: 21, 22, later described as *N. wildii*) . . **25.** ***N.*** **species 1**
(vol. 2/1 Fig. 33: 23–25, later described as *N. reichardtiana*) . . . . **26.** ***N.*** **species 2**
68a (4) Striae continuously radial (vol. 2/1 Fig. 38: 1–4) . . **42.** ***N. schroeteri***
68b Striae convergent at the ends **69**
69a Valve ends not elongated, bluntly to rarely sharply rounded (vol. 2/1 Fig. 38: 5–9) . . . . **43.** ***N. erifuga***
69b Valve ends more or less elongated **70**
70a Striae 7–10/10 μm (vol. 2/1 Fig. 37: 1–9) . . . . **45.** ***N. viridula***
70b Striae 13–22/10 μm (vol. 2/1 Fig. 38: 10–15) . . . . **44.*****N. gregaria***
71a (1) Central area only slightly lengthened transapically or not lengthened at all (vol. 2/1 Fig. 42: 1–11) . . . . **58.** ***N. capitata***
71b Central area markedly lengthened transapically **72**
72a Central area enlarged on both sides, delimited by a shortened striae (vol. 2/1 Fig. 41: 8–11) . . . . **57.** ***N. lesmonensis***
72b Central area reaching to the edge **73**
73a Ends sharply rounded, striae 7–10/10 μm (vol. 2/1 Fig. 42: 13–15) . . . . **59.** ***N. costulata***
73b Ends bluntly rounded, striae around 12/10 μm (vol. 2/1 Fig. 42: 16, 17) . . . . **60.** ***N. subcostulata***

Navicula key group B – ("Orthostichae") (vol. 2/1, p. 124)
Species with the following combination of characteristics: narrow axial area, weakly developed central area, more or less parallel striae. Species which are usually less than 20 μm follow in the separate key group C.

Key to species:
1a Valves narrowly linear-lanceolate (spindle-shaped) with broadened cross-shaped central nodule (vol. 2/1 Fig. 52: 4–6) . . . . **see "Fusiformes"(Haslea)** (vol. 2/1 p. 132) (p. 104)
1b Valves not as above **2**
2a Valves without distinctly recognisable longitudinal ribs, or more than 25/10 μm **3**

2b Valves with distinct longitudinal ribs, 8–20/10 µm (vol. 2/1 Fig. 43: 1–8); *N. perrotetii* is not found in this region . . . . **61.** ***N. cuspidata***
3a Longitudinal ribs 28/10 µm or more, still recognisable (vol. 2/1 Fig. 44: 1, 2) . . . . **62.** ***N. halophila*** (**robusta**-forms)
3b Longitudinal ribs not distinguishable **4**
4a Valves more broadly elliptical-lanceolate (vol. 2/1 Fig. 45: 13–20) . . . . **64.** ***N. accomoda***
4b Valves more narrowly lanceolate **5**
5a Central pores close to each other (vol. 2/1 Fig. 39: 6–11) . . . . **49.** ***N. duerrenbergiana***
5b Central pores more widely spaced **6**
6a Valve size usually greater than 20 µm **7**
6b Valve size usually less than 20 µm **key group C** (vol. 2/1 p. 129) (p. 103)
7a Forms in electrolyte poor water, poles without striae (this is may be difficult to determine) (vol. 2/1 Fig. 44: 12, 13) . . . . . . **63.** ***N. riparia***
7b Forms in electrolyte rich water, striae present up to the poles (vol. 2/1 Fig. 44: 1–11 to 14–18) . . . . **62.** ***N. halophila***

Navicula key group C (vol. 2/1, p. 129)

group of species with heterogeneous fine structure; striae approximately parallel along most of the length of the valve, without or with very small central area, valves elliptical to lanceolate (never strictly linear), usually less than 20 µm long.

Key to species:

1a Largest valves in a population lanceolate with more or less elongated ends **6**
1b The largest valves within the population more or less elliptical with stubby to widely rounded ends. Compare with *N. digitulus* (vol. 2/1 Fig. 77: 19–28), *N. subhamulata* (vol. 2/1 Fig. 66: 32–34), if ends short and narrowly elongated, *N. citrus* (vol. 2/1 Fig. 58: 6–8) **2**
2a Valve width 6–7 µm, striae 14–18/10 µm (vol. 2/1 Fig. 64: 26–28) . . . . **71.** ***N. kriegeri***
2b Valves narrower and/or more densely striate **3**
3a More than 25 striae /10 µm (vol. 2/1 Fig. 45: 24, 25) . . . . . **69.** ***N. fluens***
3b Usually fewer than 25 striae /10 µm **4**
4a Central area transversely broadened, at least weakly (vol. 2/1 Fig. 73: 12–15) . . . . **192.** ***N. submuralis***
4b Central area hardly developed or not widened at all **5**
5a Median rib and terminal nodule standing out conspicuously (vol. 2/1 Fig. 45: 31–33) . . . . **70.** ***N. muraliformis***
5b Not as above, valve ends only rarely bluntly rounded, striae usually distinctly radial (vol. 2/1 Fig. 76: 21–26) . . . . . **224.** ***N. subminuscula***
6a (1) Striae in the middle of the valve usually fewer than 26/10 µm . . . **7**
6b Striae almost always more narrowly spaced in the middle of the valve, at the ends to about 40/10 µm **8**
7a Valves narrowly lanceolate, forms found in more acid, electrolyte poor, clean water (vol. 2/1 Fig. 45: 26–30) . . . . **68.** ***N. submolesta***
7b Valves more broadly lanceolate, forms found in water with moderate to high electrolyte concentrations or in environments with changing humidity levels (vol. 2/1 Fig. 45: 10–12) . . . . **65.** ***N. minusculoides***
8a (6) In middle Europe so far only reported from Neusiedler See and in

brine pools in the Austrian alps(vol. 2/1 Fig. 45: 21–23) . . . . **67. *N. halophilioides***

8b Cosmopolitan with ecological preference for a-meso to moderately polysaprobic water (vol. 2/1 Fig. 45: 1–9) . . . . **66. *N. molestiformis***

"Fusiformes" (vol. 2/1, p. 132)

[*Haslea* Simonsen 1974
Typus generis: *Haslea ostrearia* (Gaillon) Simonsen 1974]

Key to species:

Valves spindle-shaped with central pores close to each other and very narrow axial area. These are predominantly marine forms. Only 2 species are found in freshwater in central Europe.

1a Around 12 striae/10µm (vol. 2/1 Fig. 52: 4) . . . . **72. *N. crucigera***
1b More than 25 striae/10µm (vol. 2/1 Fig. 52: 5, 6) . . . . . . **73. *N. spicula***

Navicula key group D – Species group around Navicula elginensis (vol. 2/1, p. 134)

In contrast to the "Lineolatae" the basic form of the valve in this group is mostly elliptical, seldom rhomboid or more or less broadly lanceolate; ends usually blunt, seldom somewhat sharply rounded, frequently more or less blunt to capitate and elongated. The valves of individual species tend to "cymbelloid" symmetry. Raphe as in the "Lineolatae", however the terminal fissures of several taxa turn to opposite sides. Principal characters differentiating this group from the "Lineolatae" are the proportions between the striae and the transapical ribs (interstriae) as well as the form of the lineolae or the foramina: usually the transapical ribs are broader than the striae.

Key to species:

1a Small valved, lanceolate to linear-lanceolate forms, usually less than 20 µm long with indistinct, very fine structure of the striae, always without isolated stigmata in the central area (see for example vol. 2/1 Fig. 35: 9–20) . . . . **see under "Lineolatae"** or other key groups
1b Average valves in the population larger and/ or outline more or less elliptical, rhomboid, broadly lanceolate, often with blunt or capitate elongated ends and often with isolated stigmata in the central area which is, however, never cross-shaped as it is in the section around *N. mutica* . **2**
2a Terminal fissures bent to opposite sides **3**
2b Terminal fissures bent to the same side **10**
3a Central area with stigmata cut off from the striae **4**
3b Central area without stigmata cut off from the striae **7**
4a Central area with 2 or more stigmata **5**
4b Central area with only 1 stigma **6**
5a The middle of the valve with numerous shortened striae alternating with longer striae (vol. 2/1 Fig. 47: 1–9) . . . . **78. *N. clementis***
5b Without striae regularly shortened in such a way (vol. 2/1 Fig. 48: 3–8) . . . . **79. *N. clementioides***
6a (4) Valves with short rostrate ends (vol. 2/1 Fig. 48: 10, 11), compare with *N. latens* with terminal fissures which are difficult to differentiate (vol. 2/1 Fig. 48: 15, 16) . . . . **80. *N. constans***
6b Valves broadly rhomboid-lanceolate with barely elongated ends (vol. 2/1 Fig. 47: 19–21) . . . . **82. *N. porifera***

7a (3) The middle of the valve with many more shortened striae inserted between the longer ones **8**
7b Without striae regularly shortened in such a way **9**
8a Valves with short rostrate ends (vol. 2/1 Fig. 48: 12–14) compare with *N. pseudanglica* (vol. 2/1 Fig. 46: 13, 14) . . **80. *N. constans* var. *symmetrica***
8b Valves broadly rhomboid-lanceolate with barely elongated ends (vol. 2/1 Fig. 47: 22–24) . . . . **82. *N. porifera* var. *opportuna***
9a (7) Valves linear-lanceolate with striae approximately parallel in the rostrate ends (vol. 2/1 Fig. 49: 1–3) . . . . **85. *N. explanata***
9b Not with the above combination of characteristics (vol. 2/1 Fig. 50: 5–8), compare with *N. pseudanglica* (vol. 2/1 Fig. 46: 13–15) . . . . **89. *N. subplacentula***
10a (2) At each pole 2 puncta or striae-like structures sit close to the raphe . **14**
10b Pole without such structures **11**
11a Central area with 1 stigma **12**
11b Central area without stigma **15**
12a Striae in the centre alternately long and short (vol. 2/1 Fig. 47: 10–18) . . . . **81. *N. decussis***
12b All striae equally shortened, 14–18/10 µm, or some individual ones shortened, but not alternately long and short **13**
13a Valve ends short and broadly elongated with flat rounded ends (vol. 2/1 Fig. 49: 7–9) . . . . **86. *N. gastrum* var. *signata***
13b Valve ends elongated, more or less rostrate to capitate (vol. 2/1 Fig. 46: 15, 18) . . . . . **75. *N. pseudanglica* var. *signata*, 76. *N. exigua* var. *signata***
14a (10) Valves small, length 10–20 µm (vol. 2/1 Fig. 41: 5–7) . **84. *N. similis***
14b Valves larger, length 24–40 µm (vol. 2/1 Fig. 48: 15, 16) . . . **83. *N. latens***
15a (11) Striae, at least in the centre, relatively coarsely punctate, mostly forms from oceans and brackish water or environments exposed to intermittently high osmotic pressure (see also key groups G and H) **26**
15b Striae more finely punctate **16**
16a Valve ends distinctly rostrate to capitate **22**
16b Valve ends not as above, mostly bluntly elongated and broadly rounded **17**
17a Central area distinctly widened transversely, rhomboid, with a large number of shortened striae inserted at the edges (vol. 2/1 Fig. 51: 3–5) . . . . **93. *N. platystoma***
17b Not with the above combination of characteristics **18**
18a The central striae markedly elongated on both sides, flanking striae shortened by comparison (vol. 2/1 Fig. 50: 9–13) . . . **90. *N. hambergii***
18b Not as above **19**
19a Valves slightly bent, cymbelloid **20**
19b Valves always completely symmetrical **21**
20a Valve width usually less than 10 µm, striae in the middle more or less the same length (vol. 2/1 Fig. 49: 10–13) . . . . **87. *N. diluviana***
20b Valve width greater than 10 µm, individual striae in the middle of the valve shortened (vol. 2/1 Fig. 49: 4–9) . . . . **86. *N. gastrum***
21a (19) Valves bluntly rounded, central area large (vol. 2/1 Fig. 46: 10–12) . . . . **74. *N. elginensis* var. *cuneata***
21b Not with the above combination of characteristics (vol. 2/1 Fig. 50: 1–4) compare with *N. diluviana* (vol. 2/1 Fig. 49: 10–13) and *N. gastrum* (vol. 2/1 Fig. 49: 4–9) . . . . **88. *N. placentula***
22a (16) Striae very distinctly convergent at the ends . . **173. *N. laterostrata***
22b Striae radial or parallel at the ends, possibly convergent at the distal extreme **23**

rows more strongly prominent, without alternating short and long striae (vol. 2/1 Fig. 67: 14, 15) . . . . **159.** *N. aboensis*
2b Not with the above combination of characteristics **3**
3a Striae very coarsely punctate, 10–16/10 µm (vol. 2/1 Fig. 59: 16–19) . . . . **112.** *N. scutelloides*
3b Puncta more dense **4**
4a Striae weakly radial **7**
4b Striae strongly radial **5**
5a Central area quite large, rounded (vol. 2/1 Fig. 59: 10, 11) . . . . **110.** *N. scutiformis*
5b Central area small or absent **6**
6a Central area almost circular-elliptical (vol. 2/1 Fig. 59: 12–15) . . . . **111.** *N. pseudoscutiformis*
6b Valves more or less rhomboid-lanceolate (vol. 2/1 Fig. 59: 2–5) . . . . **108.** *N. cocconeiformis*
7a (4) Striae in the middle 9–12/10 µm (vol. 2/1 Fig. 60: 1, 2) . . . . **113.** *N. jentzschii*
7b Striae 25–36/10 µm (vol. 2/1 Fig. 59: 6–9) . . . . **109.** *N. jaernefeltii*
8a (1) Central area barely outlined (vol. 2/1 Fig. 59: 2–5) . . . . **108.** *N. cocconeiformis*
8b Central area more or less large, approximately elliptical (vol. 2/1 Fig. 75: 29–31) . . . . **223.** *N. confervacea*

Navicula key group G – "Microstigmaticae" (vol. 2/1, p. 160)

Key to species:
1a Valve length around 200 µm (vol. 2/1 Fig. 52: 3) . . . **114.** *N. bergenensis*
1b Valves shorter **2**
2a Particular structural design in the central region as shown in vol. 2/1 Fig. 55: 11 . . . . **119.** *N. zeta*
2b Not as above **3**
3a Valves narrow lanceolate, width always less than 7.5 µm, with one isolated punctum in the small central area, which is often difficult to resolve . **7**
3b Not with the above combination of characteristics **4**
4a Valve ends with separate points and pseudosepta (vol. 2/1 Fig. 55: 1–3) . . . . **116.** *N. integra*
4b Not as above **5**
5a Valve outline lanceolate, gradually narrowed in the middle, with more or less narrow elongated ends (vol. 2/1 Fig. 54: 1–13) . . **115.** *N. crucicula*
5b Valves more elliptical to linear with ends not elongated or broadly elongated, broad to bluntly rounded ends **6**
6a Striae very weakly radial, the central striae barely more widely spaced than the others (vol. 2/1 Fig. 55: 4) . . . . **117.** *N. plicata*
6b Striae distinctly radial, the central striae distinctly more widely spaced than the others (vol. 2/1 Fig. 55: 5–10) . . . . **118.** *N. protracta*
7a (3) Valve ends rostrate (vol. 2/1 Fig. 56: 6–9) . . . . . **121.** *N. bulnheimii*
7b Valve ends not elongated, but with pointed or blunt ends **8**
8a Valve ends blunt (vol. 2/1 Fig. 56: 10) . . **121.** *N. bulnheimii* var. *belgica*
8b Valve ends pointed (vol. 2/1 Fig. 56: 1–5) . . . . **120.** *N. complanata*

Navicula key group H (vol. 2/1, p. 165)
Remaining species of the heterogeneous "Punctatae"

Key to species:

1a Valves with ends more or less abruptly elongated, rostrate or capitate 2
1b Valve ends not separated in such a manner 10
2a Valves linear 3
2b Valves with more or less distinctly convex edges 6
3a Valves broadly linear with narrow elongated ends 5
3b Valves linear with broad elongated ends 4
4a Valves a maximum of 30 µm long (vol. 2/1 Fig. 57: 5) compare with var. *capitata* which does not occur in Middle Europe (vol. 2/1 Fig. 57: 6) . . . . **124.** ***N. pusilla*** **var.** ***incognita***
4b Valves at least 40 µm long . . . . ***77. N. abiskoensis***
5a (3) Striae strongly radial with many shortened striae inserted between longer striae in the central region of the valve (vol. 2/1 Fig. 58: 1) . . . . **125.** ***N. humerosa***
5b Striae weakly radial (vol. 2/1 Fig. 58: 4–5) . . . . **127.** ***N. maculosa***
6a (2) Central area transversely broadened 7
6b Central area more or less isodiametric 8
7a Ends very sharply rostrate, striae approximately parallel, puncta around 20/10 µm (vol. 2/1 Fig. 58: 6–8) . . . . **128.** ***N. citrus***
7b Ends bluntly rounded, striae radial, puncta around 24/10 µm (vol. 2/1 Fig. 60: 10–15) compare with *N. pseudotuscula* (vol. 2/1 Fig. 82: 1–4) . . . . **131.** ***N. kotschyi***
8a (6) Valve width less than 10 µm, striae in the centre of the valve more than 20/10 µm (vol. 2/1 Fig. 57: 1–4) . . . . **123.** ***N. ordinaria***
8b Valve width over 10 µm 9
9a Central area rounded (vol. 2/1 Fig. 57: 7–9) . . . . **124.** ***N. pusilla***
9b Central area transversely broadened (vol. 2/1 Fig. 51: 1) **91.** ***N. amphibola***
10a (1) Valves linear, edges with approximately 3 undulations (vol. 2/1 Fig. 60: 3–5), compare with *N. levanderi* (vol. 2/1 Fig. 65: 16–17) . . . . **130.** ***N. pseudosilicula***
10b Not as above 11
11a Valves to around 10 µm long, rhomboid-elliptical with sharply rounded ends (vol. 2/1 Fig. 60: 6–9) . . . . **129.** ***N. ingenua***
11b Not with the above combination of characteristics 12
12a Striae distinctly convergent at the ends 13
12b Striae continuously radial, at most the extreme distal striae convergent . 14
13a Striae in the middle 6–8/10 µm (vol. 2/1 Fig. 51: 2) . . . . **92.** ***N. semen***
13b Striae considerably more closely spaced (vol. 2/1 Fig. 52: 1–2) . . . . **122.** ***N. brasiliana***
14a Central area transversely broadened 15
14b Central area rounded (vol. 2/1 Fig. 58: 2–3) . . . . **126.** ***N. lacustris*** compare with *N. pusilla* (vol. 2/1 Fig. 57: 9)
15a Valve width greater than 20 µm, striae less than 10/10 µm (vol. 2/1 Fig. 51: 1) . . . . **91.** ***N. amphibola***
15b Smaller, more densely striate forms (vol. 2/1 Fig. 60: 10–15) . . . . **131.** ***N. kotschyi***

Navicula key group I "Lyratae" (vol. 2/1, p. 170)

Key to species:

1a Striae coarse (13–16, rarely to 22/10 µm), median rib widened at the vesicular central pore, central area and lateral area connected in a bracket shape (vol. 2/1 Fig. 65: 12–13) . . . . **135.** ***N. forcipata***
1b Without the above combination of characteristics **2**
2a Central pores conspicuously separate from each other **3**
2b Not as above **4**
3a Lateral area bracket shaped (vol. 2/1 Fig. 65: 7–9) . . **133.** ***N. cryptolyra***
3b Lateral area in a more or less straight line (vol. 2/1 Fig. 65: 11) . . . . **136.** ***N. muralibionta***
4a Central area and lateral area distinctly combined with each other, central pores strongly marked (vol. 2/1 Fig. 65: 1–6') . . . . . **132.** ***N. pygmaea***
4b Not with the above combination of characteristics (vol. 2/1 Fig. 65: 10) . . . . **134.** ***N. pseudoforcipata***

Navicula key group J (vol. 2/1, p. 173)

Forms with longitudinal hyaline lines which cross the striae and can have different fine structural detail. Parts of this group, particularly the species complex around *N. tenera* and *N. insociabilis*, are more closely aligned to the Lyratae.

Key to species:

1a Central area uni- or bilateral, forming a broad transverse band reaching to the valve edge **2**
1b Not as above **3**
2a Valves transapically widened in the centre and at the ends (vol. 2/1 Fig. 66: 5–8) . . . . **140.** ***N. occulta***
2b Valve edges at most weakly convex (vol. 2/1 Fig. 66: 9–11) . . . . **141.** ***N. lucinensis***
3a (1) One side of the valve with a distinctly marked apical puncta row lying in the median rib (rarely fragments are on both sides), interrupted by the central area; also a hyaline line of striae on the periphery and a central puncta row running a parallel to the valve edge (vol. 2/1 Fig. 66: 19–23) . . . . **139.** ***N. tenera***
3b If there is a similar combination of characteristics and an isolated row of puncta, then it is not found in, but beside, the (contrasting) median rib **4**
4a Valves linear with (sometimes elongated) broadly rounded ends . . . **5**
4b Valves with distinct convex (or concave) edges, elliptical to rhomboid **8**
5a Valves relatively large (28–45 µm long), linear, usually with three undulations, striae distinctly punctate (vol. 2/1 Fig. 65: 16, 17) **142.** ***N. levanderi***
If valve ends rostrate to capitate, compare with *N. hoefleri* (vol. 2/1 Fig. 79: 28)
5b Not with the above combination of characteristics **6**
6a Central pores of the raphe separate from each other, central area quite large, rounded (vol. 2/1 Fig. 65: 11) . . . . **136.** ***N. muralibionta***
6b Not with the above combination of characters **7**
7a Each hyaline longitudinal line separates only a fine row of puncta from the striae (vol. 2/1 Fig. 66: 24–27) . . . . **160.** ***N. helensis***
7b The longitudinal lines, approximately in the middle of each side of the valve, limited to a weakly hyaline zone parallel to the median rib, terminal fissures long, hook-shaped (vol. 2/1 Fig. 66: 32–34) **162.** ***N. subhamulata***

8a (4) Valve edges concave (vol. 2/1 Fig. 75: 26–28) . . . . **230.** ***N. aerophila***
8b Not as above **9**
9a Valve ends more or less elongated or sharply rounded (vol. 2/1 Fig. 72: 21–24) . . . . . . **143.** ***N. krasskei*** see also key group M, couplets 36–38
9b Valve ends blunt or rounded, not elongated **10**
10a Axial and central area forming a more or less extensive hyaline zone, in addition to a hyaline line close to the valve edge (vol. 2/1 Fig. 66: 1–4) . . . . **138.** ***N. insociabilis***
10b A hyaline zone at most present in outline, the hyaline line in a more or less central position (vol. 2/1 Fig. 66: 12–18) . . . . **137.** ***N. monoculata***

Navicula key group K – "Annulatae" (vol. 2/1, p. 178)

Key to species:
1a Valves with undulating edges (vol. 2/1 Fig. 64: 12–15) . . . . **146.** ***N. ignota*** **var.** ***ignota***
1b Edges not undulating **2**
2a Striae strongly radial, particularly the central ones **4**
2b Middle striae weakly radial, usually with an isolated stigma lying near the middle of the central nodule **3**
3a Valve length 16–24 µm (vol. 2/1 Fig. 64: 16–19) . . . . **146.** ***N. ignota*** **var.** ***palustris***
3b Valve length 6–14 µm (vol. 2/1 Fig. 64: 20–24) . . . . **146.** ***N. ignota*** **var.** ***acceptata***
4a Central area large, nearly reaching the edges of the valve (vol. 2/1 Fig. 64: 29–32) . . . . **145.** ***N. dolomitica***
4b Central area smaller, with an irregular border (vol. 2/1 Fig. 64: 1–11) . . . . **144.** ***N. schoenfeldii***

Navicula key group L (vol. 2/1, p. 180)

Species group around *N. bryophila* and *N. subtilissima*

Key to species:
1a Striae crossed by 2 relatively wide hyaline lines in a lanceolate shape (vol. 2/1 Fig. 79: 28), compare with many forms of *N. subtilissima* . . . . **151.** ***N. hoefleri***
1b Not as above **2**
2a Central area quite large, transversely rectangular, elliptical or rhomboid **3**
2b Central area small or only weakly outlined **4**
3a Striae more than 30/10 µm, ends weakly elongated (vol. 2/1 Fig. 79: 18–21) . . . . **153.** ***N. jaagii***
3b Striae 24–27/10 µm, ends more strongly capitate (vol. 2/1 Fig. 79: 16, 17) . . . . **152.** ***N. brockmannii***
4a (2) Striae at the limits of resolution, around 42/10 µm (vol. 2/1 Fig. 79: 22–26) . . . . **150.** ***N. subtilissima***
4b Striae coarser, considerably less than 40/10 µm **5**
5a Valves narrowly linear with gradually elongated ends (vol. 2/1 Fig. 79: 9–12) . . . . **148.** ***N. suchlandtii***
5b Valves broadly linear with more distinct, usually rostrate to capitate elongated ends **6**
6a Width 2.5–4 µm(vol. 2/1 Fig. 79: 1–8') . . . . **147.** ***N. bryophila***
6b Width 4.5–5.5 µm (vol. 2/1 Fig. 79: 13–15) . . . **149.** ***N. pseudobryophila***

Navicula key group M – "Bacillares" (vol. 2/1, p. 184)

Key to species:

1a Terminal nodules transapically widened **2**
1b Not as above **3**
2a Valve ends not elongated, distinct longitudinal furrows flanking the median ribs, by EM: with conopeum (vol. 2/1 Fig. 67: 2–4) **154.** ***N. bacillum***
2b Valve ends usually more or less elongated, longitudinal furrows only line-like or visible because the ends of the striae close to the axial area appear more strongly outlined (vol. 2/1 Fig. 68: 1–21) . . . . . . **158.** ***N. pupula***
3a (1) Valve ends rostrate to more or less capitate **27**
3b Valve ends at most weakly elongated **4**
4a Valves linear with broadly rounded ends, rarely weakly concave or weakly convex in the middle **5**
4b Valves rhomboid, elliptical or lanceolate **14**
5a Valves relatively large, greater than 10 µm wide, furrows conspicuously broad, central nodule broadened elliptically, sharply outlined, raphe undulating (vol. 2/1 Fig. 67: 1), if furrows less broad compare with *N. bacillum* (vol. 2/1 Fig. 67: 2–4) . . . . **156.** ***N. americana***
5b Not with the above combination of characteristics **6**
6a Striae on the lateral area or crossed by distinct line-like structures . . . . **see key group J** (p. 110)
6b Without such structures **7**
7a Central area more or less transversely widened **8**
7b Central area not as above **10**
8a Central area reaching to the edge, forming a transverse band (vol. 2/1 Fig. 77: 2–3) . . . . **242.** ***N. muraloides***
8b Not as above **9**
9a Valve width always greater than 5 µm, raphe undulating (vol. 2/1 Fig. 67: 6–13) . . . . **157.** ***N. laevissima***
9b Valve width less than 5 µm, small forms (vol. 2/1 Fig. 69: 1–10), compare with vol. 2/1 Fig. 77: 27 under *N. digitulus* . . . . **165.** ***N. stroemii***
10a (7) Furrows flat and broad, covering $^1/_3$ – $^1/_2$ valve width, conspicuously long, hook-shaped terminal fissures (vol. 2/1 Fig. 66: 32–34) . . . . **162.** ***N. subhamulata***
10b Not with the above combination of characteristics **11**
11a Valve width less than 6 µm, striae more than 30/10 µm (vol. 2/1 Fig. 66: 35–39) . . . . **163.** ***N. lenzii***
11b Not with the above combination of characteristics **12**
12a Striae 22–26/10 µm (vol. 2/1 Fig. 66: 24–27) . . . . **160.** ***N. helensis***
12b Striae 26–32/10 µm **13**
13a Valve length greater than 14 µm (vol. 2/1 Fig. 66: 31) . . . . . . ***N. fracta***
13b Valve length less than 10 µm (vol. 2/1 Fig. 66: 40–42) . **164.** ***N. sublucidula***
14a (4) Valves more or less elliptical with broadly rounded ends **15**
14b Valves rhomboid, lanceolate, ends strongly narrowed **22**
15a Striae more or less coarsely punctate, around 25 puncta/10 µm (vol. 2/1 Fig. 67: 14–15) . . . . **159.** ***N. aboensis***
15b Puncta finer or not distinguishable **16**
16a Striae around the central nodule alternately shorter and longer, valve ends broadly rounded (vol. 2/1 Fig. 69: 11–13) . . . . **166.** ***N. rotunda***
16b Not with the above combination of characteristics **17**
17a Central area large, more or less circular, striae weakly radial, symmetrically shortened in the middle (vol. 2/1 Fig. 67: 5 does not represent this taxon, which is shown in vol. 2/1 Fig. 73: 1, 2) . . . . **155.** ***N. bacilloides***

17b Not with the above combination of characteristics **18**
18a Central area forming a transverse band covering half or more of the valve width (vol. 2/1 Fig. 73: 4–7) . . . . **190.** ***N. lapidosa***
18b Not as above **19**
19a Valves linear-elliptical to linear with considerably larger, more or less transverse elliptically widened central area (vol. 2/1 Fig. 73: 8–11) . . . . **191.** ***N. variostriata***
19b Not with the above combination of characteristics, valves elliptical . . **20**
20a Median rib with central and terminal nodules noticeably prominent, however without the outline of flanking furrows (vol. 2/1 Fig. 74: 1–38) . . . . **key group Nc (Species group around *N. atomus*)** (p. 118)
20b Median rib with the 3 nodules not prominent in such a way **21**
21a Central area of at least moderate size (vol. 2/1 Fig. 73: 1–2) . . . . **189.** ***N. weinzierlii***
This taxon is a synonym of 155. *N. bacilloides*, as we have found out recently.
21b Several middle striae in the axial area elongated, without a distinct central area (vol. 2/1 Fig. 59: 2–9) . **108.** ***N. cocconeiformis***, **109.** ***N. jaernefeltii***
22a (14) Central area distinct **23**
22b Central area very small or absent **25**
23a Central area more or less circular **24**
23b Central area forming a more or less elongated transverse band (vol. 2/1 Fig 73: 4–7) . . . . **190.** ***N. lapidosa***
24a Valve width 8–12 µm (vol. 2/1 Fig. 73: 1, 2, syn. *N. weinzierlii*) . . . . **155.** ***N. bacilloides***
24b Valve width less than 6 µm (vol. 2/1 Fig. 78: 21–25) . **187.** ***N. subadnata***
25a (22) Striae more than 30/10 µm (vol. 2/1 Fig. 72: 29–33) **188.** ***N. egregia***
25b Striae coarser **26**
26a Valve length greater than 12 µm, valve width 6–7 µm (vol. 2/1 Fig. 64: 26–28) . . . . **71.** ***N. kriegeri***
26b *Valve length less than 12 µm, width less than 6 µm (vol. 2/1 Fig. 76: 21–26) . . . . **24.** ***N. subminuscula***
27a (3) Valve ends more or less rostrate, however not capitate **28**
27b Valve ends more or less capitate **39**
28a Furrows conspicuously broad, at least more so than the zones, which do not lie in the same focal plane **36**
28b Furrows narrow or only line-like or barely outlined **29**
29a Striae around 20/10 µm **30**
29b Striae narrower, more than 22/10 µm **32**
30a Valves 7–10 µm wide, striae distinctly punctate (vol. 2/1 Fig. 70: 22–24) . . . . **173.** ***N. laterostrata***
30b Valves narrower and/or not distinctly punctate **31**
31a Ends broadly elongated and broadly rounded, middle of the valve strongly convex, central area relatively large (vol. 2/1 Fig. 71: 3–8) . . . . **176.** ***N. pseudoventralis***
31b Characteristic not present or at least less distinctly pronounced (vol. 2/1 Fig. 71: 9–13) . . . . **177.** ***N. modica***
32a (29) Several of the striae convergent at the ends, rarely parallel (vol. 2/1 Fig. 71: 15–21) . . . . **178.** ***N. absoluta***
32b Striae continuously radial, so far as can be seen **33**

---

* if a hyaline longitudinal line crosses the striae or the central area and lateral area form a more or less large hyaline region, see key group J (p. 110).

33a Ends narrowly elongated, almost sharply rounded (vol. 2/1 Fig. 72: 14–16) . . . . **185.** *N. ingrata*
33b Ends very blunt to broadly rounded **34**
34a Central irregularly widened transversely (vol. 2/1 Fig. 71: 25–31') . . . . **180.** *N. vitabunda*
34b Central area at most weakly widened elliptically **35**
35a Striae in the middle of the valve distinctly alternately short and long (vol. 2/1 Fig. 70: 19–21) . . . . **168.** *N. pusio*
35b Striae approximately equal in length (vol. 2/1 Fig. 69: 14–17) . . . . **167.** *N. detenta*
36a (28) Striae not distinguishable or barely so, in the type form around 50/10 µm (vol. 2/1 Fig. 72: 21–24) compare with *N. kuelbsii* without elongated ends (vol. 2/1 Fig. 80: 33–35) . . . . **143.** *N. krasskei*
36b Striae at least distinguishable at the ends, around or less than 30/10 µm . **37**
37a Depressions outlined by more or less distinguishable longitudinal lines . **38**
37b Longitudinal lines not present or indistinct (vol. 2/1 Fig. 72: 17–20) . . . . **186.** *N. maceria*
38a Depression broadened in the middle, lanceolate (bulging) (vol. 2/1 Fig. 72: 8–10) . . . . **184.** *N. naumannii*
38b Depression without a bulging broad area (vol. 2/1 Fig. 72: 1–7) . . . . **183.** *N. festiva*
39a (2) Striae not visible with the light microscope (vol. 2/1 Fig. 70: 14–15), this species has been placed in the genus *Nupela* . . . . **197.** *N. impexa*
39b Striae distinguishable **40**
40a Central area of at least moderate size, usually broadened transversely . **41**
40b Central area small, not distinctly set off from the axial area **49**
41a Valves 5–10 µm wide **42**
41b Valves narrower, 2–5 µm wide **43**
42a At least the central striae distinctly punctate (vol. 2/1 Fig. 70: 22–24) . . . . **173.** *N. laterostrata*
42b Striae not distinctly punctate (vol. 2/1 Fig. 71: 1–2) . . **175.** *N. ventralis*
43a (41) Striae around 30/10 µm **44**
43b Striae more widely spaced, fewer than 30/10 µm **45**
44a Valves at most 10 µm long (vol. 2/1 Fig. 70: 8–13) . **170.** *N. schmassmannii*
44b Valves greater than 10 µm long (vol. 2/1 Fig. 70: 1–7) . . . . **169.** *N. medioconvexa*
45a (43) Valves with strongly convex edges, capitate ends much narrower than the middle of the valve **46**
45b Difference in width between capitate ends and valve middle less marked **48**
46a Valves narrowly elliptical-lanceolate (vol. 2/1 Fig. 71: 22–24) . . . . **179.** *N. hustedtii*
46b Valves with strongly distended central area **47**
47a Striae more than 25/10 µm (vol. 2/1 Fig. 71: 32–38) . . . **181.** *N. schadei*
47b Striae around 30/10 µm or more (vol. 2/1 Fig. 71: 39) . . **182.** *N. glomus*
48a (45) Ends strongly constricted, central area rounded (vol. 2/1 Fig. 70: 18) . . . . **172.** *N. laticeps*
48b Ends weakly constricted, central area forming a transverse band reaching nearly out to the valve edge (vol. 2/1 Fig. 70: 16, 17) . **171.** *N. disjuncta*
49a (40) Valves with narrower, capitate constricted ends and isolated stigma in the central area (vol. 2/1 Fig. 70: 25–27) . . . . **174.** *N. declivis*
49b Not with the above combination of characteristics **50**
50a Striae in the middle of the valve alternately shorter and longer (vol. 2/1 Fig. 70: 19–21) . . . . **168.** *N. pusio*

50b Striae approximately equal in length (vol. 2/1 Fig. 69: 14–17) . . . . **167.** ***N. detenta***

Navicula key group N – "Minusculae" (vol. 2/1, p. 205)
This key group summarizes the remaining heterogeneous group of small valved forms (usually less than 20 µm long) without the combination of characters (visible with the light microscope) which characterises the other key groups.

Key to (artificial) subgroups:

1a Striae not visible with the light microscope or only very indistinct . . . . **subgroup Na** (p. 115)
1b Striae distinctly visible, at least in the middle of the valve **2**
2a Valve ends rostrate to more or less capitate . . . . **subgroup Nd** (p. 118)
2b Valve ends at most slightly elongated **3**
3a Valves more or less elliptical, median rib with the marked, puncta-like central and terminal nodules conspicuously contrasting, however without flanking longitudinal furrows . . . . **subgroup Nb Species group around** ***N. atomus*** (p. 117)
3b Without the above combination of characteristics **4**
4a Frustules in girdle view forming band-shaped aggregates (the chain can frequently be destroyed by cleaning in acid, less likely with heat preparation . . . . **subgroup Nc** (p. 118)
4b Not as above **5**
5a Forms with very small or indistinctly outlined central area or completely without central area . . . . **subgroup Ne** (p. 119)
5b Forms with more or less well formed central area and without the characteristics of other key groups or sub groups **subgroup Nf** (p. 120)

Navicula key group N – subgroup Na (vol. 2/1, p. 205)
Forms in which the striae are not distinguishable (or barely so) by light microscope.

Key to species:

1a Valves strictly elliptical with broadly rounded ends **2**
1b Valves linear, elliptical-lanceolate, rhomboid or with more or less elongated ends **6**
2a Striae distinguishable, at least in part, using special optics or oblique light **5**
2b Striae and other fine structure only visible with the EM **3**
3a Valves broadly elliptical, terminal nodules close to the poles **4**
3b Terminal nodules displaced proximally (vol. 2/1 Fig. 75: 34–37) . . . . **201.** ***N. lacunolaciniata***
4a Valves 9–12.5 µm long, 4–6.2 µm wide (vol. 2/1 Fig. 74: 37, 38) . . . . **199.** ***N. pelliculosa***
4b Valves 3.8–7.6 µm long, 2–4 µm wide (vol. 2/1 Fig. 74: 35, 36) . . . . **198.** ***N. saprophila***
5a (2) Terminal nodules displaced a long way from the poles, striae radial, around 35/10 µm (vol. 2/1 Fig. 75: 32, 33) . . . . . . **200.** ***N. nolensoides***
5b Terminal nodules only slightly displaced proximally (vol. 2/1 Fig. 74: 14–17) . . . . **216.** ***N. atomus*** **var.** ***permitis***
6a (1) Valve ends not elongated or barely so **7**
6b Valve ends more or less elongated, valve edges sometimes undulating . **11**

7a Central area still distinguishable **8**
7b No central area distinguishable **9**
8a Central area with one side more strongly developed (vol. 2/1 Fig. 80: 31) . . . . **310.** ***N. vitiosa***
8b Central area more or less similar on both sides (vol. 2/1 Fig. 76: 48 and Fig. 77: 10–12) . . . . **232.** ***N. semihyalina***
. . . . **243.** ***N. tenerrima***
9a (7) Valves with bluntly rounded ends, 5–6 µm long, 2–2.5 µm wide (vol. 2/1 Fig. 78: 28) . . . . **202.** ***N. diabolica***
9b Valves usually wider and with more broadly rounded ends **10**
10a Raphe enclosed in one conspicuously strong median rib (vol. 2/1 Fig. 80: 32–34) . . . . **213.** ***N. kuelbsii***
10b Not as above (vol. 2/1 Fig. 80: 26, 27) . . . . **209.** ***N. difficillimoides***
11a (6) Valve edge with 3 undulations (vol. 2/1 Fig. 80: 1–3) **203.** ***N. tridentula***
11b Valve edge not undulating or rarely only slightly so **12**
12a Central area visible in outline **13**
12b No central area clearly visible **14**
13a Valves with more or less strongly elongated ends, striae (by EM) 48–55/10 µm (vol. 2/1 Fig. 80: 18–20), if striae in the middle in part still distinguishable, around 40/10 µm (vol. 2/1 Fig. 80: 21, 22) . . . . **207.** ***N. gerloffii***
. . . . **205.** ***N. arvensis*** **var.** ***major***
13b Ends at most very weakly elongated, striae just visible, 36–39/10 µm (vol. 2/1 Fig. 76: 48) . . . . **232.** ***N. semihyalina***
14a (12) Valves with pole caps which appear more strongly silicified, with appropriate focus appearing as a dark section (vol. 2/1 Fig. 80: 7, 8) . . . . **204.** ***N. difficillima***
14b Not as above **15**
15a Striae around 40/10 µm, can only be resolved along the border of the valve **16**
15b Striae distinctly more than 40/10 µm, not easily resolved **17**
16a Valves mostly larger, 3.2–5.5 µm wide (vol. 2/1 Fig. 69: 18–27) . . . . **196.** ***N. minuscula***
16b Valves usually smaller, 2.5–3.5 µm wide with more strongly set off, rostrate ends (vol. 2/1 Fig. 80: 10–12), compare with *N. subarvensis* (vol. 2/1 Fig. 80: 13) . . . . **205.** ***N. arvensis*** **var.** ***arvensis***
17a Valves broadly linear, with abruptly set off ends, maximum length 10 µm (vol. 2/1 Fig. 80: 17) . . . . **206.** ***N. pseudoarvensis***
17b (15) Without the above combination of characteristics **18**
18a Valves elliptical lanceolate with more or less capitate ends **20**
18b Valve ends not capitate **19**
19a Valve width 2.5–3 µm (vol. 2/1 Fig. 80: 28–30) . . . . **211.** ***N. indifferens***
19b Valve width 3.4–5.5 µm (vol. 2/1 Fig. 72: 21–25) . . . . . **143.** ***N. krasskei***
20a (18) Valve width 3–4 µm (vol. 2/1 Fig. 80: 25 and Fig. 80: 7, 8) . . . . **197.** ***N. impexa***
. . . . **204.** ***N. difficillima***
20b Valve width 2–2.5 µm (vol. 2/1 Fig. 80: 23, 24) . . . . **208.** ***N. enigmatica***

Navicula key group Nb – species group around N. atomus (vol. 2/1, p. 214)
Valves more or less elliptical, median rib with central and terminal nodules conspicuously prominent, however without flanking longitudinal furrows.
Key to species:

1a Striae not able to be resolved by light microscope even with special lighting 2
1b Not as above 5
2a Terminal nodules very close to the poles 3
2b Terminal nodules somewhat separate from the poles 4
3a Length of the valves 9–12.5 µm, width 4–6.2 µm (vol. 2/1 Fig. 74: 37–38), compare with *N. minuscula* var. *muralis* (vol. 2/1 Fig. 69: 25–27) . . . . . **199. *N. pelliculosa***
3b Length 4–7.6 µm, width 2–4 µm (vol. 2/1 Fig. 74: 35, 36) . . . . . **198. *N. saprophila***
4a (2) Terminal nodules moderately far from the poles, striae not present, even using EM (vol. 2/1 Fig. 75: 34–37) . . . . . **210. *N. lacunolaciniata***
4b Terminal nodules quite displaced proximally from the poles, striae still distinguishable using the light microscope under optimal conditions, around 35/10 µm (vol. 2/1 Fig. 75: 32, 33) . . . . . **200. *N. nolensoides***
Several taxa with similar characteristics found only on the sea coast and/or outside middle Europe.
5a (1) Terminal nodule displaced quite a distance from the pole 6
5b Not as above 7
6a Striae around 35/10 µm, visible only under optimal conditions using the light microscope (vol. 2/1 Fig. 75: 32, 33) . . . . . **200. *N. nolensoides***
6b Striae coarsely punctate (vol. 2/1 Fig. 74: 34) . . . . . **218. *N. asellus***
7a (5) Terminal nodules often close to the more or less elongated poles, striae 30–45/10 µm, radial in the centre of the valve, abruptly convergent at the ends (vol. 2/1 Fig. 69: 18–27) . . . . . **196. *N. minuscula***
7b Combination of characteristics untrue 8
8a Striae relatively coarse, occasionally appearing punctate, around 20/10 µm 10
8b Striae more dense, occasionally barely discernible 9
9a Valves usually narrowly elliptical-lanceolate with moderately blunt to almost sharply rounded ends (vol. 2/1 Fig. 74: 1–7) . . . **214. *N. agrestis***
9b Valves broadly elliptical with more broadly rounded ends (vol. 2/1 Fig. 74: 14–17) . . . . . **216. *N. atomus* var. *permitis***
10a (8) Striae approximately radial 13
10b Striae approximately parallel, or parallel or slightly convergent at the ends 11
11a Striae approximately parallel (vol. 2/1 Fig. 45: 31–33) . **70. *N. muraliformis***
11b Not as above 12
12a Central area transversely elliptical, subpolar striae convergent (vol. 2/1 Fig. 74: 9) . . . . . **219. *N. fossaloides***
12b Central area small, rounded, striae 25/10 µm (vol. 2/1 Fig. 74: 8) . . . . . **215. *N. destricta***
13a (10) Central area not pronounced 14
13b Central area set off from the axial area or combined with it to form a more or less large hyaline zone 15
14a Valves 12–16 µm long, striae around 17/10 µm (vol. 2/1 Fig. 74: 11–13) . . . . . **216. *N. atomus* var. *excelsa***
14b Valves usually somewhat smaller, striae around 20/10 µm, median rib usually appearing less broad (vol. 2/1 Fig. 74: 10, see also 18–26) . . . . .

. . . . **216.** ***N. atomus*** **var.** ***atomus***

15a (13) Central and axial area combined in a wide hyaline area **16**

15b Central area distinctly set off from the axial area and broadened transversely **17**

16a Valve length 8–9 µm, striae around 20/10 µm (vol. 2/1 Fig. 74: 27, 28) . . . . **216.** ***N. atomus*** **var.** ***recondita***

16b Valve length 10–13 µm, striae 16–20/10 µm, appearing punctate (vol. 2/1 Fig. 74: 29–31) . . . . **217.** ***N. fossalis*** **var.** ***obsidialis***

17b (15) Valve length 9–12 m µm (vol. 2/1 Fig. 74: 32, 33) . . . . **217.** ***N. fossalis*** **var.** ***fossalis***

17b Valve length 12–16 µm, terminal nodules more separate from the poles (vol. 2/1 Fig. 74: 34), compare with *N. parsura* (vol. 2/1 Fig. 76: 8–10) . . . . **218.** ***N. asellus***

Navicula key group Nc (vol. 2/1, p. 219)
Frustules in girdle view aggregated in band-shaped chains.

Key to species:

1a Valves linear, not punctate at the edges, terminal nodule displaced away from the poles (vol. 2/1 Fig. 75: 23–25) . . . . **222.** ***N. brekkaensis***

1b Not with the above combination of characteristics **2**

2a Valves narrowly linear-lanceolate, length 10–36 µm (vol. 2/1 Fig. 75: 18–22) . . . . **221.** ***N. gallica*** **var.** ***laevissima***

2b Not with the above combination of characteristics **3**

3a Striae coarse, distinctly punctate (vol. 2/1 Fig. 75: 29–31) . . . . **223.** ***N. confervacea***

3b Striae more delicate, lines of puncta not distinguishable **4**

4a Valves linear with at most weakly concave edges, central area usually distinctly set off from the axial area, striae parallel, without puncta around the edges, raphe always present (vol. 2/1 Fig. 75: 1–5) . . **20.** ***N. contenta***

4b Not with the above combination of characteristics (vol. 2/1 Fig. 75: 6–17') . . . . **221.** ***N. gallica***

Navicula key group Nd (vol. 2/1, p. 222)
Central area very small, with an irregular outline or completely absent

Key to species:

1a Striae parallel or only very weakly radial **2**

1b Striae, at least in the middle of the valve, distinctly radial **12**

2a Valves linear to linear-elliptical **3**

2b Valves without a tendency to be linear **10**

3a Axial area in the middle with a sharp border and not widened (vol. 2/1 Fig. 45: 31–33) . . . . **70.** ***N. muraliformis***

3b Not as above **4**

4a Valves narrowly linear, striae relatively widely spaced (around 20/10 µm), the actual central area widened transversely because of the widely spaced striae, or without with a sharp border because of weakly contoured striae **7**

4b Striae more dense **5**

5a Ends broadly round **6**

5b Ends more or less cuneate (vol. 2/1 Fig. 74: 8) . . . . . **215.** ***N. destricta***

6a Raphe branches curved, central pores strongly marked (vol. 2/1 Fig. 66: 40–42) . . . . **164. *N. sublucidula***
6b Not with the above combination of characteristics (vol. 2/1 Fig. 77: 19–24, possibly 25–28) . . . . **195. *N. digitulus***
7a (4) Valves with a weak, sharply set off distended area in the middle (vol. 2/1 Fig. 78: 14–16) . . . . **227. *N. mediocris***
7b Valve edges strictly linear or with the centre of the valve gradually somewhat inflated or undulating **8**
8a Edges strictly linear or weakly concave (vol. 2/1 Fig. 78: 17–20) . . . . **228. *N. begeri***
8b Edges undulating or somewhat widened in the middle **9**
9a Striae approximately more or less parallel, never convergent at the ends (vol. 2/1 Fig. 78: 1–13) . . . . **226. *N. soehrensis***
9b Striae weakly radial in the middle, convergent at the ends (vol. 2/1 Fig. 35: 11–13) . . . . **33. *N. bremensis***
10a (3) Striae usually around 30/10 µm (vol. 2/1 Fig. 45: 24–25) **69. *N. fluens***
10b Striae usually far fewer than 30/10 µm **11**
11a Valve width 6–7 µm (vol. 2/1 Fig. 64: 26–28) . . . . **71. *N. kriegeri***
11b Valve width less than 6 µm (vol. 2/1 Fig. 76: 21–26) . **224. *N. subminuscula***
12a (1) An isolated stigma at the edge of the central nodule, marine and brackish water forms (vol. 2/1 Fig. 77: 29, 30), compare with *N. porifera* (vol. 2/1 Fig. 47: 19–24) as a freshwater form. This taxon has been found to be monoraphid using the SEM, and has been placed in the recently erected genus *Astartiella* . . . . **225. *N. bahusiensis***
12b Not as above **13**
13a Valve length only 2.7–4.5 µm, possibly dwarf forms of *Achnanthes minutissima* (vol. 2/1 Fig. 78: 26, 27) . . . . **229. *N. strenzkei***
13b Valves longer, raphe always distinctly formed **14**
14a Striae strongly radial **15**
14b Striae weakly to moderately radial **17**
15a Valves narrowly elliptical, often with elongated ends, striae at the edge of resolution (vol. 2/1 Fig. 69: 18–27) . . . . **196. *N. minuscula***
15b Valves broadly elliptical **16**
16a Striae 24–25/10 µm (vol. 2/1 Fig. 73: 26) . . . . **194. *N. utermoehlii***
16b Striae almost always fewer than 26/10 µm (vol. 2/1 Fig. 73: 16–20), compare with *N. porifera* (vol. 2/1 Fig. 47: 19–24) . . . **190. *N. subrotundata*** *N. utermoehlii* and *N. subrodundata* are conspecific, Fig. 73: 25 represents *N. mollicula* s. str., a species independent of *N. utermoehlii.*
17a (13) Valves narrow (2.5–3.5 µm), elliptical-lanceolate (vol. 2/1 Fig. 74: 1–7) . . . . **214. *N. agrestis***
17b Not as above . . . . **see couplets 2–11**

Navicula key group Ne (vol. 2/1, p. 226)
Forms with more or less beak-like (rostrate) or capitate elongate ends.

Key to species:
1a Terminal nodules distinctly displaced from the poles, resulting in more or less shortened raphe branches **2**
1b Not as above **3**
2a Valve edges appearing punctate (vol. 2/1 Fig. 75: 18–22) . . . . **221. *N. gallica* var. *laevissima***
2b Valve edges without puncta (vol. 2/1 Fig. 75: 23–25) **222. *N. brekkaensis***

3a Axial and central area combined into a lanceolate hyaline area (vol. 2/1 Fig. 75: 29–31) . . . . **223.** ***N. confervacea*** (f. **rostrata**)
3b Not as above **4**
4a Striae approximately parallel, 25–36/10 µm, frustules often visible as ribbon-shaped colonies (vol. 2/1 Fig. 75: 1–5) . . . . **220.** ***N. contenta***
4b Not with the above combination of characteristics **5**
5a Valves at most 3.5 µm wide, linear, edges triundulate or distended in the centre, raphe with central pores spaced apart, striae quite coarse in relation to the smaller valve size, around 18/10 µm in the middle of the valve (vol. 2/1 Fig. 78: 1–9) . . . . **226.** ***N. soehrensis***
5b Not with the above combination of characteristics (compare with *N. bulnheimii*, Fig. 56: 6–10 see key group M couplets 26–48)

Navicula key group Nf (vol. 2/1, p. 226)
The remaining species with more or less well developed central area, without the characteristics of other key groups or subgroups

Key to species:
1a Central area distinctly broadened transversely **2**
1b Ground form of the central area rounded or elliptical **25**
2a Valves rhomboid elliptical, maximum 9 µm long, very coarsely punctate in relation to the striae, around 24 puncta/10 µm (vol. 2/1 Fig. 60: 6–9) . . . . **129.** ***N. ingenua***
2b Not as above **3**
3a Striae relatively closely spaced, more than 24/10 µm **4**
3b Striae coarse, fewer than 24/10 µm **12**
4a Valve width greater than 4 µm **5**
4b Valve width less than 4 µm **8**
5a Central area very narrow, but strongly transapically broadened, striae approximately parallel (vol. 2/1 Fig. 77: 13–18) . . . ***Neidium alpinum***
5b Not with the above combination of characteristics **6**
6a Abrupt change between the shortened striae in the central area and the other striae (vol. 2/1 Fig. 73: 4–7) . . . . **190.** ***N. lapidosa***
6b Central striae irregularly shortened with continuous transition in length **7**
7a Valves linear to linear-elliptical (vol. 2/1 Fig. 73: 8–11) **191.** ***N. variostriata***
7b Valves with strongly convex edges (vol. 2/1 Fig. 73: 1–3) . . . . **189.** ***N. weinzierlii***
8a (4) Striae very weakly radial (vol. 2/1 Fig. 77: 2, 3) . **240.** ***N. muraloides***
8b Striae distinctly radial **9**
9a Striae 35–44 µm, only able to be resolved in the valve middle with strongly oblique light **10**
9b Striae at most 30/10 µm **11**
10a Central area distinct on both sides (vol. 2/1 Fig. 76: 48) . . . . **232.** ***N. semihyalina***
10b Central area only extending to the edge on one side (vol. 2/1 Fig. 80: 31, 32) . . . . **210.** ***N. vitiosa***
11a (9) Valve ends broadly rounded, axial area always narrow, linear (vol. 2/1 Fig. 76: 39–47) . . . . **231.** ***N. minima***
11b Valve ends more variable, mostly somewhat cuneate, blunt to almost sharply rounded, axial area in many examples broadened, lanceolate (vol. 2/1 Fig. 76: 1–7) . . . . **233.** ***N. harderi***
12a (3) Central area forming a transverse band reaching to the edge on at least one side **13**

12b Regularly shortened striae present between valve edge and central nodule **16**
13a Central area wide, valves transapically widened in the middle and at the ends (vol. 2/1 Fig. 66: 5–8) . . . . **140.** ***N. occulta***
13b Not as above **14**
14b Striae appear to be interrupted by a more or less distinct line (vol. 2/1 Fig. 66: 9–11) . . . . **141.** ***N. lucinensis***
14b Not as above **15**
15a Valve edge with one small abruptly distended area in the middle (vol. 2/1 Fig. 78: 14–15) . . . . **227.** ***N. mediocris***
15b Not as above, central nodule with a cross-shaped thickening (vol. 2/1 Fig. 77: 6–9) . . . . **242.** ***N. soodensis***
16a (12) Valve edge with an abrupt small enlarged area in the middle (vol. 2/1 Fig. 78: 14–16) . . . . **227.** ***N. mediocris***
16b Not as above. **17**
17a Valves relatively broad (5–7 µm), most of the striae very weakly radial, the narrow central area with the central nodule showing a cross-shaped thickening (vol. 2/1 Fig. 77: 6–9) . . . . **242.** ***N. soodensis***
17b Not with the above combination of characteristics **18**
18a Central area rhomboid-elliptical, particularly widened in apical direction, terminal nodule strongly marked (vol. 2/1 Fig. 76: 8–10) **238.** ***N. parsura***
18b Not as above **19**
19a Valves with broadly rounded ends, linear or linear-elliptical **20**
19b Valve ends strongly narrowed **23**
20a Valve edges strictly linear or weakly concave (vol. 2/1 Fig. 78: 17–20) . . . . **228.** ***N. begeri***
20b Valve edges more or less convex **21**
21a Valve ends weakly elongated, striae strongly radial (vol. 2/1 Fig. 76: 37, 38) . . . . **235.** ***N. joubaudii***
21b Valve ends not elongated, striae weakly radial **22**
22a Striae 17–18/10 µm, central area very narrow (vol. 2/1 Fig. 73: 12–15) . . . . **192.** ***N. submuralis***
22b Striae 18–22/10 µm, central area broader (vol. 2/1 Fig. 76: 30–36) . . . . **234.** ***N. seminulum***
23a (19) Valve ends more or less weakly elongated, striae strongly radial (vol. 2/1 Fig. 64: 29–32) . . . . **145.** ***N. dolomitica***
23b Not with the above combination of characteristics **24**
24a Valve width 3.5–4.5 µm (vol. 2/1 Fig. 76: 27–29) . . . **236.** ***N. vaucheriae***
24b Valve width 2–2.5 µm (vol. 2/1 Fig. 76: 17–20) . . . . . . **237.** ***N. obsoleta***
25a (1) Central and axial area combined into a more or less hyaline area . **26**
25b Not as above **27**
26a Valve edges constricted in the middle (vol. 2/1 Fig. 75: 26–28), compare with *N. contenta* (vol. 2/1 Fig. 75: 1–5) . . . . **230.** ***N. aerophila***
26b Valve edges not constricted (vol. 2/1 Fig. 75: 29–31), compare with *N. gallica* and *N. contenta* (vol. 2/1 Fig. 75: 1–22) . . . **223.** ***N. confervaceae***
27a (25) Valve edges strictly linear or weakly concave with broadly rounded ends, striae 17–19/10 µm (vol. 2/1 Fig. 78: 17–20) . . . . . **228.** ***N. begeri***
27b Not with the above combination of characteristics **28**
28a Striae very dense, around 35/10 µm, the actual central area is present although easily overlooked (vol. 2/1 Fig. 77: 10–12) . . **241.** ***N. tenerrima***
28b Striae more coarse **29**
29a Valves linear-elliptical, central area large, rhomboid elliptical, terminal nodule strongly marked (vol. 2/1 Fig. 76: 8–10) . . . . . **238.** ***N. parsura***
29b Not with the above combination of characteristics **30**

30a Raphe running in a strongly contrasting median rib, central pores strongly marked, central area large, striae strongly radial (vol. 2/1 Fig. 78: 21–25) . . . . **187.** ***N. subadnata***
30b Not with the above combination of characteristics **31**
31a Valves very small, length 6–10 µm, width 2.5–3 µm (vol. 2/1 Fig. 76: 11–16) . . . . **239.** ***N. evanida***
31b Valves larger (vol. 2/1 Fig. 77: 19–28) . . . . **195.** ***N. digitulus***

2. *Stauroneis* Ehrenberg 1843 (vol. 2/1, p. 236)

Typus generis: *Stauroneis phoenicentreon* (Nitzsch) Ehrenberg; [*Bacillaria phoenicentreon* Nitzsch 1817] (designated by Boyer 1927)

Key to species:

1a Puncta on the striae very fine, usually not distinguishable with the light microscope, more than 30/10 µm **2**
1b Puncta on the striae usually distinguishable with high magnification and high quality of the optics **13**
2a Found exclusively in waters with raised electrolyte concentrations . . **3**
2b Found usually in freshwater, more rarely in coastal regions **4**
3a Shorter than 40 µm (vol. 2/1 Fig. 91: 16, 17) . . . . . . **28.** ***St. wislouchii***
3b Longer than 50 µm (vol. 2/1 Fig. 91: 14, 15) . . . . **27.** ***St. salina***
4a Striae difficult to distinguish with the light microscope, more than 33/10 µm **5**
4b Striae fine, but distinct with the light microscope **8**
5a Valves shorter than 12 µm (vol. 2/1 Fig. 91: 8, 9) . . . . **24.** ***St. nana***
5b Valves longer than 13 µm **6**
6a Valve width greater than 6 µm (vol. 2/1 Fig. 90: 15) . . . **19.** ***St. recondita***
6b Valve width less than 6 µm **7**
7a Ends capitate (vol. 2/1 Fig. 90: 28–30) . . . . **21.** ***St. gracillima***
7b Ends at most somewhat set off (vol. 2/1 Fig. 91: 18, 19) . **29.** ***St. alpina***
8a (4) Ends distinctly set off, rostrate or capitate **9**
8b Ends only weakly set off or not set off at all and bluntly rounded . . **11**
9a Striae relatively coarse, 20–24/10 µm (vol. 2/1 Fig. 90: 31–34) . . . . **22.** ***St. thermicola***
9b Striae finer, 26 or more/10 µm **10**
10a Central area a broad fascia (transverse band) (vol. 2/1 Fig. 90: 21–22) . . . . **18.** ***St. agrestis***
10b Central area a narrow fascia (vol. 2/1 Fig. 90: 23–27) . . **20.** ***St. kriegeri***
11a (8) Valves finely striate, 29–32/10 µm, ends not set off, broadly rounded (vol. 2/1 Fig. 91: 10, 11) . . . . **25.** ***St. lapidicola***
11b Valves more coarsely striate, ends weakly set off **12**
12a Valves relatively broad, length/breadth ratio less than 3.6, 24–26 striae/10 µm (vol. 2/1 Fig. 91: 12, 13) . . . . **26.** ***St. tackei***
12b Valves narrow, length/breadth ratio greater than 3.7, 20–21 striae/10 µm (vol. 2/1 Fig. 91: 1–7) . . . . **23.** ***St. pseudosubobtusoides***
13a (1) Fewer than 20 striae/10 µm **14**
13b More than 20 striae/10 µm **23**
14a Pseudosepta absent or weakly distinguishable **15**
14b With pseudosepta clearly visible, at least in girdle view **19**
15a Small forms, shorter than 30 µm **16**
15b Larger forms **17**
16a Central area a moderately broad fascia (vol. 2/1 Fig. 90: 10–12) . . . . **14.** ***St. borrichii***

16b Central area a broad fascia (vol. 2/1 Fig. 90: 13) . . . **15.** ***St. laterostrata***
17a Puncta rows in the edge zone different to those in the middle region (vol. 2/1 Fig. 87: 1, 2) . . . . **5.** ***St. nobilis***
17b Puncta rows similar over the whole valve **18**
18a Sides nearly parallel, ends cuneate (vol. 2/1 Fig. 88: 5) . . **6.** ***St. dilatata***
18b Valves lanceolate or elliptical-lanceolate (vol. 2/1 Fig. 84: 1–3, 85: 1–6) . . . . **1.** ***St. phoenicentreon***
19a (14) Shorter than 30 µm (vol. 2/1 Fig. 90: 13) . . . . **15.** ***St. laterostrata***
19b Longer than 30 µm **20**
20a Raphe proximally distinctly reverse-lateral (vol. 2/1 Fig. 86: 1–6) . . . . **3.** ***St. javanica***
20b Outer raphe ends proximally straight **21**
21a Ends narrow and elongate (vol. 2/1 Fig. 85: 7–9) . **4.** ***St. lauenburgiana***
21b Ends bluntly rounded and barely elongate **22**
22a Valves rhomboid-lanceolate, more or less widened in the middle (vol. 2/1 Fig. 88: 6–9) . . . . **7.** ***St. acuta***
22b Valves linear or linear-elliptical with symmetrical convex sides (vol. 2/1 Fig. 90: 1–6) . . . . **13.** ***St. obtusa***
23a (13) Pseudosepta indistinct or absent **24**
23b Pseudosepta distinct **28**
24a Valves narrow, linear, length/breadth ratio greater than 5 (vol. 2/1 Fig. 89: 11) . . . . **11.** ***St. lundii***
24b Valves broader lanceolate to elliptical-lanceolate, length/breadth ratio less than 5 . . . . **25**
25a Striae in the middle slightly radial, the central area is almost a linear fascia (vol. 2/1 Fig. 87: 3–9; 88: 1–4) . . . . **2.** ***St. anceps***
25b Central striae more strongly radial, the fascia is more or less broader towards the outer edge **26**
26a Sides of the valves usually with three undulations (vol. 2/1 Fig. 89: 1–10) . . . . **10.** ***St. undata***
26b Sides of the valves not with three undulations **27**
27a Striae curved, at least in the middle of the valves (vol. 2/1 Fig. 90: 10–12) . . . . **14.** ***St. borrichii***
27b Striae in the middle of the valve straight (vol. 2/1 Fig. 90: 7–9) . . . . **16.** ***St. schimanskii***
28a (23) Valve ends bluntly rounded, at most weakly set off (vol. 2/1 Fig. 90: 1–6) . . . . **13.** ***St. obtusa***
28b Valve ends distinctly set off and elongate to capitate **29**
29a Valve ends sharply elongated, pseudosepta reaching to the foot of the elongated part **30**
29b Valve ends more broadly elongated to capitate, pseudosepta filling only the end of the elongated section **31**
30a Valves narrow, linear, edges parallel to undulate (with three undulations), the middle undulation about the same width as the others (vol. 2/1 Fig. 90: 16–20) . . . . **17.** ***St. prominula***
30b Valves more broadly lanceolate or, if with three undulations, then the middle undulation considerably broader than the others (vol. 2/1 Fig. 89: 16–23) . . . . **12.** ***St. smithii***
31a Sides of the valves with three strong undulations (vol. 2/1 Fig. 89: 12–15) . . . . **9.** ***St. legumen***
31b Sides of the valves convex **32**
32a Striae coarsely punctate, puncta irregularly arranged on the striae (vol. 2/1 Fig. 85: 7–9) . . . . **4.** ***St. lauenburgiana***

32b Striae finely punctate, 28–33 puncta/10 µm (vol. 2/1 Fig. 89: 1–7) . . . . **8.** ***St. producta***

3. *Anomoeoneis* Pfitzer 1871 (vol. 2/1 p. 251)

Typus generis: *Anomoeoneis sphaerophora* (Ehrenberg) Pfitzer; [*Navicula sphaerophora* Ehrenberg 1841]

Key to species:

1a Valves strongly distended in the middle, ends elongated and often somewhat capitate (vol. 2/1 Fig. 93: 4) . . . . **2.** ***A. (Brachysira) follis***
1b Valve not inflated in the middle, sides convex 2
2a Largest forms usually with fewer than 18 striae/10 µm (vol. 2/1 Fig. 92: 1–6; Fig. 93: 1–3) . . . . **1.** ***A. sphaerophora***
2b Structure more delicate, 19 or more striae/10 µm 3
3a Central nodule usually elongated (vol. 2/1 Fig. 94: 15–20) . . . . **5.** ***A. (Brachysira) styriaca***
3b Central nodule not elongated 4
4a More than 30 striae/10 µm 5
4b Fewer than 30 striae/10 µm 7
5a With distinct longitudinal ribs in the valve face, ends barely set off (vol. 2/1 Fig. 93: 8, 9) . . . . **7.** ***Brachysira aponina***
5b Longitudinal ribs absent from the valve face 6
6a 36 and more striae/10 µm, ends not set off (vol. 2/1 Fig. 103a: 10–13) . . . . **7.** ***A. (Brachysira)*** *garrensis*
6b Fewer than 36 striae/10 µm, ends almost always set off or capitate (vol. 2/1 Fig. 94: 21–28; 103a: 14) . . . . **6.** ***A. (Brachysira) vitrea***
7a (4) Mostly large forms with sharply rounded ends, valves broader than 10 µm, fewer than 23 striae/10 µm (vol. 2/1 Fig. 93: 5–7) . . . . **3.** ***A. (Brachysira) serians***
7b Small forms with bluntly rounded, bluntly cuneate or sharply rounded ends with more than 23 striae/10 µm (vol. 2/1 Fig. 94: 1–14) . . . . **4.** ***A. (Brachysira) brachysira***

4. *Frustulia* Rabenhorst 1853 nom. cons. (vol. 2/1 p. 258)

Typus generis: *Frustulia saxonica* Rabenhorst 1853 (typ. cons.)

Key to species:

1a Raphe branches with central fissures which are bent in the same direction 2
1b Raphe branches without central fissures which can be seen with the light microscope 3
2b Raphe branches accompanied by distinct axial ribs, striae with approximately 40 puncta/10 µm, very delicately punctate (vol. 2/1 Fig. 97: 12–14) . . . . **5.** ***F. weinholdii***
2b Axial ribs absent, puncta 28–30/10 µm (vol. 2/1 Fig. 97: 10, 11) . . . . **4.** ***F. creuzburgensis***
3a Central nodule very narrow, linear, the relatively coarse puncta irregularly arranged on the striae (vol. 2/1 Fig. 97: 7–9) . . . . **3.** ***F. spicula***
3b Central nodule not as above, puncta regularly arranged 4
4a Sides of the central nodule indented (concave), valves linear-lanceolate (vol. 2/1 Fig. 95: 1–7; 96: 4, 5) . . . . **1.** ***F. rhomboides***
4b Sides of the central nodule convex 5

5a Forms larger than 80 μm, linear-lanceolate (vol. 2/1 Fig. 96: 1–3) . . . . **1.** ***F. rhomboides* var. *viridula***
5b Forms smaller than 70 μm, linear-elliptical (vol. 2/1 Fig. 97: 1–6) . . . . **2.** ***F. vulgaris***

5. *Amphipleura* Kützing 1844 (vol. 2/1 p. 262)

Typus generis: *Amphipleura pellucida* (Kützing) Kützing [*Frustulia pellucida* Kützing 1833] (designated by Boyer 1927)

Key to species:
1a Valves only about 2μm broad (vol. 2/1 Fig. 98: 7, 8) . . **4.** ***A. kriegeriana***
1b Valves considerably broader **2**
2a Striae very delicate, only distinguishable in the densely packed middle region (vol. 2/1 Fig. 98: 4–6) . . . . **1.** ***A. pellucida***
2b Striae more coarse, less than 30/10 μm **3**
3a Longer than 120 μm, freshwater forms (vol. 2/1 Fig. 98: 1–3) . . . . **2.** ***A. lindheimeri***
3b Shorter than 35 μm, brackish water forms (vol. 2/1 Fig. 98: 9–11) . . . . **3.** ***A. rutilans***

6. *Neidium* Pfitzer 1871 (vol. 2/1 p. 265)

Typus generis: *Neidium affine* (Ehrenberg) Pfitzer [*Navicula affinis* Ehrenberg 1841] (designated by Boyer 1927)

Key to species:
1a Proximal raphe ends with central pores or only with a very short sideways curve (see also 14. *N. alpinum*) **2**
1b Proximal raphe ends with distinct central fissures bent to different sides **10**
2a Less than 24 striae/10 μm **3**
2b More than 24 striae/10 μm **7**
3a Raphe with central pores **4**
3b Central pores absent, raphe very slightly bent in different directions at the proximal ends **6**
4a Central area a broad transverse band, almost reaching to the edge (vol. 2/1 Fig. 100: 1, 2) . . . . **5.** ***N. ladogensis***
4b Central area of a different form **5**
5a Striae strongly radial in the middle of the valve, curved, alternately short and long around the central nodule, valves broadly lanceolate (vol. 2/1 Fig. 99: 8, 9) . . . . **2.** ***N. opulentum***
5b Striae less radial in the middle of the valve, not curved, valves linear to broadly linear-elliptical (vol. 2/1 Fig. 99: 1–7) . . . . **1.** ***N. dubium***
6a (3) Valves narrowly linear-elliptical, ends bluntly rounded (vol. 2/1 Fig. 99: 11) . . . . **4.** ***N. juba***
6b Valves broadly elliptical, ends cuneate **7**
7a Valves with two undulations, constricted in the middle **8**
7b Valves with convex sides **9**
8a Central area small, round or absent, puncta very delicate (vol. 2/1 Fig. 100: 3–5) . . . . **8.** ***N. binode***
8b Central area large, rectangular or transversely elliptical, striae coarsely punctate (vol. 2/1 Fig. 100: 6–8) . . . . **9.** ***N. binodeforme***
9a (7) Valves linear-elliptical, ends bluntly rounded, around 20 puncta/10 μm (vol. 2/1 Fig. 99: 10) . . . . **3.** ***N. testa***

9b Valves broadly lanceolate, ends cuneate, puncta 24–29/10 µm (vol. 2/1 Fig. 100: 10–13) . . . . **6.** ***N. densestriatum***
10a (1) Central fissures very long, reaching nearly to the marginal longitudinal canal **11**
10b Central fissures shorter **13**
11a Less than 7 µm broad **12**
11b Broader than 7 µm (vol. 2/1 Fig. 101: 2–4) . . . . **11.** ***N. calvum***
12a Striae distinctly punctate, 21–25/10 µm (vol. 2/1 Fig. 101: 5) . . . . **12.** ***N. minutissimum***
12b Striae very delicately punctate, nearly indistinguishable, puncta 27–32/10 µm (vol. 2/1 Fig. 101: 6, 7) . . . . **13.** ***N. javanicum***
13a (10) More than 30 striae/10 µm **14**
13b Fewer than 30 striae/10 µm **16**
14a Sides convex, not undulating, ends bluntly rounded (vol. 2/1 Fig. 101: 13–17) . . . . **14.** ***N. alpinum***
14b Ends distinctly set off **15**
15a Valves linear, sides usually with two or three undulations (vol. 2/1 Fig. 101: 8–12) . . . . **15.** ***N. septentrionale***
15b Valves linear, sides not undulating (vol. 2/1 Fig. 103a: 4, 5) . . . . **27.** ***N. affine*** var. ***longiceps***
16a (13) Striae and central fissure very diagonal (vol. 2/1 Fig. 103a: 1, 2) . . . . **24.** ***N. carteri***
16b Striae not diagonal or slightly diagonal, central fissure never diagonal . **17**
17a More than 26 striae/10 µm **18**
17b Fewer than 24 striae/10 µm **20**
18a Ends distinctly set off and broadly elongate (vol. 2/1 Fig. 106: 8–10) . . . . **27.** ***N. affine*** in part
18b Ends bluntly rounded **19**
19a Central fissures long, 1–2 distinct submarginal longitudinal canals (vol. 2/1 Fig. 103: 1–10) . . . . **22.** ***N. bisulcatum***
19b Central fissure short, marginal longitudinal canal often not distinguishable in valve view (vol. 2/1 Fig. 103: 11–16) . . . . **23.** ***N. hercynicum***
20a (17) Linear-elliptical forms, length/width ratio usually considerably less than 3, central fissures very short **21**
20b Not with the above combination of characteristics **22**
21a Striae coarsely punctate, 16–18 puncta/10 µm (vol. 2/1 Fig. 101: 1) . . . . **10.** ***N. dilatatum***
21b Striae finely punctate, 24–26 puncta/10 µm (vol. 2/1 Fig. 100: 9) . . . . **7.** ***N. apiculatum***
22a (20) Valves shorter than 36 µm **23**
22b Valves longer than 36 µm **25**
23a Puncta not in undulating longitudinal rows (vol. 2/1 Fig. 102: 1, 2) . . . . **16.** ***N. hustedtii*** (see also **27.** ***N. affine***)
23b Puncta at least marginally in undulating rows **24**
24a Puncta fine, around 30/10 µm, a curved marginal longitudinal band (vol. 2/1 Fig. 102: 7–9) . . . . **18.** ***N. perforatum***
24b Puncta coarser, several curved longitudinal bands in the valve face (vol. 2/1 Fig. 102: 3–6) . . . . **17.** ***N. bergii***
25a (22) Puncta pearl-like, conspicuously large **26**
25b Puncta not as above **27**
26a Central area absent (vol. 2/1 Fig. 102: 10, 11) **20.** ***N. distincte-punctatum***
26b Central area round or a diagonal fascia (vol. 2/1 Fig. 102: 12–15) . . . . **21.** ***N. kozlowii***
27a (25) Puncta 20/10 µm and more (vol. 2/1 Fig. 106: 8–10) . . **27.** ***N. affine***

27b Fewer than 20 puncta/10 µm **28**
28a Sides with three undulations (vol. 2/1 Fig. 107: 3) . . **29. *N. hitchcockii***
28b Sides not as above **29**
29a Ends narrowly elongated, usually somewhat capitate (vol. 2/1 Fig. 107: 4–6) . . . . **28. *N. productum***
29b Ends not as above **30**
30a Puncta organised in 3–4 longitudinal lines on each side of the valve (vol. 2/1 Fig. 103a: 3) . . . . **19. *N. decoratum***
30b Puncta only forming longitudinal lines in the region of the longitudinal canals **31**
31a Central fissures very small, ends cuneate, very large forms (vol. 2/1 Fig. 104: 1–4; 105: 1) . . . . **25. *N. iridis***
31b Central fissures longer, ends bluntly rounded to capitate (vol. 2/1 Fig. 105: 2–6; 106: 1–7; 107: 1, 2) . . . . **26. *N. ampliatum***

8. *Diploneis* Ehrenberg 1844 (vol. 2/1 p. 283)

Typus generis: *Diploneis didyma* (Ehrenberg) Ehrenberg [*Navicula (Pinnularia) didyma* Ehrenberg] (designated by Boyer 1927)

Key to species:
1a Smaller forms with more than 18 striae/10 µm **2**
1b Larger forms with fewer than 18 striae/10 µm **8**
2a Striae with double puncta (vol. 2/1 Fig. 109: 15, 16) . . . . . **8. *D. puella***
2b Striae with single puncta **3**
3a Striae very fine, barely distinguishable with the light microscope, 35–39/10 µm (vol. 2/1 Fig. 110: 6–8) . . . . **17. *D. minuta***
3b Striae easily distinguished by light microscope **4**
4a Longitudinal canals narrow **5**
4b Longitudinal canals broad **7**
5a Edges of the longitudinal canals arched outwards in the region of the central nodule (vol. 2/1 Fig. 108: 7–10) . . . . **3. *D. oblongella***
5b Outer edge of the longitudinal canal not arched outwards in the region of the central nodule **6**
6a Striae strongly contrasting, 18–20/10 µm (vol. 2/1 Fig. 110: 9–12) . . . . **18. *D. modica***
6b Striae weakly contrasting, 20–24/10 µm (vol. 2/1 Fig. 110: 13–15) . . . . **19. *D. oculata***
7a (4) Outer edge of the longitudinal canals strongly concave, both enclosed in a broad lanceolate region(vol. 2/1 Fig. 110: 16, 17) . **20. *D. petersenii***
7b Outer edges of both longitudinal canals almost parallel to each other, both enclosed in a linear-elliptical region (vol. 2/1 Fig. 110: 3–5) . . . . **16. *D. marginestriata***
8a (1) Sides of the valves concave in the middle **9**
8b Sides of the valves straight or convex **11**
9a Fresh water forms in waters with moderate electrolyte concentrations (vol. 2/1 Fig. 111: 4) . . . . **11. *D. alpina*** in part
9b Coast water forms or in inland waters with high electrolyte concentrations **10**
10a Valves with longitudinal ribs, sides weakly constricted in the middle (vol. 2/1 Fig. 112: 7) . . . . **15. *D. didyma***
10b Longitudinal ribs absent, sides strongly constricted in the middle (vol. 2/1 Fig. 112: 5, 6) . . . . **14. *D. interrupta***
11a (8) Striae with single puncta **12**

11b Striae with double puncta, at least in the border region **18**
12a Longitudinal canal or its outer side not arched outwards in the region of the central nodule **13**
12b Longitudinal canal or its outer side arched in the region of the central nodule **14**
13a Longitudinal canal very broad (vol. 2/1 Fig. 110: 1, 2) . . **10. *D. finnica***
13b Longitudinal canal narrow (vol. 2/1 Fig. 111: 5, 6) . **12. *D. domblittensis***
14a Valves linear-elliptical with nearly parallel sides **15**
14b Valves broadly elliptical with convex sides **16**
15a 14–15 striae/10 µm (vol. 2/1 Fig. 109: 10, 11) . . . . **7. *D. boldtiana***
15b 7–11 striae/10 µm (vol. 2/1 Fig. 111: 1–4) . . . . **11. *D. alpina***
16a Small forms, usually narrower than 12 µm (vol. 2/1 Fig. 109: 10, 11) . . . . **7. *D. boldtiana***
16b Valves usually considerably broader than 12 µm **17**
17a Structure coarse, longitudinal canal with a structure distinctly different to that of the striae, areolae with subdivided areolae foramina (vol. 2/1 Fig. 108: 6) . . . . **1. *D. elliptica***
17b Structure fine, longitudinal canal of the same structure as the striae, areolae with a puncta-like foramen (vol. 2/1 Fig. 108: 16) . . . . . . **2. *D. ovalis***
18a (11) Double puncta of the striae arranged in a quincunx **19**
18b Double puncta of the striae in a different arrangement **23**
19a Longitudinal canal broad **20**
19b Longitudinal canal narrow **21**
20a Longitudinal canal finely punctate (vol. 2/1 Fig. 112: 2–4) **13. *D. smithii***
20b Longitudinal canal with coarse flecks (vol. 2/1 Fig. 109: 12–14) . . . . **9. *D. mauleri***
21a Outer edge of the longitudinal canal not arched outwards in the region of the central nodule (vol. 2/1 Fig. 112: 2) . . . . **13. *D. smithii* var. *pumila***
21b Outer edge of the longitudinal canal arched outwards in the central nodule **22**
22a Central nodule small, 14–17 striae/10 µm (vol. 2/1 Fig. 109: 1–7) . . . . **5. *D. parma***
22b Central nodule large, 10–12 striae/10 µm (vol. 2/1 Fig. 109: 8–9) . . . . **6. *D. subovalis***
23a (18) Striae very finely punctate, 25–30/10 µm (vol. 2/1 Fig. 109: 10, 11) . . . . **7. *D. boldtiana***
23b Striae considerably more coarsely punctate **24**
24a Large forms, longer than 35 µm (vol. 2/1 Fig. 112: 1) . . . **10. *D. finnica***
24b Smaller forms, shorter than 32 µm **25**
25a Central nodule small, longitudinal canal not arched in the middle (vol. 2/1 Fig. 109: 15, 16) . . . . **8. *D. puella***
25b Central nodule large, longitudinal canal arched in the middle (vol. 2/1 Fig. 108: 11–13) . . . . **4. *D. pseudovalis***

9. *Pleurosigma* W. Smith 1852 nom. cons. (vol. 2/1 p. 294)

Typus generis: *Pleurosigma angulatum* (Quekett) W. Smith; [*Navicula angulata* Quekett 1848]

1a Transverse and diagonal striae a similar distance from each other and from similar structures (vol. 2/1 Fig. 113: 1, 2; Fig. 114: 1, 2) . **1. *P. angulatum***
1b Transverse striae coarser or finer than the diagonal striae **2**
2a Transverse striae coarser than the diagonal striae (vol. 2/1 Fig. 113: 3) . . . . **2. *P. salinarum***

2b Transverse striae finer than the diagonal striae (vol. 2/1 Fig. 113: 4; 114: 3) . . . . **3.** ***P. elongatum***

10. *Gyrosigma* Hassal 1843 nom. cons. (vol. 2/1 p. 295)

Typus generis: *Gyrosigma hippocampus* (Ehrenberg) Hassall 1845; [*Navicula hippocampus* Ehrenberg 1838]

1a Ends elongated **2**
1b Ends blunt or sharply rounded **4**
2a Large forms, longer than 200 µm (vol. 2/1 Fig. 116: 5) . **11.** ***G. macrum***
2b Shorter than 150 µm **3**
3a Ends moderately broadly elongate (vol. 2/1 Fig. 116: 4) . **10.** ***G. parkeri***
3b Ends very narrowly elongate (vol. 2/1 Fig. 116: 6) . . . . . **12.** ***G. fasciola***
4a (1) Longitudinal and transverse striae the same distance apart **5**
4b Transverse striae more narrowly or more widely spaced than the longitudinal striae **6**
5a Valves shorter than 200 m, 16–23 striae/10 µm, common freshwater form (vol. 2/1 Fig. 114: 4, 8) . . . . **1.** ***G. acuminatum***
5b Valves longer than 200 µm, 11–16 striae/10 µm, common brackish water form (vol. 2/1 Fig. 115: 5) . . . . **7.** ***G. balticum***
6a Transverse striae more delicate than the longitudinal striae (vol. 2/1 Fig. 114: 5, 7, 9) . . . . **2.** ***G. attenuatum***
6b Transverse striae coarser than the longitudinal striae **7**
7a Structure coarse, 11–14 striae/10 µm (vol. 2/1 Fig. 114: 6) **3.** ***G. strigilis***
7b Structure more delicate **8**
8a Raphe in the distal region very close to the convex valve side (vol. 2/1 Fig. 115: 3) . . . . **6.** ***G. peisonis***
8b Raphe in the distal region situated in the mid-line of the valve **9**
9a Striae difficult to distinguish with the light microscope (around 40/10 µm, vol. 2/1 Fig. 116: 1, 2) . . . . **8.** ***G. obscurum***
9b Striae easily distinguished with the light microscope, fewer than 30/10 µm **10**
10a More than 28 longitudinal lines/10 µm, ends usually somewhat obliquely rounded (vol. 2/1 Fig. 116: 3) . . . . **9.** ***G. scalproides***
10b 26 and fewer longitudinal lines/10 µm **11**
11a Central area not diagonal, marine and brackish water forms (vol. 2/1 Fig. 115: 2) . . . . **5.** ***G. spenceri***
11b Central area usually oblique, in inland waters with moderate to high electrolyte concentration (vol. 2/1 Fig. 115: 1) . . . . . . **4.*G. nodiferum***

11. *Cymbella* Agardh 1830 (vol. 2/1 p. 300)

Typus generis: *Cymbella cymbiformis* Agardh 1830

Key to the subgenera:

1a Terminal fissures bent towards the ventral side, stigmoid dorsal to the central nodule or absent **2**
1b Terminal fissures bent towards the dorsal side, stigmata on the ventral side **3**
2a Inner raphe fissure always with intermissio, usually only visible with the EM . . . . **1.** ***Encyonema*** (vol. 2/1 p. 302) (p. 130)
2b Intermissio absent . . . . **3.** ***Cymbopleura*** **numbers 37–43, 46** (vol. 2/1 p. 325) (p. 136)

3a Usually distinctly dorsiventral, apical porefields always present . . . . **2.** ***Cymbella*** (vol. 2/1 p. 312) (p. 133)
3b Usually naviculoid, less dorsiventral forms, as a rule apical pore field and stigmata absent . . . . **3.** ***Cymbopleura*** in part (vol. 2/1 p. 321) (p. 134)

1. *Cymbella* Subgenus *Encyonema* (Kützing 1833) Cleve-Euler 1955 (vol. 2/1 p. 302)

Typus subgeneris: *Monema prostratum* Berkeley 1832

Krammer (1997a, 1997b) has revised this taxon to the status of a genus. Some changes from this revision are inserted into the following key. The presented additions are marked by an asterisk (*1) and added as footnotes to this key.

Key to species:

1a Puncta or lineolae with apertures over 1.0 easily recognisable, in general fewer than 30 puncta/10 µm **2**
1b Puncta or lineolae even with apertures over 1.0 barely visible or not recognisable at all **16**
2a Areolae appearing as puncta with focus "deep" **3**
2b Areolae appearing as lineolae with focus "deep" **11**
3a Striae coarsely punctate (fewer than 24 puncta/10 µm **4**
3b Striae finely punctate **5**
4a Valve end bluntly rounded, length/breadth ratio greater than 4 (vol. 2/1 Fig. 118: 1–8) . . . . ***2. *C. mesiana***
4b Valve end sharply rounded, length/breadth ratio less than 4 (vol. 2/1 Fig. 122: 6–9) . . . . ***15. *C. elginensis***
5a Valves small (length/breadth ratio usually 5–8) (vol. 2/1 Fig. 120: 1–16) . . . . ***11. *C. gracilis***
5b Length/breadth ratio less **6**
6a Valves (other than the initial vegetative cell) crescent shaped, ventral side at most weakly convex, raphe strongly displaced to the ventral side . . . **7**
6b Ventral side distinctly convex, raphe less ventrally displaced **8**
7a Dorsal 10–15 striae/10 µm (vol. 2/1 Fig. 117: 1–24) . . . ***1. *C. silesiaca***
7b Dorsal 7–10 striae/10 µm (vol. 2/1 Fig. 119: 14–16) . . **4. *C. paucistriata***
8a Ends not set off (vol. 2/1 Fig. 119: 29–31) . . . . ***8. *C. brehmii***
8b Ends set off **9**
9a Ends capitate (vol. 2/1 Fig. 119: 21) . . . . **6. *C. latens***
9b Ends narrowly elongated **10**
10a Length/breadth ratio less than 3.8 (vol. 2/1 Fig. 119: 17–20) ***5. *C. obscura***
10b Length/breadth ratio greater than 4 (vol. 2/1 Fig. 121: 10–11) . . . . ***14. *C. hillardii***
11a (2) Striae finely lined, valves elliptical (vol. 2/1 Fig. 123: 1–6) . . . . ***18. *C. alpina***
11b Striae coarsely lined **12**
12a Valves naviculoid, barely dorsiventral (vol. 2/1 Fig. 124: 1–8) . . . . **19. *C. lacustris***
12b Valves strongly dorsiventral **13**
13a A distinct isolated punctum (stigmoid) situated on the dorsal side of the central nodule (vol. 2/1 Fig. 118: 1–8) . . . . ***2. *C. mesiana***
13b Isolated punctum (stigmoid) absent **14**
14a A primarily tropical species (vol. 2/1 Fig. 122: 10–15) . . ***17. *C. muelleri***
14b Species widespread in the temperate zone **15**

15a Terminal fissures lying on the valve surface (vol. 2/1 Fig. 123: 7–10) . . . . **20.** ***C. prostrata***
15b Terminal fissures lying at the mantle edge (vol. 2/1 Fig. 121: 12–16; 122: 1–5) . . . . ***16.** ***C. caespitosa***
16a (1) Valves strongly dorsiventral, ventral side straight or weakly convex, raphe strongly displaced ventrally **17**
16b Valves only moderately or very slightly dorsiventral, raphe less ventrally displaced **19**
17a More than 17 striae/10 µm (vol. 2/1 Fig. 119: 32–36) . . **9.** ***C. reichardtii***
17b Fewer than 16 striae/10 µm **18**
18a Ends not set off, or only slightly so, but partially curved a little towards the ventral side (vol. 2/1 Fig. 119: 1–13) . . . . ***3.** ***C. minuta***
18b Ends capitate (vol. 2/1 Fig. 119: 21) . . . . **6.** ***C. latens***
19a (16) Small forms with delicate striae, usually shorter than 25 µm . . . **20**
19b Forms longer than 25 µm, striae coarser **22**
20a 10–13 striae/10 µm, ends for the most part not elongate (vol. 2/1 Fig. 119: 22–27) . . . . ***7.** ***C. perpusilla***
20b 14–18 striae/10 µm, ends elongate or capitate **21**
21a Ends usually capitate, striae radial in the centre of the dorsal side (vol. 2/1 Fig. 119: 37–43) . . . . **10.** ***C. gaeumannii***
21b Ends only elongate, striae nearly parallel in the middle (vol. 2/1 Fig. 119: 28) . . . . ***7.** ***C. perpusilla*** **var.**
22a (19) Valves linear-elliptical, ends bluntly rounded (vol. 2/1 Fig. 120: 17–24) . . . . ***12.** ***C. norvegica***
22b Valves linear-lanceolate, ends sharply rounded (vol. 2/1 Fig. 112: 1–9) . . . . **13.** ***C. hebridica***

(*1) ***Encyonema silesiacum*** (Bleisch) Mann is a complex of different forms. Fig. 117: 1–13 in vol. 2/1 shows the nominate variety, which is 16–42 µm long, 5.9–9.6 µm broad and has 11–14 striae/10 µm and 28–31 puncta/10 µm. Widespread in the Alps is a more robust variety, ***E. silesiacum*** var. *lata* Krammer. With a breadth of 7–11 µm it is broader and has only 26–29 puncta/10 µm. Also not rare are two more coarsely structured variations with 23–26 puncta/10 µm. The first is ***E. silesiacum* var. *ventriforme*** Krammer with a bulbous ventral side and ***E. silesiacum* var. *distinctepunctatum*** Krammer with an outline similar to the nominate variety.
Very different in outline are two forms, shown in Fig. 117: 18 in vol. 2/1 and 117: 22–24. The latter, ***E. procerum*** Krammer, is common in small bogs of South Germany. It has a length of 18–40 µm, a breadth of 6.1–7.3 µm, 12–14 striae/10 µm and 29–32 puncta/10 µm. In comparison, ***E. lange-bertalotii*** Krammer (vol. 2/1 Fig. 117: 18) has apiculate ends, a length of 16–38 µm, a breadth of 6.2–11 µm, 14–16 striae/10 µm and 27–31 puncta/10 µm.
(*2) Description on page 304 in vol. 2/1 and Fig. 118: 1–8 in vol. 2/1 do not belong to ***Cymbella* mesiana** Cholnoky. ***Cymbella turgida*** var. *pseudogracilis* Cholnoky 1958 (vol. 2/1 Fig. 118: 5–7), receive with ***Encynema neomesianum*** Krammer 1997 a new status with a new combination (length 30–70 µm, breadth 9–12 µm, striae 7–10/10 µm, puncta 21–24/10 µm). Fig. 118: 1, 2 shows ***E. amanianum*** Krammer (length 35–66 µm, breadth 12–17 µm, striae 6–8/10 µm, puncta 15–18/10 µm). Fig. 118: 3 is a specimen of ***E. hungaricum*** Krammer (length 41–58 µm, breadth 10.5–13.5 µm, striae 7–8/10 µm, puncta 23–26/10 µm) .
(*3) Revised morphometric data for ***Encyonema minutum*** are: length 7–23 µm, breadth 4.2–6.9 µm, striae 15–18(19)/10 µm, puncta 34–38/10 µm. ***Cymbella affinis* var. *semicircularis*** Lagerstedt does not belong to *E. minutum*, it is a

synonym of *E. latens* (Krasske) Mann. The specimens in Fig. 119: 10–13 belong to ***E. ventricosum*** (Agardh) Grunow with the following morphometric data: length 9–21 µm, breadth 4.5–6.9 µm, striae 14–19/10 µm, puncta 33–36(39)/10 µm.
(*5) Revised morphometric data for ***Encyonema obscurum*** (Krasske) Mann are: length 18–32 µm, breadth 7.4–8.5 µm, striae 10–12/10 µm, puncta 27–30./10 µm. Especially in the Alps the new variety ***E. obscurum* var. *alpina*** Krammer is more common than the nominate variety. Their ends are smaller rostrate and the coarser striae have 23–26 puncta/10 µm.
(*7) Investigations have revealed that ***Encyonema bipartitum*** (A. Mayer) Krammer (vol. 2/1 Fig. 119: 22 as *Cymbella bipartita* A. Mayer) is not synonymous with ***E. perpusilla***. The differentiating characteristic of the former is the constantly larger space between the middle dorsal striae. The morphometric data for ***E. bipartitum***: length 11–26 µm, breadth 4–5 µm, striae 11–12/10 µm, puncta 38–42/10 µm.
(*8) Similar to ***Encyonema brehmii*** (Hustedt) Mann is ***E. brehmiforme*** Krammer. It is frequent in the type habitat in Teneriffe: length 10–26 µm, breadth 4.6–7.5 µm, striae 11–14/10 µm, puncta 28–32/10 µm.
(*11) ***Encyonema gracile*** is a collective name for more than 20 taxa (see Krammer 1997). *Cocconema gracile* Ehrenberg is not definable and therefore a new taxon, ***E. neogracile*** Krammer was described. Fig. 120: 1–9 in vol. 2/1 show this taxon: length 16–50 µm, breadth 4.7–6.6 µm, striae 12–15/10 µm, puncta 24–28/10 µm. A more finely structured form is ***E. neogracile* var. *tenuipunctata*** Krammer with approximately 28–32 puncta/10 µm. A larger and more coarsely punctate form, ***E. pergracile*** Krammer, is widespread in dystrophic waters in Lapland (length 26–57 µm, breadth 8–9.2 µm, striae 10–14/10 µm, puncta 22–26/10 µm). ***E. lunatum*** (W. Smith) Van Heurck is a taxon with broader rostrate ends and broader striae than *E. gracile* (vol. 2/1 Fig. 120: 10, 12, length 20–51 µm, breadth 5–7 µm, striae 9–12/10 µm, puncta 24–30/10 µm) .
(*12) ***Encynema norvegicum*** (Grunow) Mills is a complex of some varieties. The nominate variety has 38–42 puncta/10 µm and is seldom abundant. More common is ***E. norvegicum* var. *alpinum*** Krammer. It has more distinctly punctate striae (33–36 puncta/10 µm). Very distinct are the puncta in ***E. norvegicum* var. *crassipunctatum*** Krammer (28–31 puncta/10 µm). This variety is found in moor waters in Northern Europe. ***E. lapponicum*** (Cleve-Euler) Krammer has been separated from *E. norvegicum*. It differs in outline and structure: length 26–54 µm, breadth 5–8.5 µm, striae 8–10/10 µm, puncta 34–38(40)/10 µm.
(*14) Manguin (1960) described a single specimen as ***Cymbella hilliardii*** from Lake Karluk, Alaska (length 34 µm, breadth 9 µm, striae 12–13/10 µm). Foged (1971) has described a number of similar specimens from Alaska: length 26–46 µm, breadth about 9 µm, striae 12–15/10 µm).
(*15) Revised morphometric data for ***Encyonema elginense*** (Krammer) Mann are: length 26–63 µm, breadth 10–17 µm, striae 8–13/10 µm, puncta 20–24/10 µm. Two taxa which are similar in outline, found in Sweden and Finland, are ***E. perelginense*** Krammer and ***E. inarense*** Krammer. ***E. perelginense*** is very coarsely punctate (length 42–86 µm, breadth 14.3–19 µm, striae 7–8/10 µm, puncta 16–18/10 µm). ***E. inarense*** has pointed, not protracted ends (length 43–56 µm, breadth 11–12 µm, striae 8–10/10 µm, puncta 20–22/10 µm).
(*16) Revised morphometric data for ***Encyonema cespitosum*** Kützing (vol. 2/1 Fig. 121: 14) are: length 22–46 µm, breadth 9.8–12.4 µm, striae 10–12/10 µm, puncta 18–21/10 µm. A broader form, **E. cespitosum var. *comensis*** Krammer, lives in larger lakes in the Alps. It has somewhat rostrate, ventrally bent ends

(length 22–48 µm, breadth 10.5–13.5 µm, striae 10–11/10 µm, puncta 17–20/10 µm). The larger nordic-alpine ***E. cespitosum* var. *maxima*** (vol. 2/1 Fig. 121: 12, 13, 15) Krammer has only slightly protracted ends: length 30–57 µm, breadth 12–15 µm, striae 9–10/10 µm, puncta 17–20/10 µm). The similar but smaller species ***E. auerswaldii*** Rabenhorst (vol. 2/1 Fig 122: 4, 5), is differentiated from the nominate variety of *E. cespitosum* by its finer structure (length 15–37 µm, breadth 8–12 µm, striae 9–12/10 µm, puncta 20–25/10 µm) .
(*17) ***Cymbella muelleri*** Hustedt (1938, p. 425), typified with *Cymbella grossestriata* var. *obtusiuscula* O. Müller, is a taxon different from *Cymbella muelleri* sensu lato (e.g. the description on p. 311, Fig. 122: 10–15 in vol. 2/1 of this flora). That description and the figures are of ***E. vanoyei*** (Cholnoky) Krammer, a tropical species known only from Africa. In the temperate zone there is a taxon with similar outline, ***E. vulgare*** Krammer, however it is smaller in length and breadth: length 28–60 µm, breadth 9–13 µm, striae 10–13/10 µm, puncta 20–24/10 µm). Nearly all references in the literature to *Cymbella muelleri* and "*Cymbella turgida*" from the temperate zone belong to *E. vulgare*.
(*18) ***Encyonema alpinum*** (Grunow) Mann sensu lato combines two taxa, distinguished by their size and fine structure: ***E. alpinum*** (Grunow) Mann sensu stricto (vol. 2/1 Fig. 123: 1, 2, length 20–52 µm, breadth 9–12 µm, striae 5–8/10 µm, puncta 24–28/10 µm) and ***E. alpiniforme*** Krammer (vol. 2/1 Fig. 123: 3–6), length 14–37 µm, breadth 6–9 µm, striae 9–11/10 µm, puncta (26)28–29 (32)/10 µm, which is more common in nordic-alpine waters than *E. alpinum*.

2. Subgenus *Cymbella* (vol. 2/1 p. 312)

Typus subgeneris: *Cymbella cymbiformis* C. Agardh 1830

To *Cymbella* belong a number of taxa, which are complexes of species. A number of taxa which are species complexes belong to the Subgenus Cymbella. This includes: *C. affinis, C. cistula, C. arctica* and *C. cymbiformis*. We are working on new definitions for these taxa, see taxa with an asterisk (*).

Key to species:

1a More than 20 puncta or lineolae/10 µm **2**
1b Fewer than 20 puncta or lineolae/10 µm **8**
2a Valves cymbelloid with concave or straight ventral side **3**
2b Valves less cymbelloid with convex ventral side, or naviculoid **6**
3a Usually with one stigma ventral to the central nodule **4**
3b Usually with more stigmata ventral to the central nodule **5**
4a 26–28 puncta or lineolae/10 µm, striae fine, axial area narrow (vol. 2/1 Fig. 125: 1–22) . . . . ***21. *C. affinis*** (species complex)
4b 22–24 lineolae/10 µm, striae coarse, axial area with central area broadly lanceolate (vol. 2/1 Fig. 126: 1–3) . . . . **22. *C. hungarica***
5a 24–25 lineolae/10 µm, valves strongly cymbelloid (vol. 2/1 Fig. 128: 7, 8) . . . . **26. *C. arctica*** (species complex)
5b 18–21 puncta or lineolae/10 µm (vol. 2/1 Fig. 127: 8–11; 128: 1–6) . . . . **25. *C. cistula*** (species complex)
6a (2) Usually one stigma . . . . ***21. *C. affinis*** (species complex)
6b Usually two or more stigma **7**
7a Striae finely punctate, 28–32 puncta/10 µm (vol. 2/1 Fig. 126: 8–19) . . . . **24. *C. tumidula***

---

* *C. excisa* Kützing, *C. parva* (W. Smith) Kirchner and *C. nana* (Bleisch) Krammer, as well as other taxa, belong to the species complex Cymbella affinis sensu lato.

7b Striae coarsely punctate, 20–25 puncta/10 µm (vol. 2/1 Fig. 126: 4–7) . . . . **23. *C. turgidula***
8a (1) Six or more stigmata ventral to the central nodule, often difficult to see with the light microscope, large forms, usually greater than 100 µm long **9**
8b Fewer than six stigmata, always clearly visible **12**
9a Raphe distinctly reverse-lateral, stigmata very coarse (vol. 2/1 Fig. 132: 1) . . . . **34. *C. schimanskii***
9b Raphe not reverse-lateral **10**
10a Proximal raphe ends shepherd's crook to hook-shaped **11**
10b Proximal raphe ends with coarse round central pore (vol. 2/1 Fig. 131: 1) . . . . **31. *C. aspera***
11a Proximal raphe ends shepherd's crook shaped (vol. 2/1 Fig. 131: 2–2b) . . . . **32. *C. lanceolata***
11b Proximal raphe ends hook-shaped, foramina of the areolae (with the SEM) large, round (vol. 2/1 Fig. 131: 3–3b) . . . . **33. *C. helmckei***
12a (8) Proximal raphe section reverse-lateral **13**
12b Proximal raphe section not reverse-lateral **15**
13a Valves slightly dorsiventral, 14–18 lineolae/10 µm, stigma foramen slit-shaped when viewed with the EM (vol. 2/1 Fig. 129: 3) . **29. *C. simonsenii***
13b Valves more strongly dorsiventral, 18–20 lineolae/10 µm. Stigma foramen round when viewed with the EM (vol. 2/1 Fig. 129: 9) **14**
14a Usually one stigma ventral to the central nodule, central area absent from the dorsal side (vol. 2/1 Fig. 129: 2–9) . . . . **28. *C. cymbiformis*** (species complex)
14b Usually more than 3 stigmata ventral to the central nodule, central area of the dorsal side a distinct, semicircular clear area (vol. 2/1 Fig. 127: 8–11; 128: 1–6) . . . . **25. *C. cistula*** (species complex)
15a (12) With one large stigma ventral to the central nodule, with a canal which runs very diagonally through the central nodule (vol. 2/1 Fig. 130: 4–6) . . . . **30. *C. tumida***
15b With more than 3 stigmata with less diagonal stigma canals (vol. 2/1 Fig. 128: 9; 129: 1) . . . . **27. *C. proxima***

3. Subgenus *Cymbopleura* Krammer 1982 (vol. 2/1, p. 321)

Typus subgeneris: *Cymbella subaequalis* Grunow 1880

Krammer (1997a, 1997b) has revised the taxa within this subgenus which have terminal fissures curved towards the ventral margin and established the new genus *Encyonopsis* for these taxa. Some changes containing in those revision are inserted into the following key. The presented additions are marked by an asterisk (*) and added as footnotes to this key.

Key to species:
1a Terminal fissures curved towards ventral margin (the new genus *Encyonopsis*) **2**
1b Terminal fissures curved towards dorsal margin **11**
2a Outline linear-lanceolate or rhomboid-lanceolate with slightly drawn out ends **3**
2b Outline linear-elliptical **8**
3a Shorter than 30 µm, 18–22 striae/10 µm **4**
3b Longer than 30 µm **6**
4a Terminal fissures very long, running in the raphe direction (vol. 2/1 Fig. 134: 4–13). . . . . ***38. *C. cesatii***

4b Terminal fissures short **5**
5a Ends somewhat drawn out, terminal fissures comma-shaped (vol. 2/1 Fig. 134: 9, 11, 12) . . . . ***39. *C. falaisensis***
5b Ends capitate, terminal fissures semicircular (vol. 2/1 Fig. 135: 1–5) . . . . ***41. *C. descripta***
6a (3) Terminal fissures very long, running in the raphe direction (vol. 2/1 Fig. 134: 4–8, 10) . . . . ***38. *C. cesatii***
6b Terminal fissures not as above **7**
7a Valves coarsely striate, 11–14 striae/10 µm, stigmata absent, other raphe fissure slightly bent (vol. 2/1 Fig. 134: 1–3) . . . . ***37. *C. aequalis***
7b Valves more finely striate, 14–17 striae/10 µm, a stigmoid dorsal to the central node, outer raphe fissure undulating (vol. 2/1 Fig. 135: 11–13) . . . . ***42. *C. amphioxys***
8a (2) Large forms (longer than 40 µm), ends not drawn out (vol. 2/1 Fig. 136: 13–15) . . . . **46. *C. naviculacea***
8b Small forms (shorter than 35 µm) with capitate ends **9**
9a 10–12 striae/10 µm (vol. 2/1 Fig. 135: 6–10) . . . . **43. *C. similis***
9b More finely striated **10**
10a 18–21 striae/10 µm (vol. 2/1 Fig. 135: 1–5) . . . . ***41. *C. descripta***
10b 22–25 striae/10 µm (vol. 2/1 Fig. 134: 23–32) . . . . ***40. *C. microcephala***
11a (1) More strongly dorsiventral forms, dorsal side considerably more convex than ventral side **12**
11b More naviculoid forms, dorsal side a little more convex than ventral side **19**
12a Lineolae relatively coarse and easily visible **13**
12b Lineolae or puncta finer or not visible with the light microscope . . . **16**
13a Striae in the middle and at the end equally far apart, terminal nodule subterminal, long terminal fissure **14**
13b Striae at the end distinctly closer together than in the middle **15**
14a Structure very coarse (7–8 striae/10 µm, 15–16 lineolae/10 µm), valves with the dorsal side strongly convex (vol. 2/1 Fig. 133: 9) . . **36. *C. balatonis***
14b Structure somewhat more fine (8–12 striae/10 µm, 16–23 lineolae/10 µm), valves with the dorsal side moderately convex (vol. 2/1 Fig. 133: 1–8) . . . . **35. *C. helvetica***
15a (13) Usually longer than 40 µm, striae finely punctate (with high and low magnification), outer fissure of the raphe strongly undulating (vol. 2/1 Fig. 138: 1–6) . . . . **50. *C. austriaca***
15b Usually smaller than 50 µm, striae with low magnification very coarsely lined, outer raphe fissure bent, but not undulating (vol. 2/1 Fig. 143: 1–13) . . . . **61. *C. leptoceros***
16a (12) Raphe fissure and proximal raphe ends pointing dorsally (vol. 2/1 Fig. 148: 1–9) . . . . ***72. *C. pusilla***
16b Raphe fissure pointing dorsally, proximal raphe ends pointing ventrally **17**
17a Valves small, striae very fine (16–22/10 µm) (vol. 2/1 Fig. 137: 1–11) . . . . **47. *C. delicatula***
17b Valves broader, striae coarser (11–14/10 µm) **18**
18a Moderately large forms, striae finely punctate, puncta difficult to distinguish with the light microscope (vol. 2/1 Fig. 139: 4–18) . **52. *C. laevis***
18b Small forms, puncta usually distinct with the light microscope (vol. 2/1 Fig. 140: 9–17) . . . . **56. *C. hustedtii***
19a (11) Valves small or broadly lanceolate **20**
19b Valves linear, narrowly elliptical, broadly elliptical or linear-elliptical . **36**
20a Valves broadly lanceolate **21**
20b Valves linear-lanceolate **27**

21a Fewer than 22 puncta/10 µm **22**
21b More than 25 puncta/10 µm **24**
22a More than 14 striae/10 µm (vol. 2/1 Fig. 139: 1–3) . . . . . **51.** *C. hauckii*
22b Fewer than 11 striae/10 µm **23**
23a Central fissure crochet-hook to shepherd's crook shaped (vol. 2/1 Fig. 147: 3) . . . . **71.** *C. heteropleura*
23b Central fissure absent, proximal raphe ends with coarse central pore (vol. 2/1 Fig. 144: 1–6) . . . . **64.** *C. ehrenbergii*
24a (21) Valves linear elliptical-lanceolate, ends distinctly set off and widely rounded, central area large, rhomboid (vol. 2/1 Fig. 147: 1, 2) **70.** *C. tynnii*
24b Valves broadly lanceolate, ends barely set off, central area small **25**
25a Striae at the end a great deal narrower than in the middle, very delicately punctate (around 30/10 µm) (vol. 2/1 Fig. 141: 1–3) . . **57.** *C. reinhardtii*
25b Difference in density of striae between the middle and the ends smaller, striae distinctly punctate **26**
26a Raphe moderately reverse-lateral (vol. 2/1 Fig. 146: 5) **68.** *C. budayana*
26b Raphe slightly lateral (vol. 2/1 Fig. 143: 17, 18) . . . . **63.** *C. lata*
27a (20) 14 or fewer striae/10 µm **28**
27b 15 or more striae/10 µm **33**
28a Central area distinctly set off, broadly rhomboid or transversely elliptical **29**
28b Central area not as above **30**
29a Ends sharply rounded, slightly elongated, approximately 30 puncta/10 µm (vol. 2/1 Fig. 145: 4, 5) . . . . **65.** *C. hybrida* **var.**
29b Ends bluntly rounded, 23–25 puncta/10 µm (vol. 2/1 Fig. 140: 1) . . . . **53.** *C. moelleriana*
30a (28) Apical axis heteropol (vol. 2/1 Fig. 148: 18–20) . . . . . **74.** *C. ancyli*
30b Apical axis isopol **31**
31a Striae very broad, 7–10/10 µm (vol. 2/1 Fig. 137: 18) . . . **49.** *C. borealis*
31b Striae narrower, more than 12/10 µm **32**
32a Ends sharply drawn out (vol. 2/1 Fig. 143: 14–16) . . . . **62.** *C. designata*
32b Ends bluntly rounded (vol. 2/1 Fig. 140: 2–6) . . . . . . **54.** *C. rupicola*
33a (27) Central area a fascia (vol. 2/1 Fig. 140: 7–8) . **55.** *C. stauroneiformis*
33b Central area not as above **34**
34a Small narrow forms, puncta barely visible with the light microscope (vol. 2/1 Fig. 137: 1–11) . . . . **47.** *C. delicatula*
34b Larger forms, puncta delicate, but distinguishable **35**
35a Dorsal side often with 3 undulations, ends usually capitate, outline linear to elliptical-lanceolate (vol. 2/1 Fig. 135: 15–18) . . . . . **44.** *C. angustata*
35b Nearly naviculoid, rhomboid-lanceolate; ends short, sharp, set off (vol. 2/1 Fig. 137: 12–17) . . . . **48.** *C. lapponica*
36a (19) Valves linear to linear -elliptical **37**
36b Valves broadly elliptical **40**
37a Ends capitate (see also 73. *C. sinuata*, vol. 2/1 Fig. 148: 13) **38**
37b Ends not capitate **39**
38a 9–13 striae/10 µm (vol. 2/1 Fig. 145: 1–3) . . . . **65.** *C. hybrida*
38b 16–20 striae/10 µm (vol. 2/1 Fig. 135: 15–18) . . . . **44.** *C. angustata*
39a (37) With stigma between the proximal raphe ends (vol. 2/1 Fig. 148: 10–17) . . . . ***73.** *C. sinuata*
39b Without stigma **40**
40a Striae fine, but distinctly punctate **41**
40b Striae indistinctly punctate (vol. 2/1 Fig. 141: 4–19) . **58.** *C. subaequalis*
41a Raphe narrowly-lateral (vol. 2/1 Fig. 136: 1–12) . . . . . . **45.** *C. incerta*
41b Raphe broadly lateral (vol. 2/1 Fig. 142: 1, 2) . . . . . . **59.** *C. bernensis*

42a (36) Striae coarsely punctate **43**
42b Striae finely punctate **44**
43a Raphe proximally with central pore, central fissure absent (vol. 2/1 Fig. 146: 1–4) . . . . **67.** ***C. cuspidata***
43b Raphe proximally with shepherd's crook shaped central fissure (vol. 2/1 Fig. 146: 6, 7) . . . . **69.** ***C. subcuspidata***
44a (42) Puncta not distinguishable with the light microscope (vol. 2/1 Fig. 142: 3–21) . . . . **60.** ***C. amphicephala***
44b Puncta fine, but distinguishable with the light microscope **45**
45a Raphe strongly lateral, outer fissure with undulations, ends bluntly rounded (vol. 2/1 Fig. 138: 1–7) . . . . **50.** ***C. austriaca***
45b Raphe moderately lateral, outer fissure curved, but not undulating, ends drawn out or capitate (vol. 2/1 Fig. 145: 6–11) . . . **66.** ***C. naviculiformis***

(*37) ***Encyonopsis aequalis*** (W. Smith) Krammer. Revised morphometric data are length 30–55 µm, breadth 7–9 µm, striae 5–8/10 µm, puncta 36–40/10 µm.
(*38) ***Encyonopsis cesatii*** (Rabenhorst) Krammer contains a number of different taxa. The smaller ***Encyonpsis cesatii* var. *cesatii*** (vol. 2/1 Fig. 134: 6–9, 11, 12) has the following morphometric data: length 20–40 µm, breadth 5–6 µm, striae 18–19/10 µm, puncta 35–40/10 µm. The larger ***Encyonopsis cesatii* var. *geitleri*** Krammer (vol. 2/1 Fig. 134: 4, 5, 10) has a length of 30–60 µm and a breadth of 6–8 µm. In Lapland lives the larger and coarsely structured ***E. cesatiformis*** Krammer (length 26–49 µm, breadth 7–9 µm, striae 16–17/10 µm, puncta 28–32/10 µm.
(*39) Similar to the very common ***Encyonopsis falaisensis*** (Grunow) Krammer is ***Encyonopsis lanceola*** (Grunow) Krammer. In contrast to the former their ends are broader rostrate and it has more than 20 str./10 µm. The morphometrical data of known populations are: length 22–32 µm, breadth 4–4.8 µm, striae 22–26/10 µm, puncta 35–40/10 µm).
(*40) ***Cymbella microcephala*** Grunow sensu lato is a complex of taxa. On the base of Grunow's type ***Encyonopsis microcephala*** (Grunow) Krammer was revised, it corresponds with Fig. 134: 25–30 (length 10–23 µm, breadth 3.5–4.2 µm, striae 23–25/10 µm, puncta 38–42/10 µm). A taxon with more lanceolate outline and subcapitate ends is ***Encyonopsis minuta*** Krammer & Reichardt. In the SEM this taxon has characteristic foramina and terminal fissures. Morphometric data: Length 8–17 µm, breadth 2.8–3.5 µm, striae 24–25/10 µm, puncta 36–45/10 µm. In the Nördlichen Kalkalpen and the Voralpen the most common taxon from the *E. microcephala*-complex is ***Encyonopsis krammeri*** Reichardt (vol. 2/1 Fig. 134: 23). It has a linear-lanceolate outline, the ends are indistinctly capitate, shoulders are absent and the striae are very fine: Length 11.5–23.5 µm, breadth 2.6–3.8 µm, striae (27–)28–32/10 µm, puncta 35–42/10 µm.
(*41) ***Encyonpsis descripta*** (Hustedt) Krammer. A form with broader capitate ends and a irregular, asymmetric central area was described as ***Encyonpsis descripta* var. *asymmetrica*** Krammer. Reports from South Germany, perhaps also Greenland.
(*42) The Fig. of *Navicula amphioxys* Ehrenberg 1843 do not agree with *Navicula amphioxys sensu* Kützing 1844. Therefore a new name was necessary for the taxon of Kützing, ***Encyonopsis neoamphioxys*** Krammer.
(*72) Krammer placed this taxon in the new genus *Navicymbula*. It combines characters of the genera *Cymbella* and *Navicula* s. str.
(*73) Kociolek & Stoermer placed this taxon in the new genus *Reimeria*.

12. *Amphora* Ehrenberg in Kützing 1844 (vol. 2/1 p. 342)

Typus generis: *Navicula amphora* Ehrenberg 1831 (designated by Boyer 1927)

To the taxa numbers with an asterisk (*) see the footnotes.

Key to species

1a Cells without intercalary bands **2**
1b Cells with intercalary bands **8**
2a Raphe branch strongly curved and proximally turned round to the dorsal side **3**
2b Raphe branch weakly curved or straight, proximally not with a dorsal curve **6**
3a Smaller forms, dorsal side (residuum) distinctly ribbed with the light microscope (vol. 2/1 Fig. 150: 18–22) . . . . **6. *A. aequalis***
3b Larger forms, dorsal side not ribbed **4**
4a Valves with distinct dorsal central area (vol. 2/1 Fig. 149: 3–11) **2. *A. libyca***
4b Dorsal central area absent **5**
5a Dorsal side with three undulations, also with silicified, lantern-shaped cap (vol. 2/1 Fig. 153: 8, 9) . . . . **19. *A. calumetica***
5b Dorsal side convex (vol. 2/1 Fig. 149: 1, 2) . . . . **1. *A. ovalis***
6a Dorsal area absent (vol. 2/1 Fig. 150: 14–17) . . . . **5. *A. fogediana***
6b Dorsal area present **7**
7a Ends not set off, 18–25 striae/10 µm (vol. 2/1 Fig. 150: 1–13) . . . . **4. *A. pediculus***
7b Ends set off and somewhat drawn out ventrally, 15–17 striae/10 µm (vol. 2/1 Fig. 150: 1–6) . . . . **3. *A. inariensis***
8a (1) More than 20 striae/10 µm **9**
8b Fewer than 20 striae/10 µm **14**
9a Larger forms with many distinctly ornamented intercalary bands (vol. 2/1 Fig. 150: 23, 24) . . . . **7. *A. lineolata***
9b Forms shorter than 25 µm **10**
10a Valves strongly convex dorsally, small cymbelloid forms (vol. 2/1 Fig. 152: 15–18) . . . . **15. *A. thumensis***
10b Dorsal edge only moderately convex **11**
11a Central area crossed by two fortified striae on the dorsal side (vol. 2/1 Fig. 151: 18–27). . . . . **10. *A. montana***
11b Central area absent **12**
12a Dorsal side usually with three undulations, ends ventrally capitate (vol. 2/1 Fig. 152: 7, 8) . . . . **12. *A. dusenii***
12b Dorsal side not undulating, ends at most weakly drawn out **13**
13a Fewer than 26 striae/10 µm (vol. 2/1 Fig. 152: 12–14) **14. *A. spitzbergensis***
13b Around 30 striae/10 µm (vol. 2/1 Fig. 152: 19–23) . ***16. *A. delicatissima***
14a (8) Structure coarse, rib like (vol. 2/1 Fig. 152: 9–11) . **13. *A. commutata***
14b Striae not as above **15**
15a Striae very finely punctate (vol. 2/1 Fig. 151: 1–6) . . . . **8. *A. coffeaeformis* var. *coffeaeformis***

(*16) Smaller than *Amphora delicatissima* Krasske is ***Amphora hassiaca*** Krammer & Strecker (1997, Bibl. Diatomologica **37**, p. 225, Fig. 206: 1–8). This taxon is common in the brackish inland water of the Werra region in Germany. The frustules and valves have a similar outline as *Amphora delicatissima.* The frustules are 8–15.8 µm long and 4.5–6 µm broad, the breadth of the valves is 2.7–3.4 µm, striae on the dorsal side 27–32/10 µm, puncta about 30/10 µm.

15b Striae distinctly punctate **16**
16a Middle dorsal striae clearly further apart **17**
16b Middle dorsal striae not clearly further apart or dorsal central area present **18**
17a Valves narrower than 18 µm, usually more than 20 striae/10 µm (vol. 2/1 Fig. 151: 7–17) . . . . **9.** ***A. veneta***
17b Valves broader than 20 µm, fewer than 20 striae/10 µm (vol. 2/1 Fig. 153: 1–3) . . . . **17.** ***A. subcapitata***
18a (16) Central pores close together (vol. 2/1 Fig. 151: 6') . . . . **8.** ***A. coffeaeformis*** **var.** ***acutiuscula***
18b Central pores further apart **19**
19a Proximal ends of the raphe bent dorsally in a right angle, central nodule very distinct (vol. 2/1 Fig. 153: 4–7) . . . . **18.** ***A. normanii***
19b Proximal raphe ends only weakly curved dorsally, central nodule indistinct (vol. 2/1 Fig. 152: 1–6) . . . . ***11.** ***A. holsatica***

13. **Gomphonema** Ehrenberg 1832 nom. cons. (vol. 2/1 p. 352)

Typus generis: *Gomphonema acuminatum* Ehrenberg 1832 (typ. cons.)

Key to the species:**

1a Valves large with very distended centres and broad capitate poles, 2–5 apical stigmata on one side of the central area. Terminal fissure of the raphe acutely angled in the valve face (vol. 2/1 Fig. 166: 15) . . . . **15.** ***Didymosphenia*** (vol. 2/1 p. 380) only one species
1b Not with the above combination of characteristics, particularly terminal fissures appear as more or less moderately curved extensions of the raphe **2**
2a A line running apically more or less widely displaced from the valve edge, or appearing on the mantle face in girdle view (vol. 2/1 Fig. 166: 12–14) . . . . **14.** ***Gomphoneis*** (vol. 2/1 p. 379) only one species
2b Without distinct longitudinal lines on valve or mantle face **3**
3a Central area without stigma. **39**
3b Central area with 1, 2 or 4 stigmata. **4**
4a Four stigmata in a rectangular shape around the central nodule (vol. 2/1 Fig. 165: 14–18) . . . . **24** ***G. olivaceum*** **var.** ***minutissimum***
4b Central area with 1 or 2 stigmata **5**

(*11) More common in brackish inland waters and on the coast is ***Amphora subholsatica*** Krammer (1997, Bibl. Diatomologica **37,** p. 223, Fig. 205: 1–10) · In this flora the species is illustrated in vol. 2/1 Fig. 152: 4, 5. Most records in the literature of Hustedt belong to this taxon. The cells have a length of 20–38 µm and a breadth of 13–17 µm. The numerous intercalary bands, 14–16/10 µm, have 23–26 puncta/10 µm. The ends of the valves are narrowly capitate and bent ventrally. The axial area is very narrow on the dorsal side, on the ventral side the striae are only near the margin. Striae on the dorsal side 17–20/10 µm, at the middle of the ventral side about 24 striae/10 µm. Puncta about 18/10 µm. ***A. holsatica*** has (on the ventral side) only 13–15 striae/10 µm and only about 10 intercalary bands /10 µm.
** The genera *Gomphoneis, Didymosphenia* and *Rhoicosphenia*, with only one species each, are listed here. See also additional species of *Gomphonema* and some revisions in vol. 2/4, plates 73–88. For more new species see Iconographia Diatomologica vol. 8 and references given there.

5a 2 stigmata on one side of the central area (vol. 2/1 Fig. 162: 8, 9) . . . . **14. *G. bipunctatum***
5b Valves with 1 stigma in the central area, either a lengthening of the central striae (which are usually short) or in an almost central position (rarely with additional "isolated puncta" distinguishable) 6
6a Valves strongly constricted (transapically) above the midline of the valve 7
6b Valve without such a constriction, at most with slightly triundulate or flat concave outline, in the mid section a distended bulge or head end elongate, rostrate or with a distinct small knob (capitulum) 9
7a Head pole more or less cuneate or drawn out and sharply rounded (vol. 2/1 Fig. 160: 1–12) . . . . **9. *G. acuminatum***
7b Head end blunt to broadly rounded 8
8a Constricted head end small, valve narrowly sublinear (vol. 2/1 Fig. 162: 10–13) . . . . **15. *G. subtile***
8b Constricted head end larger, broadly rounded, middle striae alternately short and long (vol. 2/1 Fig. 159: 11–18) . . . . **16. *G. truncatum***
9a (6) Greatest width of the valves just below the head pole 10
9b Greatest width in the middle, or valve almost linear 13
10a Head end drawn out, rostrate or capitatate 11
10b Head end not rostrate, more or less wedge-shaped (cuneate) (vol. 2/1 Fig. 160: 1–12) . . . . **9. *G. acuminatum***
11a Head end strongly rostrate or capitate (vol. 2/1 Fig. 157: 1–8), compare with *G. pseudoaugur* (vol. 2/1 Fig. 159: 1–4) . . . . **6. *G. augur***
11b Head end rounded or with slightly drawn out point 12
12a Head end widely rounded (vol. 2/1 Fig. 159: 11–18) . **16. *G. truncatum***
12b Head end drawn out, more or less a sharply rounded, compare with *G. augur* (for example vol. 2/1 Fig. 157: 3) (vol. 2/1 Fig. 154: 21–22) . . . . **1. *G. parvulum***
(vol. 2/1 Fig. 156: 9) . . . . **4. *G. gracile***
13a (9) Valve drawn out at the head end, more or less long pointed end, point rounded or capitate 14
13b Head pole bluntly wedge-shaped or bluntly rounded 25
14a Head end or both poles rostrate or capitate 15
14b Head end rostrate, long and pointed, never capitate 17
15a Only head end capitate, head ends of sagitta forms cuneately rounded (vol. 2/1 Fig. 162: 10–13) . . . . **15. *G. subtile***
15b Both poles capitate 16
16a Central area large, apical area widened (vol. 2/1 Fig. 162: 14–18) . . . . **21. *G. helveticum***
16b Central area small, apical area not widened (vol. 2/1 Fig. 157: 10) . . . . ***G. sphaerophorum* sensu Mayer** (see under 6. *G. augur*)
17a (14) Valve long and narrow, approximately symmetrical (naviculoid) with more or less triundulate sides (vol. 2/1 Fig. 155: 22–24) **3. *G. lagerheimii***
(vol. 2/1 Fig. 156: 12–14) . . . . **5. *G. hebridense***
17b Valve not triundulate 18
18a Valve width not over 4 µm (vol. 2/1 Fig. 164: 22–24) . . . . **20. *G. pseudotenellum***
18b Valve width usually over 4 µm 19
19a Valve frequently approximately symmetrical about the transapical axis (naviculoid), one or both ends sharply rounded to quite sharply drawn out, usually more or less rhomboid-lanceolate forms (vol. 2/1 Fig. 156: 1–11; 154 : 26–27), if valve border triundulate or distended in the middle, refer to *G. hebridense* and *G. lagerheimii* . . . . **4. *G. gracile***
19b Valve form not as above 20

20a Head end drawn out and rostrate (similar to *G. augur* var. *augur*) with the widest breadth at the middle of the valve, or just the tip of the valve suddenly narrows to a wedge shape (vol. 2/1 Fig. 158: 1–6) compare with "*turris*" form of the named variety and other taxa (see vol. 2/1 Fig. 157: 7 and 159: 4) . . . . **6. *G. augur* var. *turris***
20b Head end not strongly rostrate, at most slightly drawn out **21**
21a Greatest width above the middle of the valve (vol. 2/1 Fig. 159: 1–3) compare with small forms of *G. acutiusculum* (vol. 2/1 Fig. 162: 1) . . . . **7. *G. pseudoaugur***
21b Greatest width in the middle of the valve or valves nearly linear **22**
22a Valves usually more or less oval-lanceolate, central area barely distinct, median density of the striae within a population usually greater than or equal to 12/10 µm (vol. 2/1 Fig. 154: 1–25) . . . . **1. *G. parvulum***
22b Valves with a different combination of characteristics **23**
23a Valves usually linear-lanceolate, structure on the whole more coarse, median density of the striae within a population less than 12/10 µm, central area one-sided and usually distinctly pronounced* (vol. 2/1 Fig. 155: 1–21) compare with *G. bohemicum* sensu Hustedt (vol. 2/1 Fig. 154: 29–32) if the valve borders have 3 undulations, compare with *G. lagerheimii* (vol. 2/1 Fig. : 22–24) ***2. *G. angustatum*** auct. nec. Kützing
23b Valves not with the above combination of characteristics **24**
24a Valves lanceolate with sharply rounded poles (vol. 2/1 Fig. 156: 1–11, Fig. 154: 26–27) . . . . **4. *G. gracile***
24b Valves distinctly club-shaped with strongly narrowed foot (vol. 2/1 Fig. 162: 1–3) compare with "*turris*"-form of *G. gracile* (vol. 2/1 Fig. 156: 5–10) . . . . **8. *G. acutiusculum***
25a (13) Central area relatively broad, extending on one side to the valve border **26**
25b Central area not as above **27**
26a Stigma almost in the middle point of the broadly oval club-shaped valve (vol. 2/1 Fig. 162: 6, 7) compare with small forms of *G. angustum* . . . . **22. *G. tergestinum***
26b Stigma clearly separate from the middle point, valves narrow, linear club-shaped (vol. 2/1 Fig. 164: 1–16), compare with *G. angustatum* and *G. bohemicum* sensu Hustedt (vol. 2/1 Fig. 154: 29–32) . **18. *G. angustum***
27a (25) Valves with more or less wedge-shaped head end **28**
27b Head pole not wedge-shaped **29**
28a Head end wedge-shaped or more or less conical, widest part usually above the middle of the valve. Stigma near a shortened middle striae in the central area (vol. 2/1 Fig. 160: 1–12) compare with *G. augur* var. *turris* as well as "*turris*" form of other tax . . . . **9. *G. acuminatum***
28b Head end (in critical examples) more or less wedge-shaped, with the end however widely rounded or cut off, greatest width in the centre of the valve (vol. 2/1 Fig. 163: 1–12) . . . . **12. *G. clavatum***
29a (27) Many striae in the centre alternately short and long (in the smallest forms these characteristics may disappear) **30**
29b Striae not as above **31**
30a Valves barely narrowed before a conspicuously widely rounded head end (vol. 2/1 Fig. 159: 11–18) . . . . **16. *G. truncatum***
30b Valves strongly narrowed towards both head and foot ends, central area

---

* The characteristics of *G. parvulum*, *C. micropus* and *G. angustatum* auct. nec. typus overlap considerably, so that just light microscopic examination might not distinguish these species.

round, bordered by many shortened striae, structure conspicuously coarse (vol. 2/1 Fig. 162: 4, 5) . . . . **23. *G. ventricosum***

31a (29) Striae close to the edge, so the axial and central area form a broad hyaline area (vol. 2/1 Fig. 164: 20, 21) . . . . **19. *G. clevei***

31b Striae less markedly shortened **32**

32a Valves linear-lanceolate to elliptical (regardless of the central enlarged area in larger valves), axial and central area relatively wide. The inner raphe fissure curves further to the side at the distal end, the other raphe fissure runs straight to the central pore, in the smallest forms frequently only weakly visible (vol. 2/1 Fig. 164: 1–16 representing different revised species) . . . . **18. *G. angustum***

32b Form and structure not in the combination above **33**

33a Valves slightly club-shaped **34**

33b Valves strongly club-shaped, with foot end strongly narrowed **36**

34a Striae in valve and girdle view not strongly punctate (vol. 2/1 Fig. 161: 9–11), if smaller forms are found, compare with *G. parvulum*, *G. angustatum*, G. *minutum* . . . . **13. *G. amoenum***

34b Striae distinctly punctate **35**

35a Central area one-sided, usually small (vol. 2/1 Fig. 161: 1–3) . **10. *G. affine***

35b Central area larger, often bilaterally developed (vol. 2/1 Fig. 161: 4, 5, 7, 8) . . . . **11. *G. insigne***

36a (33) Smaller forms (length 10–35 µm), puncta in the striae practically indistinguishable (double puncta), only one separate point for each striae at the valve mantle in girdle view (vol. 2/1 Fig. 159: 5–10) **17. *G. minutum***

36b Usually larger forms, puncta on the valve more or less clearly distinguishable, striae lengthened by a row of puncta in girdle view (vol. 2/1 Fig. 161: 8, 163: 12) **37**

37a Striae in the valve mantle in girdle view lengthened by small coarse puncta, valves sometimes with a tendency to triundulate form (vol. 2/1 Fig. 163: 1–12) . . . . **12. *G. clavatum***

37b Form and/or structure not as above **38**

38a Raphe always strongly lateral, in girdle view fine, dense punctate extensions of the striae at the edge (vol. 2/1 Fig. 161: 4, 5, 7, 8) **11. *G. insigne***
(vol. 2/1 Fig. 161: 1–3) . . . . **10. *G. affine***

38b Raphe usually weakly lateral, on the valve mantle in girdle view there are no individual separated puncta which lengthen striae (vol. 2/1 Fig. 164: 1–16) . . . . **18. *G. angustum***

39a (3) Frustule bent in girdle view, raphe branch on convex valve considerably shortened (vol. 2/1 Fig. 91: 20–28) . . . . ***Rhoicosphenia abbreviata*** (vol.2/1p.381)

39b Characteristics not as above **40**

40a Striae short, around the edge, occasionally transapically lengthened by an individual punctum (vol. 2/1 Fig. 166: 1–11) . . . . **27. *G. grovei***

40b Striae extending closer to raphe **41**

41a Striae very delicate and close together, more than 25/10 µm (vol. 2/1 Fig. 165: 22–24) . . . . **26. *G. tackei***

41b Fewer than 25 striae /10 µm **42**

42a Fresh and brackish water forms, central area more or less transapically widened, var. *minutissimum* with 4 stigmata arranged in a rectangle (vol. 2/1 Fig. 165: 1–18) . . . . **24. *G. olivaceum***

42b Salt water forms, without a distinct central area or central area present due to a loss of the middle striae on both sides, small narrow linear forms (vol. 2/1 Fig. 165: 19–21) . . . . **5. *G. exiguum***

17. *Caloneis* Cleve 1894 (vol. 2/1 p. 382)

Typus generis: *Caloneis amphisbaena* (Bory) Cleve 1894; [*Navicula amphisbaena* Bory 1824] (designated by Boyer 1927)

Key to species:

1a Large, broadly lanceolate or broadly elliptical forms with distinct longitudinal lines in the valve face 2
1b Valves narrower, longitudinal lines marginal, commonly not distinguishable in valve view 6
2a Axial and central area together form a large, rhomboid or lanceolate area 3
2b Central area clearly set off on at least one side 4
3a Ends bluntly rounded, central area with maculae (moon spots) (vol. 2/1 Fig. 169: 5–7) . . . . **4. *C. obtusa***
3b Ends rounded wedge-shaped to capitate, maculae absent (vol. 2/1 Fig. 168: 4, 5) . . . . **2. *C. amphisbaena***
4a Valves lanceolate with rounded wedge shaped ends (vol. 2/1 Fig. 168: 1–3; 169: 4) . . . . **1. *C. permagna***
4b Valves elliptical-lanceolate, ends bluntly rounded 5
5a 12–14 striae/10 µm (vol. 2/1 Fig. 170: 1, 2) . . . . **5. *C. westii***
5b 17–21 striae/10 µm (vol. 2/1 Fig. 169: 1–3) . . . . **3. *C. latiuscula***
6a (1) More than 30 striae /10 µm 7
6b Fewer than 30 striae /10 µm 8
7a Longitudinal striae not distinguishable with the light microscope (vol. 2/1 Fig. 173: 6–8) . . . . **12. *C. hyalina***
7b Longitudinal striae submarginal, easily distinguished with the light microscope (vol. 2/1 Fig. 173: 22–24) . . . . **19. *C. branderi***
8a With distinct maculae in the central area 9
8b Without distinct maculae in the central area, at best very weakly contrasting macula-like structures present 10
9a Valves linear to linear-elliptical, small central area (vol. 2/1 Fig. 170: 3–7) . . . . **6. *C. alpestris***
9b Valves linear to lanceolate, sides often triundulate, large central area (vol. 2/1 Fig. 171: 1–11) . . . . **7. *C. schumanniana***
10a Valves with undulating sides 11
10b Valves sides not as above 13
11a Ends broadly capitate (vol. 2/1 Fig. 175: 1–6) . . . . . . **21. *C. undulata***
11b Ends bluntly rounded or drawn out 12
12a Large, broad forms with variable outline, ends bluntly rounded (vol. 2/1 Fig. 172: 1–13) . . . . **8. *C. silicula***
12b Mostly smaller, narrow forms, ends broadly elongate (vol. 2/1 Fig. 174: 5–10) . . . . **17. *C. tenuis***
13a (10) Valves distinctly transapically inflated in the middle 14
13b Valves not inflated in the middle 15
14a Ends bluntly rounded, central area large, round (vol. 2/1 Fig. 174: 1–4) . . . . **16. *C. pulchra***
14b Ends sharply wedge-shaped, central area a broad fascia (vol. 2/1 Fig. 174: 13–15 . . . . **24. *C. leptosoma***
15a Valve sides parallel in the middle, weakly concave or convex 16
15b Valve sides moderately convex (outline linear-lanceolate or linear-elliptical) 20
16a Striae distinctly convergent in the middle of the valve (vol. 2/1 Fig. 173: 1) . . . . **9. *C. clevei***
16b Striae parallel or radial in the middle of the valve 17

17a Valves narrower than 6 µm 18
17b Valves broader than 6 µm 19
18a Striae radial in the middle of the valve (vol. 2/1 Fig. 173: 21) . . . . **14. *C. lepidula***
18b Striae parallel in the middle of the valve (vol. 2/1 Fig. 174: 11, 12) . . . . **18. *C. sublinearis***
19a Raphe strongly curved, ends rounded wedge-shaped (vol. 2/1 Fig. 173: 2, 3) . . . . **10. *C. lauta***
19b Raphe weakly curved, ends broadly rounded (vol. 2/1 Fig. 173: 25, 26) . . . . **15. *C. thermalis***
20a (15) Outline linear-lanceolate 21
20b Outline linear-elliptical 23
21a Striae distinctly radial in the middle of the valve (vol. 2/1 Fig. 174: 16–21) . . . . **20. *C. molaris***
21b Striae parallel in the middle of the valve 22
22a Ends bluntly rounded, raphe weakly curved(vol. 2/1 Fig. 173: 9–20) . . . . **13. *C. bacillum***
22b Ends rounded, wedge-shaped (cuneate), raphe strongly curved(vol. 2/1 Fig. 173: 4) . . . . **11. *C. aerophila***
23a (20) Ends capitate to broadly elongated and rounded 24
23b Ends blunt to flattened rounded wedge-shaped 25
24a Terminal fissures large (vol. 2/1 Fig. 175: 10–13) . . . . . **23. *C. budensis***
24b Terminal fissures small (vol. 2/1 Fig. 175: 7–9) . . . . **22. *C. macedonica***
25a Ends broadly rounded (vol. 2/1 Fig. 172: 1–13) . . . . **8. *C. silicula***
25b Ends flattened wedge-shaped (vol. 2/1 Fig. 174: 16–21) . . **20. *C. molaris***

18. *Pinnularia* Ehrenberg 1843 nom. cons. (vol. 2/1 p. 397)

Typus generis: *Pinnularia viridis* (Nitzsch) Ehrenberg 1843 (typ. cons.); [*Bacillaria viridis* Nitzsch 1817]

Revisions of this genus were produced by Krammer (1992a, 1992b, 2000). In the following it is impossible to present all these changes. Only the most important new species, varieties and combinations are given. The additions are marked by an asterisk (*1) and added as footnotes to this key.

Key to the artificial species groups:
1a Striae broad, far apart, not touching each other . . . . **1. group around *P. borealis*** (vol. 2/1 p. 398) (p. 145)
1b Striae and transapical ribs narrower 2
2a Striae strongly divergent in the middle and at the ends of the valves, the orientation changing abruptly 3
2b Striae somewhat divergent or not divergent 4
3a Valves broadly lanceolate or rhomboid-lanceolate . . . . **3. group around *P. acoricola*** (vol. 2/1 p.399) (p. 145)
3b Valves not as above . . . . **4. group around *P. divergentissima*** (vol. 2/1 p.400) (p. 145)
4a Valves linear with broad central area . . . . **2. group around *P. brevistriata*** (vol. 2/1 p.399) (p. 145)
4b Valves with moderately broader or narrower central area 5
5a Raphe strongly lateral, large forms . . . . **6. group around *P. viridis*** (vol. 2/1 p.402) (p. 148)

5b Raphe less lateral, small to large forms . . . . **5. group around *P. microstauron*** (vol. 2/1 p.400) (p. 146)

1. Group around *P. borealis* (Distantes)

1a Valves elliptical-lanceolate, large forms (vol. 2/1 Fig. 176: 1, 2) . **1. *P. alpina***
1b Valves linear to linear-elliptical, large and small forms **2**
2a Valves shorter than 15µm, central pores far apart (vol. 2/1 Fig. 176: 11, 12) . . . . ***4. *P. balfouriana***
2b Valves longer than 16µm **3**
3a Smaller forms with 7–10 striae/10µm **4**
3b Mostly larger forms with 6 and fewer striae/10µm **5**
4a Striae parallel in the middle, often only at the edges, marginally becoming more spread out (vol. 2/1 Fig. 176: 8–10) . . . . **3. *P. lagerstedtii***
4b Striae radial in the middle, becoming more spread out at the central area, axial area narrow (vol. 2/1 Fig. 178: 1–6) . . . . ***6. *P. intermedia***
5a Proximal raphe ends bent to one side, but not reverse-lateral (vol. 2/1 Fig. 177: 1–12; 178: 7) . . . . ***5. *P. borealis***
5b Proximal raphe ends somewhat reverse-lateral, outer fissure of the raphe displaced towards the median at first, then moving in further (vol. 2/1 Fig. 176: 3–7) . . . . **2. *P. lata***

2. Group around *P. brevistriata* (Brevistriatae)

1a Upper surface of the axial area ornamented (particularly clear with differential interference or phase contrast microscopy) **2**
1b Axial area not ornamented **4**
2a Raphe accompanied by a distinct axial rib, sides triundulate (vol. 2/1 Fig. 181: 4–10) . . . . ***14. *P. nodosa***
2b Axial ribs absent, valves not triundulate, at most distended in the middle **3**
3a Valves broader than 8µm, ends broadly capitate, ornamentation very distinct (vol. 2/1 Fig. 181: 1–3) . . . . ***13. *P. acrosphaeria***
3b Valves narrower than 6µm, ends not drawn out or barely so, ornamentation indistinct (vol. 2/1 Fig. 188: 4–8) . . . . ***40. *P. schwabei***
4a (1) Ends broadly rounded (vol. 2/1 Fig. 182: 4–7, 9) . ***16. *P. brevicostata***
4b Ends rounded, wedge-shaped (cuneate) **5**
5a 8–10 striae/10µm, longitudinal bands of the striae broad, striae parallel in the middle (vol. 2/1 Fig. 182: 1–3) . . . . ***15. *P. hemiptera***
5b 12–15 striae/10µm, longitudinal bands of the striae narrow, striae convergent in the middle (vol. 2/1 Fig. 175: 14–18) . . . . . **28. *P. schroederi***

3. Group around *P. acoricola*

1a Finely striae forms with 13–16 striae/10µm (vol. 2/1 Fig. 183: 8–12) . . . . **18. *P. acoricola***
1b More coarsely striate forms with fewer than 13 striae/10µm **2**
2a Valve width 9µm and larger (vol. 2/1 Fig. 183: 1–3) . . **17. *P. suchlandtii***
2b Valve width 8µm and smaller (vol. 2/1 Fig. 183: 4–7) . . . **19. *P. cuneola***

4. Group around *P. divergentissima*

1a Larger linear forms with bluntly rounded ends and broad, robust striae, 7–10/10µm (vol. 2/1 Fig. 186: 6–8) . . . . . . **34. *P. superdivergentissima***
1b Smaller and larger forms with finer striae, 9 or more/10µm **2**
2a Ends not set off, broadly rounded **3**
2b Ends drawn out or capitate **5**
3a Ends flatly truncated (vol. 2/1 Fig. 185: 18, 19) . ***31. *P. subrostrata*** in part
3b Ends blunt or rounded, wedge-shaped **3**

4a Outline broadly linear-elliptical (vol. 2/1 Fig. 185: 20–23) ***32. *P. obscura***
4b Outline narrowly rhomboid-lanceolate (vol. 2/1 Fig. 185: 1, 2) . . . . ***29. *P. similis***
5a (2) 9–12 striae/10µm, divergence of the groups of striae moderately large (vol. 2/1 Fig. 185: 11–17) . . . . ***31. *P. subrostrata***
5b 12–14 striae/10µm, divergence of the groups of striae large (vol. 2/1 Fig. 185: 3–10) . . . . ***30. *P. divergentissima***

5. Group around *P. microstauron*

1a Central area with distinct markings 2
1b Central area without markings, but occasionally with puncta or other weakly visible structures 4
2a Marks in the region of the central nodule 3
2b Marks in the mantle border (vol. 2/1 Fig. 179: 3–8) . . ***10. *P. divergens***
3a Small lunate markings (vol. 2/1 Fig. 178: 8–10; 179: 1) ***7. *P. stomatophora***
3b Large lunate markings usually organized in striae (vol. 2/1 Fig. 178: 11–14) . . . . ***8. *P. brandelii***
4b (1) Small forms, usually with more than 15 striae/10 µm 5
4b Larger forms with fewer than 15 striae/10 µm 8
5a Ends widely rounded and distinctly capitate 32
5b Ends rostrate out or indistinctly capitate 6
6a Striae weakly radial in centre of valve (vol. 2/1 Fig. 185: 24, 25) . . . . **33. *P. lapponica***
6b Striae distinctly radial in centre of valve 7
7a Axial area continually widens from end to middle (vol. 2/1 Fig. 185: 26) . . . . **7. *P. kneuckeri***
7b Axial area usually narrow at the valve ends and lanceolate in the middle (vol. 2/1 Fig. 193: 19–29) . . . . ***45. *P. appendiculata***
8a (4) Both terminal fissures different, a row of puncta running apically along the border of the axial area (vol. 2/1 Fig. 188: 1–3) . **37. *P. platycephala***
8b Both terminal fissures similar 9
9a Valve sides parallel, ends sharply wedge-shaped (vol. 2/1 Fig. 188: 13) . . . . **39. *P. balatonis***
9b Valve sides not parallel, ends not as above 10
10a Valve sides triundulate 11
10b Valve sides not as above 20
11a Middle undulation of valve side broader than the others 12
11b Middle undulation of valve side of similar width or narrower than the others 14
12a Middle undulation very much broader than the other undulations (vol. 2/1 Fig. 184: 5) . . . . ***26. *P. polyonca***
12b Middle undulation a little broader than the others 13
13a Axial area moderately broad (vol. 2/1 Fig. 184: 2) . ***21. *P. legumen*** in part
13b Axial area narrow (vol. 2/1 Fig. 190: 1) . . . . . ***42. *P. interrupta*** in part
14a (11) Middle undulation as broad as the others 15
14b Middle undulation narrower than the others 18
15a Longitudinal bands on the valve face formed from the inner alveolar openings 16
15b Valves without longitudinal bands, occasionally with band-like artefacts, which are formed by protruding curved edges of the valve face 17
16a Wide longitudinal bands on the striae (vol. 2/1 Fig 179: 2) . . ***9. *P. esox***
16b Narrow longitudinal bands on the striae, marginal (vol. 2/1 Fig. 184: 7–10) . . . . **22. *P. pulchra*** in part
17a Axial area moderately broad (vol. 2/1 Fig. 184: 3) . ***21. *P. legumen*** in part

17b Axial area narrow (vol. 2/1 Fig. 190: 2, 3, 5) . . ***42. *P. interrupta*** in part
18a (14) Ends rounded, wedge-shaped (vol. 2/1 Fig. 183: 17) . . . . **20. *P. infirma*** in part
18b Ends drawn out and bluntly rounded **19**
19a Valves 6–11 µm wide (vol. 2/1 Fig. 184: 6) . . . . . **22. *P. pulchra*** in part
19b Valves 15–23 µm wide (vol. 2/1 Fig. 184: 1) . . . . ***21. *P. legumen*** in part
20a (10) Valves concave in the centre **21**
20b Valve sides parallel or convex **23**
21a Ends wedge-shaped, rounded (vol. 2/1 Fig. 183: 14–16) . . . . **20. *P. infirma*** in part
21b Ends broadly rounded, capitate or drawn out **22**
22a Area very broad, lanceolate (vol. 2/1 Fig. 187: 4) . . . . . **24. *P. braunii***
22b Area narrower (vol. 2/1 Fig. 190: 6, 10) . . . . ***42. *P. interrupta*** in part (see also vol. 2/1 Fig. 206: 3 *P. lundii*)
23a (20) Ends bluntly rounded, middle striae weakly radial (vol. 2/1 Fig. 186: 9, 10) . . . ***36. *P. rupestris*** (see also 43. *P. microstauron* var. *brebissonii*)
23b Ends weakly set off **24**
24a Ends somewhat narrowed, bluntly wedge-shaped (vol. 2/1 Fig. 186: 4, 5) . . . . ***35. *P. sudetica***
24b Ends distinctly set off, elongate or capitate **25**
25a Valve broadly linear **26**
25b Valve narrowly linear **28**
26a Longitudinal bands on the valve face distinct, central area usually rhomboid, rarely a fascia, ends only very weakly separate (vol. 2/1 Fig. 188: 9–12) . . . . **38. *P. karelica***
26b Longitudinal bands absent, central area always a fascia, ends always clearly broadly set off (exception: 43. *P. microstauron* var. *brebissonii*, vol. 2/1 Fig. 191: 7–9 with ends rounded to rounded wedge-shaped, not set off) . . **27**
27a Terminal fissure comma-shaped **33**
27b Terminal fissure question mark shaped (vol. 2/1 Fig. 191: 2–7; 192: 11–16) . . . . ***43. *P. microstauron*** in part
28a (25) Central area very broad (vol. 2/1 Fig. 187: 1–3, 5) . . **24. *P. braunii***
28b Central area narrower **29**
29a Valves not wider than 7 µm (vol. 2/1 Fig. 193: 1–18) ***44. *P. subcapitata***
29b Valves wider **30**
30a Ends broadly drawn out (vol. 2/1 Fig. 191: 1, 9) . . . . ***43. *P. microstauron*** in part
30b Ends narrowly drawn out or capitate **31**
31a Ends narrowly drawn out or with capitate ends, similar in width to the rest of the valve, broad axial area (vol. 2/1 Fig. 189: 1–9; 186: 1–3) ***41. *P. gibba***
31b Ends drawn out or capitate, a great deal narrower than the valve, axial area narrower (vol. 2/1 Fig. 190: 1–11) . . . . ***42. *P. interrupta***
32a (5) Outline elliptical, capitate ends considerably narrower than the middle of the valve (vol. 2/1 Fig. 187: 9; 206: 8, 9, 11) . . . . **25. *P. krookii***
32b Outline linear, capitate ends only slightly narrower than the valve centre (vol. 2/1 Fig. 187: 6–8, 9'; 206: 12–19) . . . . **25a. *P. ignobilis***
33a (27) Capitate ends narrower than the middle of the valve, sides parallel, concave or convex (vol. 2/1 Fig. 187: 10–16; 206: 1–3) . . . **23. *P. lundii***
33b Capitate ends almost as wide as the middle of the valve, sides with a distended bulge in the centre of the valve (vol. 2/1 Fig. 206: 4–7) . . . . **23a. *P. globiceps***

6. Group around *P. viridis*

1a Ends sharply narrowing and broadly wedge-shaped, middle of the valve not distended or barely so **2**

1b Valves almost not narrowing towards the end, usually more strongly distended in the middle, ends bluntly rounded to somewhat broadened and capitate **3**

2a Central area usually relatively small, elliptical (vol. 2/1 Fig. 194: 1–4; 195: 1–6) . . . . ***46. *P. viridis***

2b Central area a single sided fascia, 4–5 striae broad (vol. 2/1 Fig. 195: 7) . . . . **47. *P. semicruciata***

3a Outer fissure of the raphe slightly undulating or not undulating **4**

3b Outer fissure of the raphe strongly undulating (complex raphe **8**

4a Valves without longitudinal bands **5**

4b Valves with longitudinal bands **6**

5a Striae strongly convergent at the ends (vol. 2/1 Fig. 180: 1, 2) . . . . **11. *P. episcopalis***

5b Striae slightly convergent to parallel at the ends (vol. 2/1 Fig. 180: 3–5) . . . . **12. *P. cardinaliculus***

6a Striae consistently parallel to weakly radial, 7–10/10µm (vol. 2/1 Fig. 197: 1, 2) . . . . ***49. *P. macilenta***

6b Striae radial in the middle of the valve, fewer than 7/10µm **7**

7a 4–5 striae/10µm, longitudinal band broad (vol. 2/1 Fig. 198: 1) . . . . ***51. *P. dactylus***

7b 5–7 Striae/10µm, longitudinal band narrow (vol. 2/1 Fig. 196: 1–4) . . . . ***48. *P. maior***

8a 6 or more striae/10µm **9**

8b Usually fewer than 6 striae/10µm **10**

9a Valves distended in the middle and at the ends (vol. 2/1 Fig. 198: 4) . . . . ***53. *P. gentilis***

9b Valves not distended in the middle and at the ends (vol. 2/1 Fig. 197: 3, 4) . . . . **50. *P. aestuarii***

10a Valves distended in the middle and at the ends (vol. 2/1 Fig. 198: 2, 3) . . . . ***52. *P. nobilis***

10b Valves linear, middle and ends not widened **11**

11a Outer fissure of the raphe strongly undulating, central area slightly developed (vol. 2/1 Fig. 199: 1–3) . . . . ***54. *P. streptoraphe***

11b Outer fissure of the raphe somewhat less undulating, central area a broad fascia (vol. 2/1 Fig. 199 ; 4, 5) . . . . **55. *P. cardinalis***

(*4) ***Pinnularia balfouriana*** has a very different valve structure to the other *Pinnularia* species. Therefore Krammer & Lange-Bertalot (2000) have established the new genus *Hygropetra* with the typus generis ***Hygropetra balfouriana***.

(*5) The ***Pinnularia borealis*** complex contains very different taxa and therefore a number of new taxa have been separated:

1. The larger forms belong to ***P. rabenhorstii*** (Grunow) Krammer, length 50–90 µm, breadth 16–18 µm, striae 4–5/10 µm. In Southern Germany a smaller variety is more common, ***P. rabenhorstii*** var. ***franconia*** Krammer (vol. 2/1 Fig. 177: 1–4), length 45–67 µm, breadth 12–13.5 µm. In La Palma, Canary Islands a form was found with cuneate ends and a broader axial area, ***P. rabenhorstii*** var. ***cuneata*** Krammer & Lange-Bertalot, length 40–61 µm, breadth 14–15 µm. Fig. 177: 5 shows ***P. subrabenhorstii*** Krammer, recorded from Java and the Alps. It is smaller and has broadly subrostrate flat ends, length

55–65 µm, breadth 11–12 µm, striae 4–6/10 µm. The smaller *borealis* forms belong to ***P. angustiborealis*** Krammer & Lange-Bertalot. However this species is very similar to *P. subrabenhorstii* in outline. It also has broadly subrostrate, flat ends and a fascia. Length 34–45 µm, breadth 7.4–8 µm, striae 5–6/10 µm.

2. The smaller forms of ***Pinnularia borealis*** are shown in vol. 2/1 Fig. 177: 7 and 10. They are 24–42 µm long, 8–10 µm broad and have 4–6 striae /10 µm. A larger nordic variety is found in Iceland, ***P. borealis* var. *subislandica*** Krammer, length 30–47 µm, breadth 9.4–11.5 µm, striae 4–5/10 µm. Fig. 177: 6 shows ***P. borealis* var. *tenuistriata*** Krammer from Saxon. The striae are smaller than the interstriae and the valve ends subrostrate, length 27–47 µm, breadth 9.2–10.2 µm, striae 4–6/10 µm. The valves of ***P. borealis* var. *sublinearis*** Krammer (vol. 2/1 Fig. 177: 9, 10) are more linear. The investigated specimens were 19–40 µm long, 7–8 µm broad with 4–6 striae/10 µm. Forms with parallel, linear sides and rounded ends belong to ***P. borealis* var. *scalaris*** (Ehrenberg) Rabenhorst. Ehrenberg's iconotype has a length of 43 µm, a breadth of 11 µm and 4–5 striae/10 µm. Vol. 2/1 Fig. 177: 12 shows ***P. borealis* var. *lanceolata*** Hustedt from the Schwarzsee near Davos with an elliptical-lanceolate outline. Vol. 2/1 Fig. 177: 11 shows the rectangular ***P. dubitabilis*** Hustedt (length 23–40 µm, breadth 6–7 µm, striae 3–5/10 µm) · The nominate variety is only known from Java and Sumatra. A similar smaller form with very short, marginal striae, ***P. dubitabilis* var. *minor*** Krammer (length 20–25 µm, breadth 6.5–6.8 µm, striae 5/10 µm), was found in Northern Germany (vol. 2/1 Fig. 178: 7) · The larger ***P. angulosa*** Krammer also has a rectangular outline. The striae are not only present around the margin, however they are relatively short (length 42–53 µm, breadth 9.7–13.3 µm, striae 3–4/10 µm).

(*6) ***Pinnularia schimanskii*** Krammer is similar to ***P. intermedia*** (Lagerstedt) Cleve, but smaller. The valve outlines are linear-elliptical with parallel to slightly concave sides, ends broadly cuneiform, rounded; the fascia is not expanded towards the valve margin; length 26–32 µm, breadth 5–5.5 µm, striae 9–10/10 µm.

(*7) The description of ***Pinnularia stomatophora*** (Grunow) Cleve on vol. 2/1 p. 406 belongs to a complex of a number of taxa. In the nominate variety crescent-shaped markings are present, forming more or less compact structures visible with the LM, divided only by striae-like lines. The ends are obtusely rounded, length 55–115 µm, breadth 9.5–12.5 µm, striae 11–14/10 µm.(vol. 2/1 Fig. 178: 8–10).

P. stomatophora* var. *erlangensis (A. Mayer) Krammer is a larger form with broadly rounded ends and very rugged crescent-like markings in the central area. Length 90–115 µm, breath 13–14 µm, striae more robust, 9–10/10 µm.

P. graciloides Hustedt has capitate ends. The nominate variety is a tropical form. ***P. graciloides* var. *triundulata*** (Fontell) Krammer (vol. 2/1 Fig. 179: 1) lives in the temperate zone · Its sides are slightly undulate, the undulations equal in width or the middle undulation a little smaller; undulations in small valves nearly absent; fascia a relatively broad widening of the large rhombic central area. Length 82–105 µm, breadth 11–13 µm, 10–12 striae/10 µm.

(*8) Only vol. 2/1 Fig. 178: 13, belongs to ***P. brandelii*** Cleve. Fig. 178: 11, 12 in vol. 2/1 are specimens of ***P. brandeliformis*** Krammer. Its valve outline is rhombic-lanceolate, often weakly triundulate with the central swelling more pronounced. Poles indistinctly offset, capitate and broadly rounded. Very robust, lunate patterning is restricted to a band that reaches far into the axial area. Length 50–78 µm, breath 10–13.6 µm, striae more robust, 9–11/10 µm.

(*9) ***Pinnularia esox*** Ehrenberg is a broadly lanceolate *Caloneis*, therefore Patrick in Patrick & Reimer (1966) created the name ***Pinnularia clevei*** Patrick

P. schoenfelderi Krammer differs from the two above mentioned taxa by the very narrow axial area. Valves linear-elliptical to linear-lanceolate, continuously narrowed to the ends, ends obtusely rounded, length 19–37 µm, breadth 5–7 µm, striae 13–16/10 µm. The narrow axial area becomes somewhat lanceolate towards the middle; central area forming a moderately broad, often asymmetric fascia. Striae radiate in the middle becoming convergent at the ends.
(*16) ***Pinnularia brevicostata*** is, in the description on page 410 (vol. 2/1), a complex taxon and a revised concept is necessary. All members of this complex show a similarity to the specimens in Fig. 182: 4–9 (vol. 2/1). The outline is linear with nearly parallel sides, the ends are broadly rounded and the axial area is relatively broad. ***P. brevicostata*** Cleve is the largest taxon of the complex, length 80–150 µm, breadth 19–22 µm, striae 6–7/10 µm. The axial area and lateral raphe are both very broad.
P. cruxarea Krammer (vol. 2/1 Fig. 182: 5) also has a very wide axial area. Sides parallel to slightly undulate in larger forms or tapering smoothly to weakly protracted, broadly rounded ends; length 67–100 µm, breadth 12–14 µm, striae 8–10/10 µm. Lateral raphe very broad, two longitudinal lines may be visible, outer fissure slightly undulate, curved, proximally bent reverse laterally, central pores small, round to drop-shaped, terminal fissures large question mark-shaped. Axial area 1/2–3/4 the breadth of the valve; central area elliptical with its greatest length in the apical axis, very large and nearly dilated towards the margins of the valve face, not distinctly differentiated from the axial area. Frequently in larger specimens a small, very asymmetric fascia is present., slightly radiate in the middle.
P. crucifera Cleve-Euler (vol. 2/1 Fig. 182: 4, 6–9) has a smaller axial area; length 80–140 µm, breadth 12–14.5 µm, striae 8–10/10 µm. Lateral raphe very broad, three longitudinal lines may be visible, outer fissure slightly, undulate curved, proximally bent reverse laterally, central pores small, round, terminal fissures large, sickle-shaped. Axial area 1/4–1/3 the breadth of the valve, linear, tapering at the ends, central area a small fascia, distinctly differentiated from the axial area. Striae parallel in the middle, parallel to very slightly convergent towards the ends, a longitudinal line is visible.
P. spitsbergensis Cleve is a similar taxon with crescent-shaped markings in the central area; length 50–105 µm, breadth 8–11 µm, striae 14–15/10 µm. Its lateral raphe is broad, three longitudinal lines may be visible, outer fissure slightly undulate, curved, proximally bent reverse laterally, central pores small, round, terminal fissures large, bayonet-shaped. Axial area 1/3 the breadth of the valve, central area rhomboidal with a marginal fascia.
P. ivaloensis Krammer is much smaller; length 40–78 µm, breadth 6.5–7 µm, striae 9–10/10 µm. The axial area is 1/4–1/3 the breadth of the valve, broadening from the ends to the centre of the valve, central area a small fascia, distinctly differentiated from the axial area. Striae slightly radiate in the middle, moderately convergent towards the ends, a small marginal longitudinal band is clearly visible.
P. isostauron (Grunow) Cleve is similar, however it has a semicomplex raphe and very large central pores; length 30–70 µm, breadth 8–10 µm, striae 9–11/10 µm. Axial area 1/3 the breadth of the valve, linear, central area a small fascia, distinctly differentiated from the axial area.
Fig. 182: 8 (vol. 2/1) shows a specimen of ***P. conifera*** (Brun & Héribaud) Krammer with cuneate protracted ends, length 50–55 µm, breadth 7–8 µm, striae 11–12/10 µm. Axial area 1/4 the breadth of the valve, tapering at the ends, central area a small symmetric or asymmetric fascia.
(*21) ***P. legumen*** (Ehrenberg) Ehrenberg is a fossil taxon. Neither of Ehren-

berg's three iconotypes from diverse locations in America agrees with *P. legumen* sensu lato in the diatom literature. The oldest valid described taxon which agrees with our "*P-legumen*"-forms is ***P. undula*** Schumann (vol. 2/1 Fig. 184: 3). The valves are linear, sides parallel to more or less triundulate, the central undulation narrower or as broad as the others, ends subcapitate to capitate, in small species truncate, much narrower than the valve body; length 54–135 µm, breadth 14–22 µm, striae 8–10/10 µm. Axial area 1/5–1/3 the valve breadth, linear or slightly widening from the poles towards the middle; central area large, round, often almost reaching the valve margin and distinct from the axial area, often irregularly ornamented. ***P. undula* var. *major*** (A. Schmidt) Krammer (vol. 2/1 Fig. 184: 1) from many fossil samples from the USA has triundulate valves, the central undulation is narrower than the other two, the ends capitate, length 100–135 µm, breadth 20–22 µm.

P. legumiformis Krammer is a similar taxon with straighter sides. Its ends are broadly rounded, always offset and broadly protracted or capitate; length 58–110 µm, breadth 16–18 µm, striae 8–10/10 µm. Axial area 1/4 to 1/3 the valve breadth, central area very large, rhombic to round, sometimes reaching the margin.

(*26) ***Pinnularia polyonca*** (Brébisson) W. Smith has triundulate sides. ***P. polyonca*** var. **similis** Krammer is more common in Middle Europe. Sides not undulate, nearly parallel or weakly convex, 8–9 striae/10 µm.

P. mayeri Krammer is similar, however the outline is not triundulate but linear with nearly parallel to slightly convex sides, the ends are broadly capitate, length 40–60 µm, breadth 6.5–8 µm, striae 9–10/10 µm. Axial area lanceolate, tapering from the end to the large rhombic fascia, striae radiate in the middle, convergent towards the ends.

The nordic taxon ***P. lange-bertalotii*** Krammer is biundulate with concave sides. Valves linear, biundulate with concave sides, ends broadly capitate, rounded; length 43–72 µm, breadth 6.5–8.7 µm, striae 10–13/10 µm. Axial area linear, 1/5–1/4 of the valve width, central area rhombic, broadening into a fascia almost as wide as the valve. Mottling originating from solitary puncta in the central area.

(*29) ***Pinnularia similis*** Hustedt is a tropical form from Java, 50–75 µm long, 7.5–9 µm broad and with 10–12 striae/10 µm and so far not found in the temperate zone.

P. similiformis Krammer is a similar taxon from the Northern Europe (vol. 2/1 Fig. 185: 1, 2) · It has a striking resemblance to *P. similis* in valve outline, the broad cell girdle and the arrangement of the striae, however the dimensions are much smaller. Length 32–60 µm, breadth 5.3–6.5µm.

P. carteri Krammer is a smaller form from Scotland. The valve outline is linear rhombic-elliptical, slightly tumid at mid-valve, ends capitate, rounded; length 36–40 µm, breadth 4.8–5.2 µm, striae 10–11/10 µm. Axial area narrow, linear to slightly lanceolate; central area a broad fascia, widening to the margins. Striae strongly radiate in the middle becoming strongly convergent towards the ends.

P. pseudosimilis Krammer was found in Lake Inari, Finnish Lapland. Valves rhombic-lanceolate to rhombic-elliptical, ends not or very slightly offset and obtusely rounded; length 38–60 µm, breadth 8.6–9.8 µm, striae 14–16/10 µm. Raphe filiform to weakly lateral, outer fissure almost straight; central pores small, tear-shaped, close together; terminal fissures question mark-shaped. Axial area 1/4–1/3 the valve width, widening continuously from the ends to the valve middle; central area usually united with the axial area, elongated-rhombic, widened to a broad fascia. Striae radiate in the middle, becoming weakly convergent towards the ends; longitudinal bands absent.

(*30) ***Pinnularia divergentissima*** (Grunow) Cleve includes a number of varieties, all with striae extremely divergent with a sudden change in direction halfway to the ends, strongly radiate towards the valve centre, strongly convergent towards the ends, with an acute angle being formed between the two striae groups where they meet. The valves of the nominate variety (vol. 2/1 Fig. 185: 5, 7) are linear to linear-lanceolate, sides weakly convex, ends obtusely rounded to broadly subrostrate, length 25–46 µm, breadth 5.8–6.3 µm, striae 12–14/10 µm. ***P. divergentissima*** **var.** ***subrostrata*** Cleve-Euler (vol. 2/1 Fig. 185: 15–17) is similar in size to the nominate variety, the ends in moderate and larger sized specimens are broadly capitate. Length 23–40 µm, breadth 5.7–6.5 µm, 12–15 striae/10 µm. ***P. divergentissima*** **var.** ***triundulata*** Krammer (vol. 2/1 Fig. 185: 3, 6) has triundulate sides, the undulation at mid valve is more tumid than the others, ends broadly capitate. ***P. divergentissima*** **var.** ***minor*** Krammer (vol. 2/1 Fig. 185: 8, 9) is similar to *P. divergentissima* var. *subrostrata*, but smaller. Length 8–28 µm, breadth 4.4–5.4 µm, 14–16 striae/10 µm.

(*31) ***P. divergentissima*** **var.** ***subrostrata*** Cleve-Euler see under *30.

Forms like Fig. 185: 18 (vol. 2/1) belong to the nordic taxon ***P. krammeri*** Metzeltin. Its valve outline is linear to linear-elliptical, the sides in small individuals slightly concave in the middle; ends broadly wedge-shaped, not suddenly offset, pole broadly rounded; length of the individuals found so far, 26–45 µm, breadth 6.7–7.8 µm, striae 11–13 / 10 µm. Axial area very narrow, linear, lanceolate in larger individuals; central area variable in shape and size, in small individuals it is a fascia reaching to the valve margin, in larger specimens, it is rhombic with a fascia of similar width, striae strongly radiate in the middle, strongly convergent at the ends.

The valves of ***P. marchica*** Ilka Schönfelder are subcapitate, similar to Fig. 185: 14 (vol. 2/1) but less capitate; the ends are relatively long, rostrate or subcapitate; length 22–37 µm, breadth 4.7–6.3 µm, striae 11–14/10 µm. Axial area narrow, central area a broad rhombic fascia. Striae strongly radiate in the middle, becoming strongly convergent towards the ends, 11–14/10 µm.

(*32) ***P. obscura*** Krasske (vol. 2/1 Fig. 185: 20–23) has rounded or weakly subrostrate ends.

The similar species ***P. obscuriformis*** Krammer has valves with straight, parallel sides, ends cuneate and acutely narrowly rounded; length 15–34 µm, breadth 5.2–6.7 µm, striae 10–12/10 µm. It is common in the waters of the lower mountain region in Europe.

Pinnularia acidophila Hofmann & Krammer is similar in outline, but smaller. In girdle view the frustules are very broad due to the broad girdle bands. Valves linear-lanceolate with straight to weakly convex sides. Ends cuneiform and not offset, length 12–22 µm, breadth 3–3.3 µm, striae 13–16/10 µm. Axial area narrow, fascia broad.

(*35) ***P. persudetica*** Krammer is similar to ***P. sudetica*** (Hilse) M. Peragallo, but broader with linear elliptical-lanceolate outline. Sides slightly convex, narrowing toward the cuneate rounded ends; length 63–97 µm, breadth 12–15 µm, striae 8–9/10 µm. Lateral raphe broad in larger specimens, outer fissure almost curved to undulate, central pores large drop-shaped, not close to each other, bent laterally; terminal fissures question mark-like. Axial area 1/4 to more than 1/3 the breadth of the valve, linear to lanceolate, widening from the end to the centre; central area large, roundish-rhombic, 1/2–2/3 the breadth of the valve, differentiated from the axial area, sometimes a unilateral fascia is present. A distinct broad longitudinal band is present on the distal side of the striae.

(*36) Additional investigations on the type material of ***P. rupestris*** Hantzsch shows the following morphometric data: Length 40–90 µm, breadth 9–12.4,

striae 12–13/17 µm, a longitudinal band on the stria is absent. The following taxa are similar to the former species:
P. frequentis Krammer has more cuneate ends, a small rhombic central area and smaller central pores; length 30–63 µm, breadth 8.7–11 µm, striae 9–11/10 µm. Axial area narrow, up to 1/4 the breadth of the valve, widened from the ends to the centre, central area round to rhombic-roundish, 1/3–1/2 the breadth of the valve. A small, always distinct, longitudinal band is present.
Unlike *P. rupestris*, ***P. subrupestris*** Krammer has a lanceolate outline, a broader axial area and a small marginal longitudinal band on the striae. Length 35–77 µm, breadth 8.7–12 µm, 9–13 striae/10 µm.
The common ***P. subcommutata*** Krammer has a smaller length-to-breadth ratio and always has a small semifascia. Its outline is linear-elliptical to linear-lanceolate, sides slightly convex, narrowing towards the broadly rounded ends; length 32–83 µm, breadth 10–13.4, µm striae 9–12/10 µm. Lateral raphe narrow, outer fissure slightly curved, proximally bent laterally, central pores small, elongated, moderately closely spaced and bent laterally; terminal fissures large. Axial area narrow, up to 1/5 the breadth of the valve, linear, tapering, lanceolate at the ends, central area roundish-rhombic or orbicular, 1/3–1/2 the breadth of the valve, differentiated from the axial area, in the nominate variety 1–2 striae on one side are absent or strongly shortened, forming a semifascia, slightly radiate in the middle, parallel to slightly convergent at the ends, broad submarginal longitudinal bands are present.
(*40) The description and Fig. 188: 4–7 (8?) (vol. 2/1) are of ***Pinnularia kriegeriana*** Krasske. *P. schwabei* from Chile is another taxon and therefore the two are not synonymous. Both have a structure very different to that of *Pinnularia* and therefore Krammer (2000) has established the genus ***Pulchella*** for these taxa.
(*41) ***Pinnularia gibba*** Ehrenberg (only Fig. 189: 1, vol. 2/1) has a rhombic-lanceolate outline, gradually or slightly curved, tapered to the broadly rounded, capitate ends; length 60–110 µm, breadth 10–13.5 µm, striae 8–11/10 µm. Axial area 1/2–2/3 the valve breadth, continuously widened from the ends to the central area, which is not clearly differentiated; axial and central area together form a large rhombic area with a small symmetric or asymmetric fascia in the middle. The proximal part of the area is accompanied by large markings, with a different structure on each side, larger on the ventral side (a character of nearly all taxa in the *gibba*-complex). *P. macilenta* and. *P. subgibba* are similar.
In the past the concept of ***P. macilenta*** Ehrenberg sensu Cleve (1895) was very different from Ehrenberg's type (Reichardt 1995) · This is also true for the description and figures on p. 429 (vol. 2/1) of this flora. Larger specimens of *P. macilenta* have a linear outline, the sides are nearly parallel, weakly convex or, in large specimens, slightly swollen in the median or in both the median portion and at the ends. Ends broadly rounded or slightly capitate and broadly rounded; length 60–134 µm, breadth 12–15.5 µm, striae 8–10/10 µm. Axial area 1/4–1/2 valve breadth, linear, central area irregularly rhombic, 2/3 to 3/4 the valve breadth, a small asymmetric transverse fascia always is present. Striae radiate in the middle, convergent at the ends; longitudinal bands are absent. Similar to *P. gibba* the central areae are accompanied by four indistinct, large markings. Fig. 197: 1, 2 (vol. 2/1) shows *P. socialis* var. *debesii* Hustedt (see under *49) and not *P. macilenta*.
The smaller and more common ***P. subgibba*** Krammer (vol. 2/1 Fig. 189: 2, 7, 8) has an identical outline to *P. macilenta*. Its sides are very slightly convex, ends subcapitate to broadly subrostrate, length 60–100 µm, breadth 10–12 µm, 8–9 striae/10 µm. Post initial cells are swollen in the middle and at the ends. The ends are more cuneately rounded and the axial area is smaller.

Pinnularia mesogongyla sensu Cleve (*P. gibba* var. *mesogongyla* (Ehrenberg) Hustedt, (see. vol. 2/1 p. 424, only Fig. 186: 1 not, 2, 3,) has absolutely no similarity with the protologue of *P. mesogongyla* Ehrenberg. Therefore a new name was created, ***Pinnularia erratica*** Krammer: outline rhombic-lanceolate to elliptical, sides weakly convex, ends broadly capitate, swollen; length 50–90 µm, breadth 9–12 µm, striae 9–11/10 µm. Raphe lateral, the outer fissure weakly curved and proximally bent to the ventral side, central pores small with lateral appendices in the LM, terminal fissures very large, bayonet-shaped. Axial area 1/5–1/3 the valve breadth, linear, central area large and clearly differentiated from the axial area, without a fascia, sometimes one submarginal stria is absent on one or two sides. Striae strongly radiate in the middle, strongly convergent towards the ends, longitudinal lines absent. Only two large or some irregularly arranged markings on each side of the central nodule.
There is much confusion in the literature connected with the epithet "*acrosphaeria*". The names *Frustulia acrosphaeria* sensu Brébisson; *Navicula acrosphaeria* sensu Kützing and *Pinnularia acrosphaeria* sensu Rabenhorst do not belong to *P. acrosphaeria* W. Smith. They are synonyms of ***P. gibbiformis*** Krammer, best distinguished by its outline, length-to-breadth ratio and areae. Its outline is narrowly linear-lanceolate to rhombic-lanceolate, tapered continuously from the sometimes swollen middle to the capitate and broadly rounded ends; length 34–84 µm, breadth 6.7–8.5 µm, striae 10–13/10 µm, length-to-breadth ratio in medium size valves about 10–12. Lateral raphe with weakly curved outer fissure, central pores very small, round, closely spaced, terminal fissures question mark-shaped. Axial area narrow, linear, sometimes broadened, lanceolate in the middle; the elongated-elliptical round central area may be almost clearly differentiated, 1/2, in large specimen up to 2/3, the breadth of the valve. Striae radiate in the middle, convergent towards the ends, longitudinal lines absent.
(*42) Most forms designated as ***Pinnularia interrupta*** W. Smith belong to ***P. biceps*** Gregory (vol. 2/1 Fig. 190: 2–4). These are linear forms, the capitate ends relatively small, sides straight to weakly undulate, if undulate then the central undulation a little narrower than the other two; length 48–85 µm, breadth 11–13 µm, striae 9–13/10 µm, radiate in the middle, convergent towards the ends, longitudinal lines are absent. ***P. biceps* var. *gibberula*** (Hustedt) Krammer (not Fig. 190: 1–11) has slightly triundulate sides, the middle undulation hardly or not broader than the other two, or sides weakly convex.
Fig. 190: 1 (vol. 2/1) shows another taxon, ***P. mareifelana*** Krammer, which was in the past confused with *P. biceps* var. *gibberula*. Outline rhombic-lanceolate with a central inflation, ends capitate, clearly differentiated from the valve body by shoulders, length 58–72 µm, breadth 12–14.7 µm, striae 10–12/ 10 µm. Axial area narrow, linear, clearly differentiated from the rhombic central area, in the valve middle one or more submarginal striae on one or both sides are absent or irregularly arranged; central area 1/2–2/3 the valve breadth. Striae radiate in the middle, convergent towards the ends, longitudinal lines are absent.
P. latarea Krammer has a relatively wide axial area. It has an outline similar to *P. biceps* however with parallel to slightly concave sides; length 35–64 µm, breadth 8–10 µm, striae 9–11/10 µm. Axial and central area together form a wide, lanceolate space with a broad fascia. In hot-house waters of some Botanical gardens (Marburg, Munich) there is a variety with distinctly concave sides and very wide areae and fasciae: ***P. latarea* var. *thermophila*** Krammer.
Only ***P. anglica*** Krammer has protracted subcapitate to rostrate narrower ends, parallel sides, small axial areae and large rhomboid central areae; length 30–70 µm, breadth 10–13 µm, striae 9–11/10 µm.
P. falaiseana Krammer has a large broad outline. Its outline is broadly linear-

elliptical, the ends are narrowly capitate, less than half the valve width, length 61–72 µm, breadth 14.7–15 µm, striae 11–13/10 µm, average length/breadth ratio of the valve body without ends is 3. Raphe weakly lateral; central pores bent laterally, terminal fissures indistinct. Axial area narrow, central area rhombic to rounded, usually about 1/2 the valve width, frequently a little asymmetrical and with irregularly arranged submarginal striae.
P. subfalaiseana Krammer is very similar to *Pinnularia falaiseana* but smaller; length 45–55 µm, breadth 10.7–11.5 µm, striae 11–12/10 µm.
P. pisciculus Ehrenberg has a similar outline, but it is smaller; length 22–50 µm, breadth 6.0–8.3 µm, striae 10.5–12/10. Axial area narrow or lanceolate, central area relatively large, rhombic, widened into a fascia, striae radiate in the middle, convergent at the ends, often somewhat curved; longitudinal lines absent.
P. ferrophila Krammer lives in springs with a high iron content. Its outline is linear, sides in larger specimens slightly triundulate, in smaller biundulate with concave sides; ends in larger specimens broadly capitate, in smaller truncate and flatly rounded, length 30–62 µm, breadth 8.8–10 µm, striae 9–10/10 µm. Axial area moderately broad, 1/4–1/5 the breadth of the valve, widening from the end to the very large, rhomboid central area, in larger specimens widened to a small or moderately broad fascia. Striae very robust, radiate in the middle, convergent at the ends, longitudinal lines absent.
The common and widespread ***P. grunowii*** Krammer is a smaller taxon than *P. biceps.* In the literature it is erroneous named *P. mesolepta* (Ehrenberg) W. Smith. However the protologue of Ehrenberg does not agree with the latter concept. *P. grunowii* (vol. 2/1 Fig. 190: 5, 6) has valves with margins moderately triundulate, the central undulation of similar breadth or narrower than the other two, ends capitate, somewhat attenuated but broad and obtusely rounded, sharply offset from the valve body; length 27–55 µm, breadth 6.5–9 µm, striae 11–14 /10 µm. Axial area narrow, linear, clearly differentiated from the rhombic central area, combined with a small or moderately broad fascia. Striae radiate in the middle becoming strongly convergent towards the ends.
P. turbulenta (Cleve-Euler) Krammer is similar in size to *P. grunowii.* Its valves are strongly triundulate, the central undulation is similar in breadth to the other two, ends capitate to subcapitate, sharply offset from the valve body; length 30–40 µm, breadth 6–7 µm, striae 11–12 /10 µm. Axial area narrow, linear, central area a fascia, occupying the space formed by the middle undulation. Striae strongly radiate in the middle, nearly parallel in the distal undulations, becoming strongly convergent towards the ends.
P. schroeterae Krammer is similar in outline to *P. biceps*, but much smaller. Its valves are linear to elliptic-lanceolate with straight to weakly convex margins, in large specimens weakly triundulate and, if so, then the middle undulation slightly broader than the other two, ends capitate with broad shoulders, smaller specimens only truncate; length 13–28 µm, breadth 4.3–5.4 µm, striae 13–16/10 µm. Axial area very narrow, central area in larger specimens a large, rhombic fascia, in smaller specimens a small linear fascia, sometimes asymmetric. Striae radiate to strongly radiate in the middle, becoming convergent to strongly convergent towards the ends, in larger specimens a sudden transition between the radial and the convergent striae. *P. schroeterae* lives in sphagnetum in moors. ***P. streckerae*** Krammer, which is very similar, is found in waters with very high electrolyte content. A third small form, ***P. subinterrupta*** Krammer & Schroeter, has parallel sides and a smaller roundish central area.
There are many additional taxa in the *P. biceps* complex, see Krammer (2000).
(*43) ***Pinnularia microstauron*** (Ehrenberg) Cleve and ***Pinnularia brebissonii*** (Kützing) Rabenhorst are both very comprehensive complexes and therefore only some taxa are discussed. The main differences between the two complexes

are the valve outline (in *microstauron* more linear, in *brebissonii* more lanceolate) and the form of the ends (in *microstauron* always broadly offset and protracted, in *brebissonii* wedge-shaped and broadly rounded or slightly cuneate).

1. ***Pinnularia brebissonii*** (Kützing) Rabenhorst (vol. 2/1 Fig. 191: 7–9) has linear-lanceolate to linear-elliptical valves with broadly rounded to wedge-shaped ends, length 28–60 µm, breadth 9–12 µm, 10–12 striae/10 µm. The central area is a more or less broad fascia in most specimens in a population, rarely with short striae at the fascia edge. A number of varieties differ primarily in the form of the valve ends. ***P. brebissonii* var. *acuta*** Cleve-Euler has acutely rounded ends. In ***P. brebissonii* var. *bicuneata*** Grunow sides are almost straight and parallel, ends distinctly obtuse or acutely wedge-shaped. In ***P. brebissonii* var. *lancarea*** Krammer the outline is similar to that of the nominate variety, however the axial area is broadly lanceolate. ***P. brebissonii* var. *minuta*** Krammer has the outline of the nominate variety, however it is smaller (length 20–36 µm, breadth 6.3–7.5 µm, striae usually 11–13/10 µm. Central area mostly a variously sized rhombus, a fascia present in initial and post-initial cells).

Much larger than the above varieties of the *brebissonii*-group is ***P. brebissonii-formis*** Krammer from the Soosmoor in Bohemia. The outline is similar those of *P. brebissonii*; length 65–100 µm, breadth 15.2–18 µm, striae 9–10/10 µm, strongly radiate in the middle, weakly convergent at the ends. Axial area moderately broad, lanceolate, broadening from the ends to the rhombic-roundish central area, with a fairly broad transverse fascia on one or both sides.

2. ***Pinnularia microstauron*** (Ehrenberg) Cleve (vol. 2/1 Fig. 192: 1–5) has weakly convex sides, ends in larger specimens broadly rostrate and wedge-shaped, hardly differentiated in small specimens. Fascia narrow to moderately broad, usually asymmetric, often absent in many individuals in a population. Length 30–78 (100) µm, breadth 10–12.4 µm, striae 9–11 (15)/10 µm. ***P. microstauron* var. *angusta*** Krammer has a similar outline as the nominate variety, the fascia is broader and the size smaller, length 25–47 µm, breadth 6.5–8.0 µm, 10–12 striae/10 µm. ***P. microstauron* var. *rostrata*** Krammer has narrower rostrate ends and is smaller than the nominate variety, length 30–50 µm, breadth 6–7 µm, 10–11 striae/10 µm. In ***P. microstauron* var. *nonfasciata*** Krammer (vol. 2/1 Fig. 192: 11–13) a fascia is absent, length 20–60 µm, breadth 8–11 µm, striae 11–13/10 µm.

P. rhombarea Krammer (vol. 2/1 Fig. 191: 1–5) is also a species complex. The broadly hatchet-like rostrate and flat rounded ends distinguish this complex from the *P. microstauron* complex. Outline linear, sides parallel to weakly convex, concave or weakly undulate, length 40–100 µm, breadth 10.5–16 µm, striae 9–11/10 µm. Axial area narrow, linear with acute ends; central area a large rhombus, widened to an almost asymmetrical fascia. This taxon has a number of varieties. An important variety from brackish water is ***P. rhombarea* var. *halophila*** Krammer with a more compact outline, sides weakly convex, ends (only in larger specimens) weakly protracted, length 40–67 µm, breadth 12–15 µm, 10–11 striae/10 µm. ***P. rhombarea* var. *undulata*** Krammer has triundulate sides. The ends are barely protracted and broadly rounded. The axial area wider as in the nominate variety, nearly 1/3 the breadth of the valve, central area large, rhombic, nearly reaching the valve margin. Length 58–87 µm, breadth 11.5–13.4, µm, 8–9 striae/10 µm. ***P. rhombarea* var. *brevicapitata*** Krammer contains stocky forms with broad, shorts capitate ends. Length 40–55 µm, breadth 10.5–12.5, 10–11 striae/10 µm.

In comparison to the large forms of the *microstauron* complex there is the very small species, ***P. submicrostauron*** Schroeter. This oblong taxon has until now been lumped with *P. subcapitata* (*P. hilseana*) · Its sides are parallel, weakly concave, convex or undulate; ends bluntly wedge-shaped, weakly to moderately rostrate and broadly rounded; length 27–57 µm, breadth 5.6–7.6 (8) µm, striae 12–14/10 µm. Axial area very narrow, only moderately narrow in larger individuals, linear or a little widened towards the middle; central area distinct, large, round, often reaching the valve margin. Striae radiate to strongly radiate in the middle, convergent to strongly convergent at the ends; longitudinal bands not visible.

The valves of ***P. renata*** Krammer have convex sides. The outline is broadly elliptical-lanceolate, ends not or only slightly protracted and flatly rounded. Length 26–52 µm, breadth 7.3–11 µm, striae 10–12/10 µm Axial area very narrow, central area large, rhombic, extending nearly to the margins of the valve face. Striae strongly radiate in middle, slightly convergent at the ends.

(*44) ***Pinnularia subcapitata*** Gregory, syn. *P. hilseana* Janisch (only vol. 2/1 Fig. 193: 14, 16, 17) is a capitate to subcapitate form. Length-to-breadth ratio in average individuals less than 6; axial area narrow; length 20–43 µm, breadth 4–6 µm; striae 11–13/10 µm. A larger capitate form is ***P. subcapitata* var. *elongata*** Krammer (vol. 2/1 Fig. 193: 15–18), length 27–57 µm, breadth 5.4–6.6 µm, striae 10–11/10 µm. ***Pinnularia subcapitata* var. *subrostrata*** Krammer (vol. 2/1 Fig. 193: 1–3) has broadly subrostrate ends, length 29–57 µm, breadth 4.7–6.7 µm; striae 10–13/10 µm, usually c. 12/10 µm; length-to-breadth ratio often over 8 in average individuals; central area a narrow to moderately broad, often asymmetric fascia.

P.sinistra Krammer (vol. 2/1 Fig. 193: 4–11) was in the past usually confused with the very different *P. subcapitata. P. sinistra* has an outline with slightly convex, more rarely straight or weakly concave sides, ends distinctly differentiated and broadly protracted, almost as wide as the valve; length 17–52 µm, breadth 4–6.5 µm, striae 11–13(14)/10 µm. Axial area linear, in large individuals lanceolate; central area an often slightly asymmetric fascia. Striae parallel to radiate in the middle, weakly to moderately convergent at the ends.

(*45) Fig. 193: 19–29 (vol. 2/1) does not show ***Pinnularia appendiculata*** (Agardh) Cleve (see below). P. *appendiculata* sensu typo is a more robust form like Fig. 193: 12. The smallest forms are almost elongate-elliptical, the ends broadly protracted; length 16–46 µm, breadth 4.7–6.1 µm, striae 16–18/10µm. Raphe filiform; central pores close to each other. Axial area moderately broad, widening from the end to the centre, usually continuous with the rhombic central area, widening into a very variable broad fascia. Striae radiate in the middle, convergent at the ends. ***P. appendiculata* var. *amaniana*** Krammer (vol. 2/1 Fig. 193: 19, 21) is a more rhomboid variety with a broad fascia.

Pinnularia kuetzingii Krammer is more linear with slightly convex sides, a narrow axial area and a smaller fascia (vol. 2/1 Fig. 193: 20) · Length 18–46 µm, breadth 3.8–6.7 µm, striae 18–20/10 µm.

P. silvatica Petersen is also similar. Outline elliptical to rhombic-lanceolate, not capitate or subcapitate, length 16–25 µm, breadth 3–4.2 µm, striae (19)20–24(25)/10 µm. Axial area very narrow, usually broadening towards the middle, ending in a large rhombic central area, sometimes forming a fascia nearly reaching the margin. Striae radiate in the middle, convergent at the ends.

P. irrorata (Grunow) Hustedt has a similar outline to *P. silvatica*. However it has a small rhombic central area and always has a fascia, length 18.9–24 µm, breadth 3.7–4.2 µm, striae 22–24/10 µm.

P. perirrorata Krammer (vol. 2/1 Fig. 193: 22–28) is most common in moorland waters, where it can be widespread and is often in huge numbers. Outline rhombic-lanceolate, ends rounded, in large specimens weakly offset or a little

subcapitate; length 17–30 µm, the majority in the range 20–25 µm, breadth 4.2–4.4 µm, striae 16–18/10 µm. Axial area relatively broad, lanceolate, central area large, rhombic, widening into a broad fascia reaching the valve margin. Striae, radiate in the middle, convergent at the ends.
(*46) *P. viridis* Nitzsch sensu *Bacillaria viridis* is an unknown species, the description of Nitzsch (1817) is meaningless, taxonomically *Bacillaria viridis* a nomen nudum (not in the sense of the ICBN) · Nitzsch described a diatom with green chloroplasts. Type material is absent. However *P. viridis* is the conserved typus generis and therefore Krammer (2000) has produced a neotype. This concept (similar to Fig. 195: 1, vol. 2/1) has a linear outline, sides parallel or slightly convex, narrowing toward the rounded ends, length 90–182 µm, breadth 21–30 µm, striae about 6–7/10 µm. Raphe broadly lateral to semicomplex, in which case the outer fissures are undulate, three longitudinal lines are visible, proximally bent laterally, central pores small, round. Axial area 1/5–1/4 the breadth of the valve, linear, tapering lanceolate at the ends; central area variable, roundish or irregular, a little wider than the axial area, almost asymmetrical. Striae moderately radiate in the middle, slightly convergent towards the ends. Distinct narrow to moderately wide longitudinal bands are present.
However the most common form of the *P. viridis* sensu lato complex is ***P. viridiformis*** Krammer (e.g. vol. 2/1 Fig. 194: 1, 2) · The characters of this infraspecific complex: outline linear, sides parallel, slightly convex or triundulate, ends rounded, length 67–145 µm, breadth 14–21 µm, striae about 7–9/10 µm. Raphe broadly lateral, occasionally slightly semicomplex, proximally bent laterally, central pores small, round, close to each other. Axial area 1/5–1/4 the breadth of the valve, linear, tapering on the ends; central area roundish, a little wider than the axial area, almost asymmetrical. Striae slightly to moderately radiate in the middle, parallel to slightly convergent towards the ends. Distinct narrow to moderate wide longitudinal bands are present, sometimes their marginal limits are indistinct.
P. rhomboelliptica Krammer (e.g. vol. 2/1 Fig. 194: 4) is a complex with a large number of varieties. Infraspecific complex has the following characters: outline rhombic-elliptic, sides regularly narrowing toward the rounded ends, length 70–160 µm, breadth 17–24 µm, striae 7–8/10 µm. Raphe broadly lateral with undulate outer fissure, in some populations slightly semicomplex, proximally strongly bent laterally, central pores round, distinct, terminal fissures *viridis*-like. Axial area linear, tapering on the ends, 1/4–1/3 the breadth of the valve, central area rounded, large, 1/2 the breadth of the valve, or smaller, asymmetrically rounded. Striae radiate in the middle, convergent towards the ends, crossed by a narrow to moderately broad longitudinal band. ***P. rhomboelliptica* var. *inflata*** Krammer (vol. 2/1 Fig. 196: 4) is distinctly inflated in the middle.
P. rathsbergiana Krammer has parallel sides, cuneate ends and a semicomplex raphe; length 83–120 µm, breadth 18–21 µm, striae about 6–8/10 µm. Axial area 1/4–1/3 the breadth of the valve, linear, tapering on the ends; central area large, roundish or rhombic, 1/2–2/3 the breadth of the valve, almost asymmetrical.
P. bleischii and ***P. rhenohassiaca*** are large taxa with linear outline and broad axial areae. The valves of ***P.bleischii*** Krammer have triundulate sides and cuneate ends, length 93–158 µm, breadth 20–25 µm, striae about 7–7.5/10 µm. Raphe broadly semicomplex. Axial area 1/3 the breadth of the valve, linear, tapering on the ends; central area hardly wider than the axial area, almost asymmetrical. Both areae have an irregular surface structure. Narrow longitudinal bands on the middle of the striae.
The valves of ***P. rhenohassiaca*** Krammer have parallel sides, slightly swollen in the middle, and broadly rounded ends, length 100–145 µm, breadth 22–24 µm, striae about 7–8/10 µm. Raphe semicomplex to slightly complex. Axial area 1/3

the breadth of the valve, linear, central area large, roundish with its largest extension in the apical direction. Distinct narrow to moderately wide longitudinal bands are present.

P. halophila and ***P. oriunda*** also have linear valves, however their autecology is unusual in the genus *Pinnularia*. ***P. halophila*** Krammer lives in puddles with high electrolyte content and is associated with the coastal brackish waters of the Baltic sea. The sides are slightly convex, not gibbous in the middle portion, ends broadly rounded, length 70–134 µm, breadth 17–22 µm, striae 8–9/10 µm. Lateral to broad semicomplex raphe, axial area 1/4–1/2 the breadth of the valve, linear, central area rhombic to rhombic-roundish, in smaller specimens 1/2, in larger 2/3–3/4 the breadth of the valve, differentiated from the axial area. A broad longitudinal band is good visible.

P. oriunda Krammer (in the literature frequently designated *Pinnularia viridis* var. *leptogongyla* (Ehrenberg) Cleve) also lives in biotopes with middle to higher electrolyte content. Outline linear-elliptic, larger specimens in the centre slightly gibbous, ends broadly rounded, length 55–105 µm, breadth 12–17 µm, striae 8–10/10 µm. Lateral raphe moderately broad, axial area 1/4–1/3 the breadth of the valve, tapering from the central area towards the small terminal areae, lanceolate, central area rhombic to irregularly roundish, 1/2 up to nearly the breadth of the valve. Longitudinal band moderately broad, frequently indistinctly visible.

(*48) ***Pinnularia major*** Rabenhorst 1853 has priority for the epithet *major* in the genus *Pinnularia*. Rabenhorst relates his taxon to *Pinnularia viridis* sensu Ehrenberg and not to *Navicula major* Kützing. Therefore a new name was necessary for *P. major* auct. nonull: ***Pinnularia neomajor*** Krammer (vol. 2/1 Fig. 196: 1–3), large linear forms, length 160–230 µm, breadth 24–30 µm, about 7 striae/10 µm, sides parallel to very slightly undulate, lateral raphe moderately broad, longitudinal bands narrow. ***P. neomajor* var. *intermedia*** (Cleve) Krammer is smaller variety with similar outline and areae, length 90–170 µm, breadth 15–19 µm, about 8–9 striae/10 µm.

A rare taxon in Europe (common in New Zealand) is ***Pinnularia transversa*** (A. Schmidt) Mayer. It has a large length-to-breadth ratio of 10–12. Outline linear, sides parallel, almost slightly gibbous in the middle and sometimes at the rounded ends, length 160–246 µm, breadth 17–23 µm, striae 8–9/10 µm. Raphe strongly lateral (leading to the name), both lines slightly curved, proximally bent laterally, central pores moderately large, round, terminal fissures hook-shaped. Axial area 1/3–1/2 the breadth of the valve, linear, tapering on the ends; central area a little wider than axial area, almost asymmetrical. Striae with distinct, narrow longitudinal bands.

(*49) ***P. macilenta*** Ehrenberg is another taxon (see under *41). The chain-forming taxon erroneously described on page 429 (vol. 2/1) as *P. macilenta* is ***Pinnularia socialis* var. *debesii***(Hustedt) Krammer (vol. 2/1 Fig. 197: 1, 2) · The valves are joined on the edges with a groin of linking spines. Length 110–140 µm, breadth 19–21 µm, 7–8 striae/10 µm.

(*51) The valid name for *Pinnularia dactylus* is ***P. gigas*** Ehrenberg. The former is a smaller form, very different from the latter.

(*53) This is a revised description of ***Pinnularia gentilis*** (Donkin) Cleve: outline linear, valve sides slightly swollen in the central portion and at the rounded ends, length 140–220 µm, breadth 26–33 µm, striae about 6–8/10 µm. Raphe semicomplex, two straight and one undulate longitudinal lines are always visible, proximally bent laterally, central pores small, round, moderately close to each other. Axial area 1/4–1/3, linear, tapering on the ends; central area roundish, 1/2 the breadth of the valve, distinctly differentiated from the axial area. Distinct broad longitudinal bands cross the striae.

(*54) Revised morphometric data for ***Pinnularia streptoraphe*** Cleve are length 140–280 µm, breadth 26–35 µm, striae 4.8–5.8/10 µm. The following species also have a similar outline (linear, sides parallel, ends broadly rounded) and complex raphes have also the following species:
The more abundant ***P. substreptoraphe*** Krammer (syn. *Pinnularia streptoraphe* var. *minor* (Cleve) Cleve 1895, p. 93, pro parte) has the following characteristics: length 75–148 µm, breadth 19–23 µm, striae 6–8/10 µm. Axial area linear or tapering lanceolate to the ends, 1/4–1/2 the breadth of the valve; central area a large, roundish, asymmetrical widening of the axial area, 1/2–2/3 the breadth of the valve. Striae slightly radiate to radiate in the middle, slightly to moderately convergent towards the ends, crossed by a small, sometimes indistinct, longitudinal band.
Similar to the latter, but smaller, is ***P. ilkaschoenfelderae*** Krammer. Length 75–105 µm, breadth 13.4–17 µm, striae 6–7/10 µm. Axial area linear, tapering on the ends, 1/5–1/3 the breadth of the valve; central area large, roundish, 1/2–2/3 the breadth of the valve. Striae slightly radiate in the middle, slightly convergent towards the ends, crossed by a broad, distinct longitudinal band.

19. *Mastogloia* Thwaites in W. Smith 1856 (vol. 2/1 p.432)

Typus generis: *Mastogloia elliptica* var. *dansei* (Thwaites) Cleve; [*Dickieia dansei* Thwaites 1848]

Key to species:

1a Striae with double puncta (vol. 2/1 Fig. 202: 3–5) . . . . . **6. *M.grevillei***
1b Striae with single puncta rows **2**
2a Lateral area and central area forming an H-shaped pattern **3**
2b Lateral area absent **4**
3a Valves broad, 14–28 µm, raphe undulating (vol. 2/1 Fig. 200: 1–5) . . . . **1. *M. braunii***
3b Valves narrower, 6–10 µm, raphe filiform (vol. 2/1 Fig. 200: 6, 7) . . . . **2. *M. pumila***
4a Axial area accompanied by strong silicious ribs (vol. 2/1 Fig. 200: 8–10) . . . . **3. *M. baltica***
4b Silicious ribs absent **5**
5a Raphe filiform, at most with the centre of the raphe somewhat curved (vol. 2/1 Fig. 201: 1–9) . . . . **4. *M. smithii***
5b Outer raphe fissure in the centre strongly turned outwards (vol. 2/1 Fig. 201: 10–14; 202: 1, 2) . . . . **5. *M. elliptica***

20. *Diatomella* Greville 1855 nom. cons. (vol. 2/1 p.436)

Typus generis: *Diatomella balfouriana* Greville 1855

Only 1 species, *Diatomella balfouriana.*

21. *Oestrupia* Heiden ex Hustedt 1935 (vol. 2/1 p.436)

Typus generis: *Oestrupia powellii* (Lewis 1861) Heiden ex Hustedt (designated in Patrick & Reimer 1966)

Key to species:

1a Valves linear to linear-elliptical, sides weakly convex (vol. 2/1 Fig. 202: 6–8) . . . . **1. *Oe. zachariasii***

1b Valves linear with triundulate sides (vol. 2/1 Fig. 202: 9–12) . . . . **2. *Oe. bicontracta***

22. *Entomoneis* Ehrenberg 1845 (vol. 2/1 p.438)

Typus generis: *Entomoneis alata* (Ehrenberg) Ehrenberg 1845; [*Navicula alata* Ehrenberg 1840]

Key to species:

1a Keel with strong ribs (vol. 2/1 Fig. 203: 5; 204: 1) . . . . . . **2. *E. costata***
1b Keel without ribs **2**
2a Separation lines very strongly undulating . . . . **4. *E. ornata***
2b Separation lines less undulating **3**
3a Keel distinctly punctate, particularly at the separation lines, 15–17 striae/10 µm on the valve (vol. 2/1 Fig. 203: 1–4) . . . . **1. *E. alata***
3b Keel not punctate, striae on the valve very delicate, around 24/10 µm, the valves are usually very delicately silicified (vol. 2/1 Fig. 204: 2–4; 205: 9) . . . . **3. *E. paludosa***

The families (4–6) with canal raphes (Bacillariaceae, Epithemiaceae, Surirellaceae) (vol 2/2)

Key to the families with canal raphes:

1a Rib system, with alveoli, situated on the inside of the valve, raphe canal running from pole to pole and mostly laterally to the median line* . . **2**
1b Valves without alveolate rib system on the inside of the valve (in *Stenopterobia*, the ribs lie on the outside of the valve), but undulating in a different way; there is one raphe canal on both sides of the valve at the border between the surface of the valve and the valve mantle, which is usually interrupted at the valve poles . **6. *Surirellaceae*** (vol. 2/2 p.166) (p. 184)
2a Fibulae extended and forming transapical walls, in between each one there are several rows of alveoli; valvocopula with septa arranged at the edge and/or ladder-like . . . . **5. *Epithemiaceae*** (vol. 2/2 p.135) (p. 182)
2b Fibula not forming transapical walls; valvocopula without septa or absent . . . . **4. *Bacillariaceae*** (vol. 2/2 p. 6) (p. 163)

4. Family Bacillariaceae Ehrenberg 1840 (vol. 2/2 p. 6)

Family type: *Bacillaria* Gmelin 1791

Key to the genera of the Bacillariaceae:

1a Fibulae absent, the raphe canal lies distally on a wing and is linked by an alar canal with the inside of the valve; monotypic genus with two subspecies currently described (vol. 2/2 Fig. 84: 13–19) . . . . **6. *Simonsenia*** (vol. 2/2 p. 135)
1b Fibulae always present, wing (lobe) and alar canals absent **2**
2a Valves twisted two or three times around the apical axis, raphe canal therefore like a screw thread (vol. 2/2 Fig. 87: 3) . . . . **5. *Cylindrotheca*** (one species, vol. 2/2 p. 134)
2b Not as above **3**

* if the character combinations in couplets 1a and 1b do not apply, go to 2b

3a Apical axis heteropolar as in *Gomphonema* (vol. 2/2 Fig. 92: 12–14), no description given* . . . . ***Gomphonitzschia*** (vol. 2/2 p. 133)
3b Apical axis isopolar **4**
4a Transapical axis of the valve usually heteropolar, as a result the valves are dorsiventral **5**
4b Transapical axis of the valves usually isopolar or only very weakly heteropolar, for example because of constrictions in the raphe canal in the middle (or at certain points) **6**
5a Valves very small, outline similar to *Cymbella* (vol. 2/2 Fig. 92: 10,.11) . . . . **4. *Cymbellonitzschia*** (one species, vol. 2/2 p. 133)
5b Valve outline not as above, ends always more or less cuneate, narrowed and with variations in the separation of the ends, keel and canal raphe always on the same side of the frustule . . . .
3. *Hantzschia* (vol. 2/2 p.126) (p. 181)
6a (4) Raphe only on one side of the frustule, without exception in all valves in a population . . . . **3. *Hantzschia*** (vol. 2/2 p. 126) (p. 181)
6b Raphe on different sides of the frustule in at least half of the population (or always central) **7**
7a Frustules combined in table like bands, each individual is able to slide against the others with its central keel, so that this formation can separate and contract; isolated individuals with characteristics similar to those shown in vol. 2/2 Fig. 87: 4 . . . . . **1. *Bacillaria*** (vol. 2/2 p. 6) (p. 164)
7b Not with the above combination of characteristics . . . . **2. *Nitzschia*** (vol. 2/2 p. 8) (p. 164)

1. *Bacillaria* Gmelin 1781 (vol. 2/2 p. 8)

Typus generis: *Bacillaria paradoxa* Gmelin 1791

Only 1 species, *Bacillaria paradoxa.*

2. *Nitzschia* Hassall 1845 nom. cons. (vol. 2/2 p. 8)

Typus generis: *Bacillaria sigmoidea* Nitzsch 1817; (*Nitzschia sigmoidea* (Nitzsch) W. Smith 1853; *Nitzschia elongata* Hassal nom. illeg.)

Key to the species groups (sections):

1a Raphe keel on the outside of the valve covered by a canopy, recognisable directly (SEM, vol. 2/2 Fig 1: 1–5) or recognisable by an adjoining line which runs parallel to it (light microscope vol. 2/2 Fig 6: 5 and 8: 6); raphe never with a central node, raphe approximately central to slightly eccentric, valves linear or convex in the centre, never concave, at the most sigmoidly curved . . . . **("subgenus *Nitzschia*") 13**
1b Canopy not positively distinguishable **2**
2a Frustules more or less sigmoidly curved in girdle and/or valve view . **14**
2b Frustules not curved in this way in girdle and valve view **3**
3a Transapical ribs run far into the valve face; these are elongated fibula or run as independent elements beside the short fibula **10**
3b No such rib structures recognisable **4**

* Without diagnosis: the taxon is discussed here and elsewhere in the text, but not under a specific number in the text.

4a Valve with conspicuously elongated rostrate ends . . . . **"Nitzschiellae"** (vol. 2/2 p. 121) (p. 180)
. . . . **"Lanceolatae"** in part (vol. 2/2 p. 76) (p. 173)
4b Valve ends at most short, rostrate 5
5a Valves with more or less obvious longitudinal folds, puncta on the striae usually relatively coarse or number of striae (transapical ribs) corresponding to the number of fibulae, points sometimes together in double rows (Tribes shown in Fig. 35: 7–38: 11) . . . . **"Tryblionellae"** in part (vol. 2/2 p. 35) (p. 169)
5b Not with the above combination of characters 6
6a Valves with more or less obvious longitudinal fold, which either results in an optical interference of the continuous striae or sometimes appears as a hyaline zone, and thus interrupts the striae; the actual striae usually very closely spaced, delicate, often difficult to separate and obscured by linear supporting ribs; raphe always with a central node (vol. 2/2 Fig. 28: 1–35; 6 and 38: 12–39: 9) . . . . **"Tryblionellae"** in part, **"Apiculatae"**, ... **"Panduriformes"** (vol. 2/2 p. 35) (p. 169)
6b Valves without obvious longitudinal fold, at least not in combination with the characteristics shown in the figures given under 6a 7
7a Frustules in girdle view often quite wide because of numerous bands, the valves usually lie in this position following preparation, so that the more or less eccentric raphe keel appears marginal. Valve in this position more or less bracket or boat-shaped, the edge with a raphe keel more strongly constricted than the edge without the keel, central nodes always present . . . . **"Bilobatae"**,**"Dubiae"** in part. (vol. 2/2 p. 53) (p. 170)
7b Not with the above combination of characters or it is not possible to be certain that the characters are present 8
8a Frustules and valves with characters of the taxa grouped in vol. 2/2 Figures 55: 1–58: 15; this is a remainder group of the historical **Lineares** and other related taxa which are difficult to define . . . .
"Lineares" (vol. 2/2 p. 69) (p. 172)
8b Not as above 9
9a Raphe keel recognisable as slightly to somewhat eccentric in suitable valve position, central nodes always absent, shown in vol. 2/2 Fig. 11: 1–14, 17a . . . . **"Dissipatae"** (vol. 2/2 p. 18) (p. 167)
9b Raphe keel strongly eccentric, positioned at the edge between valve surface and mantle . . . . **"Lanceolatae"** (vol. 2/2 p. 76) (p. 173)
. . . . **"Dubiae"** in part (vol. 2/2 p. 53) (p. 170)
10a (3) Raphe without central node; ribs, which do not extend to the opposite edge of the valve, are extensions of the fibulae, striae relatively far apart and with coarse puncta 1
10b Raphe with central nodes, striae very closely spaced **12**
11a Valves very long as shown in vol. 2/2 Fig. 25: 1–4, compare also . . . . **"Insignes"** (vol. 2/2 Fig. 26: 1–6)
. . . . **"Scalares"** (vol. 2/2 p. 34) (one species, vol. 2/2 p. 34)
11b Valves much shorter as shown in vol. 2/2 Fig. 39: 10 to 40: 8, compare also *Denticula kuetzingii* . . . . . . **"Grunowia"** (one species, vol. 2/2 p. 52)
12a (10) Ribs extending at least part way to the opposite valve edge; in the group occurring in central Europe there are elements which run next to the fibula; striae very dense, around 40/10 µm or more (vol. 2/2 Fig. 40: 9–17) . . . . **"Epithemioideae"** (one species, vol. 2/2 p. 50)
12b If fibulae are more or less extended and rib-like, but do not extend as far into the surface of the valve, striae at the most around 30/10 µm, then compare with *N. bremensis*, *N. palustris*, *N. homburgiensis*, *Hantzschia* spp.

compare with *N. vidovichii* (vol. 2/2 Fig. 18: 6, no description given) . . . . **15. *N. obtusa***

2b Length and/or width less 3

3a Sigmoid curve in valve view only suggested by the asymmetric outlines of the poles, poles not short rostrate or capitate 4

3b Not with the above combination of characteristics 5

4a Ends narrowed very close to the poles, strongly asymmetrical, resulting in a conspicuous scalpel-shape (vol. 2/2 Fig. 18: 2–5), compare also *N. vidovichii* (vol. 2/2 Fig. 18: 6, no description given) **16. *N. scalpelliformis***

4b Ends gradually narrowed and less asymmetrically constricted (vol. 2/2 Fig. 19: 7 to Fig. 20: 7) . . . . **18. *N. filiformis***

5a (3) Ends relatively broad at the poles (not constricted in any way) and sigmoidally curved (vol. 2/2 Fig. 17: 4–8), compare with *N. subcohaerens* var. *scotica* (vol. 2/2 Fig. 20: 8–12) . . . . **17. *N. nana***

5b Not with the above combination of characteristics 6

6a Striae of smaller valves comparatively widely spaced, 24–28/10 µm (vol. 2/2 Fig. 18: 8–10, no description given) . . . . ***N. aremonica***

6b Without the above combination of characteristics 7

7a Valve ends rostrate and more or less bent to opposite sides, 38–42 striae/ 10 µm (vol. 2/2 Fig. 19: 1–7) . . . . **24. *N. clausii***

7b Not with the above combination of characteristics 8

8a The middle fibulae noticeably separated from each other, somewhat longer than the others and usually only at a slight angle to each other (vol. 2/2 Fig. 21: 5–7), see also *N. submarina*, which may be conspecific (vol. 2/2 Fig. 21: 8) . . . . **25. *N. amplectens***

8b Not with the above combination of characteristics 9

9a Valves usually more or less concave in the centre, narrowed, and ends more or less rostrate 10

9b Without the above combination of characteristics, although there might be a suggestion of the characters in some examples (vol. 2/2 Fig. 20: 8–12) . . . . **19. *N. subcohaerens***

10a Valve length of the population within relatively narrow limits (vol. 2/2 Fig. 22: 1–6) . . . . **22. *N. brevissima***

10b Valve length of the population less narrowly limited (vol. 2/2 Fig. 22: 7–11) . . . . **23. *N. terrestris***

Nitzschia "Sigmata" sensu Grunow 1879 (vol. 2/2 p. 32)

Key to species:

1a The two middle fibulae usually more widely spaced than the others, small valved forms (vol. 2/2 Fig. 70: 25–27), compare with *N. lorenziana* (vol. 2/2 Fig. 86: 6–10) as well as "false *N. sigma* tribes" (vol. 2/2 Fig. 24: 3–7) and also sporadically occurring sigmoid populations found in other sections with and without central nodules (for example vol. 2/2 Fig. 59: 5; 70: 8, 26) . . . . **28. *N. austriaca***

1b Middle fibulae not widely spaced 2

2a Fibulae of the slightly eccentric raphe keel noticeably extending into the valve face (vol. 2/2 Fig. 22: 12–14), see also "**Insignes**" (vol. 2/2 Fig. 26: 1–6) . . . . **27. *N. fasciculata***

2b Keel strongly eccentric without conspicuously elongated fibulae, valve size and density of the striae extremely variable in different tribes, which are probably not likely to be conspecific (vol. 2/2 Fig. 23: 1–9; 24: 1–2B) . . . . **26. *N. sigma*** s. lato

Nitzschia **"Tryblionellae"** sensu lato Hustedt 1930 (vol. 2/2 p. 35)

"Tryblionellae", **"Pseudotryblionella"** pro parte(?), **"Apiculatae"**, **"Circumsutae"** (all sensu Grunow 1880)

Key to species:

1a Raphe continuous, central nodule absent, the number of fibulae always corresponds to the number of transapical ribs, therefore they are difficult to differentiate **16**

1b Central nodule present, although careful focussing may be necessary to distinguish it **2**

2a The arrangement of the puncta on the striae in quincunx form* on the whole valve face (or on parts of it), apical and transapical rows occasionally may seem to be interlaced giving a honeycomb appearance (vol. 2/2 Fig. 39: 1–9), there is only one species with description given in this group (vol. 2/2 Fig. 38: 13–15A) . **"Panduriformes"**(vol. 2/2 p. 49) ***N. coarctata***

2b Puncta not in quincunx arrangement **3**

3a Number of the fibulae is the same as the number of transapical ribs, therefore it is difficult to differentiate them **15**

3b Fibulae easily differentiated **4**

4a Striae (in the narrowest sense of the definition) formed from very fine lines of puncta, which are occasionally broken in places. In some cases the striae are very difficult to resolve, in some tribes they may be overlaid by many coarser lines (of second order) which correspond to the transapical folds and alway follow the supporting ribs **8**

4b Striae simple (not including the longitudinal folds), "normally" marked, not broken or diffuse, occasionally puncta are not distinguishable because they are made up of two or more rows of areolae **5**

5a Puncta very coarse, both sides of the valves strongly concave in the centre (vol. 2/2 Fig. 38: 13–15A) . . . . **48. *N. coarctata***

5b Not with the above combination of characteristics **6**

6a Width of the valves 9 µm or more (vol. 2/2 Fig. 33: 6), if the valves have longitudinal folds which appear to be "diffuse", see also *N. plana* (vol. 2/2 Fig. 33: 1–3) . . . . **42. *N. marginulata***

6b Width of the valves reaching 9 µm at most **7**

7a Striae 16–20/10 µm (vol. 2/2 Fig. 34: 1–3) . . . . **37. *N. hungarica***

7b Striae 24–30/10 µm (vol. 2/2 Fig. 51: 7–15) . . . . **41. *N. aerophila*** . . . . **40. *N. parvula***

8a (4) Supporting ribs on the valve face may appear stepped, continuous or interrupted in the median area by longitudinal folds or displaced from each other **9**

8b Supporting ribs not visible at all or only visible at the edge or on one side or only distinguishable in outline by focussing up and down **11**

9a Valves width 8–11 µm, relatively narrow in relation to length, with relatively narrow and very dense supporting ribs, 9–17 ribs/10 µm (vol. 2/2 Fig. 30: 1–5) . . . . **33. *N. calida***

9b Not with the above combination of characteristics **10**

10a Supporting ribs always dominating very strongly over the finely punctate striae, numerous tribes difficult to define because of their common characteristics (vol. 2/2 Fig 28: 1 to Fig. 29: 5) . . . . **31. *N. levidensis***

* A quincunx is an arrangement of 5 dots in a rectangle, one at each corner and one in the centre.

10b Supporting ribs less strongly dominant, valves always relatively large, 50–180 µm long, 16–35 µm wide, not abruptly narrowed towards the ends but much more gradually wedge-shaped (cuneate) (vol. 2/2 Fig. 27: 1–4) . . . . **30. *N. tryblionella***
11a (8) Valves relatively small, length almost always less than 30 µm, width less than 7 µm (vol. 2/2 Fig. 27: 5–11) . . . . **32. *N. debilis***
11b Valves larger **12**
12a Valve width approximately 50–65 µm (vol. 2/2 Fig. 32: 1–4) . . . . **36. *N. circumsuta***
12b Valve width much less **13**
13a Striae 16–22/10 µm (vol. 2/2 Fig. 33: 1–3) . . . . **35. *N. plana***
13b Striae more closely spaced **14**
14a Valve width 12–30 µm, striae 30–38/10 µm (vol. 2/2 Fig. 30: 6 to Fig. 31: 5) . . . . **34. *N. littoralis***
14b Valves broader, 35–40 µm (vol. 2/2 Fig. 33: 4, 5 no description) **N.** species
15a (3) Width of the valves greater than 12 µm (vol. 2/2 Fig. 34: 4–6) . . . . **39. *N. acuminata***
15b Width of the valves less than 9 µm (vol. 2/2 Fig. 35: 1–6) . **38. *N. constricta***
16a (1) Puncta on the striae around the edge, in double rows (vol. 2/2 Fig. 35: 7, 8) . . . . **43. *N. navicularis***
16b Not as above **17**
17a Valves more or less elliptical with very bluntly rounded poles, striae very coarsely punctate in the middle of the valve, double punctate at least near the raphe keel (vol. 2/2 Fig. 35: 9–13) . . . . **44. *N. granulata***
17b Not with the above combination of characteristics **18**
18a Striae coarsely punctate as seen in vol. 2/2 Fig. 37: 1–10 . . . . **45. *N. compressa*** in part
18b Striae more finely punctate **19**
19a Valves comparatively long, linear, 4–12 µm wide (vol. 2/2 Fig. 36: 1–5). Vol. 2/2 Fig 36: 2, 3 represent the more recently described *N brunoi* . . . . **46. *N. angustata***
19b Valves linear-lanceolate, lanceolate or elliptical (if approximately linear, then only about 4 µm wide) **20**
20a Striae not able to be resolved as single rows of puncta (vol. 2/2 Fig. 36: 6–10), see also *N. siliqua* (vol. 2/2 Fig. 36: 11–13) . . **47. *N. angustatula***
20b Striae formed of single puncta, ends very short, rostrate to tubercle-like (vol. 2/2 Fig. 38: 1–8) . . . . **45. *N. compressa*** in part

Nitzschia "Dubiae" and "Bilobatae" (as in Grunow 1880) **(vol. 2/2 p. 53)**

Key to species:
1a Striae with puncta arranged in a quincunx* (vol. 2/2 Fig. 54: 1–3 and Fig. 54: 4–5A) . . . . **69. *N. littorea*** . . . . **70. *N. lacunarum***
1b Puncta not in a quincunx arrangement **2**
2a Fibulae narrow, all the same width or narrowed into a single structure like the root of a tooth, more or less extended into the valve face **17**
2b Not as above, fibulae may be slightly extended **3**
3a Raphe keel comparatively narrow at the central nodule, usually con-

* A quincunx is an arrangement of 5 spots in a rectangle, one on each corner and one in the centre.

stricted at some point, valves therefore mostly linear, fibulae very slim, usually only combined with one stria (more precisely one transapical costa), striae 30/10 µm or more **15**

3b Not with the above combination of characteristics **4**

4a All fibulae very wide, thick, each combined with three or more striae **14**

4b Fibulae combined only with one, two or at most three (irregularly thickened) striae, in any case not particularly wide **5**

5a Frustules appear very wide in girdle view because of a large number of intercalary bands, valve edge with the raphe keel and central nodule strongly constricted and concave, valves "naviculoid" or "bracket-shaped" in girdle view with bent, rostrate, asymmetric ends; if the valve is in a more favourable position (which occurs rarely) it is possible to distinguish the slightly eccentric position of the raphe keel; fibulae always narrow and combined with only one stria **12**

5b Not with the above combination of characteristics or characters difficult to distinguish **6**

6a Frustules and valves as described in couplet 5a, but raphe keel more strongly eccentric and less strongly constricted at the central nodule (vol. 2/2 Fig. 41: 1, 2) . . . . **51.** ***N. dubia***

6b Not with the above combination of characteristics **7**

7a Valves always particularly short in relation to their width, as shown in vol. 2/2 Fig. 65: 1–2A . . . . **73.** ***N. laevis***

7b Valves not as above **8**

8a Valves linear or wider to narrower linear-lanceolate, each fibula narrow and only combined with one stria, striae comparatively dense, 28 to more than 40/10 µm (vol. 2/2 Fig. 55: 1–10) . . . . **71.** ***N. linearis***

8b Not with the above combination of characteristics **9**

9a All fibulae short and combined with more than one stria, striae 24–30/10 µm (vol. 2/2 Fig. 51: 1–6A) . . . . **65.** ***N. umbonata***

9b Not with the above combination of characteristics **10**

10a Proximal fibulae combined with two, three or more striae, distal fibulae combined with only one; in individuals in brackish water the striae are less dense, 19–22/10 µm, in electrolyte-rich freshwater to 24/10 µm (vol. 2/2 Fig. 43: 1–4) . . . . **54.** ***N. commutatoides***

10b Not with the above combination of characteristics **11**

11a Valves with central nodule difficult to distinguish, almost unmarked and weakly constricted, with the middle fibulae narrow and only combined with one stria, striae around 20/10 µm (vol. 2/2 Fig. 42: 7,.8) . **53.** ***N. gisela***

11b Valves less narrowly linear, with raphe keel more strongly constricted in the centre, proximal fibulae often broader than distal fibulae, the two central ones distinctly separated (vol. 2/2 Fig. 42: 1–6) **52.** ***N. commutata***

12a (5) Striae widely spaced, 16–20/10 µm, however there are problems differentiating this species (vol. 2/2 Fig. 46: 1,.2) . . . . **59.** ***N. bilobata***

12b Striae greater than 20/10 µm **13**

13a Frustules very like *N. bilobata*, however usually smaller or less robust and striae more dense, 21–27/10 µm (vol. 2/2 Fig. 46: 3 to Fig. 47: 3) . . . . **60.** ***N. hybrida***

13b Frustules commonly still smaller and/or narrower, or appearing less robust, striae around 30–40/10 µm (vol. 2/2 Fig. 47: 4 to Fig. 48: 9; Fig. 44: 8–10), compare with those forms listed here under couplet 58 as "critical heterogeneous tribes around *N. thermaloides*, *N. normanii* and *N. dubiiformis*" (vol. 2/2 Fig. 45: 1–16) . . . . **61.** ***N. pellucida***
. . . . **57.** ***N. dubiiformis***

14a (4) Frustules very large, width in girdle view around 40 µm, striae 17–21/10 µm (vol. 2/2 Fig. 52: 1, 2) . . . . **66. *N. kittlii***
14b Frustules smaller, striae much more dense (vol. 2/2 Fig. 52: 3 to 53: 8), see also *N. polaris* (vol. 2/2 Fig. 56: 9) . . . . **67. *N. dippelii***
vol. 2/2 Fig. 56: 8, 8A . . . . **68. *N. vasta***
15a (3) Valve length/breadth ratio compact, broadly linear with blunt cuneate elongated ends, striae disrupted by a narrow longitudinal fold more or less in the middle of the valve (vol. 2/2 Fig. 43: 5–7) . . . . . **55. *N. normanii***
15b Not with the above combination of characteristics **16**
16a The two centre fibulae distinctly separated; variable tribes from freshwater; these tribes might be confused with *N. thermaloides*, however they have scarcely more than 12 fibulae/10 µm (vol. 2/2 Fig. 55: 1–10) . . . . **71. *N. linearis***
16b Fibulae around 16/10 µm or more, the middle two comparatively less widely spaced; tribes predominantly on the seacoast, rarely also in saline inland waters (vol. 2/2 Fig. 44: 1–7), see also critical heterogeneous tribes of *N. thermaloides* (vol. 2/2 Fig. 45: 1–6) . . . . **56. *N. thermaloides***
17a (2) Valves comparatively large, length greater than 60 µm, width more than 6 µm (vol. 2/2 Fig. 49: 1–5) . . . . **62. *N. bremensis***
17b Valves usually smaller **18**
18a Fibulae on average more widely spaced, 6–10/10 µm and (with suitable focus) greater than half the valve width (vol. 2/2 Fig. 50: 1–3) . . . . **63. *N. palustris***
18b Fibulae more closely spaced, 9–15/10 µm, mostly shorter and less markedly lengthened (vol. 2/2 Fig. 50: 4–9) . . . . **64. *N. homburgiensis***

Nitzschia – a group of heterogeneous species which are difficult to recognise or not affiliated with a defined section, and another remaining group which cannot be justifiably placed in the "Lineares" (vol. 2/2 p. 69)

Key to species:
1a Raphe with central nodule, the centre two fibulae usually more widely spaced than the adjacent ones 7
1b Not as above 2
2a Fibulae broad to very broad, combined with three or more striae . . . 3
2b Fibulae narrower, often very narrow 5
3a Fibulae 2–3/10 µm (vol. 2/2 Fig. 57: 1–4) . . . . **75. *N. peisonis***
3b Fibulae 4–8/10 µm 4
4a Valves comparatively large, striae coarsely punctate, around 20/10 µm (vol. 2/2 Fig. 56: 1, 2) . . . . **74. *N. vitrea*** in part
4b Valves smaller, striae considerably more closely spaced (vol. 2/2 Fig. 56: 3–7) . . . . **74. *N. vitrea*** in part
. . . . **74A. *N. ebroicensis***
5a (2) Valves around 100 µm or longer, fibulae 5–7/10 µm (vol. 2/2 Fig. 57: 5–8) . . . . **78. *N. monachorum***
5b Valves shorter, more like the "Lanceolatae", fibulae narrower and more closely spaced 6
6a Valves linear-lanceolate to linear, striae 34–38/10 µm, easily distinguishable with oblique lighting (vol. 2/2 Fig. 58: 10–15) . . . . . **76. *N. sublinearis***
6b Valves on average somewhat narrower, linear lanceolate to lanceolate, striae 40–50/10 µm, difficult to distinguish with the light microscope (vol. 2/2 Fig. 58: 1–9) . . . . **77. *N. pura***

7a (1) Marine tribes, perhaps belonging to the Lanceolatae, but valves with unusual proportions, short and broad with widely spaced central fibulae as shown in vol. 2/2 Fig. 65: 1–2A . . . . **73. *N. laevis***

7b Not with the above combination of characteristics **8**

8a Fibulae wide, appearing thick, the middle two particularly widely separated (vol. 2/2 Fig. 56: 8, 8A), see also examples of *N. polaris* (vol. 2/2 Fig. 56: 9) . . . . **68. *N. vasta***

8b Without the above combination of characteristics, in particular fibulae much narrower **9**

9a Marine tribes, shape only known from one individual, shown in vol. 2/2 Figure 56: 10. . . . . **79. *N. pseudocommunis***

9b Richly variable tribes from fresh water, all characterised by narrow fibulae, each combined with one stria (vol. 2/2 Fig. 55: 1–10) . . . **71. *N. linearis*** . . . . **72. *N. subtilis***

Nitzschia "Lanceolatae" (group section) **(vol. 2/2 p. 76)**

Key to species groups:

1a The centre fibulae usually equidistant, combined with this the characteristic of a continuous raphe uninterrupted from pole to pole (this can definitely only be confirmed with the electron microscope) . . . . **key group A** (vol. 2 page 78) (p. 173)

1b Raphe is usually interrupted in the centre between the poles by a central nodule or a 'fold' in the silica, which appears to be (with the electron microscope) somewhat more heavily silicified; the two fibulae in the middle bordering the central nodule are usually somewhat more widely spaced than the others . . . . **key group B** (vol. 2 page 82) (p. 177)

Key group A
Key to species:

1a Fibulae very narrow, elongated, wedge-shaped or denticle-like; (if striae are visible) each combined with only one stria (= transapical rib), sometimes only visible with careful focussing **55**

1b Fibulae wider, if striae are not visible because they are quite dense, particularly if they appear to be broad and don't seem to narrow, then the fibula must be linked with more than one stria **2**

2a Striae comparatively more widely spaced, fewer than 15 to 25/10 µm (also visible without using oil immersion), usually distinguishable as punctate **3**

2b Striae more dense, more than 25/10 µm **15**

3a Striae coarse, 17–20/10 µm, puncta not distinguishable, because they are in quincunx* or in double rows arranged close together **4**

3b Not with the above combination of characteristics **5**

4a Ends bluntly rounded (vol. 2/2 Fig. 84: 1–8) . . . . **136. *N. valdecostata***

4b Ends sharply rounded (vol. 2/2 Fig. 84: 13–19). . . . . ***Simonsenia delognei*** (vol. 2/2 pg. 135)

5a (3) Striae fewer than 15/10 µm, coarsely punctate, fibulae somewhat lengthened into the valve face (vol. 2/2 Fig. 78: 27–29) incorrectly identified as . . . . *N. amphibioides* = ***N. semirobusta***

* A quincunx is an arrangement of 5 dots, one at each corner of a square, and one in the middle.

spaced in the middle of the valve (approximately 10/10 µm) than at the ends (vol. 2/2 Fig. 70: 10–13) . . . . **95. *N. desertorum***

36b Width 2.5–4 µm, striae 25–34/10 µm, fibulae also closely spaced in the centre of the valve (vol. 2/2 Fig. 70: 14–21) . . . . . . **93. *N. supralitorea***

37a (15) Valves very short (less than 20 µm), with conspicuously capitate or short rostrate ends (vol. 2/2 Fig. 83: 10–19) . . . . **133. *N. microcephala***

37b Not with the above combination of characteristics **38**

38a Ends markedly blunt to widely rounded **39**

38b Ends sharply rounded or somewhat capitate or elongate rostrate, compare with *N. recta* (vol. 2/2 Fig. 12: 1–11) **44**

39a Populations with valves always short and relatively broadly elliptical . **40**

39b Valves mostly linear-elliptical or more narrowly elliptical **42**

40a Valves 4.5–6.6 µm broad, fibulae 12–16/10 µm (vol. 2/2 Fig. 79: 7–11) . . . . **117. *N. ovalis***

40b Valves 2.5–5 µm broad, fibulae on average more narrowly spaced . . . **41**

41a Valves quite delicately silicified, by EM: areolae in quincunx arrangement (vol. 2/2 Fig. 80: 16–21) . . . . **121. *N. aurariae***

41b Silicification "normal", as with similar "Lanceolatae", areolae not in quincunx, presumably two heterospecific tribes in freshwater and salt water (vol. 2/2 Fig. 79: 12–15) . . . . **118. *N. pusilla*** in part

42a Valves somewhat broader in comparison, 4–5 µm, striae 35–40/10 µm (vol. 2/2 Fig. 80: 10–15) . . . . **120. *N.* (?) *bergii***

42b Striae always more than 43/10 µm **43**

43a Population does not show any tendency to elongate poles, brackish water forms (vol. 2/2 Fig. 80: 1–9) . . . . **119. *N. perspicua***

43b Larger valves showing a tendency to elongate poles, two possibly hetero-specific tribes in fresh and brackish water (vol. 2/2 Fig. 79: 12–15, 16–28) . . . . **118. *N. pusilla***

44a (38) Valves comparatively short, more or less lanceolate, less than 5 m wide with sharply rounded to weakly capitate poles **45**

44b Not with the above combination of characteristics **51**

45a Striae 35–40/10 µm, found in predominantly oligo to mesotrophic waters with moderate electrolyte levels (vol. 2/2 Fig. 78: 1–6) . **112. *N. lacuum***

45b Not as above, particularly with striae more closely spaced **46**

46a Ends sharply rounded, without tending towards capitate, fibulae appearing somewhat broader at the base than at the valve edge, found only in fresh water (vol. 2/2 Fig. 83: 1–9) . . . . **132. *N. sociabilis***

46b Not with the above combination of characteristics **47**

47a Tribes from fresh water with more or less moderate electrolyte levels . **49**

47b Tribes in brackish water **48**

48a Valves shorter, up to approximately 20 µm, and therefore appearing more widely lanceolate (vol. 2/2 Fig. 81: 17–21) . . . . . . **126. *N. rosenstockii***

48b Valves longer, greater than 30 µm (vol. 2/2 Fig. 82: 9–11A) . **128. *N. agnita***

49a (47)Valves with longer, elongate rostrate ends, more than 50 striae /10 µm (vol. 2/2 Fig. 81: 14–16) compare with part of *N. palea*, with around 40 striae /10 µm (vol. 2/2 Fig. 59: 23) . . . . **125. *N. pumila***

49b Ends possibly slightly elongated **50**

50a Fibulae very thin, with a line-like appearance (vol. 2/2 Fig. 58: 1–9) . . . . **77. *N. pura***

50b Fibulae rather short and point-like, more than 46 striae /10 µm (vol. 2/2 Fig. 81: 10, 12, ?13) compare with part of *N. palea* with about 40 striae /10 µm . . . . **124. *N. archibaldii***

51a (44) Fibulae somewhat line-like (vol. 2/2 Fig. 58: 10–15) compare with *N. pura* (vol. 2/2 Fig. 58: 1–9) . . . . **76. *N. sublinearis***

51b Fibulae rather point-like, short **52**
52a Valves lanceolate to linear lanceolate, not narrowly linear **53**
52b Valves generally narrow, at least some valves within the population linear **54**
53a Ends distinctly capitate, fibulae quite thick (vol. 2/2 Fig. 65: 11–13) . . . . **86.** ***N. pseudofonticola***
53b Combination of characteristics not present or usually not present in an entire population (vol. 2/2 Fig. 59: 1 to Fig. 60: 7) . . **80.** ***N. palea*** in part
54a (52) Length always less than 40 µm, striae 34–37/10 µm, nordic-alpine, probably always in electrolyte poor waters (vol. 2/2 Fig. 66: 12–16) . . . . **88.** ***N. suchlandtii***
54b Not with the above combination of characteristics (vol. 2/2 Fig. 66: 1–11, Fig. 67: 1–3) . . . . **87.** ***N. gracilis*** . . . . **129.** ***N. acicularioides***
55a (1) Fibulae appear relatively long and narrow, however at the base leading into more than one stria and showing a trace of one delicate intersecting longitudinal line (vol. 2/2 Fig 12: 1–11) . . . . **11.** ***N. recta***
55b Not with the above combination of characteristics **56**
55a Fibulae line-like, fewer than 25 striae /10 µm (vol. 2/2 Fig. 13: 1–5) . . . . **13.** ***N. heufleriana***
56b Not as above **57**
57a Fibulae shorter and pointed or with a 'denticle' combined with each stria, striae 24–28/10 µm (vol. 2/2 Fig. 71: 1–12) . . . . **96.** ***N. solita***
57b Fibulae not as above and striae considerably more closely spaced . . . **58**
58a Fibulae particularly small both in relation to the valve size and absolute terms, more than 20/10 µm, often appearing to be irregularly grouped together (vol. 2/2 Fig. 13: 6, 7) . . . . **14.** ***N. fibulafissa***
58b Not with the above combination of characteristics **59**
59a Striae around 35/10 µm, valves on average somewhat broader with a strong tendency to a linear outline (vol. 2/2 Fig. 58: 10–15) . **76.** ***N. sublinearis***
59b Striae about 40/10 µm, valves comparatively somewhat narrower with a stronger tendency to lanceolate outline (vol. 2/2 Fig. 58: 1–9) **77.** ***N. pura***

Key group B
Key to species:

1a Fibulae very narrow, on the whole long or cuneate, narrowed like a tooth-root, usually combined with only one stria (transapical rib) (this may only be distinguishable with careful focussing) **26**
1b Fibulae broader; if the striae cannot be resolved because they are too closely but the fibulae do not appear thin, then the fibulae must be combined with more than one stria **2**
2a Striae comparatively widely spaced, less than 15–25/10 µm, usually distinguishable as punctate **3**
2b Striae more dense, more than 25/10 µm **10**
3a Valves always small, striae coarse, 16–19/10 µm, puncta nevertheless not distinguishable because they lie in closely packed double rows (vol. 2/2 Fig. 84: 9–12) . . . . **135.** ***N. valdestriata***
3b Not with the above combination of characteristics **4**
4a Striae always fewer than 20/10 µm and fibulae (visible by focussing) sharply merging with the transapical ribs (vol. 2/2 Fig. 78: 13–26), compare with "*N. amphibioides*" (vol. 2/2 Fig. 78: 27–29), as well as *N. gisela* (vol. 2/2 Fig. 42: 7, 8) . . . . **114.** ***N. amphibia***
4b Not with the above combination of characteristics **5**

5a Valves comparatively elongate and linear, fibulae 18–21/10 µm, fibulae merging with one transapical rib (vol. 2/2 Fig. 42: 7, 8) . . . **53. *N. gisela***
5b Without the above combination of characteristics **6**
6a Valves more or less lanceolate, striae 18–21/10 µm, tribes in fresh water with moderate electrolyte levels (vol. 2/2 Fig. 76: 8–16) . . **107. *N. fossilis***
6b Not with the above combination of characteristics **7**
7a Striae around 20/10 µm, tribes in very electrolyte rich water, particularly saline, (by EM) areolae rows not double in the raphe keel region (vol. 2/2 Fig. 68: 11–17) . . . . **89. *N. frustulum*** var. ***bulnheimiana***
7b Not with the above combination of characteristics **8**
8a Largest valves usually linear, striae 20–26/10 µm (by EM) double areolae rows in the raphe keel region, tribes predominantly in electrolyte poor waters (vol. 2/2 Fig. 73: 9–18) . . . . **99. *N. hantzschiana***
8b Without the above combination of characteristics, in particular found under different water conditions **9**
9a Valves longer in comparison, lanceolate or linear-lanceolate with pointed to slightly capitate rounded ends (vol. 2/2 Fig. 75: 1 to 76: 7) . . . .
. . . . **104. *N. fonticola*** in part
. . . . **105. *N. tropica*** in part
9b Valves always short, ends bluntly rounded (vol. 2/2 Fig. 69: 6–13). Note: vol. 2/2 Fig. 69: 1–5 shows *N. inconspicua* syn. *N. frustulum* . . . .
. . . . **90. *N. abbreviata*** in part
10a (2) Frustules forming stellate clusters (vol. 2/2 Fig. 77: 1–5. ?6) . . . . . .
. . . . **108. *N. incognita***
10b Without the above combination of characters, or characters not clearly identifiable. **11**
11a Valves quite broad and always particularly short, middle two fibulae widely spaced (vol. 2/2 Fig. 65: 1–2A), compare with *N. pseudocommunis* (vol. 2/2 Fig. 56: 10) . . . . **73. *N. laevis***
11b Not with the above combination of characteristics **12**
12a Fibulae line-like (vol. 2/2 Fig. 44: 1–7) . . . . **56. *N. thermaloides***
12b Fibulae point-like or broader **13**
13a Longer valves in the population linear, not lanceolate, valve ends short and cuneate **14**
13b The longest valves in the population tending to lanceolate or linear-lanceolate shape, valve ends appear to be more elongate cuneate **16**
14a Valve width 3–6 µm, tribes in electrolyte rich water **15**
14b Valves 2.5–3 µm wide and usually considerably shorter than the last group, tribes in predominantly electrolyte poor water (vol. 2/2 Fig. 73: 1–8), compare with the very similar *N. incognita* in electrolyte rich water (vol. 2/2 Fig. 77: 1–5) . . . . **98. *N. acidoclinata***
15a Striae 26–29/10 µm, difficult to resolve without oblique lighting (need to be viewed in relief) (vol. 2/2 Fig. 77: 11–14) . . . . **109. *N. gessneri***
15b Striae 29 to greater than 35/10 µm, more difficult to resolve (vol. 2/2 Fig. 63: 8 to Fig. 64: 16), compare with *N. capitellata* (vol. 2/2 Fig. 62: 1–13A) . . . . **84. *N. tubicola***
16a (13) Valves always short, elliptical to linear-elliptical as shown in Fig. 69: 6–13, Note: Fig. 69: 1–5 shows *N. inconspicua* syn. *N. frustulum* . . . . .
. . . . **90. *N. abbreviata*** in part
16b Not with the above combination of characteristics, particularly valve pole less blunt, pointed rather than rounded capitate **17**
17a Valves distinctly lanceolate with sharply rounded poles . . . .
. . . . **104.“*N. fonticola* in part”**
17b Not with the above combination of characteristics **18**

18a (16) Tribes with strictly lanceolate valves, usually much shorter than 40 µm, found in freshwater with moderate to moderately raised electrolyte concentrations; (using EM) double rows of areolae in the raphe keel region (vol. 2/2 Fig. 75: 1–23), compare with *N. radicula*, which has valves which are on average longer and narrower (vol. 2/2 Fig. 77: 7–10) . . . . **104.** ***N. fonticola***

18b Not with the above combination of characteristics, particularly areolae rows are always simple **19**

19a Valves lanceolate to linear-lanceolate, barely reaching 30 µm, usually considerably shorter and rarely exceeding 3 µm wide, ends not rounded capitate (vol. 2/2 Fig. 68: 1–8 and Fig. 68: 20–24A) . . . . **89.** ***N. frustulum*** in part
. . . . **91.** ***N. leistikowii***

19b Not with the above combination of characteristics **20**

20a Valves narrow lanceolate, never concave on both sides, width 2.5–3 µm, striae comparatively easy to resolve (vol. 2/2 Fig. 77: 7–10 and Fig. 77: 1–5, ?6) . . . . **106.** ***N. radicula***
. . . . **108.** ***N. incognita***

20b Valves broader, 3.5–6.5 µm, often with both sides somewhat concave in the centre, striae more difficult to resolve **21**

21a Fibulae broader, irregularly distributed (vol. 2/2 Fig. 63: 8 to Fig. 64: 16) . . . . **84.** ***N. tubicola*** (complex)

21b Fibulae narrower, more evenly spaced (vol. 2/2 Fig. 62: 1 to Fig. 63: 3), compare with *N. calcicola* (vol. 2/2 Fig. 63: 4–6) as well as *N. paleaeformis* "Tribe from Ireland" (vol. 2/2 Fig. 65: 9, 10) . . . . **83.** ***N. capitellata*** (complex including *N. subinvicta)*

21c With other combination of characters **22**

22a Valves narrowly lanceolate to narrowly linear-lanceolate or ends elongated and rostrate, width very seldom over 3 µm, striae can be resolved under the light microscope with strong contrast **23**

22b Not with the above combination of characteristics, striae comparatively easily resolved with oblique lighting **24**

23a Ends elongated, rostrate (vol. 2/2 Fig. 81: 8, 9) . . . **123.** ***N. graciliformis***

23b Not as above, outline variable (vol. 2/2 Fig. 81: 1–7) . **122.** ***N. paleacea***

24a (22) Fibulae broader, unevenly distributed (vol. 2/2 Fig. 63: 8 to Fig. 64: 16) . . . . **84.** ***N. tubicola*** (complex) part

24b Fibulae narrower, more evenly distributed **25**

25a Tribes in water with moderate and raised electrolyte concentrations (vol. 2/2 Fig. 62: 1 to Fig. 63: 3) . . . **83.** ***N. capitellata*** (species complex) part

25b Tribes in water with lower electrolyte concentrations (vol. 2/2 Fig. 65: 3–10) . . . . **85.** ***N. paleaeformis***

26a (1) Fibulae not line-like, but relatively dense, somewhat narrowed, cuneate and root-like leading into the transapical ribs; valves always comparatively small with particularly coarse striae, fewer than 20/10 µm (vol. 2/2 Fig 78: 13–26) compare with *N. semirobusta* which is identified here as *N. amphibioides* (vol. 2/2 Fig. 78: 27–29) . . . . **114.** ***N. amphibia***

26b Not with the above combination of characteristics **27**

27a Striae around 20/10 µm (vol. 2/2 Fig. 42: 7, 8) . . . . **53.** ***N. gisela***

27b Striae considerably more dense **28**

28a Valves around 8 µm wide with narrower longitudinal folds (vol. 2/2 Fig. 43: 5–7) . . . . **55.** ***N. normanii***

28b Not with the above combination of characteristics **29**

29a Valve width about 4 µm (vol. 2/2 Fig. 55: 1–10) . . . . . . **72.** ***N. subtilis***
. . . . **71.** ***N. linearis*** part.

29b Valves wider **30**
30a Fibulae very dense, more than 15/10 µm, the centre two fibulae only slightly apart, marine tribes (vol. 2/2 Fig. 44: 1-.7) . . **56.** ***N. thermaloides***
30b Fibulae not particularly dense, (here) fewer than 15/10 µm, a distinct space between the two middle fibulae, predominantly freshwater tribes (vol. 2/2 Fig. 55: 1–10) . . . . **71.** ***N. linearis*** in part

Nitzschia "Nitzschiellae" (vol. 2/2 p. 121)

(Genus: *Nitzschiella* Rabenhorst 1864)

Key to species:
(also included are "Lanceolatae" tribes, with which they might be confused)

1a Apical axis sigmoid or (rarely) sickle-shaped **2**
1b Apical axis straight **6**
2a Striae coarse, usually fewer than 20/10 µm (vol. 2/2 Fig. 86: 6–10) . . . . **141.** ***N. lorenziana***
2b Striae very numerous and more closely spaced **3**
3a Fibulae quite wide, moderately or slightly eccentric raphe keel, the middle two fibulae equidistant (vol. 2/2 Fig. 86: 1–4) . . . . **142.** ***N. behrei***
3b Not with the above combination of characteristics **4**
4a Frustules very weakly silicified, fine hair-like ends, easily bent or broken in preparation; mostly sickle-shaped, occasionally sigmoid, frustules often combined in dense aggregates; predominantly in marine plankton, occasionally in saline water (vol. 2/2 Fig. 87: 1, 2) . . . . **140.** ***N. closterium***
4b Not as above, particularly more strongly silicified than other *Nitzschia* with similar valve width **5**
5a Characteristics as shown in vol. 2/2 Fig. 85: 7–10 . . . . **139.** ***N. reversa***
5b If valves are sickle-shaped, possibly *N. acicularis* var. *closterioides*; if valves sigmoid, possibly other marine taxa not listed here.
6a (1) Raphe with central nodule, a space between the middle two fibulae **7**
6b Not as above, all fibulae in the centre equidistant **9**
7a Valves with spindle-shaped 3–4.5 µm wide centre section and abruptly elongated rostrate ends, fibulae 18–21/10 µm, more than 50 striae/10 µm (vol. 2/2 Fig. 85: 5, 6) . . . . **138.** ***N. draveillensis***
7b Not with the above combination of characteristics **8**
8a Ends less abruptly elongated from the narrow middle section (less than 3 µm wide), striae 45–60/10 µm, or at the limits of resolution with the light microscope (vol. 2/2 Fig. 81: 8, 9) . . . . **123.** ***N. graciliformis***
8b Valves wider and striae more widely spaced . . . . **83.** ***N. capitellata*** in part (***tenuirostris*-forms**)
9a (6) Ends abruptly elongated, rostrate, middle section of the valve broader, spindle-shaped, striae not able to be resolved by light microscope (vol. 2/2 Fig. 85: 1–4) . . . . **137.** ***N. acicularis***
9b Not with the above combination of characteristics **10**
10a Striae around 30/10 µm, comparatively easily seen **11**
10b Striae more closely spaced, possibly visible with oblique lighting . . . **12**
11a Width of the valve at most about 3 µm (vol. 2/2 Fig. 67: 4–10) . . . . **130.** ***N. subacicularis***
11b Valves broader (vol. 2/2 Fig. 67: 11) also compare with *N. longirostris* (Lange-Bertalot and Krammer 1987) . . . . **31.** ***N. rostellata***
12a (10) Striae around 40/10 µm, still quite easily visible with oblique lighting . . . . **87.** ***N. gracilis***
. . . . **129.** ***N. acicularioides***

. . . . **80.** ***N. palea*** part

12b Striae still more closely spaced, at least more difficult to distinguish (vol. 2/2 Fig. 81: 14–16), compare also with *N. agnita* and others (vol. 2/2 Fig. 82: 9–14) and additional tribes in Lange Bertalot & Krammer (1987) . . . . **125.** ***N. pumila***

3. *Hantzschia* Grunow 1877 nom. cons. (vol. 2/2 p. 126)

Typus generis: *Eunotia amphioxys* Ehrenberg 1843 typ. cons.; (*Hantzschia amphioxys* (Ehrenberg) Grunow in Cleve & Grunow 1880)

Note: Since 1988 – when the first edition of vol. 2/2 was published – the numbers of known *Hantzschia* species has more than doubled. See "Additions and Corrections" in the second and third edition of this vol. and elsewhere in comprehensive books on diatoms.

Key to species:

1a Number of fibulae and number of transapical ribs similar, striae with double rows of puncta in quincunx, which are more or less distinct (vol. 2/2 Fig. 93: 1–6) . . . . **10.** ***H. marina***

1b Not with the above combination of characteristics 2

2a Raphe running from pole to pole without interruption, middle fibulae equidistant 3

2b Raphe interrupted in the middle, quite a wide space between the two centre fibulae 4

3a Frustules with a weak sigmoid curve, valves without the typical *Hantzchia*-like dorsiventral form (vol. 2/2 Fig. 91: 1–4) . . . **11.** ***H. spectabilis***

3b Valves typically *Hantzschia*-like, dorsiventral, frustules not sigmoid (vol. 2/2 Fig. 91: 5, 6) . . . . **4.** ***H. vivax***

4a (2) Striae comparatively dense, 30 or more/10 µm, marine (vol. 2/2 Fig. 92: 1–4; 5–7; 8, 9) . . . . **5.** ***H. weyprechtii***
. . . . **6.** ***H. petitiana***
. . . . **7.** ***H. baltica***

4b Not with the above combination of characteristics 5

5a Fibulae narrow along their entire length, always merging with only one transapical rib, predominantly marine, rarely in brackish inland water 6

5b Fibulae at least in part broader and merging with more than one transapical rib, predominantly freshwater 7

6a Striae markedly coarsely punctate in relation to the moderately large valves, number of the striae less than double the number of the fibulae, which are barely elongated (vol. 2/2 Fig. 88: 8–10) **9.** ***H. distinctepunctata***

6b Number of striae more than double the number of the fibulae, which are very long (vol. 2/2 Fig. 90: 1–8) . . . . **8.** ***H. virgata***

7a (5) Valves remarkably narrow (10–14 µm in relation to the length (230–430 µm), (vol. 2/2 Fig. 89: 1–2A) . . . . **2.** ***H. elongata***

7b Proportion less markedly linear 8

8a Valve always very large, length 170–265 µm, breadth 17–25 µm (vol. 2/2 Fig. 89: 3–5) . . . . **3.** ***H. rhaetica***

8b Valve proportions extremely variable, but hardly ever reaching the dimensions of *H. rhaetica* (vol. 2/2 Fig. 88: 1–7) . . . .
. . . . **1.** ***H. amphioxys*** (species complex)

5. Family Epithemiaceae sensu
Karsten in Engler & Prantl 1928 (vol. 2/2 p. 135)

Family type genus: *Epithemia* Brébisson ex Kützing 1844

Key to the genera of the Epithemiaceae:

1a Transapical axis heteropol, therefore valves are dorsiventral **2**
1b Transapical axis isopol, valves not dorsiventral . . . . **1.** ***Denticula*** (vol. 2/2 p.137) (p. 182)
2a The raphe, and least in the distal region, is running more strongly on the ventral side of the valve, and proximally rises more or less into the valve face . . . . **2.** ***Epithemia*** (vol. 2/2 p.145) (p. 145)
2b Valves with narrower, steeply sloping dorsal side and broader, less sloping ventral side, the raphe is running on the ridge between the dorsal and ventral sides . . . . **3.** ***Rhopalodia*** (vol. 2/2 p. 157) (p. 157)

1. *Denticula* Kützing 1844 (vol. 2/2 p. 137)

Typus generis: *Denticula elegans* Kützing 1844 (designated by Boyer 1927)

Key to species:

1a Double rows of areolae between the transapical ribs which are easily visible with light microscope at high magnification (vol. 2/2 Fig. 99: 1–10) . . . . **7.** ***D. thermalis***
1b Always with a single row of areolae between the transapical ribs . . . **2**
2a Septa form relatively large capitate areas on the fibular partitions, particularly noticeable in girdle view **3**
2b No capitate septa in girdle view or capitate areas very small **5**
3a 15–20 striae/10 µm **4**
3b Around 12 striae/10 µm (vol. 2/2 Fig. 98: 8–18) . . . . **5.** ***D. eximia***
4a On average more than 40 fibulae/100 µm (if possible, measure a number of examples, to exclude irregularities), capitate septa moderately large (vol. 2/2 Fig. 96: 10–33; 97: 1–5) . . . . **3.** ***D. elegans***
4b Fewer than 40 fibulae/100 µm, capitate septa very large, very robustly built valves (vol. 2/2 Fig. 97: 9–17; 98: 1–7) . . . . **4.** ***D. valida***
5a (2) Fibular partitions very low and seen only with careful focussing, humps on the fibular partitions along the raphe side, septa short, around the edge, structure coarse (vol. 2/2 Fig. 99: 11–23; 100: 1–4, 18–22) . . . . **6.** ***D. kuetzingii***
5b Fibular partitions across most of the valve width as high as the valve mantle, septa with complete transapical bridges, structure relatively delicate (more than 22 striae/10 µm) **6**
6a Raphe canal marginal, usually difficult to see (vol. 2/2 Fig. 96: 1–9) . . . . **2.** ***D. subtilis***
6b Raphe canal somewhat to one side of the median line, always easily seen by light microscope (vol. 2/2 Fig. 95: 4–25) . . . . **1.** ***D. tenuis***

2. *Epithemia* Brébisson ex Kützing 1844 (vol. 2/2 p. 145)

Epithema Brébisson 1838 (non *Epithema* Blume 1826)
Typus generis: *Eunotia turgida* Ehrenberg 1838 (designated by Boyer 1927)

Key to species:

1a Capitate septa on the fibular partitions distinctly seen in girdle view (see also 4. *E. cistula* vol. 2/2 Fig. 105: 7–11) **2**

1b Capitate septa on the fibular partitions not easily seen in girdle view . **3**
2a Central portion of the raphe usually rising towards the dorsal area over half of the valve, the breaks in the septa near the ventral edge of the valve (vol. 2/2 Fig. 102: 1–9; 103: 1–5) . . . . **1.** ***E. argus***
2b Central portion of the raphe not rising towards the dorsal edge over half of the valve, fibular partitions parallel, the breaks in the septa nearer the dorsal edge (vol. 2/2 Fig. 103: 6–9) . . . . **2.** ***E. goeppertiana***
3a (1) Raphe running along on the ventral valve edge, only the central pores distinguishable, fibular partitions parallel (see also *E. adnata*) (vol. 2/2 Fig. 104: 1–7) . . . . **3.** ***E. frickei***
3b Raphe curving towards the dorsal edge or lying in the valve face . . . **4**
4a Raphe branch almost straight, completely on the valve face near the dorsal side of the valve (vol. 2/2 Fig. 105: 7–11) . . . . **4.** ***E. cistula***
4b Raphe branch more or less curved, at least in the central region running on the ventral edge of the valve **5**
5a On average (within one valve) 3 or more striae between each pair of fibulae **6**
5b 3 or fewer striae between each pair of fibulae **7**
6a The raphe branch running on the valve face over its entire length and curving towards the middle of the dorsal side (vol. 2/2 Fig. 105: 1–6) . . . . **5.** ***E. smithii***
6b The raphe branch along the ventral edge of the valve for almost the entire length, only visible in the centre of the valve, but barely reaching the middle of the valve (vol. 2/2 Fig. 107: 1–11; 108: 1–3) . . . . **6.** ***E. adnata***
7a (5) Valves with 5 or more fibulae/10 µm and more than 10 striae/10 µm. Dorsal side strongly convex, the raphe branch reaching the dorsal edge in the middle the valve (vol. 2/2 Fig. 106: 1–14) . . . . **7.** ***E. sorex***
7b Valves with usually fewer than 5 fibulae/10 µm and 9 or less areolae/10 µm, dorsal side less convex, the raphe branch curving dorsally only in the middle of the valve **8**
8a Ends broad and bluntly rounded, valve sides nearly parallel, valves strongly curved (vol. 2/2 Fig. 104: 8–10) . . . . **8.** ***E. hyndmanii***
8b Ends narrower, often elongated, valves more weakly curved (vol. 2/2 Fig. 108: 4–8; Fig. 109: 1–7) . . . . **9.** ***E. turgida***

3. *Rhopalodia* O. Müller 1895 (vol. 2/2 p. 157)

Typus generis: *Navicula gibba* Ehrenberg 1830 (designated by Boyer 1927)

Key to species:
1a Valves bracket-shaped, bent ventrally at the ends and sharply rounded (vol. 2/2 Fig. 111: 1–13; 111A: 1–7) . . . . **1.** ***Rh. gibba***
1b Valve outline not as above **2**
2a Puncta distinguishable under the light microscope with high magnification, fewer than 30 puncta/10 µm on the striae **3**
2b Puncta difficult to distinguish or indistinguishable by light microscope with high magnification, always more than 30 puncta/10 µm **7**
3a Valve outline broad and sickle-shaped, fewer than 16 puncta/10 µm on the striae, valves with strongly convex dorsal edge and compact outline (vol. 2/2 Fig. 114: 1–11) . . . . **4.** ***Rh. musculus***
3b Structure finer **4**
4a Ventral edge concave, outline of the valve sickle-shaped **5**
4b Ventral edge straight or slightly convex, valve outline a segment of a circle or lanceolate **6**

5a Keel weakly developed, raphe running along the valve edge, barely distinguishable in valve view (vol. 2/2 Fig. 112: 1–6; 113: 4–6) . . . . **2. *Rh. gibberula***

5b Keel and dorsal side of the valve more distinct, canal raphe almost always easily seen in valve face view (vol. 2/2 Fig. 112: 7–10; 113: 1–3) . . . . **3. *Rh. acuminata***

6a Large forms with straight ventral edge and double puncta irregularly arranged on the striae, usually appearing as individual puncta with the light microscope (vol. 2/2 Fig. 113A: 1–6) . . . . **5. *Rh. constricta***

6b Smaller, delicate forms with straight or concave ventral edge; in the ventral part of the valves there are distinctly contrasting individual puncta on the striae (vol. 2/2 Fig. 113: 7–13; 113A: 7–12) . . . . **6. *Rh. brebissonii***

7a More than 50 puncta/10 µm on the striae, length/breadth ratio of the ventral valve face greater than 5.5 (vol. 2/2 Fig. 115: 1–8) **7. *Rh. rupestris***

7b Fewer than 45 puncta/10 µm on the striae, length/breadth ratio of the ventral valve face less than 5 (vol. 2/2 Fig. 115: 9–12) . **8. *Rh. operculata***

6. Family Surirellaceae Kützing 1844 (vol. 2/2 p. 166)

Family type genus: *Surirella* Turpin

Key to the genera of the Surirellaceae:

1a Valves curved in a saddle shape, the apical axis of hypotheca and epitheca cross at right angles . . . . . . **4. *Campylodiscus*** (vol. 2/2 p.211) (p. 188)

1b Valve otherwise formed, apical axis of the hypotheca and eiptheca run parallel **2**

2a Valves elongated, linear, apical axis S-shaped or straight, median area narrowly linear, but always distinctly bordered, real transapical ribs on the outside of the valve, with several rows of small areolae between . . . . **3. *Stenopterobia*** (vol. 2/2 p.207) (p. 188)

2b Not with the above combination of characteristics

3a Transapical undulations and striae interrupted, at least in the region of the median line . . . . **2. *Surirella*** (vol. 2/2 p.172) (p. 185)

3b Transapical undulations and striae not interrupted in the region of the median line . . . . **1. *Cymatopleura*** (vol. 2/2 p.168) (p. 184)

1. *Cymatopleura* W. Smith 1851 (vol. 2/2 p. 168)

Typus generis: *Surirella solea* Brébisson (designated by Boyer 1927)

Key to species:

1a The fibulae continue transapically in narrow undulations, which reach to the axial area. Structure of the troughs and crests of the waves of the apical undulations is the same, valve outline almost always panduriform (vol. 2/2 Fig. 117: 1–5; 118: 1–8) . . . . **1. *C. solea***

1b Valves not as above, structure of the wave troughs and crests differ . . **2**

2a Outline linear, lanceolate or rhomboid (vol. 2/2 Fig. 119: 1–4; 120: 1–6; 121: 1–3) . . . . **2. *C. elliptica***

2b Outline broadly elliptical to nearly round (vol. 2/2 Fig. 122: 1, 2) . . . . **3. *C. brunii***

2. *Surirella* Turpin 1828 (vol. 2/2 p. 172)

Typus generis: *Surirella striatula* Turpin 1828

Key to species groups:

1a Valves with wings and alar canals, often with loops visible along the valve edge, outline often linear-oval, usually fresh water . . . . **2. Robustae group** (vol. 2/2 p. 191) (p. 186)

1b Without the above characteristics, outline often broadly oval, rarely linear, portulae connecting the raphe canal with the inside of the cell, freshwater and marine 2

2a Species characterised by infundibulae, window-like fibulae, and a central crown shaped field. Exclusively marine (no species in this group covered) . . . . **3. Fastuosae group**

2b Usually small species with pseudoinfundibula, a central field absent or shaped like a gable-roof or a hyaline area, or it consists only of a narrow, undulating keel running to the apex. The fibulae are canal like structures, bordered by the interfibular area . . . . **1. Pinnatae group** (vol. 2/2 p.176) (p. 185)

1. Group Pinnatae (vol. 2/2 p. 176)

Moderate to large forms can be identified with this key. Isolated small forms from this group cannot be identified with certainty with the light microscope.

Key to species:

1a Apical axis isopolar 2

1b Apical axis heteropolar 5

2a Margin of the valve with very distinct pseudoinfundibula, limited to the valve face (vol. 2/2 Fig. 138: 1–5; 139: 1–8) . . . . **14.** ***S. amphioxys***

2b Narrow, rib-like fibulae visible only in the edge zone 3

3a Large, robust, linear forms with fewer than 17 striae/10 µm (vol. 2/2 Fig. 136: 1–4) . . . . **13.** ***S. gracilis***

3b Delicate, finely-structured, linear forms with more than 22 striae/10 µm 4

4a Shorter forms, length/breadth ratio of the girdle side around 5:1, transapical undulations run to the median line (vol. 2/2 Fig. 133: 6–13; 134: 1, 6–10) . . . . **11.** ***S. angusta***

4b Forms with long, narrow valves, length/breadth of the girdle side up to 15:1 (vol. 2/2 Fig. 135: 15–17) . . . . **12.** ***S. lapponica***

5a (1) Transapical undulations running parallel to the median line, length/breadth ratio up to 5:1, 60–80 fibulae/100 µm (vol. 2/2 Fig. 127: 14; 134: 2, 11, 12; 135: 1–14) . . . . **10.** ***S. minuta***

5b Not with the above combination of characteristics 6

6a Valves broad, pear-shaped 7

6b Valves oval to linear-oval 9

7a Valves with only weak concentric undulations or no concentric undulations at all (vol. 2/2 Fig. 129: 1–5) . . . . **4.** ***S. crumena***

7b Valves with strong concentric undulations 8

8a Valves usually broader than 40 µm, fibulae and pseudoinfundibula form a narrow marginal ring (vol. 2/2 Fig. 131: 1–3) . . . . **6.** ***S. peisonis***

8b Valves usually narrower than 40 µm, fibulae and pseudoinfundibula form a broad marginal ring (vol. 2/2 Fig. 132: 1–8; 133: 1–4) . **7.** ***S. brightwellii***

9a (6) Valves twisted around the apical axis, valve face undulating transapically (vol. 2/2 Fig. 130: 1–8) . . . . **5.** ***S. hoefleri***

9b Valves not twisted **10**
10a Both valve poles wedge-shaped (only distinct in large forms) **11**
10b One or both valve poles broadly rounded in moderate to large forms . **13**
11a Linear-elliptical forms usually with fewer than 35 fibulae/100 µm (vol. 2/2 Fig. 137: 1–9) . . . . **9.** ***S. patella***
11b Broadly lanceolate forms with more than 30 fibulae/100 µm **12**
12a Pseudoinfundibula with indistinct borders, valves with weak concentric undulations (vol. 2/2 Fig. 125: 1–71; 126: 1) . . . . **1.** ***S. ovalis***
12b Pseudoinfundibula with distinct borders, valves with strongly concentric undulations (vol. 2/2 Fig. 132: 1–8; 133: 1–4) . . . . . . **7.** ***S. brightwellii***
13a (10) Transapical undulations or striae distinct **14**
13b Transapical undulations or striae indistinct **15**
14a Valves always shorter than 50 µm (vol. 2/2 Fig. 128: 1–10) . **3.** ***S. subsalsa***
14b Valves longer than 70 µm **16**
15a Valve outline linear-elliptical (vol. 2/2 Fig. 130: 9, 10; 134: 3–5) . . . . **8.** ***S. visurgis***
15b Valve outline oval (vol. 2/2 Fig. 123: 4, 5; 126: 2–11; 127: 1–13) . . . . **2.** ***S. brebissonii***
16a (14) 20–30/100 µm narrow ribs as extensions of the fibulae running in to a narrow median rib (vol. 2/2 Fig. 140: 1–3) . . . . **16.** ***S. gemma***
16b 6–15 transapical undulations/100 µm, wave crests nearly as broad as the wave troughs (vol. 2/2 Fig. 140: 4, 5) . . . . **15.** ***S. striatula***

2. Group Robustae (vol. 2/2 p. 191)

Key to species:

1a Frustule strongly twisted (rotated) around the apical axis (vol. 2/2 Fig. 168: 1–7) . . . . **35.** ***S. spiralis***
1b Frustule not twisted around the apical axis or only slightly so **2**
2a Apical axis isopol* **3**
2b Apical axis heteropol* **13**
3a Delicate valves with finer structure, 40–75 alar canals/ 100 µm, ends clearly narrowing and usually elongated (vol. 2/2 Fig. 173: 1–8) [see vol. 2/2 p. 210] . . . . ***Steneropterobia delicatissima***
3b Ends not elongated **4**
4a Outline lanceolate **5**
4b Outline lanceolate to linear-lanceolate **8**
5a More than 35 alar canals/100 µm (vol. 2/2 Fig. 154: 1–5, 155 : 1) . . . . **21.** ***S. constricta***
5b Fewer than 30 alar canals /100 µm **6**
6a Undulations extending from the alar canals appearing as round circles at the margins (vol. 2/2 Fig. 148: 1–4) . . . . **19.** ***S. birostrata***
6b Valves with a different structure, distinctly punctate **7**
7a Median area narrowly lanceolate, edges of the wave crest with distinct spinules (vol. 2/2 Fig. 145: 2–4; 146: 1–4; 147: 1–5; 150: 4–6) . **18.** ***S. bifrons***
7b Median area broadly lanceolate, undulations not visible at the margins, spinules very distinct and irregularly arranged in the central region (vol. 2/2 Fig. 152: 1–5) . . . . **20.** ***S. turgida***

---

* Almost all species with isopolar apical axes also have forms with weak heteropolar apical axes. In particular this is so for *S. constricta*, *S. bifrons*, *S. biseriata* and less so for most other species. *S. biseriata* is therefore also with the heteropol species in the key.

8a (4) 30–50 alar canals/100 µm **9**
8b Fewer than 32 alar canals/100 µm **10**
9a Freshwater species (vol. 2/2 Fig. 148: 5–9) . . . . **23.** ***S. roba***
9b Brackish water species (vol. 2/2 Fig. 154: 1–5; 155: 1) . . **21.** ***S. constricta***
10a (8) Transapical undulations visible in the median area with the light microscope **11**
10b Transapical waves only just visible along the edge with the light microscope **12**
11a Smaller delicate forms with more than 20 alar canals/100 µm (vol. 2/2 Fig. 149: 1–9; 150: 1; 151: 1–4) . . . . **22.** ***S. linearis***
11b Large robust forms with fewer than 20 alar canals/100 µm, often with numerous very delicate spinules on the top of the undulations (vol. 2/2 Fig. 141: 1–3; 142: 1–5; 143: 1–9, 144: 1–3, 145: 1) . . . . **17.** ***S. biseriata***
12a (10) Pole with an indentation visible in valve view (vol. 2/2 Fig. 153: 5–8) . . . . **25.** ***S. barrowcliffia***
12b Pole without such an indentation (vol. 2/2 Fig 153: 1–4) . **24.** ***S. didyma***
13a (2) Valve with strong spines just in front of the broad pole and occasionally also just in front of the narrow pole (vol. 2/2 Fig. 166: 1–4; 167: 1–4) . . . . **32.** ***S. capronii***
13b Typical spines not present **14**
14a More than 35 alar canals/100 µm, transapical undulations barely visible at the edges, or only slightly so* **15**
14b Fewer than 35 alar canals/100 µm, transapical undulations also distinct in the mid region of the valve **16**
15a 85–100 alar canals/100 µm, valve face nearly hyaline (vol. 2/2 Fig. 151: 5–7) . . . . **33.** ***S. suecica***
15b 35–50 alar canals/100 µm, transapical waves only outlined at the margin-(vol. 2/2 Fig. 151: 8, 9) . . . . **34.** ***S. tenuis***
16a (14) Alar projection distinct **17**
16b Alar projection not as clear or indistinct, large forms (vol. 2/2 Fig. 160: 5; 161: 1, 2: 162: 1–7; 163: 1–4) . . . . **31.** ***S. elegans***
17a More than 20 alar canals/100 µm (see also *S. splendida*) **18**
17b Fewer than 20 alar canals/100 µm **19**
18a Larger forms with distinct middle rib (vol. 2/2 Fig. 164: 1–4; 165: 1–3) . . . . **29.** ***S. tenera***
18b Smaller forms, middle rib narrow to absent (vol. 2/2 Fig. 155: 2–9) . . . . **30.** ***S. bohemica***
19a (17) Valve face slightly twisted (vol. 2/2 Fig. 160: 1, 2) . . **27.** ***S. astridae***
19b Valve face not twisted **20**
20a 9–15 (usually 10 or less) alar canals/100 µm, transapical undulations very distinct (vol. 2/2 Fig. 156: 1–5; 157: 1–4) . . . . **26.** ***S. robusta***
20b Transapical undulations more delicate **21**
21a Alar projections small and indistinct, usually around 15 alar canals/100 µm (vol. 2/2 Fig. 158: 1–3; 159: 1–6; 160: 3, 4) . . . . **28.** ***S. splendida***
21b Alar projections always distinct (compare also 11b) . . . **17.** ***S. biseriata***

* if it is not possible to separate the species on the number of alar canals, then try both parts of the couplet.

3. *Stenopterobia* Brébisson in litt. ex Habirshaw *et al.* 1878 (vol. 2/2 p. 207)

Typus generis: *Surirella anceps* Lewis 1863

Key to species:

1a Apical axis sigmoid **2**
1b Apical axis straight (vol. 2/2 Fig. 173: 1–8; 174: 1–12) **4. *St. delicatissima***
2a Supporting ribs marginal, median area occupying at least 1/3 of the valve width (North American species) (vol. 2/2 Fig. 171: 1–4) . . **1. *St. anceps***
2b Median area very narrow, but distinct **3**
3a Fewer than 24 striae/10 µm (vol. 2/2 Fig. 171: 5–9; 172: 1–3) **2. *St. curvula***
3b More than 26 striae/10 µm (vol. 2/2 Fig. 172: 4, 6) . . . **3. *St. densestriata***

4. *Campylodiscus* Ehrenberg 1840 (vol. 2/2 p. 211)

Lectotype: *Cocconeis* (?) *clypeus* Ehrenberg 1938 (designated by Boyer 1927)

Key to species:

1a Valve face dotted with breaks through the wall, marginally arranged in irregular radial rows and somewhat scattered in the middle of the valve (vol. 2/2 Fig. 175: 1, 2; 176: 1–3) . . . . **1. *C. echeneis***
1b Valve structure not as above **2**
2a Wave troughs very narrow, with the appearance of narrow, radial, sharply contoured ribs **3**
2b Radial undulations irregular and with broader wave troughs **5**
3a Valves strongly saddle-shaped and twisted around the apical axis, radial undulations 13–18/100 µm (vol. 2/2 Fig. 181: 4–6) . . . . **6. *C. levanderi***
3b Valves saddle-shaped, but not twisted around the apical axis **4**
4a Radial undulations 20–30/100 µm, often reaching the centre of the valves (vol. 2/2 Fig. 182: 1–5) . . . . **2. *C. noricus***
4b Radial undulations 10–20/100 µm, always with a large area free of undulations in the centre of the valve (vol. 2/2 Fig. 175: 5; 179: 1–4; 180: 1–7; 181: 1–3) . . . . **3. *C. hibernicus***
5a (2) Marginal radial undulations interrupted by a hyaline concentric ring (vol. 2/2 Fig. 175: 3, 4); 177: 1–5) . . . . **4. *C. clypeus***
5b Marginal radial undulations not interrupted, the centre of the valve with the same pattern as the marginal region (vol. 2/2 Fig. 178: 1–6) . . . . **5. *C. bicostatus***

II B Clefs françaises

(Les numéros de taxa marqués d'un *, comparez les chefs de détermination en anglais).

A. Ordre Centrales (vol. 2/3)

Clef pour les genres

1a Cellules ayant de nombreuses bandes intercalaires ouvertes et lignes d'imbrication **2**
1b Cellules de structure différente **3**
2a Cellules ayant à chaque extrémité 1 longue soie . . . . **13. *Rhizosolenia*** (vol. 2/3 p. 84) (p. 197)
2b Cellules ayant à chaque extrémité 2 longues soies . . . . **11. *Acanthoceras*** (vol. 2/3 p. 83) (one species, vol. 2/3 p. 83)
3a Cellules aux manteaux valvaires élevés et le plus souvent des ceintures nettes formant en général de longues chaînes fermées **4**
3b Cellules en forme de disque ou de tambour, manteaux valvaires plus courts, ceintures en général peu visibles, cellules uniques ou en chaînes courtes **8**
4a Cellules très faiblement siliceuses, si traitées à l'acide, en général plus ou moins déformées . . . . **10. *Skeletonema*** (vol. 2/3 p. 81) (p. 197)
4b Cellules plus fortement siliceuses **5**
5a Disque avec centre net et structuré différemment **6**
5b Disques autrement structurés **7**
6a Disques nettement aréolés . . . . . **2. *Orthoseira*** (vol. 2/3 p. 12) (p. 190)
6b Sur les disques, seules les côtes de jonction sont reconnaissables . . . . **3. *Ellerbeckia*** (vol. 2/3 p. 17) (p. 191)
7a Sous microscope optique, les aréoles sur le manteau sont difficilement reconnaissables . . . . **1. *Melosira*** (vol. 2/3 p. 7) (p. 190)
7b Sous microscope optique, les aréoles sur le manteau sont distinctes (cfr aussi *M. lineata* morphotype *juergensii*) . . . . **4. *Aulacoseira*** (vol. 2/3 p. 19) (p. 191)
8a Cellules à valves elliptiques avec de longues soies courbées latéralement aux extrémités apicales . . . . **12. *Chaetoceros*** (vol. 2/3 p. 84) (one species, vol. 2/3 p. 84)
8b Valves de construction différente **9**
9a Valves avec 2 grands "yeux" . . . . **14. *Pleurosira*** (vol. 2/3 p. 86) (one species, vol. 2/3 p. 86)
9b Valves sans "yeux" **10**
10a Aréoles ordonnées dans des secteurs structurés **11**
10b Aréoles non ordonnées dans des secteurs structurés **13**
11a Valves avec des aréoles en sandwich (double paroi de la cellule) rangées en fascicules (secteurs structurées), sans côtes radiales ou différences de niveau (pour des cellules plus petites, les fascicules sont nettement moins reconnaissables) **12**
11b Valves avec des côtes radiales nettes . . . . **9. *Stephanocostis*** (vol. 2/3 p. 80) (p. 197)
12a Un pseudonodule peut être vu lors d'une résolution optimale au bord de la valve, sans processus renforcé au centre de la face valvaire . . . . **15. *Actinocystus*** (vol. 2/3 p. 88) ***Actinocyclus*** (one species, vol. 2/3 p. 88)

12b Absence de pseudonodule; au centre un ou plusieurs processus renforcés . . . . **8. *Thalassiosira*** (vol. 2/3 p. 77)
13a Rangées d'aréoles/stries se multiplient du centre jusqu'au bord de la valve **14**
13b Rangées d'aréoles seulement aux bords, le centre est différemment structuré . . . . **5. *Cyclotella*** (vol. 2/3 p. 40) (p. 192)
14a Interstries marginalement fourchues . . . . **6. *Cyclostephanos*** (vol. 2/3 p. 61) (p. 195)
14b Interstries non fourchues . . . **7. *Stephanodiscus*** (vol. 2/3 p. 65) (p. 195)

1. Genre ***Melosira*** Agardh 1827 nom. cons. (vol. 2/3 p. 6)

Typus generis: *Melosira nummuloides* (Dillwyn) Agardh

Clef pour les espèces

1a Valves avec carène circulaire en marge du disque (vol. 2/3 Fig. 8: 1–8) . . . . **4. *M. nummuloides*** vol. 2/3 p. 11
1b Valves sans carène circulaire en marge du disque **2**
2a Centre du disque avec quelques points plus gros, de nombreuses cellules avec des valves internes (vol. 2/3 Fig. 9: 1–13) . **5. *M. disckiei*** vol. 2/3 p. 12
2b Disque différemment structuré **3**
3a Manteau et ceinture à peine structurés sous microscope optique **4**
3b Manteau et ceinture nettement structurés sous microscope optique . . **5**
4a Cellules rectangulaires en vue connective (vol. 2/3 Fig. 4: 1–8) . . . . **1. *M. varians*** vol. 2/3 p. 7
4b Cellules oblongues-octogonales en vue connective (vol. 2/3 Fig. 6: 1–5) . . . . **2. *M. moniliformisvar. octagona*** vol. 2/3 p. 9
5a Sur disque, manteau et ceinture, structure relativement grossière (vol. 2/3 Fig. 5: 1–7) . . . . **2. *M. moniliformis*** vol. 2/3 p. 8
5b Membrane cellulaire délicatement et régulièrement ponctuée (vol. 2/3 Fig. 7: 1–9) . . . . **3. *M. lineata*** vol. 2/3 p. 10

2. Genre ***Orthoseira*** Thwaites 1849 (vol. 2/3 p. 12)

Typus generis: *Melosira americana* Kützing 1844

Clef pour les espèces

1a Disques sans carinoportules au centre **2**
1b Disques avec carinoportules nettes au centre **3**
2a Manteau presque absent (vol. 2/3 Fig. 13: 1–8) . . . . ***5. *Melosira arentii*** vol. 2/3 p. 15
2b Manteau relativement élevé (vol. 2/3 Fig. 6: 6–8) . . . . **6. *Melosira undulata*** vol. 2/3 p.16
3a A règlage "haut", 4–6 grandes taches claires dans la marge du disque (vol. 2/3 Fig. 11: 6–9) . . . . **3. *O. dendrophila*** vol. 2/3 p.14
3b Pas de taches claires **4**
4a Bandes connectives ayant aux bords des rangées nettes de points, carinoportules très petites (vol. 2/3 Fig. 12: 8–12) **4. *O. circularis*** vol. 2/3 p. 15
4b Bandes connectives irrégulièrement ponctuées **5**
5a Epines de jonction courtes (vol. 2/3 Fig. 10: 1–11; 11: 1–4) . . . . **1. *O. roeseana*** vol. 2/3 p. 13
5b Epines de jonction longues (vol. 2/3 Fig. 12: 1–7) . . . . **2. *O. dentroteres*** vol. 2/3 p 14

3. Genre *Ellerbeckia* Crawford 1988 (vol. 2/3 p. 17)

Typus generis: *Melosira* Moore ex Ralfs 1843
Une seule espèce, *Ellerbeckia arenaria.*

4. Genre *Aulacoseira* Thwaites 1848 (vol. 2/3 p. 21)

Typus generis: *Melosira crenulata* Kützing 1844

Clef pour les espèces

1a Manteau sans aréoles ou une seule rangée oblique aux bords et à l'extrémité du manteau 2
1b 2 et plus d'aréoles sur les stries périvalvaires 3
2a Manteau sans aréoles (vol. 2/3 Fig. 33: 12–17) . . . . **14.** ***A. perglabra*** vol. 2/3 p. 37
2b A chaque fois, une rangée d'aréoles distales et proximales au manteau (vol. 2/3 Fig. 36: 1, 2) . . . . **15.** ***A. lirata var. biseriata*** vol. 2/3 p. 37
3a 2–3 aréoles sur les stries périvalvaires 4
3b Toujours plus de 3 aréoles sur les stries périvalvaires 11
4a Disque avec un seul anneau marginal de petites aréoles (vol. 2/3 Fig. 32: 1–9) . . . . **12.** ***A. tethera*** vol. 2/3 p. 36
4b Disque aréolé d'une autre façon 5
5a "Aréolation" très variables, à côté des disques avec 2–3 anneaux marginaux d'aréoles, on trouve aussi des disques complètement aréolés, le plus souvent avec une aire centrale sans aréoles entièrement formées (vol. 2/3 Fig. 23: 12–17) . . . . **14.** ***A. perglabra*** vol. 2/3 p. 37
5b Toute la surface du disque aréolée 6
6a Toutes les valves ont 2–3 aréoles sur les stries périvalvaires . . . . **9.** ***A. distans var. tenella*** vol. 2/3 p. 32
6b La plupart des valves avec > 3 aréoles sur les stries périvalvaires 7
7a Disque aréolé irrégulièrement avec des aréoles relativement grandes (vol. 2/3 Fig. 29: 1–22) . . . . **9.** ***A. distans*** vol. 2/3 p. 32
7b De grandes aréoles forment des structures régulières sur le disque . . . 8
8a Sulcus court (vol. 2/3 Fig. 30: 2–7) . **9.** ***A. distans var. nivalis*** vol. 2/3 p. 32
8b Sulcus long (vol. 2/3 Fig. 30: 1–9) . . . . . **13.** ***A. pfaffiana*** vol. 2/3 p. 36
9a Points (aréoles) sur le manteau très grands, en forme de perles, le plus souvent < 10 points/10 µm sur les stries périvalvaires ou points situés très irrégulièrement sur les stries 10
9b Aréoles du manteau différentes 11
10a Hauteur du manteau de presque toutes les cellules de l'échantillon inférieure au diamètre (vol. 2/3 Fig. 34: 1–12) . . **15.** ***A. lirata*** vol. 2/3 p. 37
10b Pour la plupart des cellules, hauteur du manteau supérieure au diamètre (vol. 2/3 Fig. 37: 1–10) . . . . **18.** ***A. crassipunctata*** vol. 2/3 p. 39
11a Stries périvalvaires parallèles à l'axe périvalvaire, très légèrement obliques ou seulement sur quelques cellules une chaîne légèrement oblique à l'axe périvalvaire 12
11b Stries périvalvaires pour toutes les cellules, nettement obliques par rapport à l'axe périvalvaire sauf pour les cellules de séparation 19
12a Points sur stries périvalvaires du manteau du moins partiellement plus ou moins allongées 13
12b Points sur stries périvalvaires du manteau circulaires ou carrées au microscope optique 15
13a < 15 points/10 µm sur les stries périvalvaires (vol. 2/3 Fig. 26: 1–9; 27: 1–12) . . . . **7.** ***A. crenulata*** vol. 2/3 p. 30

13b > 18 points/10 µm sur les stries périvalvaires **14**
14a Disques grossièrement aréolés, le centre le plus souvent sans aréoles ou juste une indication d'aréoles (vol. 2/3 Fig. 35: 1–13) . . . . **16.** *A. lacustris* vol. 2/3 p. 38
14b Toute la surface du disque finement aréolée (vol. 2/3 Fig. 36: 3–18) . . . . **17.** *A. tenuis*
15a Chaînes de cellules de séparation avec des aréoles déviantes et ayant de longues épines (vol. 2/3 Fig. 20: 1–9) . . **2.** *A. muzzanensis* vol. 2/3 p. 24
15b Cellules de séparation sans aréoles déviantes **16**
16a Stries périvalvaires très finement aréolées, > 20 stries/10 µm (vol. 2/3 Fig. 31: 16, 17) . . . . **11.** *A. laevissima* vol. 2/3 p. 35
16b Moins de 17 stries périvalvaires/10 µm **17**
17a Valves presque toujours plus longues que larges (vol. 2/3 Fig. 22: 1–12) . . . . **4.** *A. islandica* vol. 2/3 p. 26
17b Valves toujours plus larges que hautes **18**
18a Col relativement court (vol. 2/3 Fig. 29: 1–23) **9.** *A. distans* vol. 2/3 p. 32
18b Col presque aussi long que le reste du manteau valvaire (vol. 2/3 Fig. 23: 1–10) . . . . **13.** *A. pfaffiana* vol. 2/3 p. 36
19a Cellules avec cellules de séparation présentant des rangées de grandes aréoles périvalvaires parallèles et quelques épines de jonction particulièrement longues (vol. 2/3 Fig. 18: 1–12; 19: 1–9) **1.** *A. granulata* vol. 2/3 p. 22
19b Absence de telles cellules de séparation **20**
20a Epines de jonction très grandes **21**
20b Epines de jonction plus petites **22**
21a Paroi relativement fine, liteau étroit (vol. 2/3 Fig. 24: 1, 3–6; 25: 1–15) . . . . **6.** *A. italica* vol. 2/3 p. 29
21b Paroi épaisse, liteau large (vol. 2/3 Fig. 28: 1–11) **8.** *A. valida* vol. 2/3 p. 32
22a Surface du disque finement ponctuée, épines de jonction (dans REM) avec ancres bifidiens (vol. 2/3 Fig. 21: 1–16) . . . . **3.** *A. ambigua* vol. 2/3 p. vol. 2/3 p. 25
22b Surface de disque tout au plus avec "bague" de points marginale, épines de jonction (dans REM) formées autrement **23**
23a Epines de jonction pointues, sans ancre, sortant de deux côtes périvalvaires (vol. 2/3 Fig. 23: 1–11) . . . . **5.** *A. subarctica* vol. 2/3 p. 28
23b Epines de jonction larges, courtes, avec ancre ramifié, sortant chacune d'une côte périvalvaire (vol. 2/3 Fig. 2: 6; 31: 1–15; 32: 10–16) . . . . **10.** *A. alpigena* vol. 2/3 p. 34

5. Genre *Cyclotella* (Kützing) Brebisson 1838 nom. cons. (vol. 2/3 p. 41)

Typus generis: *Cyclotella tecta* Håkansson & Ross 1984 (=*Cyclotella distinguenda* Hustedt 1927)

Clef pour les espèces

1a Bord valvaire sans lignes d'ombre **2**
1b Bord valvaire avec lignes d'ombre **26**
2a Champ central à structure peu nette ou non structurée (voyez aussi nr. 22a) **3**
2b Champ central à structure nette **12**
3a Champ central à ondulation tangentielle **4**
3b Champ central à structure déviante **9**
4a Champ central ovale **5**
4b Champ central différent **6**

5a Pas de processus renforcés dans le champ central (vol. 2/3 Fig. 46: 6–8; ?9–11) . . . . **8.** ***C. iris*** vol. 2/3 p. 47
5b De nombreux processus renforcés dans le champ central (vol. 2/3 Fig. 46: 9–11? 12–13) . . . . **9.** ***C. michigeniana*** vol. 2/3 p. 48
6a Champ central présentant des creux et des bosses bien réparties **7**
6b Champ de structure différent, voir description vol. 2/3 p. 44 (vol. 2/3 Fig. 44: 1–10) . . . . **3.** ***C. meneghiniana*** vol. 2/3 p. 45 voir description . . . . (vol. 2/3 Fig. 44: 11a, b) . . . . **4.** ***C. gamma*** vol. 2/3 p. 45
7a Généralement diamètre > 20 µm (vol. 2/3 Fig. 45: 1–8) . . . . **5.** ***C. striata*** vol. 2/3 p. 46
7b Diamètre < 20 µm **8**
8a Zone marginale finement structurée (vol. 2/3 Fig. 46: 1a, b) . . . . **6.** ***C. caspia*** vol. 2/3 p. 47
8b Zone marginale à structure plus grossière (vol. 2/3 Fig. 46: 2–5) . . . . **7.** ***C. hakanssoniae*** vol. 2/3 p. 47
9a Espèce relativement grande, > 25 µm (Fig 43: 12–14) . . . . **2.** ***C. plitvicensis*** vol. 2/3 p. 44
9b Espèces relativement petites, < 20 µm **10**
10a Les cellules forment des colonies (vol. 2/3 Fig. 52: 3) . . . . **22.** ***C. delicatula*** vol. 2/3 p. 53
10b Le plus souvent des cellules individuelles **11**
11a Structure radiale ondulée, rangée en secteurs, dans le champ central (vol. 2/3 Fig. 52: ?1, 2, 4–6, ?7–9) . . . . **23.** ***C. comensis*** vol. 2/3 p. 53
11b Champ central relativement plat (vol. 2/3 Fig. 51: ?7, 10–14) . . . . **20.** ***C. cyclopuncta*** vol. 2/3 p. 52
12a Champ central à structure étoilée **13**
12b Champ central avec structure déviante **16**
13a Des formes petites, < 10 µm **14**
13b Formes plus grandes, > 10 µm **15**
14a Les cellules constituent des colonies (vol. 2/3 Fig. 49: 11) . . . . **17.** ***C. glomerata*** vol. 2/3 p. 51
14b Cellules individuelles (vol. 2/3 Fig. 49: 5–8) . . . . **16.** ***C.pseudostelligera*** vol. 2/3 p. 51
15a Les éléments d'une structure étoilée dans le champ central sont non-structurés (vol. 2/3 Fig. 49: 1a-4, ?9) . . . **15.** ***C. stelligera*** vol. 2/3 p. 50
15b Les éléments d'une structure étoilée dans le champ central sont aréolés (vol. 2/3 Fig. 64: 1–8) . . . . **35.** ***C. rossii*** vol. 2/3 p. 60
16a De petites formes, de 3 à 10 µm **17**
16b Formes > 25 µm **19**
17a Champ central des valves avec papilles circulaires **18**
17b Champ central avec des papilles non circulaires (vol. 2/3 Fig. 50: 12) . . . . **19.** ***C. trichonidea*** vol. 2/3 p. 52
18a Champ central plat, cellules circulaires (vol. 2/3 Fig. 50: 1–11, 13, 14; 51: 1–5) . . . . **18.** ***C. ocellata*** vol. 2/3 p. 51
18b Champ central ondulé (vol. 2/3 Fig. 48: 4a-7) . . . . **14.** ***C. tripartita*** vol. 2/3 p. 49
19a Champ central strié radialement **20**
19b Champ central de structure différente **22**
20a Cellules vivant seules (vol. 2/3 Fig. 47: 1) . . . **10.** ***C. elgeri*** vol. 2/3 p. 48
20b Cellules vivant en colonies **21**
21a Colonies plutôt en forme de chaînes (vol. 2/3 Fig. 56: 1a-2; 64: 9–11) . . . . **32.** ***C. planctonica*** vol. 2/3 p. 59

21b Colonies plutôt irrégulières (vol. 2/3 Fig. 64: 12–14) . . . . **33.** *C. socialis* vol. 2/3 p. 59
22a Pas de structure dans le champ central **23**
22b Structure nette dans le champ central **24**
23a Champ central ovale (vol. 2/3 Fig. 65: 1–3) . . . . **36.** *C. schumannii* vol. 2/3 p. 60
23b Champ central plutôt circulaire (vol. 2/3 Fig. 43: 1–11; 51: 6–8, 16, 18) . . . . **1.** *C. distinguenda* vol. 2/3 p. 43
24a Champ central sans structures en forme de cratère **25**
24b Champ central à structure plus ou moins ordonnée, visiblement pointillée (vol. 2/3 Fig. 65: 5, 6) . . . . **37.** *C. krammeri* vol. 2/3 p. 60
25a Grand champ central présentant des creux et des bosses bien réparties (vol. 2/3 Fig. 47: 2) . . . . **11.** *C. areolata* vol. 2/3 p. 49
25b Champ central relativement petit à ponctuation irrégulière (vol. 2/3 Fig. 47: 3a-4) . . . . **12.** *C. fottii* vol. 2/3 p. 49
26a Champ central à structure radiale **27**
26b Champ central à structure différente **32**
27a Champ marginal à structure fine **28**
27b Champ marginal à structure plus grossière **30**
28a Valves à forte ondulation radiale (vol. 2/3 Fig. 54: 3a-4b; 55: 1, ?2–7b; 56: ?3a-5; 57: ?4, 5; 61: ?1–5b) . **24.** *C. bodanica* var. *lemanica* vol. 2/3 p. 55
28b Valves sans ondulation radiale **29**
29a Champ central très finement structuré. Les points “passent” dans les rayures radiales de la zone marginale (vol. 2/3 Fig. 53: 1–6; 54: 1–2; 57: 1–3) . . . . **24.** *C. bodanica* var. *bodanica* vol. 2/3 p. 54
29b Champ central à structure très relâchée, pas de passage des points centraux dans les rayures radiales de la zone marginale (vol. 2/3 Fig. 58: 1–4b) . . . . **24.** *C. bodanica* var. *affinis* vol. 2/3 p. 54
30a Les cellules constituent des colonies à 4 ou plus de cellules (vol. 2/3 Fig. 60: 7–10) (voir description page 59) . . . . **30.** *C. praetermissa* vol. 2/3 p. 58
(vol. 2/3 Fig. 59: ?1–3; 60: ?1–5) . . . . **31.** *C. quadrijuncta* vol. 2/3 p. 58
30b Cellules vivant seules **31**
31a Valves très délicatement structurées (vol. 2/3 Fig. 63: 1–7) . . . . **34.** *C. glabriuscula* vol. 2/3 p. 59
31b Valves à structure très nette (vol. 2/3 Fig. 62: 1–6, 9–12) . . . . **29.** *C. radiosa* vol. 2/3 p. 57
32a Petites formes < 7 µm **33**
32b Grandes formes > 7 µm **34**
33a Structure marginale visible, grand champ central (vol. 2/3 Fig. 51: 19–21) . . . . **21.** *C. atomus* vol. 2/3 p. 53
33b Structure reconnaissable sur toute la valve, petit champ central (vol. 2/3 Fig. 65: 8a-10) . . . . **38.** *C. wuethrichiana* vol. 2/3 p. 61
34a Dans le champ central, structure étoilée de grandes aréoles (vol. 2/3 Fig. 47: 5, 6; 48: 1a-3) . . . . **13.** *C. antiqua* vol. 2/3 p. 49
34b Champ central de structure différente **35**
35a Valves elliptiques (vol. 2/3 Fig. 61: 6a-10) . **28.** *C. austriaca* vol. 2/3 p. 56
35b Valves rondes **36**
36a Champ central radialement ponctué (vol. 2/3 Fig. 59: 4) . . . . **25.** *C. styriaca* vol. 2/3 p. 56
36b Champ central à ondulation tangentielle **37**
37a Structure marginale plutôt fine (vol. 2/3 Fig. 59: 6) . . . . **26.** *C. stylorum* vol. 2/3 p. 56
37b Structure marginale plutôt grossière (vol. 2/3 Fig. 59: 5) . . . . **27.** *C. baikalensis* vol. 2/3 p. 56

6. Genre *Cyclostephanos* Round in Theriot et al. 1987 (vol. 2/3 p. 62)

Typus generis: *Stephanodiscus* (*bellus* A. Schmidt var.?) *novae zeelandiae*, Cleve

Clef pour les espèces

1a Valves ondulées **2**
1b Valves plates **5**
2a Stries radiales, côtes radiales en bordure externe de la face valvaire (vol. 2/3 Fig. 67: 8a-9b) . . . . **6. *C. dubius*** vol. 2/3 p. 64
2b Valves à structure différente **3**
3a Stries radiales grossièrement aréolées, interstriae fourchues (vol. 2/3 Fig. 66: 1a-3) . . . . **1. *C. novaezeelandiae*** vol. 2/3 p. 62
3b Stries radiales finement aréolées, interstriae fourchues **4**
4a Valves > 15 µm (vol. 2/3 Fig. 66: 4a,b; 67: 1a-2) **2. *C. damasii*** vol. 2/3 p. 63
4b Valves < 15 µm (vol. 2/3 Fig. 67: 5) . . **4. *C. costatilimbus*** vol. 2/3 p. 64
5a Valves à interstriae non fourchues (vol. 2/3 Fig. 67: 6a,b) . . . . **5. *C. tholoformis*** vol. 2/3 p. 64
5b Valves à interstriae fourchues (vol. 2/3 Fig. 67: 3, 4) . . . . **3. *C. invisitatus*** vol. 2/3 p. 63

7. Genre *Stephanodiscus* Ehrenberg 1846

Typus generis: *Stephanodiscus niagarae* Ehrenberg 1846 (Boyer 1927)

Clef pour les espèces

1a Les cellules nettement plus longues que larges forment en général de longues chaînes (vol. 2/3 Fig. 74: 10–11) **11. *S. binderanus*** vol. 2/3 p. 72
1b Cellules nettement plus courtes que larges **2**
2a Structure irrégulière en bordure de la valve **3**
2b Structure régulière en bordure de la valve **7**
3a Diamètre des plus grandes valves dans le cycle de division toujours > 60 µm (vol. 2/3 Fig. 68: 1–3, 5; 69: 1a,b; 70: 1) **1. *S. niagarae*** vol. 2/3 p. 67
3b Diamètre des plus grandes valves dans le cycle de division toujours < 60 µm **4**
4a Diamètre < 12 µm (vol. 2/3 Fig. 71: 6; 72: 1–2b) . . . . **4. *S. agassizensis*** vol. 2/3 p. 69
4b Diamètre > 12 µm **5**
5a Position variée des processus renforcés sur les deux valves d'un frustule (soit dans le champ central soit sur les bords) (vol. 2/3 Fig. 73: 4a-5b) . . . . **7. *S. galileensis*** vol. 2/3 p. 70
5b Position des processus renforcés déviante **6**
6a Dans le champ central de la valve, 1 ou plusieurs processus renforcés (vol. 2/3 Fig. 68: 4a,b; 69: 4, 5; 71: 1–2b) . . . . **2. *S. rotula*** vol. 2/3 p. 68
6b Champ central de la valve sans processus renforcé (vol. 2/3 Fig. 69: 3; 70: 3; 71: 3a-5b) . . . . **3. *S. neoastraea*** vol. 2/3 p. 68
7a Diamètre < 12 µm **8**
7b Diamètre > 12 µm **13**
8a Valves plates **9**
8b Valves à ondulation concentrique **10**
9a Processus renforcé sur la face valvaire (vol. 2/3 Fig. 74: 1–4) . . . . **9. *S. parvus*** vol. 2/3 p. 71

9b Pas de processus renforcé sur la face valvaire (vol. 2/3 Fig. 75: 4–11; 76: 1–3) . . . . **13.** ***S. hantzschii*** vol. 2/3 p. 73
10a Structure fine sur la face valvaire **11**
10b Structure plus grossière sur la face valvaire **12**
11a Structure des rangées d'aréoles nette seulement en bordure (vol. 2/3 Fig. 74: 8a-9) . . . . **10.** ***S. vestibulis*** vol. 2/3 p. 72
11b Structure des rangées d'aréoles nette partout (vol. 2/3 Fig. 74: 5–7) . . . . **8.** ***S. minutulus*** vol. 2/3 p. 71
12a Récente (vol. 2/3 Fig. 75: 1a,b?; 2a-3b) . . . **12.** ***S. medius*** vol. 2/3 p. 73
12b Fossile (récent ?) (vol. 2/3 Fig. 76: 7a, b) **15.** ***S. oregonicus*** vol. 2/3 p. 75
13a Valves plates (vol. 2/3 Fig. 754–11; 76: 1–3) **3.** ***S. hantzschii*** vol. 2/3 p. 73
13b Valves à ondulation concentrique **14**
14a Récente **16**
14b Fossile **15**
15a A chaque fois, 4 rangées d'aréoles en faisceau en bordure de la valve (vol. 2/3 Fig. 73: 1a-2) . . . . **6.** ***S. aegyptiacus*** vol. 2/3 p. 70
15b Toujours 2–3 rangées d'aréoles en faisceau en bordure de la valve (vol. 2/3 Fig. 76: 7a,b) . . . . **15.** ***S. oregonicus*** vol. 2/3 p. 75
16a Toujours 4–6 rangées d'aréoles en faisceau en bordure de la valve (vol. 2/3 Fig. 76: 4a-6) . . . . **14.** ***S. transsylvanicus*** vol. 2/3 p. 75
16b Au plus, 3 rangées d'aréoles en faisceau en bordure de la valve **17**
17a Structure des bords régulière (vol. 2/3 Fig. 72: 3a-4) . . . . **5.** ***S. alpinus*** vol. 2/3 p. 70
17b Structure des bords légèrement irrégulière (vol. 2/3 Fig. 75: 1a,b•, 2a-3b) . . . . **12.** ***S. medius*** vol. 2/3 p. 73

8. Genre *Thalassiosira* Cleve 1873 (emend. Hasle) (vol. 2/3 p. 78)

Typus generis: *Thalassiosira nordenskioeldii* Cleve 1873

Clef pour les espèces

1a Valves avec des rangées ordonnées en faisceau **2**
1b Valves avec des rangées ordonnées plus ou moins radialement **4**
2a Diamètre des valves > 25 µm **3**
2b Diamètre des valves < 25 µm (vol. 2/3 Fig. 77: 5a,b) . . . . **2.** ***T. virsurgis*** vol. 2/3 p. 78
3a Valves avec un seul processus renforcé central (vol. 2/3 Fig. 77: 1, 2) . . . . **1.** ***T. nordenskioeldii*** vol. 2/3 p. 78
3b Valves à plusieurs processus renforcés centraux (vol. 2/3 Fig. 77: 6a,b) . . . . **3.** ***T. baltica*** vol. 2/3 p. 79
4a Valves avec un cercle de processus renforcés centraux (vol. 2/3 Fig. 77: 3, 4) . . . . **4.** ***T. weissflogii*** vol. 2/3 p. 79
4b Valves avec un processus renforcé central, voir description page 79 (vol. 2/3 Fig. 78: 1–3) . . . . **5.** ***T. proschkinae*** vol. 2/3 p. 79
. . . . . Voir description page 80 (vol. 2/3 Fig. 60: 6a,b) **6.** ***T. pseudonana*** . . . . vol. 2/3 p. 80

9. Genre *Stephanocostis* Genkal & Kuzmin 1985 (vol. 2/3 p. 81)

Typus generis: Genkal & Kuzmin 1985
Une seule espèce, *Stephanocostis chantaicus.*

10. Genre *Skeletonema* Greville 1865 (vol. 2/3 p. 81)

Typus generis: *Skeletonema barbadense* Greville 1865 (vol. 2/3 p. 81)

Clef pour les espèces

1a Pseudosulcus manquant chez des cellules serrées (vol. 2/3 Fig. 84:15–10; 85: 1–3) . . . . **1. *Skeletonema subsalsum*** vol. 2/3 p. 82
1b Pseudosulcus visible chez des cellules serrées (vol. 2/3 Fig. 85: 4–8) . . . . **2. *Skeletonema potamos*** vol. 2/3 p. 82

13. Genre *Rhizosolenia* Ehrenberg 1843; emend Brightwell 1858 (vol. 2/3 p. 85)

Typus generis: *Rhizosolenia americana* Ehrenberg 1843

Clef pour les espèces

1a Soies plus ou moins au milieu des calyptrae – sutures des bandes intercalaires et ligne d'imbrication peu nettes sous microscope optique (vol. 2/3 Fig. 86: 1–4) . . . . **1. *Rh. longiseta*** vol. 2/3 p. 85
1b Soies sur les côtés des calyptrae – sutures des bandes intercalaires et ligne d'imbrication déjà visibles en préparation aqueuse (vol. 2/3 Fig. 86: 5–8) . . . . **2. *Rh. eriensis*** vol. 2/3 p. 85

B. Ordre des Pennales (vol. 2/3 p. 90)

Raphé présent . . . . **I. Sous-ordre Araphideae** (p. 197)
Raphé absent . . . . **II. Sous-ordre Raphidineae** (p. 205)

I. Sous-Ordre Araphidineae

Famille Fragilariaceae Hustedt 1930 (vol. 2/3 p. 90)

Typus familiae: *Fragilaria* Lyngbye

Clef pour les genres

1a Bandes intercalaires des cellules à septums **2**
1b Bandes intercalaires des cellules sans septums **3**
2a Valves ayant des cloisons transversales nettement visibles . . . . **1. *Tetracyclus*** (vol. 2/3 p. 90) (p. 198)
2b Valves sans cloisons transversales **5. *Tabellaria*** (vol. 2/3 p. 104) (p. 199)
3a Valves toujours avec des cloisons transversales nettes **4**
3b Valves sans cloisons transversales (mais cf. aussi Fig. 118: 1–10) **5**
4a Valves hétéropolaires . . . . **3. *Meridion*** (vol. 2/3 p. 101) (p. 199)
4b Valves isopolaires . . . . **2. *Diatoma*** (vol. 2/3 p. 93) (p. 199)
5a Cellules presque toujours hétéropolaires, le plus souvent en forme étoilée, formant rarement des colonies en zigzag . . . . **4. *Asterionella*** (vol. 2/3 p. 102) (p. 199)
5b Caractéristiques différentes **6**

6a Valves à alvéoles partiellement fermées (reconnaissables uniquement sous microscope électronique), espèces d'eau saumâtre (voir Fig. 136: 8, 9) . . . . **6. *Synedra*** (vol. 2/3 p. 111) (one species, vol. 2/3 p. 111)
6b Caractéristiques différentes 7
7a Cellules sans exception hétéropolaires, espèces d'eau saumâtre, structure fine des aréoles (voir Fig. 118: 20). . . . . **8. *Opephora*** (vol. 2/3 p. 165) (one species, vol. 2/3 p. 165)
7b Caractéristiques différentes . . . . **7. *Fragilaria*** (vol. 2/3 p. 113) (p. 200)

1. Genre *Tetracyclus* Ralfs 1843 (vol. 2/3 p. 41)

Typus generis: *Tetracyclus lacustris* Ralfs 1843

Clef pour les espèces

1a Longueur valvaire < 30 µm, contour linéaire-elliptique à rond (vol. 2/3 Fig. 89:8–20) . . . . **3. *T. rupestris*** vol. 2/3 p. 93
1b Longueur valvaire > 30 µm, élargissement transapical au milieu, cruciforme, rarement rhomboédriques 2
2a Gonflement central en élargissement, valves larges-cruciformes, rarement rhomboédriques, les bombements ne sont pas spécialement bordés (vol. 2/3 Fig. 87: 1–8; 88: 1–8; 89: 1–6) . . . . **1. *T. glans*** vol. 2/3 p. 91
2b Gonflement central très distinct avec en plus des bords convexes (vol. 2/3 Fig. 89: 8–20) . . . . **2. *T. emarginatus*** vol. 2/3 p. 92

2. Genre *Diatoma* Bory 1824 nom.cons. (vol. 2/3 p. 94)

Typus generis: *Diatoma vulgaris* Bory de Saint-Vincent 1824

Clef pour les espèces

1a Généralement plus de 6 parois de séparation/10 µm; les valves constituent souvent des bandes en zigzag 2
1b Généralement moins de 5 parois de séparation/10 µm; les valves constituent le plus souvent des bandes fermées 5
2a Valves ayant aux deux extrémités des processus labiés, les stries transapicales sont très délicates, invisibles sous microscope optique, silhouette conique ou linéaire, sans extrémités très marquées (vol. 2/3 Fig. 92: 6; 96: 11–21) . . . . **4. *D. moniliformis*** vol. 2/3 p. 98
2b Valves avec un processus labié à une seule extrémité 3
3a Largeur valvaire 5 µm et moins, valves étroites et longues (vol. 2/3 Fig. 96: 1–9, 10) . . . . **3. *D. tenuis*** vol. 2/3 p. 97
3b Valves de construction plus robuste, largeur > 6 µm 4
4a Valves linéaires avec des extrémités bien marquées et capitées (vol. 2/3 Fig. 92: 5; 95: 5–14) . . . . **2. *D. ehrenbergii*** vol. 2/3 p. 97
4b Valves le plus souvent larges, linéaires ou elliptiques-lancéolées, si linéaires alors normalement sans extrémités bien marquées (vol. 2/3 Fig. 91: 2, 3; 93: 1–12; 94: 1–13; 95: 1–7; 97: 3–5) . . . . **1. *D. vulgaris*** vol. 2/3 p. 95
5a Valves linéaires avec extrémités marquées, le plus souvent capitées (vol. 2/3 Fig. 102: 4–10) . . . . **7. *D. anceps*** vol. 2/3 p. 100
5b Extrémités non capitées, tout au plus légèrement marquées 6
6a Valves très robustes, le plus souvent longues de > 40 µm, stries nettes, 18–22/10 µm (vol. 2/3 Fig. 97: 6–10; 98: 1–6) **5. *D. hyemalis*** vol. 2/3 p. 99
6b Valves plus délicates, le plus souvent < 40 µm de long, stries très délicates, 22–35/10 µm (vol. 2/3 Fig. 91: 1; 92: 1–4; 98: 7; 99: 1–12) . . . . **6. *D. mesodon*** vol. 2/3 p. 100

3. Genre *Meridion* Agardh 1824 (vol. 2/3 p. 101)

Typus generis: *Echinella circularis* Greville 1823
Une seule espèce *Meridion circulare* (Greville) Agardh

4. Genre *Asterionella* Hassall 1850 (vol. 2/3 p. 102)

Typus generis: *Asterionella formosa* Hassall 1850

Clef pour les expèces

1a Vue connective à extrémités triangulaires élargies (vol. 2/3 Fig. 103: 1–9; 104: 9, 10) . . . . **1.** ***A. formosa*** vol. 2/3 p. 103
1b Vue connective faiblement élargie du côté pied (vol. 2/3 Fig. 104: 1–8) . . . . **2.** ***A. ralfsii*** vol. 2/3 p. 103

5. Genre *Tabellaria* Ehrenberg 1840 (vol. 2/3 p. 104)

Typus generis: *Tabellaria trinodis* Ehrenberg 1840 (*T. fenestrata*) (Lyngbye) Kützing 1844

Clef pour les espèces

1a Valves concaves ou presque elliptiques au centre (vol. 2/3 Fig. 105: 9–16) . . . . **5.** ***T. binalis*** vol. 2/3 p. 110
1b Valves convexes au centre et aux extrémités **2**
2a Bords valvaires (au microscope) sans spinules, bandes intercalaires (isolées) ouvertes (pas toujours) à une extrémité et sans septae rudimentaires. Autres caractéristiques combinées à celles-ci: aire axiale également étroite, linéraire au centre, extrémités nettement étranglées capitées, rimoportula net, +/- proche du point central du renflement central. 4 bandes intercalaires avec septae chez des cellules différenciées et prêtes à se diviser (vol. 2/3 Fig. 105: 1–4) . . . . **1.** ***T. fenestrata*** vol. 2/3 p. 106
2b Bords valvaires avec spinules délicates à grossières, caractéristiques différentes **3**
3a Spinules relativement grossières, lg > 0,5 µm (microscope optique). Bords valvaires parallèles entre les renflements, toujours 4 bandes intercalaires avec septae pour cellules différenciées, rimoportula dans l'aire axiale, mais nettement transposé distalement vers les bords du renflement central (vol. 2/3 Fig. 105: 5–8) . . . . **2.** ***T. quadriseptata*** vol. 2/3 p. 107
3b Combinaison de caractéristiques déviante dans 1 ou plusieurs points, particulièrement le nombre de bandes intercalaires avec septae qui est plus grand **4**
4a Valves mesurant 10–16 µm de large au milieu, un rimoportula près de chaque extrémité. Reconnaissables au microscope électronique comme traits transapicaux courts et réguliers ou comme des points nets (vol. 2/3 Fig. 107: 1–6) . . . . **4.** ***T. ventricosa*** vol. 2/3 p. 109
4b Valves mesurant moins de 10 µm de large, seulement un rimoportula près du milieu valvaire (vol. 2/3 Fig. 106: 1–13)(ensemble de formes) . . . . **3.** ***T. flocculosa*** vol. 2/3 p. 108

7. Genre *Fragilaria* Lyngbye 1819 (vol. 2/3 p. 113)

Typus generis: *Fragilaria pectinalis* (O.F. Müller) Lyngbye 1819 (?*Fragilaria capucina* Desmazières 1825)

Clef pour les espèces

1a Valves tripolaires en étoile **59**
1b Valves bipolaires **2**
2a Valves fortement à faiblement hétéropolaires **55**
2b Valves isopolaires **3**
3a Axe apical légèrement courbé en faucille ou en banane, ou valves longues > 100 µm et enflées plus fortement au milieu et irrégulièrement courbées à pliées **53**
3b Axe apical régulièrement droit **4**
4a Frustules pélagiques, liées en agrégats étoilés **51**
4b Caractéristiques différentes **5**
5a Valves comparables à celles de *Diatoma tenuis*, avec des côtes transapicales irrégulièrement disposées, ressemblant à des bandes et plus ou moins visibles (vol. 2/3 Fig. 118: 1–7) . . . **25. *F. incognita*** part. vol. 2/3 p. 142
5b Caractéristiques différentes **6**
6a Les agrégats de cellules en forme de crête; en vue connective les frustules ne se touchent qu'au centre ou, en outre, aux extrémités (vol. 2/3 Fig. 116) **50**
6b Caractéristiques différentes **7**
7a Les stries sur les frustules en vue connective s'étendent largement sur le manteau valvaire – les surfaces hyalines ne sont donc pas ou peu visibles vol. (vol. 2/3 Fig. 128: 1–10) . . . . **20. *F. nitzschioides***
7b Caractéristiques différentes **8**
8a Valves ayant des stries singulièrement courtes en bordure. L'aire axiale est donc élargie à une grande aire centrale plate (ne sont pas concernées les espèces ayant des valves étroites plus ou moins en forme d'aiguilles) . **42**
8b Caractéristiques différentes **9**
9a Aire centrale symétrique, séparée brusquement de l'aire axiale et limitée comme par un bourrelet **41**
9b Aire centrale absente ou différente **10**
10a Grandes valves (longueur 120–500 µm) avec des extrémités élargies, coniques-capitées (vol. 2/3 Fig. 123: 1–3) . . . **30. *F. dilatata*** vol. 2/3 p. 147
10b Caractéristiques différentes ou apparaissant beaucoup plus faiblement chez quelques formes de *F. ulna et F. biceps* **11**
11a Valves en aiguille, rétrécissement progressif du centre vers les extrémités **12**
11b Valves plus ou moins elliptiques, plus largement lancéolées ou au milieu convexes ou concaves ou encore linéaires et rétrécies seulement aux extrémités **13**
12a Valves en moyenne au moins 20 × plus longues que larges **38**
12b Si en moyenne les valves sont plus courtes que larges, comparer avec: 6. *F. delicatissima part.*, 5. *F. tenera part.*, 4. *F. famelica part.*, 1. *F. capucina*-groupe d'espèces part., 26. *F. ulna*-groupe d'espèces part.
13a Valves plus ou moins concaves au milieu **14**
13b Caractéristique différente **17**
14a Largeur valvaire maximale > 8 µm, en cas de plusieurs ondulations, celle du milieu est la plus nettement marquée (vol. 2/3 Fig. 128: 11–14; 129: 1, 6). Cf. aussi *F. lata* (vol. 2/3 Fig. 129: 5) et formes individuelles du groupe d'espèces *F. ulna* . . . . **21. *F. constricta*** vol. 2/3 p. 140

14b Largeur valvaire régulièrement < 8 µm **15**
15a Les valves larges de 2,5–3 µm sont légèrement entourées sur une plus longue distance (vol. 2/3 Fig. 111: 25–28) . **23.** ***F. alpestris*** vol. 2/3 p. 141
15b Valves plus larges et/ou concaves seulement dans l'aire centrale **16**
16a Valves à deux ondulations, grossièrement ponctuées (vol. 2/3 Fig. 130: 19, 20) . . . . **43.** ***F. robusta*** vol. 2/3 p. 164
16b Si ces caractéristiques ne sont pas présentes, comparer avec: 5. *F. capucina* var. *mesolepta*, 12. *F. parasitica* var. *subconstricta*, 8. *F. crotonensis*, 1. *F. capucina*-groupe d'espèces part. (de nombreuses populations), 2. *F. bidens* part., 4. *F. famelica* part., 34. *F. construens* part., 40. *F. oldenburgiana* part., 22. *F. lata* part., 15. *F. virescens* part., 26. *F. ulna*-groupe d'espèces part.
17a Au milieu, valves plus ou moins renflées **18**
17b Valves linéaires, elliptiques, lancéolées mais en général non renflées . . **24**
18a Valves étroites et linéaires (3–3,5 µm de large), légèrement renflées au centre (vol. 2/3 Fig. 134: 26–31) . . . **40.** ***F. oldenburgiana*** vol. 2/3 p. 162
18b Valves plus courtes et plus larges avec renflement plus prononcé . . . **19**
19a Stries grossières, < 12/10 µm et points discrets (vol. 2/3 Fig. 133: 33–42) . . . . **37.** ***F. leptostauron*** part. vol. 2/3 p. 159
19b Stries plus fines, régulièrement > 12/10 µm **20**
20a Silhouette plus ou moins rhomboédrique-lancéolée, extrémités plus étirées et arrondies en plus pointu, (sous microscope optique) sans épines de jonction (vol. 2/3 Fig. 130: 1–8) . . . . **12.** ***F. parasitica*** part.
20b Caractéristiques différentes **21**
21a Aire axiale non différenciée ou à peine reconnaissable comme telle (vol. 2/3 Fig. 129: 3–5) . . . . **22.** ***F. lata*** part. vol. 2/3 p. 140
21b Aire axiale plus ou moins nettement différenciée **22**
22a Stries apparaissant grossièrement ponctuées (vol. 2/3 Fig. 130: 25–30) . . . . **42.** ***F. pseudoconstruens*** part. vol. 2/3 p. 163
22b Stries toujours délicatement ponctuées **23**
23a Valves régulièrement très étirées et rostrées vers les extrémités (vol. 2/3 Fig. 116: 8–10) . . . . **11.** ***F. heidenii*** part. vol. 2/3 p. 132
23b Extrémités moins étirées et plus arrondies (vol. 2/3 Fig. 132: 1–32) . . . . **34.** ***F. construens*** part. vol. 2/3 p. 153
24a Bords des valves pluriondulés, très rarement aussi chez *F. pinnata* (vol. 2/3 Fig. 132: 23–27) . . . . **34.** ***F. construens*** part. vol. 2/3 p. 153
24b Caractéristique différente **25**
25a Aire axiale de très étroite à à peine différenciable **26**
25b Aire axiale de plutôt étroite à plutôt large, mais toujours nettement différenciable **28**
26a Valves avec une largeur > 5 µm (vol. 2/3 Fig. 126: 1–10) . . . . **15.** ***F. virescens*** vol. 2/3 p. 135
26b Valves jusqu'à 5 µm de large **27**
27a Stries régulières, 18–21/10 µm (vol. 2/3 Fig. 126: 11–20) . . . . **17.** ***F. exigua*** vol. 2/3 p. 137
27b Stries 13–17/10 µm, souvent irrégulières (vol. 2/3 Fig. 118: 11–16). Si stries plus fines, cf. aussi *F. incognita* sans formation de côtes en bandes ressemblant à *Diatoma* (vol. 2/3 Fig. 118: 1–6) . **24.** ***F. bicapitata*** vol. 2/3 p. 141
28a Aire centrale plus ou moins prononcée hémisphériquement avec une longueur valvaire régulièrement < 50 µm (vol. 2/3 Fig. 108/109); s'y ajoutent d'autres espèces ou des exemplaires individuels de ce groupe . . . . ***F. capucina* var. *vaucheriae* et var. *perminuta*** vol. 2/3 p. 124
28b Caractéristiques différentes **29**
29a Stries grossières, environ 5–12/10 µm **30**
29b Stries régulièrement > 12/10 µm **32**

30a Formes marines polyhalobes (vol. 2/3 Fig. 136: 12, 13) . . . . **33.** ***F. investiens*** vol. 2/3 p. 151
30b Espèces d'eau douce, rares en eau saumâtre **31**
31a Longueur de croissance est le plus souvent limitée à moins de 40 µm, stries non ponctuées mais apparaissant toujours lignées (vol. 2/3 Fig. 133: 1–42) cf. aussi les espèces à rayures grossières du groupe *F. capucina* . . . . **36.** ***F. pinnata*** part. vol. 2/3 p. 156
31b Longueur le plus souvent nettement supérieure à 40 µm et/ou largeur de 5 µm et plus; mais cf. aussi espèces à rayures grossières de *F. capucina* et les espèces grossiérement ponctuées de *F. minuscula* **61**
32a Stries plus ou moins nettement ponctuées **33**
32b Points des stries difficilement ou pas résolvables **36**
33a Valves linéaires à linéaires-lancéolées, le plus souvent avec une aire centrale au moins esquissée séparée de l'aire axiale (vol. 2/3 Fig. 111: 4–17) . . . . **4.** ***F. famelica*** part. vol. 2/3 p. 128
33b Valves circulaires à linéaires-elliptiques. Aire centrale non séparée de l'aire axiale **34**
34a Valves le plus souvent linéaires-elliptiques avec des extrémités larges à platement arrondies et aire axiale très étendue (vol. 2/3 Fig. 129: 10–13). Cf. aussi vol. 2/3 p. 165 *Delphineis karstenii* (vol. 2/3 Fig. 129: 16, 17) . . . . **44.** ***F. zeilleri*** part.
34b Valves courbées plus fortement en convexe ou rondes **35**
35a Valves largement elliptiques jusqu'à plus ou moins rondes (vol. 2/3 Fig. 130: 31–42). Cf. aussi les formes de Fig. 130: 21–23vol. 2/3 p. 155 . . . . **35.** ***F. elliptica*** part.
35b Valves linéaires-elliptiques aux extrémités tronquées à modérément larges-arrondies (vol. 2/3 Fig. 132: 17–22)(espèces subsalina) . . . . **34.** ***F. construens*** part. vol. 2/3 p. 153
36a Valves sans aire centrale nettement distincte de l'aire axiale **37**
36b Esquisse de l'aire centrale plus ou moins reconnaissable Fig. 108–112. Cf. aussi *F. famelica* part. d'espèces part. **1.** ***F. capucina***-groupe vol. 2/3 p. 120
37a Valves (sauf les plus petites) strictement linéaires avec des extrémités rétrécies coniquement (vol. 2/3 Fig. 127: 1–5a) . . . . **16.** ***F. neoproducta*** part. vol. 2/3 p. 136
37b Silhouette valvaire variable, rarement strictement linéaire (vol. 2/3 Fig. 132: 1–32). Cf. *F. oldenburgiana*, *F. famelica*, *F. capucina*-groupe d'espèces, *F. investiens* groupe d'espèces **34.** ***F. construens*** vol. 2/3 p. 153
38a Largeur valvaire maximale 1,5–2 µm. Stries à peine différenciables, > 22/10 µm (vol. 2/3 Fig. 115: 15, 16) . . . . . **7.** ***F. nanana*** vol. 2/3 p. 130
38b Largeur valvaire maximale au moins 2 µm et/ou stries plus nettement différenciables, jusqu'à environ 22/10 µm **39**
39a Aire axiale relativement large par rapport aux petites largeurs de 2–3 µm (vol. 2/3 p.e. Fig. 115: 8, 9), cf. 5. *F. tenera* part., *F. delicatissima* part., 1.13 *F. capucina* var. *Amphicephala*, et le groupe d'espèces autour de 32. *F. fasciculata* part.
39b Aire axiale plus étroite **40**
40a Longueur régulièrement < 100 µm en général, cf. *F. tenera* part., 1. *F. capucina*-groupe d'espèces part., 4. *F. famelica* part.
40b Longueur autour ou supérieure à 100 µm en général **61**
41a Stries grossièrement ponctuées (vol. 2/3 Fig. 136: 1–7) . . . . **31.** ***F. pulchella*** vol. 2/3 p. 148
41b Stries apparaissant faiblement ou pas ponctuées du tout (vol. 2/3 Fig. 111: 18–22) . . . . **2.** ***F. bidens*** vol. 2/3 p. 127
. . . . ***F. capucina*-grouped'espèces** part. vol. 2/3 p. 120

42a Valves rhomboédriques à largement lancéolées ou à étirement relativement long vers les pôles à partir d'un renflement à l'aspect ventral **43**
42b Valves linéaires, elliptiques, plus étroitement lancéolées **45**
43a Stries très grossières, max. 12/10 µm, pourtant jamais ponctuées mais en tous les cas lignées (vol. 2/3 Fig. 133) . . . . **36.** ***F. pinnata*** part. . . . . **37.** ***F. leptostauron*** part. vol. 2/3 p. 156 & vol. 2/3 p. 159
43b Caractéristiques différentes **44**
44a Extrémités étirées étroitement, avec un arrondi plutôt pointu (vol. 2/3 Fig. 130: 1–8) . . . . **12.** ***F. parasitica*** vol. 2/3 p. 133
44b Extrémités plus largement étirées, avec un arrondi plus tronqué (vol. 2/3 Fig. 130:9–17). Cf. aussi *F. pseudoconstruens* (vol. 2/3 Fig. 130: 24–30) et *F. construens* part . . . . **41.** ***F. brevistriata*** vol. 2/3 p. 162
45a Valves linéaires à linéaires-elliptiques **46**
45b Bords valvaires plus fortement courbés **47**
46a Stries nettement ponctuées, aire centrale toujours d'une largeur modérée cf. aussi *F. zeilleri* (vol. 2/3 Fig. 129: 10–15), *F. construens* part et *Delphineis karstenii* (vol. 2/3 Fig. 129). . . . . **35.** ***F. elliptica*** part.
46b Stries apparaissant non-ponctuées (vol. 2/3 Fig. 134: 1–8). Cf. aussi certains exemplaires de *F. investiens* . . . . **38.** ***F. lapponica*** vol. 2/3 p. 161 . . . . **36.2.** ***F. pinnata*** **var.** ***intercedens*** **vol. 2/3 p. 157**
47a Longueur valvaire rarement > 25 µm, stries le plus souvent résolvables comme 1–3 points **48**
47b Longueur valvaire rarement < 25 µm, stries non ponctuées, sous microscope électronique sans spinules de jonction (vol. 2/3 Fig. 135: 1–18). Cf. aussi *F. investiens* vol. 2/3 p. 150 (vol. 2/3 Fig. 136: 12, 13) . . . . **32.** ***F. fasciculata*** part.
48a Stries marginales, régulièrement non résolvables en deux ou trois points (vol. 2/3 Fig. 130: 1–16) . . . . **41.** ***F. brevistriata*** vol. 2/3 p. 162
48b Sous microscope optique, on peut résoudre régulièrement deux à trois points sur les stries **49**
49a Valves linéaires-elliptiques (vol. 2/3 Fig. 129: 14, 15; 130: 17) . . . . **44.** ***F. zeilleri*** var. ***elliptica*** vol. 2/3 p. 165
49b Valves plutôt rondes (vol. 2/3 Fig. 130: 21–23) **43.** ***F. robusta*** vol. 2/3 p. 164 . . . . **42.** ***F. pseudoconstruens*** vol. 2/3 p. 163 . . . . **41.** ***F. brevistriata*** var. ***elliptica*** vol. 2/3 p. 164
50a Valves relativement courtes et d'une largeur > 5 µm au milieu (vol. 2/3 Fig. 116: 8–10) . . . . **11.** ***F. heidenii*** vol. 2/3 p. 132
50b Valves longues et d'une largeur < 5 µm au milieu (vol. 2/3 Fig. 117: 1–4) . . . . **8.** ***F. crotonensis*** vol. 2/3 p. 130
51a Longueur valvaire autour ou dépassant largement les 100 µm surtout "*Synedra acus* var. *ostenfeldii*" Krieger . . . . **26. Groupe d'espèces** ***F. ulna*** vol. 2/3 p. 143
51b Longueur régulièrement < 50 µm **52**
52a Valves étroitement lancéolées aux extrémités en pointes arrondies . . . . **(vol. 2/3 Fig. 111: 23, 24)** . . . . **3.** ***F. utermoehlii*** vol. 2/3 p. 127
52b Valves elliptiques à linéaires avec des extrémités arrondies tronquées, souvent avec un léger renflement au centre (vol. 2/3 Fig. 134: 21–25) . . . . **39.** ***F. berolinensis*** vol. 2/3 p. 161
53a Côté ventral des valves plus ou moins renflé dans l'aire centrale (vol. 2/3 Fig. 117: 8–14) . . . . **14.** ***F. arcus*** part. vol. 2/3 p. 134
53b Côté ventral sans un tel renflement localement aussi limité **54**
54a Longueur des valves > 100 µm, irrégulièrement courbées, avec un renfle-

ment central des deux côtés de l'aire centrale (vol. 2/3 Fig. 116: 6, 7). Cf. aussi les auxospores et les cellules initiales des autres taxons . . . . **9.** ***F. montana*** vol. 2/3 p. 131

54b Caractéristiques différentes (vol. 2/3 Fig. 117: 15–16). Individus in situ sont positionnés sur du zooplancton . . **13.** ***F. cyclopum*** vol. 2/3 p. 134

55a Stries environ ou > 20/10 µm **56**

55b Stries < 16/10 µm **57**

56a Espèces d'eau douce peu acide (vol. 2/3 Fig. 126: 11–20) . . . . **17.** ***F. exigua*** part. vol. 2/3 p. 137

56b Espèces d'eau saumâtre des côtes marines (vol. 2/3 Fig. 127: 9–15) . . . . **18.** ***F. subsalina*** vol. 2/3 p. 139

57a Stries 13–16/10 µm, finement ponctuées, aire axiale extrêmement étroite (vol. 2/3 Fig. 127: 16–21) . . . . **19.** ***F. schulzii*** vol. 2/3 p. 138

57b Stries le plus souvent plus espacées, non ponctuées mais lignées ou hyalines, aire axiale moins étroite (espèces correspondant à *Opephora sensu auct. nonnull.*) **58**

58a Espèces d'eau saumâtre des côtes marines, spinules obliques visibles sur les stries mais non sur les côtes transapicales (vol. 2/3 Fig. 134: 9–20, voir 32, 33) . . . . ***F. pacifica*** vol. 2/3 p. 166
. . . . ***F. olsenii*** vol. 2/3 p. 166

58b Espèces vivant en grande majorité en eau douce, sans spinules ou si spinules, celles-ci se trouvent le plus souvent insérées sur les côtes transapicales (vol. 2/3 Fig. 133: 12–17, voir 28–30), *Opephora martyi sensu auct. nonnull.*, cf. aussi ***F. berolinensis*** (vol. 2/3 Fig. 134: 22, 23) . . . .
. . . . **36.4.** ***F. pinnata*** var. ***subsolitaris*** vol. 2/3 p. 158
. . . . **37.3.** ***F. leptostauron*** var. ***martyi*** vol. 2/3 p. 160

59a "Bras" longuement étirés, environ 20–40 µm et une largeur d'environ 2 µm, stries 20 et plus/10 µm (vol. 2/3 Fig. 117: 1, 2) . . . . **10.** ***F. reicheltii*** vol. 2/3 p. 167

59b Caractéristiques différentes **60**

60a Pôles arrondis capités. Stries > environ 13/10 µm (vol. 2/3 Fig. 117: 4–7a) . . . . **34.5.** ***F. construens*** f. ***exigua*** vol. 2/3 p. 154

60b Stries nettement plus espacées et pôles non arrondis capités (vol. 2/3 Fig. 117: 3) . . . . **36.3.** ***F. pinnata*** var. ***trigona*** vol. 2/3 p. 157

61a Stries non résolvables comme lignes ponctuées même avec des aides optiques. Sous microscope électronique, aréoles en rangées doubles régulières (vol. 2/3 Fig. 120: 1–5). Cf. aussi *F. goulardii* (vol. 2/3 Fig. 123: 4) et d'autres taxons du groupe autour de *F. ulna* . . . . **29.** ***F. lanceolata*** vol. 2/3 p. 147

61b Aréoles en rangées simples **62**

62a Pôles largement arrondis ou plus ou moins capités, souvent environ en forme de cuillère. Aire centrale généralement non prononcée. Longueur des valves en moyenne > 250 µm, largeur 5–10 µm. Frustules facultatives enchaînées en agrégats en forme de bandes ou ayant des spinules aux bords valvaires (vol. 2/3 Fig. 121: 1–5). Il s'agit probablement d'un groupe hétérogène . . . . **28.** ***F. biceps*** vol. 2/3 p. 146

62b Caractéristiques différentes **63**

63a Frustules régulièrement enchaînées en agrégats en forme de bandes et ayant des spinules aux bords valvaires. Valves de 7–10 µm de large et en moyenne < 200 µm de long (vol. 2/3 Fig. 121: 6–8) . . . . **27.** ***F. ungeriana*** vol. 2/3 p. 145

63b Caractéristiques différentes. Caractéristiques en combinaisons variables surtout dans les rapports largeur/longueur, la forme des pôles et de l'aire centrale (vol. 2/3 Fig. 119, 122) . . . . **26.** *F. ulna*-groupe **d'espèces** vol. 2/3 p. 143

II. Sous-Ordre Raphidineae

Les six familles de ce sous-ordre rencontrées dans les eaux continentales se distinguent par les caractéristiques suivantes

Clef pour les familles

1a Raphé présent sur une seule valve du frustule **2. Achnanthaceae** (p. 213)
1b Raphé présent sur les deux valves du frustule **2**
2a Branches du raphé très courtes situées aux extrémités de la valve, le plus souvent relativement peu recourbées dans la face valvaire . . . . **1. Eunotiaceae** (p. 205)
2b Branches du raphé le plus souvent s'étendant sur toute la longueurdes valves **3**
3a Valves avec raphé médian . . . . **3. Naviculaceae** (p. 224)
3b Valves avec canal raphéen **4**
4a Valves sans carène, canaux alaires absents . . . **5. Epithemiaceae** (p. 303)
4b Valves avec carène, canaux alaires nets **5**
5a Canal raphéen dans la face valvaire . . . . **4. Bacillariaceae** (p. 282)
5b Canal raphéen faisant le tour de la valve en longeant le bord du manteau . . . . **6. Surirellaceae** (p. 306)

1. Famille Eunotiaceae Kützing 1844 (vol. 2/3 p. 169)

Typus familiae: *Eunotia* Ehrenberg 1837

Clef pour les espèces

1a Valves droites coniques . . . . **3.** ***Peronia*** (vol. 2/3 p. 229) (one species, vol. 2/3 p. 229)
1b Valves régulièrement non coniques et droites mais plus ou moins courbées vers l'axe apical et donc dorsiventrales **2**
2a Dans une population, toutes les valves sont régulièrement hétéropolaires . . . . **2.** ***Actinella*** (vol. 2/3 p. 229) (one species, vol. 2/3 p. 229)
2b Dans une population, ou bien toutes les valves sont isopolaires ou une partie seulement est hétéropolaire . . **1.** ***Eunotia*** (vol. 2/3 p. 169) (p. 205)

1. Genre *Eunotia* Ehrenberg 1837 (vol. 2/3 p. 168)

Typus generis: *Eunotia arcus* Ehrenberg 1837

Clef pour les genres

1a Extrémités du raphé dans la face valvaire courbé vers le milieu de la valve sur une courte distance (vol. 2/3 pl. 137–140) . . **Groupe-clef C** (p. 212)
1b Caractéristique différente **2**
2a Extrémités valvaires étirées en forme nasale, extrémités distales du raphé difficilement reconnaissables au microscope, ou si les extrémités valvaires n'ont pas de forme nasale, il n'y a pas du tout de raphé reconnaissable **3**
2b Extrémités distales du raphé se terminant en arc plus ou moins long sur la face valvaire (vol. 2/3 pl. 141–160), extrémités valvaires non étirées en forme nasale . . . . **Groupe-clef A** (p. 206)
3a Sous microscope optique, on ne peut reconnaître ni le tracé du raphé ni les nodules terminaux. Valves fortement courbées jusqu'à une position environ hémisphérique (vol. 2/3 Fig. 166: 8–11). Cf. aussi *E. eruca* (vol. 2/3 Fig. 166: 6, 7) . . . . **53.** ***E. hemicyclus*** vol. 2/3 p. 227

3b Le raphé se termine exclusivement dans le manteau valvaire (les fissures terminales seules peuvent brièvement se courber dans la face valvaire) donc reconnaissable en vue valvaire seulement après réglage en profondeur. Nodules terminaux plus ou moins éloignés des pôles et donc extrémités valvaires étirées en forme nasale. Si difficile à déterminer, l'espèce est également répertoriée sous A (vol. 2/3 pl. 161–164 et 166 part.) . . . . **Groupe-clef B** (p. 211)

Groupe-clef A

1a Valves plus grandes en longueur et/ou largeur, au moins une partie des valves d'une population de longueur > 30 µm et/ou d'une largeur > 6 µm **2**

1b Valves plus petites, d'une longueur maximale < 30 µm et d'une largeur < 6 µm **4**

2a Bord dorsal ayant 2 ou plus de bosses plus ou moins prononcées . . . . **Sous-groupe Aa** (p. 206)

2b Bord dorsal simplement courbé ou à ondulations peu élevées, ou ayant des protubérances légères plus ou moins pointues, ou, avant les extrémités, avec des épaules convexes **3**

3a Valves avec des renflements variables au bord dorsal ou ventral . . . . **Sous-groupe Ab** (p. 207)

3b Bord dorsal courbé simplement . . . . **Sous-groupe Ac** (p. 208)

4a Le bord dorsal a, avant les extrémités, une épaule bien prononcée, cf. aussi *E. auriculata* (vol. 2/3 Fig. 160: 14) . . . . **41.** ***E. bactriana*** vol. 2/3 p. 218

4b Bord dorsal simplement courbé ou ayant un ou plusieurs renflements **5**

5a Bord dorsal ayant au moins un renflement ou ondulation prononcé . . . . **Sous-groupe Ad** (p. 209)

5b Bord dorsal simplement courbé . . . . **Sous-groupe Ae** (p. 210)

Sous-groupe Aa

1a Valves ayant deux bosses respectivement deux ondulations, parfois présentant chacune une épaule à arrondi pointu avant les extrémités (vol. 2/3 Fig. 150: 8, 9) parfois également le bord ventral à deux bosses (vol. 2/3 Fig. 160: 4, 5) . . . . **41.** ***E. bactriana*** vol. 2/3 p. 218
. . . . **42.** ***E. gibbosa*** vol. 2/3 p. 222

1b Valves ayant trois jusqu'à de nombreuses bosses (exceptionnellement plus de 20), si valves avec un côté dos relativement faiblement ondulé (vol. 2/3 Fig. 141: 1–7), cf. aussi *E. siberica* (vol. 2/3 Fig. 141: 8–10) . . . . **10.** ***E. pectinalis*** vol. 2/3 p. 193

2a Valves d'une population présentant constamment trois bosses (vol. 2/3 Fig. 146: 6–9) . . . . **44.** ***E. triodon*** vol. 2/3 p. 220

2b Valves présentant plus de trois bosses (vol. 2/3 Fig. 146: 1–5). Cf. aussi *E. muelleri* (vol. 2/3 Fig. 146: 10, 11) . . . . **43.** ***E. serra*** vol. 2/3 p. 219

3a Valves, et surtout bosses, fortement renflées **4**

3b Bosses moins renflées **5**

4a Chute très raide des bosses vers les extrémités, jusqu'à environ un angle de 90° (vol. 2/3 Fig. 149: 3–6). Cf. aussi *E. papilio* (vol. 2/3 pl. 160) ainsi que d'autres taxons de régions tropiques à subtropiques . . . . (espèces appartenant à)
. . . . **7.** ***E. praerupta papilio***

4b Si la chute est moins raide et le creux entre les bosses moins profond (vol. 2/3 pl. 149, 150), cf. autres taxons dans le groupe autour de *E. praerupta* ainsi que d'autres taxons de régions tropiques à subtropiques

5a Bord dorsal rentré de façon prononcée avant les extrémités qui ont donc un aspect courbé et rostré (vol. 2/3 Fig. 150: 1–7) . . . . **6.6.** ***E. praerupta*** **var.** ***bigibba*** vol. 2/3 p. 188

5b Extrémités d'aspect arrondi-pointu ou plat-renforcé **6**
6a Bord dorsal visiblement rentré vers les extrémités **7**
6b Bord dorsal non ou faiblement rentré **8**
7a Extrémités arrondies (vol. 2/3 Fig. 149: 8–19). Cf. aussi *E. circumborealis* (vol. 2/3 Fig. 143) . . . . **7.** ***E. diodon*** vol. 2/3 p. 191
7b Extrémités renforcées (vol. 2/3 Fig. 148: 9, 11, 12) . . . . **6.4.** ***E. praerupta bidens*** **(espèces appartenant à)** vol. 2/3 p. 187
8a Extrémités étroites, arrondies, parfois avec de petites dents sur le bord dorsal (vol. 2/3 Fig. 149: 8–19) . . . . **7.** ***E. diodon*** vol. 2/3 p. 191
8b Extrémités renforcées ou arrondies en largeur **9**
9a Valves relativement étroites par rapport à la longueur, environ 6–14 fois plus longues que larges **11**
9b Valves plus larges, d'aspect plus compact **10**
10a Extrémités à renforcement plat (vol. 2/3 Fig. 148: 9, 11, 12) . . . . **6.4.** ***E. praerupta bidens*** **(espèces appartenant à)** vol. 2/3 p. 187
10b Extrémités arrondies en largeur (vol. 2/3 Fig. 158: 4–6). Cf. aussi *E. circumborealis* (vol. 2/3 Fig. 143: 16–23), *E. implicata* (vol. 2/3 Fig. 143: 1–9a) ainsi que *E. zygodon* (vol. 2/3 Fig. 159: 8, 9) . . . . **30.2.** ***E. monodon*** **var.** ***bidens*** **vol. 2/3 p. 210)**
11a Bord dorsal moins convexe, plus linéaire, stries > 12/10 µm (vol. 2/3 Fig. 159: 1) . . . . **45.** ***E. ruzickae*** vol. 2/3 p. 221
11b Bord dorsal plus convexe, moins linéaire, stries < 12/10 µm (vol. 2/3 Fig. 147: 18) . . . . **5.** ***E. arcus bidens*** (espèces) vol. 2/3 p. 185

Sous-groupe Ab

1a Bord ventral ayant un renflement au milieu **2**
1b Bord ventral droit à légèrement courbé ou à faible ondulation sans renflement central **8**
2a Bord dorsal rentré au centre **3**
2b Caractéristique différente **4**
3a Extrémités élargies en forme de “pagaie” prononcée (vol. 2/3 Fig. 141: 8–10) . . . . **12.** ***E. siberica*** vol. 2/3 p. 195
3b Extrémités ayant une autre forme (vol. 2/3 Fig. 141: 1–7) . . . . **10.** ***E. pectinalis*** vol. 2/3 p. 193
4a Bord dorsal ayant des bosses ou des ondulations au centre **5**
4b Caractéristiques différentes **6**
5a Extrémités plus ou moins coniques, raphé contournant largement le pôle (vol. 2/3 Fig. 152: 8–12a) . . . . **29.** ***E. formica*** vol. 2/3 p. 209
5b Extrémités arrondies en largeur jusqu'à plus ou moins pointues. Raphé plus court, bord dorsal parfois à plusieurs ondulations légères (vol. 2/3 Fig. 141: 1–7) . . . . **10.** ***E. pectinalis*** vol. 2/3 p. 193
6a Extrémités plus ou moins coniques (vol. 2/3 Fig. 152: 8–12a) . . . . **29.** ***E. formica*** vol. 2/3 p. 209
6b Extrémités arrondies plus en largeur **7**
7a Grandes valves. Largeur > 10 µm (vol. 2/3 Fig. 158: 1–3) . . . . **30.** ***E. monodon*** vol. 2/3 p. 210
7b Valves plus délicates, notamment plus étroites (vol. 2/3 Fig. 141: 1–7) . . . . **10.** ***E. pectinalis*** vol. 2/3 p. 193
8a Extrémité du raphé ne se courbant que peu dans la face valvaire (vol. 2/3 Fig. 161: 1–7) . . . . **49.** ***E. sudetica*** vol. 2/3 p. 224
8b Extrémité du raphé se courbant plus dans la face valvaire **9**
9a Extrémités plus ou moins coniques (vol. 2/3 Fig. 152: 8–12a) . . . . **29.** ***E. formica*** vol. 2/3 p. 209
9b Extrémités arrondies-tronquées jusqu'à largement renforcées **10**

10a Largeur 4 µm, stries 16–20/10 µm, le bord dorsal renfoncé au milieu (vol. 2/3 Fig. 156: 35–40) . . . . **38. *E. silvahercynia*** vol. 2/3 p. 216
10b Caractéristiques différentes **11**
11a Stries moins espacées > 12/10 µm, d'aspect délicat (vol. 2/3 Fig. 159: 1) . . . . **45. *E. ruzickae*** vol. 2/3 p. 221
11b Stries le plus souvent > 12/10 µm avec ponctuation nette (vol. 2/3 Fig. 141: 1–7) . . . . **10. *E. pectinalis*** vol. 2/3 p. 193

Sous-groupe Ac

1a Largeur de la valve autour ou > 10 µm **17**
1b Largeur de la valve < 10 µm **2**
2a Valves, plus particulièrement leur bord central, courbées très fortement en arc; dans un cas extrème allant jusqu'à une forme hémisphérique (vol. 2/3 Fig. 157: 1–12) . . . . **34. *E. arculus*** vol. 2/3 p. 213
. . . . **33. *E. elegans*** vol. 2/3 p. 212
2b Valves moins fortement courbées **3**
3a Bord ventral plus ou moins fortement courbé **4**
3b Bord ventral légèrement courbé à droit **7**
4a Valves étroites jusqu'à 4 µm maximum. Extrémités du raphé légèrement recourbées en arrière, parfois difficilement reconnaissables (vol. 2/3 Fig. 137/140). Cf. aussi *E. subarcuatoides* ainsi que *E. steineckei, E. exigua, E. tenella, E. paludosa* (vol. 2/3 Fig. 153–155), *E. arculus* (vol. 2/3 Fig. 157: 4–12) . . . . **1. *E. bilunaris*** vol. 2/3 p. 179
. . . . **2. *E. naegelii*** vol. 2/3 p. 182
4b Largeur valvaire > 4 µm **5**
5a Bords ventral et dorsal parallèles jusqu'aux extrémités (vol. 2/3 Fig. 152: 1–3) . . . . **28. *E. parallela*** var. ***angusta*** vol. 2/3 p. 209
5b Bord dorsal plus ou moins rétréci avant les extrémités **6**
6a Extrémités arrondies en largeur (vol. 2/3 Fig. 157: 13–18). Cf. aussi *E. rostellata* (vol. 2/3 Fig. 159: 4, 5) . **35. *E. septentrionalis*** vol. 2/3 p. 213
6b Extrémités plates renforcées (vol. 2/3 Fig. 147) **5. *E. arcus*** vol. 2/3 p. 184
7a Bord ventral faiblement mais nettement courbé **8**
7b Bord ventral droit ou presque invisiblement courbé **14**
8a Structure des valves relativement délicate **9**
8b Structure plus grossière, les stries apparaissant nettement ponctuées . . **13**
9a Extrémités arrondies en largeur ayant des nodules respectivement des aires terminales d'une forme labiale singulière (vol. 2/3 Fig. 151: 11–13) . . . . **32. *E. lapponica*** vol. 2/3 p. 212
9b Caractéristiques différentes **10**
10a Valves très rétrécies vers les extrémités rostrées. Bord dorsal avec ou sans spinules si vivant en haut marécage (vol. 2/3 Fig. 157: 19–28) . . . . **25. *E. denticulata*** vol. 2/3 p. 205
10b Caractéristiques différentes **11**
11a Stries > 16/10 µm. Extrémités rostrées et courbées vers le côté dorsal (vol. 2/3 Fig. 154: 31–34) . . . . **19. *E. nymanniana*** vol. 2/3 p. 201
11b Caractéristiques différentes **12**
12a Bord dorsal rentré avant les extrémités, donc pôles plus ou moins étranglés (vol. 2/3 Fig. 147) . . . . **5. *E. arcus*** vol. 2/3 p. 184
12b Pôles non étranglés (vol. 2/3 Fig. 142: 7–15). Cf. aussi *E. intermedia* (vol. 2/3 Fig. 143: 10), *E. tenella* (vol. 2/3 Fig. 154: 23–30) et *E. bilunaris* (vol. 2/3 Fig. 137, 138) . . . . **14. *E. minor*** vol. 2/3 p. 196
13a Extrémités terminales du raphé s'étendant à peu près jusqu'au bord dorsal (vol. 2/3 Fig. 151: 1–10a) . . . . **27. *E. glacialis*** vol. 2/3 p. 207

13b Extrémités terminales du raphé plus courtes (vol. 2/3 Fig. 142: 1–6) . . . . **11.** ***E. soleirolii*** vol. 2/3 p. 194
14a Extrémités arrondies **15**
14b Extrémités plates renforcées **16**
15a Valves selon plusieurs figures de la Fig. 137 . **1.** ***E. bilunaris*** vol. 2/3 p. 179
15b Valves d'autres formes, stries plus espacées (vol. 2/3 Fig. 142: 1–6), cf. aussi *E. pectinalis* (vol. 2/3 Fig. 141) . . . . **11.** ***E. soleirolii*** vol. 2/3 p. 194
16a Rapport largeur/longueur 1/3–7 (vol. 2/3 Fig. 148) . . . . **6.** ***E. praerupta*** vol. 2/3 p. 186
16b Rapport largeur/longueur 1/6–15 (vol. 2/3 Fig. 147) . . . . **5.** ***E. arcus*** vol. 2/3 p. 184
17a Largeur valvaire maximum 10 µm **18**
17b Largeur valvaire > 10 µm **21**
18a Extrémités ayant des nodules (aires) terminaux singulièrement grands. Stries délicates environ 20/10 µm (vol. 2/3 Fig. 151: 11–13) . . . . **32.** ***E. lapponica*** vol. 2/3 p. 212
18b Caractéristiques différentes **19**
19a Bords dorsal et ventral parallèles jusqu'aux extrémités (vol. 2/3 Fig. 10: 14–18) . . . . **28.** ***E. parallela*** vol. 2/3 p. 208
19b Bord dorsal rentré avant les extrémités ou valves distales progressivement étirées **20**
20a Extrémités arrondies-tronquées ou gonflées (vol. 2/3 Fig. 151: 1–10a) . . . . **27.** ***E. glacialis*** vol. 2/3 p. 207
20b Si extrémités arrondies en largeur à plates renforcées, cf. *E. praerupta* et *E. arcus* (Fig. 147, 148)
21a Largeur valvaire 22–28 µm (vol. 2/3 Fig. 158: 7, 8) . . . . **31.** ***E. clevei*** vol. 2/3 p. 211
21b Largeur valvaire < 18 µm **22**
22a Bord dorsal en parallèle jusqu'aux extrémités (vol. 2/3 Fig. 152: 1–7) . . . . **28.** ***E. parallela*** vol. 2/3 p. 208
22b Bord dorsal plus ou moins fortement rentré, forme variable des extrémités (vol. 2/3 Fig. 148) . . . . **6.** ***E. praerupta*** vol. 2/3 p. 186

Sous-groupe Ad

1a Bord dorsal à faible ondulation, le creux étant le plus profond au centre de la valve (vol. 2/3 Fig. 156: 35–40) . . **38.** ***E. silvahercynia*** vol. 2/3 p. 216
1b Caractéristique différente **2**
2a Bord dorsal ayant un renflement plus ou moins pointu au centre . . . **3**
2b Deux ou plusieurs bosses arrondies **4**
3a Bord ventral ondulé (vol. 2/3 Fig. 156: 27–36) . . . . **24.** ***E. microcephala*** vol. 2/3 p. 205
3b Bord ventral droit (vol. 2/3 Fig. 155: 22–37) . . . . **23.2,** ***E. paludosa*** var. ***trinacria*** vol. 2/3 p. 204
4a Bord dorsal ayant plus de deux bosses **5**
4b Bord dorsal ayant deux bosses **6**
5a Bosses réparties relativement régulièrement, arrondies-plates (vol. 2/3 Fig. 156: 1–22). Cf. aussi les formes *tridentula* de *E. exigua* (vol. 2/3 Fig. 153: 21–27) . . . . **39.** ***E. muscicola*** vol. 2/3 p. 216
5b Bosses réparties irrégulièrement, d'aspect billot (vol. 2/3 Fig. 156: 23–26) . . . . **40.** ***E. crista-galli*** vol. 2/3 p. 219
6a Extrémités courbées-rostrées vers le côté dorsal ou capitées (vol. 2/3 Fig. 153: 19, 20, 24; 154: 18–22) **17.** ***E. exigua*** **var. *bidens*** vol. 2/3 p. 199 . . . . **22** ***E. rhynchocephala*** var. ***satelles*** vol. 2/3 p. 203
6b Extrémités ni capitées ni courbées-rostrées **7**

7a Extrémités arrondies-tronquées (vol. 2/3 Fig. 143: 1–9a) . . . . **15.** ***E. implicata*** vol. 2/3 p. 197
7b Extrémités arrondies en largeur jusqu'à plates ou renforcées obliquement. Stries le plus souvent un plus plus espacées que *E. implicata* (vol. 2/3 Fig. 143: 16–23) . . . . **16.** ***E. circumborealis*** vol. 2/3 p. 197

Sous-groupe Ae

1a Extrémités du raphé se courbant toujours très légèrement par-dessus le bord ventral dans la face valvaire (vol. 2/3 Fig. 164: 12–20). Cf. aussi *E. siolii* (vol. 2/3 Fig. 165: 1–10) et autres espèces de régions tropiques et subtropiques . . . . **48.** ***E. rhomboidea*** vol. 2/3 p. 223
1b Extrémités du raphé bien reconnaissables, aussi lors d'une haute focalisation **2**
2a Valves étroites, environ 1–3 µm **3**
2b Valves régulièrement plus larges que 3 µm **5**
3a Stries espacées, < 15/10 µm (vol. 2/3 Fig. 150: 10–24). Cf. aussi *E. tenella* (vol. 2/3 Fig. 154: 23–30) . . . . **26.** ***E. fallax*** vol. 2/3 p. 206
3b Stries moins espacées, > 14/10 µm **4**
4a Bord ventral plus fortement courbé (vol. 2/3 Fig. 137, 138). Cf. aussi les formes dites *falcata* et *subarcuata*, *E. subarcuatoides* et plusieurs espèces du groupe autour de *E. exigua* (pl. 153, 154) . . . . **1.2.** ***E. bilunaris*** var. ***mucophila*** vol. 2/3 p. 180
4b Bord ventral légèrement courbé ou droit. Largeur valvaire variable selon les variétés, également avec une bosse centrale courte, plutôt pointue (vol. 2/3 Fig. 155). Cf. aussi les taxons mentionnés sous 4a qui peuvent porter à confusion . . . . **23.** ***E. paludosa*** vol. 2/3 p. 203
5a Largeur valvaire de 3 à environ 5 µm **6**
5b Largeur valvaire > 5 µm **23**
6a Valves plus ou moins en forme de haricot. Bord dorsal pas du tout rentré avant les extrémités (vol. 2/3 Fig. 143: 10–15) . . . . **37.** ***E. intermedia*** vol. 2/3 p. 215
6b Caractéristiques différentes **7**
7a Bord ventral fortement courbé (vol. 2/3 Fig. 157: 4–12). Cf. aussi *E. elegans* (vol. 2/3 Fig. 157: 1–3) . . . . **34.** ***E. arculus*** vol. 2/3 p. 213
7b Bord ventral courbé modérément à faiblement ou pas du tout **8**
8a Si bord dorsal peu renflé et extrémités plus ou moins prononcées rostrées, cf. groupe autour de *E. exigua* (vol. 2/3 Fig. 155, 156) et *E. arculus* (vol. 2/3 Fig. 157: 4–12)
8b Caractéristiques différentes **9**
9a Bord ventral droit ou très faiblement courbé **21**
9b Bord ventral encore nettement concave **10**
10a Bord dorsal fortement renflé **14**
10b Bord dorsal modérément renflé **11**
11a Extrémités rostrées à capitées (pl. 153, 154, 138) . . . .
. . . . **17.** ***E. exigua*** vol. 2/3 p. 199
. . . . **20.** ***E. meisteri*** vol. 2/3 p. 202
. . . . **36.** ***E. subarcuatoides*** vol. 2/3 p. 214
11b Extrémités pas ou très peu prononcées, arrondies-tronquées à larges . **12**
12a Bord dorsal plus fortement rentré avant les extrémités (vol. 2/3 Fig. 153, 154). Cf. aussi *E. septentrionalis* (vol. 2/3 Fig. 157:13–18) et *E. denticulata* (vol. 2/3 Fig. 157: 19–28) . . . .
. . . . **17.** ***E. exigua*** vol. 2/3 p. 199
. . . . **21.** ***E. tenella*** vol. 2/3 p. 202
12b Bord dorsal peu ou pas rentré **13**

13a Bord dorsal encore nettement rentré, extrémités légèrement prononcées. (vol. 2/3 Fig. 142: 7–15). Cf. aussi la variante à une bosse de *E. implicata* (vol. 2/3 Fig. 143: 1–9) . . . . **14. *E. minor*** vol. 2/3 p. 196
13b Bord dorsal non rentré avant les extrémités. Raphé légèrement recourbé en arrière mais cela est difficilement reconnaissable (vol. 2/3 Fig. 137, 138) . . . . **1. *E. bilunaris*** formes falcata et subarcuata vol. 2/3 p. 179
14a Bord dorsal progressivement descendant du centre aux extrémités de façon à ce que la largeur valvaire diminue en continu **15**
14b Bords dorsal et ventral plus ou moins parallèles dans la partie centrale de la valve et rentrés peu avant les extrémités **19**
15a Bord dorsal légèrement ondulé **16**
15b Bord dorsal non ondulé **17**
16a Bord dorsal distal très fortement rentré et donc extrémités capitées et pédonculées (vol. 2/3 Fig. 154: 11–22) . . . . **22. *E. rhynchocephala*** vol. 2/3 p. 203
16b Extrémités différentes, arrondies en largeur (vol. 2/3 Fig. 156: 1–22) . . . . **39. *E. muscicola*** vol. 2/3 p. 216
17a Nodules (aires) terminaux très prononcés, bord dorsal avec ou sans petites dents (vol. 2/3 Fig. 157: 19–28) . . . . . **25. *E. denticulata*** vol. 2/3 p. 205
17b Nodules (aires) terminaux peu prononcés **18**
18a Stries moins espacées, environ 20 et >/10 µm (vol. 2/3 Fig. 153) . . . . **17. *E. exigua*** vol. 2/3 p. 199
18b Stries plus espacées (vol. 2/3 Fig. 154: 23–30) **21. *E. tenella*** vol. 2/3 p. 202
19a Nodules (aires) terminaux très prononcés (vol. 2/3 Fig. 157: 19–28) . . . . **25. *E. denticulata*** vol. 2/3 p. 205
19b Nodules (aires) terminaux peu prononcés, très petits **20**
20a Extrémités apparaissant comme capitées et pédonculées. Bord dorsal le plus souvent faiblement ondulé (vol. 2/3 Fig. 154: 11–22) . . . . **22. *E. rhynchocephala*** vol. 2/3 p. 203
20b Caractéristiques différentes (vol. 2/3 Fig. 157: 13–18). Cf. aussi *E. minor* (vol. 2/3 Fig. 142: 7–15) . . . . . . **35. *E. septentrionalis*** vol. 2/3 p. 213
21a Bord dorsal très courbé, rarement à deux ondulations (vol. 2/3 Fig. 154: 1–10). Cf. aussi *E. rhynchocephala* (vol. 2/3 Fig. 154: 11–22) . . . . **20. *E. meisteri*** vol. 2/3 p. 202
21b Bord dorsal moins fortement courbé ou presque en parallèle au bord ventral **22**
22a Bord dorsal très légèrement rentré aux extrémités (vol. 2/3 Fig. 154: 23–30) . . . . **21. *E. tenella*** vol. 2/3 p. 202
22b Bord dorsal fortement rentré, en outre le bord ventral est lui aussi nettement rentré (vol. 2/3 Fig. 154: 11–22) . . . . **22. *E. rhynchocephala*** vol. 2/3 p. 203
23a Extrémités des valves élargies capitées et/ou arrondies en largeur jusqu'à plates-renforcées **24**
23b Extrémités des valves autrement formées . . **14. *E. minor*** vol. 2/3 p. 196
24a Rapport largeur/longueur 1/3–7 (vol. 2/3 Fig. 148: 4–8) . . . . **Espèces de 6.3. *E. praerupta* curta** vol. 2/3 p. 187
24b Rapport largeur/longueur 1/6–15 (pl. 147) . . **5. *E. arcus*** vol. 2/3 p. 184

Groupe-clef B

1a Bord dorsal avec deux ou plus de bosses prononcées **10**
1b Bord dorsal sans multiples bosses, toujours faiblement ondulé **2**
2a Largeur de la valve 5–8 µm **3**
2b Largeur de la valve 2–6 µm **6**
3a Bord dorsal descendant vers les extrémités avec épaule plus ou moins

prononcée (vol. 2/3 Fig. 161: 1–4). Cf. aussi *E. veneris*, *E. pirla*, *E. carolina* (pl. 163) ainsi que *E. convexa* et *E. siolii* (pl. 165) . . . . **49. *E. sudetica*** vol. 2/3 p. 224

3b Bord dorsal descendant sans épaules vers les extrémités **4**

4a Valves avec un bord dorsal fortement convexe, souvent en forme de haricot (fève). Cf. aussi *E. sudetica* (vol. 2/3 Fig. 161: 5–7) et *E. incisa* (pl. 161, 163) . . . . **50. *E. faba*** vol. 2/3 p. 225

4b Valves plus étirées, paraissant souvent avoir des méandres **5**

5a Raphé non visible à réglage haut en vue valvaire (vol. 2/3 Fig. 164: 1–10) . . . . **50. *E. faba*** vol. 2/3 p. 225

5b Extrémités du raphé extrêmement distales, encore légèrement visibles à réglage haut (vol. 2/3 Fig. 152: 1–7) . . . **28. *E. parallela*** vol. 2/3 p. 208

6a Largeur de la valve environ 4–6 µm **7**

6b Largeur de la valve environ 2–4 µm **8**

7a Nodules terminaux nettement éloignés des pôles, extrémités très prononcées en dents de requin (vol. 2/3 Fig. 161: 12–19; 163: 1–7) . . . . **46. *E. incisa*** vol. 2/3 p. 221

7b Nodules terminaux plus proches des pôles, extrémités faiblement prononcées en forme nasale (vol. 2/3 Fig. 164: 11–20). Cf. aussi *E. intermedia* (vol. 2/3 Fig. 143: 10–15) . . . . **48. *E. rhomboidea*** vol. 2/3 p. 223

8a En réglage haut, extrémités du raphé non nettement reconnaissables (vol. 2/3 Fig. 164: 11–20). Cf. aussi *E. siolii* (vol. 2/3 Fig. 165: 1–9) . . . . **48. *E. rhomboidea*** vol. 2/3 p. 223

8b En réglage haut, extrémités du raphé faiblement visibles en arc **12**

9a Stries plus espacées, environ 13/10 µm. Bord dorsal rentré avant les extrémités (vol. 2/3 Fig. 154: 23–30). Cf. aussi *E. fallax* (vol. 2/3 Fig. 150: 10–24) et *E. siolii* (vol. 2/3 Fig. 165: 1–9) . . **21. *E. tenella*** vol. 2/3 p. 202

9b Stries moins espacées, environ 20/10 µm. Bord dorsal non rentré (vol. 2/3 Fig. 155). Cf. aussi espèces autour de *E. exigua* (vol. 2/3 Fig. 153, 154) . . . . **23. *E. paludosa*** vol. 2/3 p. 203

10a Bord dorsal ayant (4) 5–10 bosses (vol. 2/3 Fig. 166: 1–4) . . . . **52. *E. hexaglyphis*** vol. 2/3 p. 227

10b Bord dorsal ayant 2 bosses **11**

11a Largeur des valves aux extrémités diminuant à environ ¼ de la largeur maximale. Nodules terminaux nettement distincts des extrémités (vol. 2/3 Fig. 161: 21–25) . . . . **51. *E. bidentula*** vol. 2/3 p. 226

11b Si valves moins larges et réduites à environ ½ à ⅓ et nodules terminaux avec extrémités du raphé se trouvant plus proches des extrémités, cf. *E. diodon* et les espèces autour de *E. praerupta* (vol. 2/3 Fig. 149, 150)

Groupe-clef C

1a Grandes valves, largeur le plus souvent 5–10 µm. Longueur le plus souvent > 50 µm (vol. 2/3 Fig. 140: 7–18) **2**

1b Valves en général plus délicates. Largeur le plus souvent < 5 µm (vol. 2/3 Fig. 137, 138, 140: 1–6) **3**

2a Valves plus larges au centre que distales (vol. 2/3 Fig. 140: 7) . . . . **4. *E. pseudopectinalis*** vol. 2/3 p. 184

2b Valves de largeur régulière ou aux extrémités plus ou moins élargies (vol. 2/3 Fig. 140: 8–18) . . . . **3. *E. flexuosa*** vol. 2/3 p. 182

3a Valves beaucoup plus longues que larges, largeur 2–4 µm **4**

3b Valves d'une largeur de 3 à > 5 µm (vol. 2/3 Fig. 137 part.) . . . . **1. *E. bilunaris*** vol. 2/3 p. 179

4a Valves non rétrécies vers les extrémités (vol. 2/3 Fig. 140: 8–18) . . . . **3. *E. flexuosa*** vol. 2/3 p. 182

4b Valves progressivement rétrécies vers les extrémités **5**

5a Extrémité du raphé revenant en arrière, toujours reconnaissable comme un arc court (vol. 2/3 Fig. 137–139). Cf. aussi *E. subarcuatoides* . . . . 1. ***E. bilunaris*** vol. 2/3 p. 179

5b Extrémité du raphé revenant en arrière, visible après réglage. Valves le plus souvent moins courbées que *E. bilunaris* (vol. 2/3 Fig. 140: 1–6) . . . . **2. *E. naegelii*** vol. 2/3 p. 182

2. Famille Achnanthaceae Kützing 1844 (Vol. 2/4)

Typus familiae: *Achnanthes longipes* Agardh 1824 syn. *Conferva armillaris* O.M. Müller 1783

Clef pour les genres

1a Valves du raphé à valvocopulae clairement visibles au microscope optique également en vue connective, reliées à la face valvaire par des fimbriae . . . . **2. *Cocconeis*** (vol. 2/4 p. 83) (p. 223)

1b Caractéristiques différentes . . . . . **1. *Achnanthes*** (vol. 2/4 p. 1) (p. 213)

1. Genre *Achnanthes* Bory 1822

Typus generis: *Achnanthes adnata* Bory 1822

Clef pour les sous-genres

1a Faces valvaires et manteaux valvaires grossièrement ponctués; en vue connective large, les bandes isolées sont également ponctuées; valve sans raphé en général avec aire axiale étroite plus ou moins excentrique (pl. 1 et 2) . . . . ***Achnanthes*** (vol. 2/4 p. 1) (p. 213)

1b Caractéristiques différentes; si exceptionnellement la face valvaire est grossièrement ponctuée, le manteau valvaire a toujours une seule couronne circulaire de points, les bandes isolées de la ceinture semblent hyalines; l'aire axiale de la valve sans raphé est en position médiane, souvent élargie de forme elliptique à lancéolée . . ***Achnanthidium*** (vol. 2/4 p. 6) (p. 214)

1. Sous-genre *Achnanthes* Bory 1822 (vol 2/4 p. 2)

Typus subgeneris: *Achnanthes adnata* Bory 1822

Clef pour les espèces dans le sous-genre *Achnanthes*

1a Stries constituées de 2 (plus rarement 3) lignes de points (vol. 2/4 Fig. 1: 1) . . . . **4. *A. longipes*** vol. 2/4 p. 5

1b Stries constituées d'une seule ligne de points **2**

2a Valves renflées au milieu et aux extrémités (vol. 2/4 Fig. 2: 9, 10) . . . . **6. *A. inflata* var. *inflata*** vol. 2/4 p. 6

2b Contour valvaire différent **3**

3a Valves aux extrémités +/- étirées et +/- concaves au milieu (vol. 2/4 Fig. 2: 1–8) . . . . **5. *A. coarctata*** vol. 2/4 p. 5

3b Caractéristiques différentes **4**

4a Pôles valvaires cunéiformes, régulièrement 7–8 stries transapicales/10 µm (vol. 2/4 Fig. 1: 2, 3) . . . . **1. *A. brevipes*** vol. 2/4 p. 3

4b Pôles valvaires largement arrondis non spécifiquement cunéiformes, > 8 str/10 µm **5**

5a Aire axiale des valves sans raphé limitée aux deux extrémités valvaires par une structure tavelée (orbiculus) (vol. 2/4 Fig. 1: 11–15) . . . . **3.** *A. parvula* vol. 2/4 p. 5
5b Valves sans raphé sans orbiculus **6**
6a Valves rhomboédriques-lancéolées, jamais concaves au milieu (vol. 2/4 Fig. 2: 11, 12) . . . . . **6.** ***A. elata* syn. *A. inflata*** var *elata* vol. 2/4 p. 6
6b Contour valvaire différent, populations au moins partiellement à valves faiblement étranglées dans l'aire centrale **7**
7a Stries 13–15/10 µm (vol. 2/4 Fig. 1: 9, 10) . . . . ***A. islandica*** (sans diagnostic) vol. 2/4 p. 4
7b Stries plus éloignées les unes des autres, 9–12/10 µm (vol. 2/4 Fig. 1: 4–8) . . . . **2.** ***A. subsessilis*** syn. *A. brevipes* var. *intermedia* vol. 2/4 p. 4

2. Sous-genre *Achnanthidium* (Kützing 1844) Hustedt 1933 non sensu Hustedt (vol. 2/4 p. 6)

Typus subgeneris: *Achnanthidium microcephalum* Kützing 1844 (Reimer in Patrick & Reimer 1966)

Clef pour les groupes dans le sous-genre *Achnanthidium*

1a Valve sans raphé ou, rarement valves sans raphé et à raphé, avec "une tache en forme de fer à cheval" (pl. 41–48) . . . . . . **Groupe-clef H** (p. 223)
1b Valve sans raphé sans "tache en forme de fer à cheval" **2**
2a Valves sans et à raphé montrant des différences significatives dans la densité et les angles d'inclinaison des stries (vol. 2/4 Fig. 16: 1–21) . . . . **Groupe-clef A** (p. 215)
2b Valves sans et à raphé légèrement différentes: les aires peuvent être de grandeurs différentes, les stries interrompues par une aire hyaline sur une valve ou elles ne se différencient que par la densité ou par l'angle d'inclinaison des stries **3**
3a Stries apparaissant grossières et éloignées les unes des autres, moins de 20/10 µm; toutefois non résolvables en lignes de points au microscope optique (au microscope électronique mise en évidence de lignes doubles ou multiples d'aréoles ou alvéoles pas du tout articulées par des aréoles) . . . . **Groupe-clef B** (p. 215)
3b Stries plus proches ou décomposables en lignes de points simples . . . **4**
4a Stries pratiquement non résolvables au microscope optique, environ 40/10 µm, extrémités rostrées, très étirées (vol. 2/4 Fig. 22: 31) . . . . **75.** ***A. gracillima*** vol. 2/4 p. 57
4b Caractéristiques différentes **5**
5a Stries sur face valvaire interrompues des deux côtés de la médiane par une étroite zone hyaline. Souvent cette interruption n'apparaît que sur la valve sans raphé . . . . **Groupe-clef C** (p. 217)
5b Stries non-interrompues ou seulement en bordure du manteau **6**
6a Raphé en position plus ou moins diagonale, sigmoïde ou presque droite; seules ses extrémités distales sont courbées en sens opposé . . . . **Groupe-clef D** (p. 218)
6b Extrémités du raphé tout à fait droites, ou extrémités tournées dans le même sens ou tout au moins non visiblement courbées en sens opposé **7**
7a Contour des valves elliptique ou elliptico-lancéolé ou proche de rhomboédrique. Extrémités larges-arrondies et non ou à peine étirées . . . . **Groupe-clef E** (p. 218)
7b Contour différent. Chez la majorité des individus, la longueur dépasse le

double de la largeur (contour à tendance linéaire), ou les extrémités sont très nettement étirées **8**

8a Contour valvaire (moyenne de la population) linéaire, linéaire-elliptique ou linéaire-lancéolé, largeur valvaire < 5 µm (donc des formes minces, étirées selon pl. 32–37) . . . . **Groupe-clef F** (p. 220)

8b Caractéristiques différentes; valves plus larges ou plus elliptiques à lancéolées, souvent avec des extrémités capitées ou rostrées . . . . **Groupe-clef G** (p. 221)

Groupe-clef A

(Espèces à différence de structure considérable sur valve à raphé ou sans raphé)

1a Individus d'une longueur < 10 µm, valve sans raphé sans stries (vol. 2/4 Fig. 15: 39–44) . . . . **26. *A. kuelbsii*** vol. 2/4 p. 27

1b Les deux valves à structures **2**

2a Valve à raphé possédant un point isolé près du nodule central, stries sur valve sans raphé très proches, difficilement différenciables (vol. 2/4 Fig. 29:1–9) **3**

2b Caractéristiques différentes **4**

3a Stries sur valve sans raphé > 40/10 µm (vol. 2/4 Fig. 29: 7–9) . . . . **69. *A. bremeyeri*** vol. 2/4 p. 50

3b Stries sur valve sans raphé, 30–36/10 µm (vol. 2/4 Fig. 29: 1–6) . . . . **68. *A. bahusiensis*** vol. 2/4 p. 50

4a Valves aux extrémités plus ou moins étirées, cunéiformes ou le plus souvent rostrées **7**

4b Extrémités largement arrondies **5**

5a Aire centrale sur valve à raphé en forme de bande transversale **6**

5b Aire centrale moins typique (vol. 2/4 Fig. 18: 20, 21) . . . . ***A. reversa*** (sans diagnostic)

6a Valve sans raphé à aire centrale large et ne possédant donc des stries qu'aux bords (vol. 2/4 Fig. 16: 15–21) . . . . **31. *A. lutheri*** vol. 2/4 p. 29

6b Valve sans raphé à aire centrale plus étroite (vol. 2/4 Fig. 16: 1–14) . . . . **30. *A. oblongella*** vol. 2/4 p. 29

7a Stries sur valve à raphé nettement ponctuées (vol. 2/4 Fig. 21: 1–17) . . . . **45. *A. clevei***

7b Caractéristique différente ou difficile à reconnaître **8**

8a Valve à raphé à aire transversale similaire à un stauros (vol. 2/4 Fig. 23: 1–27) . . . . **49. *A. exigua part.***

8b Caractéristique différente **9**

9a Valve sans raphé à stries nettement ponctuées; valve à raphé sans aire centrale bien marquée (vol. 2/4 Fig. 21: 18–27) . . . . **46. *A. laterostrata*** vol. 2/4 p. 36

9b Caractéristiques différentes (vol. 2/4 Fig. 23: 39, 40) . . . . **76. *A. dispar*** vol. 2/4 p. 54

Groupe-clef B

(Valves à raphé et sans raphé à stries grossières)

1a Contour valvaire de la moyenne de la population à tendance plutôt linéaire, largeur < 5 µm (vol. 2/4 Fig. 30: 1–19; 36: 38–41; 37: 1–8) . . . **25**

1b Caractéristiques différentes **2**

2a Extrémités valvaires largement arrondies, non étirées **3**

2b Extrémités valvaires plus ou moins étirées **12**

3a Valve à raphé avec une aire centrale similaire à un stauros, allant jusqu'aux bords, et légèrement élargie de façon asymétrique (vol. 2/4 Fig. 19:1–15) . . . . **41. *A. hungarica*** part. vol. 2/4 p. 33

3b Caractéristique différente **4**
4a Valve sans raphé à petite aire centrale **8**
4b Valve sans raphé à aire centrale grande, surtout avec aire centrale plus large, donc stries plus sur les bords **5**
5a Valves rhomboédriques-elliptiques à rhomboédriques-lancéolées, généralement de longueur < 14 µm (vol. 2/4 Fig. 17: 22–34) . . . . **34.** ***A. montana*** vol. 2/4 p. 31
5b Valves plus grandes et/ou elliptiques sans tendance rhomboédrique . . **6**
6a Valve à raphé à aire axiale étroite, linéaire **7**
6b Valve à raphé à aire axiale élargie lancéolée (vol. 2/4 Fig. 17: 14–21) . . . . **33.** ***A. rupestris*** vol. 2/4 p. 30
7a Valve à raphé avec une aire centrale étroite à très étroite entre les deux stries centrales, et allant le plus souvent jusqu'aux bords de la valve (vol. 2/4 Fig. 16: 22–33) . . . . **28.** ***A. conspicua*** vol. 2/4 p. 28
7b Aire centrale plus large, n'atteignant généralement pas les bords (vol. 2/4 Fig. 17: 35–42) . . . . **35.** ***A. rupestoides*** vol. 2/4 p. 31
8a Valve à raphé sans aire centrale bien marquée, forme marine (vol. 2/4 Fig. 40: 23–26) . . . . **90.** ***A. polaris*** vol. 2/4 p. 69
8b Caractéristiques différentes **9**
9a Valve à raphé à aire centrale étroite à très étroite entre les deux stries centrales allant le plus souvent jusqu'aux bords valvaires (vol. 2/4 Fig. 16: 22–33) . . . . **28.** ***A. conspicua*** vol. 2/4 p. 28
9b Caractéristiques différentes **10**
10a Valve à raphé avec une aire centrale plus grande, plus ou moins ronde (vol. 2/4 Fig. 39; 40) . . . . **92.** ***A. delicatula*** part.
10b Aire centrale plus petite, transversalement élargie par raccourcissement de 1–2 stries **11**
11a Formes d'eaux douce et saumâtre riches en électrolytes, stries sur valve à raphé courbées, plutôt radiales (vol. 2/4 Fig. 39; 40) . . . . **92.** ***A delicatula*** part. vol. 2/3 p. 71
11b Formes d'eau douce plus pauvre en électrolytes, stries sur valve à raphé plus faiblement radiales (vol. 2/4 Fig. 18: 1–13). Si:
contour valvaire elliptico-lancéolé . . . . . **36.** ***A. distincta*** vol. 2/4 p. 32
strictement elliptiques . . . . **37.** ***A. stewartii*** vol. 2/4 p. 32
12a Extrémités terminales du raphé courbées en sens opposé, extrémités étirées tronquées (vol. 2/4 Fig. 18: 14–17) . . . . . . **38.** ***A. holstii*** vol. 2/4 p. 33
12b Extrémités terminales du raphé courbées dans la même direction ou droites **13**
13a Extrémités à forme rostrée **14**
13b Extrémités cunéiformes plus tronquées ou plus pointues ou étirées tronquées **23**
14a Stries de la valve sans raphé interrompues des deux côtés par des aires hyalines (vol. 2/4 Fig. 22: 5–12), cf. aussi quelques formes de *A. ploenensis* var. *gessneri* vol. 2/4 p. 37 (Fig. 22: 25–28) . . . . **47.** ***A kolbei***
14b Caractéristique différente **15**
15a Grandes valves; forme marine de 30 à > 40 µm de long, str. sur valve sans raphé composées de rangées de points doubles reconnaissables au microscope optique (vol. 2/4 Fig. 40: 21, 22) . . . . **91.** ***A. linkei*** vol. 2/4 p. 69
15b Caractéristiques différentes **16**
16a Stries sur valve avec raphé très fortement radiales au milieu (vol. 2/4 Fig. 23: 39, 40) . . . . **76.** ***A. dispar*** vol. 2/4 p. 54
16b Caractéristique différente **17**
17a Stries sur valve sans raphé fortement radiales (vol. 2/4 Fig. 15: 45–48) . . . . **29.** ***A. lacunarum*** vol. 2/4 p. 28

17b Stries sur valve sans raphé approx. parallèles à plus faiblement radiales . **18**
18a Valves de 12 µm de long ou plus petites **19**
18b Valves plus grandes **21**
19a Valves elliptiques à linéaires-elliptiques, à extrémités abruptes largement étirées (vol. 2/4 Fig. 38: 13–24) . . . . **63.** ***A. daui*** vol. 2/4 p. 47
19b Valves aux extrémités étirées plus étroites à rostrées tronquées **20**
20a Extrémités étirées de forme rostrée étroite (vol. 2/4 Fig. 26: 31–40) . . . . 60 ***A. lemmermannii*** vol. 2/4 p. 44
20b Extrémités courtes et larges (vol. 2/4 Fig. 38: 1–12) . . . . **62.** ***A. grana*** vol. 2/4 p. 45
21a Valves aux bords peu convexes, à tendance linéaire-elliptique; alvéoles inarticulées (sous microscope électronique) (vol. 2/4 Fig. 22: 13–30) . . . . **48.** ***A. ploenensis*** vol. 2/4 p. 37
21b Bords plus convexes, alvéoles (sous microscope électronique) à 2 ou plus de rangées de points **22**
22a Stries (sous microscope électronique) généralement constituées de deux rangées de points (vol. 2/4 Fig. 26: 47–49)(sans diagnostic) . . . . ***A.*** **species "de Bonn"**
22b Stries constituées de plus de 2 rangées de points . . . . **92.** ***A. delicatula*** part. vol. 2/4 p. 72
23a Valves toujours < 12 µm de long, formes d'eau douce pauvre en électrolytes (vol. 2/4 Fig. 38: 1–12) . . . . **62.** ***A. grana*** vol. 2/4 p. 45
23b Valves (moyenne de la population) nettement > 12 µm de long **24**
24a Extrémités arrondies tronquées ou cunéiformes, contour elliptique à rhomboédrique-lancéolé (vol. 2/4 Fig. 39; 40) . . . . **92.** ***A. delicatula*** **part** vol. 2/4 p. 71
24b Extrémités subcapitées, graduellement élargies, contour valvaire à tendance linéaire-elliptico-lancéolé (vol. 2/4 Fig. 40: 14–20) . . . . **93.** ***A. pericava*** vol. 2/4 p. 73
25a Aire centrale n'atteignant jamais les bords valvaires; valves strictement linéaires et pas plus longues que 15 µm (vol. 2/4 Fig. 37: 1–8) . . . . **86.** ***A. nodosa*** vol. 2/4 p. 67
25b Caractéristiques différentes **26**
26a Valves strictement linéaires, largeur <= 3µm, peu élargies au centre, aire centrale atteignant généralement les deux bords valvaires (vol. 2/4 Fig. 36: 38–41) . . . . **85.** ***A. kriegeri*** vol. 2/4 p. 66
26b Valves linéaires-lancéolées. En moyenne, largeur de la valve toujours > 3 µm; aire centrale n'atteignant pas le bord ou alors d'un seul côté (vol. 2/4 Fig. 30: 1–19) . . . . **73.** ***A. thermalis*** part. vol. 2/4 p. 52

Groupe-clef C
(Stries interrompues par une aire hyaline)
1a Valves elliptiques aux extrémités larges-arrondies (vol. 2/4 Fig. 28: 9–20) . . . . **66.** ***A. suchlandtii*** vol. 2/4 p. 49
1b Valves aux extrémités étirées **2**
2a Stries < 15/10 µm (vol. 2/4 Fig. 22: 5–12), cf. aussi qq formes de *A. ploenensis* var. *gessneri* (vol. 2/4 Fig. 22: 21–28) **47.** ***A. kolbei*** vol. 2/4 p. 37
2b Stries > 15/10 µm **3**
3a Partie centrale de la valve le plus souvent linéaire-elliptique, formes marines, d'eau saumâtre ou d'eau douce riche en électrolytes (vol. 2/4 Fig. 26: 7–23) . . . . **59.** ***A. amoena*** vol. 2/4 p. 44
3b Partie centrale de la valve elliptique avec le plus souvent des bords plus fortement convexes; formes d'eau douce moins riche en électrolytes (sans diagnostic) (vol. 2/4 Fig. 26: 24–29) . . . . ***A. nitidiformis***

Groupe-clef D

(Raphé diagonal, sigmoïde ou extrémités distales courbées en sens opposé)

1a Raphé fortement oblique ou à développement sigmoïde **2**
1b Raphé droit pour la majorité, courbé en sens opposé aux extrémités . **5**
2a Faces valvaires fortement ondulées en 4 parties, si bien que la structure des stries semble floue au microscope optique à réglage constant (vol. 2/4 Fig. 9: 1–10) . . . . **7.** ***A. flexella*** vol. 2/4 p. 16
2b Caractéristique différente **3**
3a Aire centrale de la valve sans raphé présente d'un seul côté; forme d'eau saumâtre (vol. 2/4 Fig. 29: 16) . . . . **40.** ***A. pseudobliqua*** vol. 2/4 p. 33 (sans diagnostic)
3b Caractéristiques différentes **4**
4a Valves elliptiques aux extrémités larges-arrondies (vol. 2/4 Fig. 12: 1–9) . . . . **11.** ***A. bioretii*** part.
4b Valves largement lancéolées, aux extrémités effilées cunéiformes (vol. 2/4 Fig. 18: 18, 19) . . . . **39.** ***A. obliqua*** vol. 2/4 p. 33
5a Stries grossières, < 15/10 µm **6**
5b Stries plus rapprochées **7**
6a Valves aux extrémités nettement étirées (vol. 2/4 Fig. 18: 14–17) . . . . **38.** ***A. holstii*** vol. 2/4 p. 33
6b Extrémités peu ou pas étirées (vol. 2/4 Fig. 19: 1–8) . . . . **36.** ***A. distincta*** vol. 2/4 p. 32
7a Valves plus ou moins linéaires aux extrémités larges arrondies ou effilées coniques. Aire centrale sur la valve à raphé constituant une bande transversale atteignant les bords des deux côtés (vol. 2/4 Fig. 19: 1–15) . . . . **41.** ***A. hungarica*** vol. 2/4 p. 33
7b Caractéristiques différentes **8**
8a Aire centrale sur la valve à raphé atteignant les bords des deux côtés (vol. 2/4 Fig. 23: 1–27) . . . . **49.** ***A. exigua*** part. vol. 2/4 p. 38
8b Aire centrale asymétrique, plus nettement marquée d'un côté (vol. 2/4 Fig. 9: 14–22; 10: 1–11) . . . . **9.** ***A. laevis*** vol. 2/4 p. 17

Groupe-clef E

(Contours elliptiques ou elliptico-lancéolées ou rhomboédriques, toujours avec des extrémités largement arrondies)

1a Valves concaves dans la partie centrale (vol. 2/4 Fig. 10: 28–33) . . . . **16.** ***A. didyma*** vol. 2/4 p. 22
1b Caractéristiques différentes **2**
2a Stries très nettement ponctuées (aussi sans effet d'éclairage particulier) **3**
2b Stries pas nettement ponctuées **7**
3a Aire centrale nettement marquée d'un côté sur la valve sans raphé ou sur les deux valves **4**
3b Aire centrale plus ou moins symétrique **5**
4a Valves larges-elliptiques, formes d'eau douce pauvre en électrolytes (vol. 2/4 Fig. 14: 11–18), cf. aussi *A. lapidosa* (vol. 2/4 Fig. 27: 1–14) . . . . **22.** ***A. hintzii*** vol. 2/4 p. 25
4b Valves plus lancéolées aux extrémités très effilées, vivant en eau saumâtre (vol. 2/4 Fig. 29: 17–22) . . . . **70.** ***A. punctulata*** vol. 2/4 p. 51 . . . . **71.** ***A. pseudopunctulata*** vol. 2/4 p. 51
5a Valves elliptiques aux extrémités largement arrondies, formes d'eau douce plus pauvres en électrolytes **6**
5b Valves aux extrémités coniques rétrécies et arrondies tronquées, formes d'eau riche en électrolytes à saumâtre (vol. 2/4 Fig. 28: 1–8) . . . . **67.** ***A. subsalsa*** vol. 2/4 p. 49

6a Aire centrale plus ou moins identique sur valves à raphé et sans raphé. Raphé plus ou moins diagonal chez une partie des individus d'une population (vol. 2/4 Fig. 12: 1–9) . . . . **11.** ***A. bioretii*** part. vol. 2/4 p. 19
6b Caractéristiques différentes (vol. 2/4 Fig. 10: 12–27) . . . . **10.** ***A. helvetica*** part. vol. 2/4 p. 18
7a Valves des populations au moins partiellement à contour rhomboédrique-lancéolé **8**
7b Tendance à contour rhomboédrique non reconnaissable **10**
8a Valves très petites, longueur < 10 µm. Stries non résolvables sans effet d'éclairage particulier (vol. 2/4 Fig. 11: 14–17) . . . . **77.** ***A. carissima*** vol. 2/4 p. 55
8b Caractéristiques différentes **9**
9a Valve sans raphé à grande aire centrale. Stries se trouvant donc aux bords (vol. 2/4 Fig. 17: 22–34), cf. aussi *A. rupestris* (vol. 2/4 Fig. 17: 14–21) . . . . **34.** ***A. montana*** part. vol. 2/4 p. 31
9b Valve sans raphé avec une aire centrale prononcée d'un seul côté (vol. 2/4 Fig. 14: 19–26). Si cette caractéristique ne s'applique pas et que les frustules sont courbés en forme de cuillère en vue connective, cf. *altaica* part.(vol. 2/4 Fig. 20: 24–32), *A. rossii* (vol. 2/4 Fig. 20: 1–12), *A. rechtensis* part. (vol. 2/4 Fig. 20: 13–23) . . . . **23.** ***A. semiaperta*** part. vol. 2/4 p. 26
10a Valves sans raphé sans aucune structure. Par contre, sur la valve à raphé, on peut résoudre, au moins en éclairage "rasant", des stries radiales (vol. 2/4 Fig. 15: 39–44) . . . . **26.** ***A. kuelbsii*** part. vol. 2/4 p. 27
10b Caractéristique différente **11**
11a Valves sans raphé à très grande aire centrale. Stries sur les bords. Distalement pas d'aire axiale plus étroite **12**
11b Aire centrale plus petite, distalement il y a encore une aire axiale parce que les stries ne sont pas ou à peine raccourcies **14**
12a Stries grossières, < 24/10 µm (vol. 2/4 Fig. 15: 29–38), cf. aussi *A. conspicua* part. (vol. 2/4 Fig. 16: 22–33) . . . . . **27.** ***A. holsatica*** part. vol. 2/4 p. 27
12b Stries plus rapprochées **13**
13a Frustules moins courbés en vue connective. Longueur régulièrement > 10 µm entre les rangées d'aréoles. Cf. aussi *A. levanderi* (vol. 2/4 Fig. 15: 8–18), *A. daonensis* (vol. 2/4 Fig. 12: 10–20) . . . . **19.** ***A. marginulata*** vol. 2/4 p. 23
13b Caractéristiques différentes (vol. 2/4 Fig. 11: 9–13) . . . . **15.** ***A. scotica*** vol. 2/4 p. 22
14a Valves comparativement très petites, longueur régulièrement < 7 µm (vol. 2/4 Fig. 11: 1–8) . . . . **14.** ***A. curtissima*** vol. 2/4 p. 21
14b Valves en moyenne plus grandes **15**
15a Valves strictement elliptiques, sans tendance linéraire-elliptique **16**
15b Les plus grands exemplaires ont tout au moins une tendance linéaire-elliptique **20**
16a Stries très rapprochées, env. 30/10 µm (vol. 2/4 Fig. 14: 1–10) . . . . **21.** ***A. subatomoides*** vol. 2/4 p. 24
16b Stries plus éloignées, env. 25/10 µm **17**
17a Raphé courbé en sens opposé aux pôles. Reconnaissable au moins chez les exemplaires plus grands, en faisant varier la mise au point (vol. 2/4 Fig. 10: 12–27) . . . . **10.** ***A. helvetica*** part. vol. 2/4 p. 18
17b Extrémités du raphé toujours droites **18**
18a Aire centrale formant des deux côtés, par le raccourcissement de plusieurs

stries, une bande transversale (vol. 2/4 Fig. 18: 22–23) (sans diagnostic) . . . . *A. johncarteri*

18b Aire centrale petite sur valve à raphé, une ou max. 2 stries sont raccourcies **19**

19a Une seule strie raccourcie d'un seul côté (vol. 2/4 Fig. 15: 1, 2) . . . . **17. *A. lacusvulcani*** vol. 2/4 p. 22

19b Populations possédant aussi des individus à 2 stries raccourcies ou à stries raccourcies des 2 côtés (vol. 2/4 Fig. 15: 8–18), cf. aussi *A. saccula* à largeur valvaire plus réduite (env. 4 µm) (vol. 2/4 Fig. 15: 19–28) . . . . **18. *A. levanderi*** vol. 2/4 p. 23

20a Largeur valvaire régulièrement <= 4 µm **21**

20b Largeur valvaire régulièrement > 4 µm **22**

21a Stries distales nettement plus rapprochées que les proximales (vol. 2/4 Fig. 34: 25–34); cf. aussi les petits individus des formes à larges valves de *A. minutissima* . . . . **83. *A. kranzii*** part. vol. 2/4 p. 64

21b Stries distales seulement un peu plus rapprochées que les proximales (vol. 2/4 Fig. 15: 19–28); cf. aussi les formes plus courtes de *A. petersenii* (vol. 2/4 Fig. 37: 28–39), *A. pusilla* (vol. 2/4 Fig. 37: 9–18), *A. nodosa* (vol. 2/4 Fig. 37: 1–8) *A. rosenstockii* (vol. 2/4 Fig. 36: 32–37) . . . . **25. *A. saccula*** vol. 2/4 p. 26

22a Raphé courbé en sens opposé aux pôles, observables au moins chez les individus plus grands en faisant soigneusement varier la mise au point (vol. 2/4 Fig. 10: 12–27) . . . . **10. *A. helvetica*** vol. 2/4 p. part. vol. 2/4 p. 18

22b Extrémités distales du raphé toujours droites **23**

23a Stries éloignées, < 24/10 µm (vol. 2/4 Fig. 17: 1–13) . . . . **32. *A. kryophila*** vol. 2/4 p. 30

23b Stries plus rapprochées, > 26/10 . . . . **24**

24a Sur valve sans raphé, l'aire centrale rectangulaire transversale est nettement distincte de l'aire axiale (vol. 2/4 Fig. 12: 21–31) . . . . **13. *A. chlidanos*** vol. 2/4 p. 20

24b Sur valve sans raphé, l'aire axiale s'élargit en forme lancéolée en une aire centrale non clairement définie (vol. 2/4 Fig. 12: 10–20) . . . . **12. *A. daonensis*** vol. 2/4 p. 20

Groupe-clef F (*A. minutissima* et espèces similaires)
(Complexe des forme d'*Achnanthes*)

1a Aire centrale de la valve à raphé rhomboédrique ou rhomboédrique-elliptique, extrémités proximales du raphé éloignées (vol. 2/4 Fig. 33: 23–31) . . . . **79. *A. exilis*** vol. 2/4 p. 61

1b Caractéristiques différentes **2**

2a Si la largeur valvaire est en moyenne > 5 µm, **cf. Groupe-clef E** (p. 218)

2b Largeur valvaire en moyenne nettement < 5 µm **3**

3a Contour valvaire variable, raphé courbé d'un seul côté en zone distale **4**

3b Développement du raphé en zone distale droit (en cas de doute, à observer au microsc. électron. ou comparer avec les figures des taxons suivants) **7**

4a Stries proximales également très rapprochées, > 30/10 µm. Aire axiale large et lancéolée sur valve à raphé. Aire centrale atteignant souvent les bords d'un seul côté (vol. 2/4 Fig. 28: 21–31) . **65. *A. silvahercynia*** vol. 2/4 p. 48

4b Caractéristiques différentes **6**

5a Stries proximales 25–30/10 µm, stries distales 35–40/10 µm. Aire centrale sur valve à raphé formant une large bande transversale; formes d'eau pauvre en électrolytes (vol. 2/4 Fig. 34: 25–34) **83. *A. kranzii*** vol. 2/4 p. 64

5b Caractéristiques différentes **6**

6a Aire axiale large et lancéolée sur valve sans raphé. Aire centrale atteignant le plus souvent les bords d'un seul côté (vol. 2/4 Fig. 30: 1–19) . . . . **73.** ***A. thermalis*** vol. 2/4 p. 52
6b Caractéristiques différentes (vol. 2/4 Fig. 36: 1–31) . . . . **81.** ***A. biasolettiana*** vol. 2/4 p. 62 (ensemble des formes)
7a Valves linéaires, largeur > 3 µm. Extrémités valvaires peu rétrécies. Stries distales peu ou pas plus étroites que les proximales. **8**
7b Caractéristiques différentes **10**
8a Stries distantes, rarement > 20/10 µm (vol. 2/4 Fig. 37: 9–18), cf. aussi *A. nodosa* (vol. 2/4 Fig. 37: 1–8) et *A. kriegeri* (vol. 2/4 Fig. 36: 38–41) . . . . **87.** ***A. pusilla*** vol. 2/4 p. 67
8b Stries > 24/10 µm **9**
9a Stries env. 30/10 µm (vol. 2/4 Fig. 37: 28–39) . . . . **88.** ***A. petersenii*** vol. 2/4 p. 67
9b Stries env. 26/10 µm (vol. 2/4 Fig. 37: 19–39) . . . . **89.** **"*A. linearis*"** sensu auct. nonnull.
10a Frustules en vue connective courbés (non pliés) en forme de cuillère et liés pour constituer des colonies en bandes (vol. 2/4 Fig. 34: 23, 24) . . . . **80.** ***A. catenata*** part. vol. 2/4 p. 62
10b Caractéristiques différentes (vol. 2/4 Fig. 32–35) . . . . **78.** ***A. minutissima*** vol. 2/4 p. 55 (ensemble des formes)

Groupe-clef G

1a Valves aux extrémités étirées rostrées ou capitées **7**
1b Valves dans tous les cas aux extrémités très faiblement étirées **2**
2a Valves toujours < 10 µm de long. Stries non-résolvables sans réglage particulier (vol. 2/4 Fig. 11: 14–17) . **77.** ***A. carissima*** part. vol. 2/4 p. 55
2b Caractéristiques différentes **3**
3a Stries ponctuées **4**
3b Stries non ponctuées ou alors caractère reconnaissable seulement à l'aide d'effets lumineux spéciaux **5**
4a Petite aire centrale, plus nettement prononcée d'un seul côté de la valve sans raphé (vol. 2/4 Fig. 27: 1–14), cf. aussi *A. punctuata* et *A. pseudo-punctuata* (vol. 2/4 Fig. 29: 17–22) . . **64.** ***A. lapidosa*** part. vol. 2/4 p. 47
4b Aire centrale lancéolée, grande (vol. 2/4 Fig. 28: 1–8) . . . . **67.** ***A. subsalsa*** part. vol. 2/4 p. 49
5a Valves le plus souvent linéaires. Aire centrale étroite et en forme de stauros, orientée transversalement **6**
5b Valves elliptiques. Aire centrale plus large, non en forme de stauros (vol. 2/4 Fig. 20: 24–32) . . . . **42.** ***A. altaica*** part. vol. 2/4 p. 34
6a Valves larges > 6 µm. Stries env. 20/10 µm (vol. 2/4 Fig. 19: 1–15) . . . . **41.** ***A. hungarica*** part. vol. 2/4 p. 33
6b Valves larges d'env. 4 µm. Stries env. 30/10 µm (vol. 2/4 Fig. 23: 28–32) . . . . **51.** ***A. subexigua*** part. vol. 2/4 p. 40
7a Stries distantes, env. 20/10 µm. Aires centrale et axiale non-différenciées (vol. 2/4 Fig. 22: 13–30); cf. aussi *A. species "de Bonn"* (vol. 2/4 Fig. 22: 47–49) . . . . **48.** ***A. ploenensis*** vol. 2/4 p. 37
7b Caractéristiques différentes **8**
8a Valves elliptiques à linéaires-elliptiques aux extrémités étirées en forme capitée et à stries difficilement résolvables, autour de jusqu' à largement au-dessus de 30/10 µm **9**
8b Caractéristiques différentes **13**
9a Valves plus largement elliptiques, longueur > 14 µm. Stries > 40/10 µm à peine résolvables même avec réglage spécial **10**

9b Caractéristiques différentes **11**
10a Valve sans raphé possédant une structure en forme de point dans le nodule central (vol. 2/4 Fig. 25: 7–12 sans diagnostic) . **A. *impexa*** vol. 2/4 p. 42
10b Valve sans raphé ne possédant pas de point (vol. 2/4 Fig. 25: 1–6) . . . . **55. *A. impexiformis*** vol. 2/4 p. 41
11a En vue connective, frustules nettement courbées en forme de cuillère, constituant des agrégats en forme de bandes (vol. 2/4 Fig. 34: 23, 24) . . . . **80. *A. catenata*** part. vol. 2/4 p. 62
11b Caractéristiques différentes **12**
12a Valves < 10 µm de longueur (vol. 2/4 Fig. 24: 23–27) . . . . **53. *A. stolida*** vol. 2/4 p. 41
12b Valves > 10 µm de longueur (vol. 2/4 Fig. 24: 1–7); cf. aussi formes capitées de l'ensemble des formes autour d'*A. minutissima* (vol. 2/4 Fig. 32–35) . . . . **54. *A. pseudoswazi*** vol. 2/4 p. 41
13a Partie centrale de la valve large-elliptique jusqu'à presque carrée à extrémités étirées de forme capitée large. Aire centrale petite sur valve à raphé, très grande sur valve sans raphé (vol. 2/4 Fig. 26: 1–6) . . . . **58. *A. submarina*** vol. 2/4 p. 43
13b Caractéristiques différentes **14**
14a Aire centrale sur les deux valves très petite à absente. Aire axiale étroitement linéaire à lancéolée (vol. 2/4 Fig. 25: 13–20), cf. aussi *A. amoena* (vol. 2/4 Fig. 26: 7–23) . . . . **56. *A. ingratiformis*** vol. 2/4 p. 42
14b Caractéristique différente **15**
15a Valves tri-ondulées à cause d'extrémités largement arrondies capitées et du centre valvaire largement bombé **16**
15b Caractéristique différente **17**
16a Valves >= 20 µm de longueur (vol. 2/4 Fig. 19: 17–22) . . . . **57. *A. trinodis*** vol. 2/4 p. 43
16b Valves < 14 µm de longueur (vol. 2/4 Fig. 36: 32–37) . . . . **84. *A. rosenstockii*** part. vol. 2/4 p. 65
17a Valves > 20 µm de long avec une aire centrale large et stries finement ponctuées, env. 30/10 µm (vol. 2/4 Fig. 31: 1–10) . . . . **74. *A. imperfecta*** vol. 2/4 p. 53
17b Caractéristiques différentes **18**
18a Stries des deux côtés régulières env. 20/10 µm. Aire centrale sur valve à raphé constituant une bande transversale très étroite (vol. 2/4 Fig. 23: 33–38) . . . . **50. *A. ziegleri*** (vol. 2/4 p. 39)
18b Caractéristiques différentes **19**
19a Aire centrale sur valve à raphé en forme de bande transversale atteignant les bords **20**
19b Caractéristique différente **21**
20a Stries sur les deux valves plus ou moins d'égale densité, env. 30/10 µm (vol. 2/4 Fig. 23: 28–32) . . . . **51. *A. subexigua*** part. vol. 2/4 p. 40
20b Stries sur valve sans raphé plus éloignées, < 26/10 µm (vol. 2/4 Fig. 23: 1–27) . . . . **49. *A. exigua*** **part** vol. 2/4 p. 38
21a Stries 28–33/10 µm. Raphé faiblement courbé en zone distale, formes . . . . . vivant de préférence dans des eaux stagnantes (vol. 2/4 Fig. 20: 1–12) . . . . **43. *A. rossii*** part. vol. 2/4 p. 34
21b Stries en moyenne un peu plus éloignées. Raphé plus ou moins droit dans sa partie distale, forme vivant de préférence dans des eaux courantes (vol. 2/4 Fig. 20: 13–23) str. 25–30/10 µm . **44. *A. rechtensis*** part. vol. 2/4 p. 35

Groupe-clef H
(Valves sans raphé à zone en forme de fer-à-cheval. Abbréviation utilisée par la suite: FC)

1a FC également présente sur la valve à raphé (vol. 2/4 Fig. 46: 1–6) . . . . **99.** ***A. calcar*** vol. 2/4 p. 82
1b FC régulière uniquement sur la valve sans raphé 2
2a Stries grossières sur la valve sans raphé, beaucoup plus rapprochées sur la valve à raphé 3
2b Stries sur les deux valves plus ou moins à la même distance 5
3a Valves aux extrémités étirées rostrées 4
3b Extrémités faiblement étirées (vol. 2/4 Fig. 48: 1–16) . . . . **97.** ***A. oestrupii*** part. vol. 2/4 p. 81
4a Valves < 20 µm (vol. 2/4 Fig. 48: 19–26) . **98.** ***A. peragalli*** vol. 2/4 p. 82
4b Valves plus longues (vol. 2/4 Fig. 48: 17, 18) . . . . **97.** ***A. oestrupii* var. *pungens*** vol. 2/4 p. 81
5a Stries >= 20/10 µm, FC relativement faiblement marquée (vol. 2/4 Fig. 14: 27–34). D'autres espèces n'appartenant pas à l'ensemble de formes autour de *A. lanceolata* au sens le plus large: voir sous *A. hungarica*, *laevis* var. *diluviana*, *A. delicatula*-ensemble de formes, *A. fonticolanceolata*, *A. semiaperta*, *A. thermalis* L. . . . . **24.** ***A. lauenburgiana*** vol. 2/4 p. 26
5b Caractéristiques différentes 6
6a Valves linéaires et brusquement faiblement renflées dans la partie centrale (vol. 2/4 Fig. 45: 1–11) . . . . **95.** ***A. fragilarioides*** vol. 2/4 p. 80
6b Caractéristiques différentes 7
7a Valves relativement larges elliptiques, aux extrémités largement arrondies (non rhomboédriques), FC possède un double contour. Stries aux extrémités très fortement radiales (vol. 2/4 Fig. 46: 7–17) . . . . **96.** ***A. joursacense*** vol. 2/4 p. 81
7b Caractéristiques différentes. FC à contour simple ou double; cf. aussi plusieurs espèces "exotiques" qu'on ne retrouve pas dans cette région ensemble des formes . . . . **94.** ***A. lanceolata*** vol. 2/4 p. 73

2. Genre *Cocconeis* Ehrenberg 1838 (vol. 2/4 p. 83)

Typus generis: *Cocconeis scutellum* Ehrenberg 1838 (Boyer 1927)

Clef pour les espèces

1a Aire de la valve sans raphé lancéolée; petites formes délicates 2
1b Aire de la valve sans raphé linéaire, le plus souvent des formes plus robustes 5
2a Valves sans raphé grossièrement ponctuées ; points le + souvent ronds 3
2b Valves sans raphé ponctuées plus finement 4
3a Valves sans raphé à 6–9 str./10 µm, valve à raphé à env. 22 stries/10 µm (vol. 2/4 Fig. 56: 1–13) . . . . **3.** ***C. disculus*** vol. 2/4 p. 89
3b Valves sans raphé à 10–12 stries/10 µm, valve sans raphé à > 30 stries/ 10 µm (vol. 2/4 Fig. 56: 14–17) . . **6.** ***C. pseudothumensis*** vol. 2/4 p. 92
4a Sur les stries des valves sans raphé points ou traits délicats, ronds ou transversaux-elliptiques. Stries le plus souvent > 20/10 µm (vol. 2/4 Fig. 57: 8–31) . . . . **5.** ***C. neothumensis*** vol. 2/4 p. 91
4b Sur les stries des valves sans raphé points ou traits plus grossiers, ronds; stries 11–14/10 µm (vol. 2/4 Fig. 55: 1–4; 56: 18–32) . . . . **4. C.** ***neodiminuta*** vol. 2/4 p. 90
5a Valves sans raphé avec un anneau marginal de rangées doubles d'aréoles (vol. 2/4 Fig. 50: 4, 6; 57: 5–7; 58: 1–13) . . **7.** ***C. scutellum*** vol. 2/4 p. 93

5b Stries des valves sans raphé constituées jusqu'au bord de simples points ou de bandes **6**
6a Frustules avec valves fortement ondulées (vol. 2/4 Fig. 55: 5–8; 57: 1–4) . . . . **2.** ***C. pediculus*** vol. 2/4 p. 89
6b Valves relativement plates (vol. 2/4 Fig. 49: 1–3; 50: 1, 2, 5; 51: 1–9; 52: 1–13; 53: 1–19; 54: 1–11) . . . . **1.** ***C. placentula*** vol. 2/4 p. 86

Variétés de *Cocconeis placentula* (vol. 2/4 p. 86)
1a Aire axiale sur valves à et sans raphé oblique à l'axe apical L.: 12–55 µm (vol. 2/4 Fig. 51: 6–9) . . . . **1.5. var. *klinoraphis*** vol. 2/4 p. 87
1b Aire axiale pas oblique **2**
2a Valves le plus souvent > 40 µm de longueur **3**
2b Valves le plus souvent < 40 µm de longueur **6**
3a Stries des valves sans raphé finement ponctuées avec 18–22 points/10 µm (vol. 2/4 Fig. 51: 1–5) . . . . **1.1. var. *placentula*** vol. 2/4 p. 86
3b Stries des valves sans raphé plus grossièrement ponctuées **4**
4a Valves sans raphé avec stries > 16/10 µm, à aréoles transapicales fines et allongées (vol. 2/4 Fig. 52: 1–13) . . . . . . **1.6. var. *lineata*** vol. 2/4 p. 87
4b Valves sans raphé avec stries < 15/10 µm, aréoles apparaissant comme des points grossiers ou des petits points ronds **5**
5a Valve sans raphé à points grossiers, 7–8/10 µm . . . . **1.2. var. *intermedia*** vol. 2/4 p. 87
5b Valve sans raphé à points plus fins, 10–12/10 µm . . . . **1.3. var. *rouxii*** vol. 2/4 p. 87
6a Aréoles sur valve sans raphé très grandes et transapicalement élargies. A réglage moyen, le centre clair est entouré d'une marge foncée. Les aréoles constituent des rangées longitudinales plus ou moins régulières (vol. 2/4 Fig. 54: 3–11) . . . . **1.8. var. *pseudolineata*** vol. 2/4 p. 87
6b Points plus fins **7**
7a Valves sans raphé avec stries > 23/10 µm, finement ponctuées **8**
7b Valves sans raphé avec stries < 22/10 µm, ponctuées ± grossièrement . **9**
8a Valve sans raphé à 30–38 str./10 µm . **1.4. var. *tenuistriata*** vol. 2/4 p. 87
8b Valve sans raphé à 24–26 str./10 µm (vol. 2/4 Fig. 51: 1–5), str. 20–23/10 µm valve à raphé, str. 24–26/10 µm valve sans raphé . . . . **1.1. var. *placentula*** vol. 2/4 p. 86
9a Sur chaque strie 3–5 aréoles larges, apicalement ordonnées en rangées longitudinales (vol. 2/4 Fig. 53: 1–19) . . **1.7. var. *euglypta*** vol. 2/4 p. 87
9b Sur chaque strie de nombreuses aréoles, apicalement ordonnées en zigzag (vol. 2/4 Fig. 52: 1–13) . . . . **1.6. var. *lineata*** vol. 2/4 p. 87

3. Famille Naviculaceae Kützing 1844 (sensu Simonsen 1979) (vol. 2/1)

Typus familiae: *Navicula* Bory.

Clef pour les genres
1a Axe apical ou (et) transapical hétéropolaire **2**
1b Axes apical et transapical isopolaires **7**
2a Axe apical isopolaire. Axe transapical hétéropolaire. Valves à symétrie dorso-ventrale **3**
2b Axe apical hétéropolaire. Axe transapical isopolaire **4**
3a Manteau valvaire dorsal légèrement plus large que le manteau valvaire ventral, residuum et bandes intercalaires absents . . . . **11.** ***Cymbella*** (vol. 2/1 p. 300) (p. 266)
3b Manteau valvaire dorsal en général beaucoup plus large que le manteau

ventral: un residuum et souvent de nombreuses bandes intercalaires donnent à la frustule une forme large-elliptique . . . . 12. ***Amphora*** (vol. 2/1 p. 342) (p. 270)

4a Raphé bien développé sur une seule valve. L'autre valve ne possède que de courts rudiments de raphé aux pôles . . . . 16. ***Rhoicosphenia*** (vol. 2/1 p. 381) (one species, vol. 2/1 p. 381)

4b Raphé bien développé sur les deux valves 5

5a Alvéoles fermées au moins dans la zone du manteau. Une ligne longitudinale est visible en vue valvaire ou connective . . . . 14. ***Gomphoneis*** (vol. 2/1 p. 379) (one species, vol. 2/1 p. 379)

5b Alvéoles ouvertes au niveau de la valve et du manteau 6

6a Les deux fissures polaires identiques et nettement coudées en arrière . . . . 15. ***Didymosphenia*** (vol. 2/1 p. 380) (one species, vol. 2/1 p. 380)

6b Les deux fissures polaires généralement construites différemment . . . . 13. ***Gomphonema*** (vol. 2/1 p. 352) (p. 271)

7a Cellules avec bandes intercalaires et septa nets 8

7b Cellules sans bandes intercalaires ni septa 9

8a Bandes intercalaires à bords alvéolés . . . . 19. ***Mastogloia*** (vol. 2/1 p. 432) (p. 281)

8b Bandes intercalaires à septa étirés sans bords alvéolés . . . . 20. ***Diatomella*** (vol. 2/1 p. 436) (p. 281)

9a Valves ou raphé en forme de S 10

9b Valves ou raphé non en forme de S 13

10a Raphé en position oblique sur la valve et en forme de S. Valves non en forme de S 11

10b Valves et raphé en forme de S 12

11a Raphé ne reposant pas sur une carène . . . . 7. ***Scoliopleura*** (vol. 2/1 p. 282) (p. 261)

11b Raphé reposant sur une carène . 22. ***Entomoneis*** (vol. 2/1 p. 438) (p. 282)

12a Points (aréoles) agencés en lignes apicales et transapicales . . . . 10. ***Gyrosigma*** (vol. 2/1 p. 295) (p. 263)

12b Points (aréoles) agencés en une ligne transapicale et deux obliques . . . . 9. ***Pleurosigma*** (vol. 2/1 p. 294) (p. 263)

13a Valves avec côtes longitudinales saillantes nettes . . . . 21. ***Oestrupia*** (vol. 2/1 p. 436) (p. 282)

13b Pas de côtes longitudinales saillantes dans la face valvaire 14

14a Points des stries transapicales non résolvables au microscope optique . 15

14b Points des stries transapicales résolvables par dispositif optique approprié 16

15a Alvéoles toujours fermées au moins au centre, à 1–2 ouvertures étroites en marge ou dans le secteur du manteau 17. ***Caloneis*** (vol. 2/1 p. 382) (p. 275)

15b Alvéoles ouvertes ou fermées avec ouvertures intérieures plus ou moins larges dans le domaine valvaire . 18. ***Pinnularia*** (vol. 2/1 p. 397) (p. 276)

16a Aire centrale avec des cornes ou en forme de stauros 17

16b Aire centrale ronde, ovale ou allongée 18

17a Aire centrale avec apophyses en forme de cornes . . . . 8. ***Diploneis*** (vol. 2/1 p. 283) (p. 261)

17b Aire centrale formant un stauros 2. ***Stauroneis*** (vol. 2/1 p. 236) (p. 255)

18a Raphé bifurqué aux pôles. Valves avec une ou plusieurs lignes longitudinales . . . . 6. ***Neidium*** (vol. 2/1 p. 265) (p. 259)

18b Raphé non bifurqué aux pôles. Pas de lignes longitudinales 19

19a Raphé accompagné de côtes axiales intérieures puissantes 20

19b Raphé non accompagné de côtes axiales intérieures 21

20a Valves fusiformes. Aire centrale prolongée, d'où extrémités proximales du raphé éloignées l'une de l'autre 5. ***Amphipleura*** (vol. 2/1 p. 262) (p. 258)

20b Valves lancéolées. Aire centrale peu prolongée. . . . . **4.** ***Frustulia*** (vol. 2/1 p. 258) (p. 258)
21a Stries lignées, peu nombreuses dans le sens transapical . . . . **3.** ***Anomoeoneis*** (vol. 2/1 p. 251) (p. 257)
21b Valves autrement structurées . . . . **1.** ***Navicula*** (vol. 2/1 p. 84) (p. 226)

1. Genre *Navicula* Bory de St. Vincent 1822 (vol. 2/1 p. 85)

Typus generis: *Navicula tripunctata* (O.F. Müller) Bory 1824

Clef pour les groupes d'espèces (sections)

1a Valves ayant, en plus de l'aire centrale et axiale, des aires latérales liées à l'aire centrale en forme de H (vol. 2/1 Fig. 65: 1–13). Dans les plus petites valves (longueur < 20 µm) une ligne longitudinale hyaline croisant les bandes transapicales court des deux côtés entre l'aire axiale et le bord de la valve (vol. 2/1 Fig. 66: 1–34) **2**
1b Valves autrement structurées **3**
2a Valves avec aires centrale et latérale nettement liées en forme de (Houtre *N. pygmaea*, peu d'espèces d'eau salée et saumâtre sont concernées) . . . . **Groupe-clef I. "Lyratae"** (vol. 2/1 p. 170) (p. 241)
2b Aires latérales, côtes longitudinales ou structures similaires en sens apical ne sont indiquées que par des lignes où il y a entre les aires longitudinale et latérale une seule ligne de points. Il n'y a donc pas de forme de lyre nettement définie. . . . . **Groupe-clef J**: formes voisines ***N. monoculata*** (vol. 2/1 p. 173) (p. 242)
3a Aux pôles, deux stries distales ou plus sont interrompues par une ligne (aire) hyaline +/- circulaire ou semicirculaire. Chez des individus plus petits, cette structure peut être réduite au point que n'apparaissent plus qu'une ou plusieurs paires de points renforcés immédiatement sur l'aire axiale (vol. 2/1 Fig. 64: 1–25) . . . . **Groupe-clef K "Annulatae"** (vol. 2/1 p. 178) (p. 243)
3b Caractéristiques différentes **4**
4a Stries comparativement très grossières (env. 9/10 µm) mais sans structure plus fine reconnaissable. Représentée ici par une seule espèce (vol. 2/1 Fig. 82: 7, 8) . . . . **"Laevistriatae"** . . . . **247.** ***N. elegans*** vol. 2/1 p. 236
4b Caractéristiques différentes **5**
5a Stries ponctuées formées par trois systèmes qui se croisent. Représentée ici par une seule espèce (vol. 2/1 Fig. 82: 5, 6) . . . . **"Decussatae"**, ... **245.** ***N. placenta*** (vol. 2/1 p. 235)
5b Caractéristiques différentes **6**
6a Pores centraux du raphé avec apophyses filamenteuses courbées latéralement semblables aux fissures terminales. Représentée ici par une seule espèce (vol. 2/1 Fig. 65: 14, 15) . . . . **"Fistulatae"** . . . . **246.** ***N. gibbula*** vol. 2/1 p. 235
6b Caractéristiques différentes (cf. toutefois *N. mutica* sous microscope électronique, Fig. 53: 8) **7**
7a Branches du raphé présentant des fissures internes et externes se croisant +/- au centre des branches. Stries formées par des points grossiers alignés +/- transapicalement. Représentée ici par deux espèces seulement (vol. 2/1 Fig. 82: 1–4). Chez *N. tuscula*, les stries au bord de la valve sont à résoudre comme des lignes pointillées doubles (vol. 2/1 Fig. 81: 1–7) . . . . **Groupe d'espèces autour de 243. *N. tuscula*** . . . . **244.** ***N. pseudotuscula*** (vol. 2/1 p. 234)

7b Caractéristiques différentes **8**
8a Stries lignées réunies. Chez certaines formes, surtout à petites valves, résolvables en microscopie optique seulement sous lumière oblique, (ou dans le champ obscur) ou stries généralement parallèles **9**
8b Stries formées +/- nettement de points ou structure fine non différenciable **12**
9a Stries +/- radiales (souvent convergentes aux extrémités) . . . . **Groupe-clef A "Lineolatae"** (vol. 2/1 p. 88) (p. 228)
9b Stries et côtes transapicales parallèles ou faiblement radiales. Chez les formes plus grandes, l'agencement régulier des stries lignées ou des rangées de points avec les côtes longitudinales forme un système +/- nettement reconnaissable de lignes se croisant à angle droit **10**
10a Valves fusiformes avec nodule central étiré en forme de stauros. Représentée ici par deux espèces seulement (vol. 2/1 Fig. 52: 4–6) . . . . **"Fusiformes"** *[Haslea]* (vol. 2/1 p. 132) (p. 235)
10b Nodule central pas en forme de stauros. Air centrale petite, ronde à lancéolée **11**
11a Longueur valvaire régulièrement > 20 µm . . . . **Groupe-clef B "Orthostichae"** (vol. 2/1 p. 124) (p. 234)
11b Longueur valvaire régulièrement < 20 µm . . . . **Groupe-clef C** (vol. 2/1 p. 129) (p. 234)
12a Stries ponctuées +/- clairement, reconnaissables mais non combinées avec des sillons longitudinaux le long de la côte médiane, ni avec côte médiane respectivement aire axiale très contrastée **13**
12b Stries ne pouvant être résolvables comme lignes de points ou alors autres caractéristiques **17**
13a Nodule central à prolongement stauroïde avec stigma +/- excentrique . . . . **Groupe-clef E, espèces autour de *N. mutica*** (vol. 2/1 p. 147) (p. 237)
13b Caractéristiques différentes **14**
14a Formes larges-elliptiques aux extrémités arrondies, larges et peu ou pas étirées (vol. 2/1 Fig. 59: 1–19 ; 60: 1, 2) . . . . **Groupe-clef F, espèces autour de *N. pseudoscutiformis*** . . . . (vol. 2/1 p. 158) (p. 239)
14b Caractéristiques différentes **15**
15a Frustules avec bandes intercalaires marquées en vue connective . . . . **Groupe-clef G, "Microstigmaticae"**(vol. 2/1 p. 160) (p. 240)
15b Frustules sans bandes intercalaires marquées **16**
16a Valves avec stries radiales relativement grossières (régulièrement > 24/10 µm) correspondant aux "Lineolatae", cependant avec foramina +/- en forme de point relativement rapproché au lieu de lignes. C'est la raison pour laquelle les côtes transapicales semblent plus larges que les stries. Longueur valvaire moyenne > 20 µm ou légèrement inférieure mais alors avec une forme nettement large-lancéolée. Souvent 1, rarement plusieurs stigmates dans le secteur du nodule central . . . . **Groupe-clef D, formes voisines de *N. elginensis*** (vol. 2/1 p. 134) p. 235)
16b Du moins au centre, stries formées par des points grossiers relativement éloignés (régulièrement < 24/10 µm). Espèces restantes des "Punctatae" avec formes valvaires très variables. Pour la plupart, formes d'eau de mer ou saumâtre ou vivant dans des conditions d'oscillations élevées de pression osmotique . . . . **Groupe-clef H, espèces autour de *N. pusilla, N. semen,N. humerosa*** . . . . (vol. 2/1 p. 165) (p. 240)
17a Valves le plus souvent étroites, aux bords linéaires à lancéolées, avec extrémités étirées +/- rostrées à capitées. Raphé avec fissures terminales à

angle abrupt. Dans la partie centrale, stries fortement radiales, à convergence abrupte avant les extrémités, le plus souvent très rapprochées (> 30/10 µm jusqu'à non résolvables sous microscope optique, cf. Fig. 79: 1–28) . **Groupe-clef L, espèces voisines *N. bryophila et N. subtilissima*** . . . . (vol. 2/1 p. 180) (p. 243)

17b Caractéristiques différentes **18**

18a Côte médiane très prononcée par des sillons longitudinaux "fortement marqués" (probablement groupe homogène mais souvent non-identifiable autour de *N. bacillum* sous microscope optique) . . . . **Groupe-clef M, "Bacillares"** (vol. 2/1 p. 184) (p. 243)

18b Sillons longitudinaux non reconnaissables. Formes petites à (rarement) moyennes avec structures fines variables, sans les caractéristiques des autres groupes-clefs . . . . **Groupe-clef N (groupe hétérogène résiduel)"Minusculae"** . . . . (vol. 2/1 p. 205) (p. 247)

Groupe-clef A – Naviculae lineolatae (vol. 2/1 p. 89) **(Subgenus *Navicula*)**

1a Valves déviant des caractéristiques habituelles des Lineolatae, avec des stries grossières et larges par rapport à la petite taille des valves (presque toujours < 30 µm). Le plus souvent pôles sans stries et apparaissant renforcés par la présence de pseudo-septa (cf. Fig. 16: 1–11 et Fig. 15: 8–11) **71**

1b Caractéristiques différentes **2**

2a Pôles également sans stries, aire centrale grande, stries 10–13/10 µm et plus fortement radiales (vol. 2/1 Fig. 34: 12, 13) L: 35–75 / l: 15–10 /str. 10–13/10 µm . . . . **29. *N. globulifera*** vol. 2/1 p. 110
En cas d'aire centrale petite, str. env. 20/10 µm, très faiblement radiales . . . . (vol. 2/1 Fig. 44: 12–13) . . . . **63. *N. riparia*** vol. 2/1 p. 127

2b Pôles pas entièrement dépourvus de stries **3**

3a Stries fortement coudées aux extrémités. Formes à valves relativement grandes. (vol. 2/1 Fig. 41: 2) L: 70–220 / l: 12–14 /str. 6–9/10µm . . . . **55. *N. oblonga*** vol. 2/1 p. 121

3b Stries non coudées **4**

4a Pores centraux des branches du raphé ou des extrémités proximales des branches courbés remarquablement d'un seul côté. Régulièrement, le nodule central apparaît également renforcée d'un seul côté (cf. Fig. 37: 1–8et Fig. 38: 1–15). Ne sont pas concernées ici les formes dont les branches du raphé se courbent d'abord au niveau du nodule central (cf. Fig. 39: 12, 13) **68**

4b Extrémités du raphé avec les pores centraux dans la médiane du nodule central; sinon pas de combinaison de caractéristiques comme 4a **5**

5a Stries (régulièrement) radiales jusqu'aux extrémités **6**

5b Stries (régulièrement) convergentes aux extrémités **15**

6a Aire centrale s'étendant jusqu'au bord de la valve. Stries apparaissant grossières par rapport à la petite taille de la valve, 7–10/10 µm (vol. 2/1 Fig. 42: 13–15) (comparé aussi 60. *N. subcostala* et *N. lesmonensis*) L: 12–35 / l: 4–7,5 /str. 7–10/10 µm . . . . **59. *N. costulata*** vol. 2/1 p. 124

6b Aire centrale n'allant pas jusqu'au bord de la valve **7**

7a Stries considérablement plus écartées au niveau du nodule central que les autres stries. Formes de grande taille (cf. Fig. 26: 1) L: 60–130 / l: 13–30 /str. 8–10/10 µm au centre . . . . **40. *N. hasta*** vol. 2/1 p. 114

7b Stries pas ou moins fortement écartées au niveau du centre **8**

8a Pores centraux très fortement rapprochés. Formes d'eaux salées, saumâtres ou douces très alcalines **9**
8b Caractéristiques différentes **10**
9a Valves étroites et lancéolées. Stries lignées à peine différenciables (vol. 2/1 Fig. 39: 6–11) L: 20–60 / l: 4–6 / str. 14–15/10 µm . . . . **49.** ***N. duerrenbergiana*** vol. 2/1 p. 119
9b Valves plus largement lancéolées ou avec des bords linéaires à faiblement concaves. Stries lignées clairement différenciables (vol. 2/1 Fig. 39: 4, 5) L: 30–80 / l: 5–10 / str. 12–14/10 µm . . . **48.** ***N. pavillardii*** vol. 2/1 p. 118
10a Branches du raphé avec pores centraux du nodule central courbés d'un seul côté (vol. 2/1 Fig. 38: 1–4) L: 30–55 / l: 5–9 / str. 12–16/10 µm . . . . **42.** ***N. schroeteri*** vol. 2/1 p. 115
10b Pores centraux +/- droits **11**
11a Stries du centre alternativement plus courtes et plus longues. Extrémités largement arrondies (vol. 2/1 Fig. 40: 1, 2); cf. *N. splendicula* (vol. 2/1 Fig. 33: 1–3) dont quelques exemplaires peuvent avoir des stries radiales continues **38**
11b Les stries du centre peuvent être plus courtes mais n'alternent pas avec des stries plus longues. Extrémités non larges arrondies **12**
12a Extrémités faiblement à nettement rostrées **14**
12b Extrémités graduellement plus étroites, arrondies-pointues, non rostrées **13**
13a Largeur valvaire 6–9 µm. Aire centrale légèrement élargie transapicalement et non nettement asymétrique (vol. 2/1 Fig. 36: 8) L: 25–50 / l: 6–9 /str. 9–12/10 µm . . . . **38.** ***N. pseudolanceolata*** vol. 2/1 p. 113
13b Largeur valvaire 9–12 µm. Aire centrale +/- élargie lancéolée et en général nettement asymétrique (vol. 2/1 Fig. 36: 10–12) L: 40–75 / l: 9–12 / str. 8–11 µm . . . . **39.** ***N. concentrica*** vol. 2/1 p. 113
14a Aire centrale assez grande et ronde (vol. 2/1 Fig. 35: 1–4), cf. aussi *N. expecta* (vol. 2/1 Fig. 31: 15). Si l'aire centrale est élargie transversalement et les stries lignées sont très nettes (env. 20/10 µm), il peut s'agir éventuellement d'une forme qui rejoint la *N. pseudolanceolata* avec extrémités faiblement étirées selon Fig. 36: 9 . . . . . **30.** ***N trivialis*** vol. 2/1 p. 110
14b Aire centrale étroite lancéolée (vol. 2/1 Fig. 36: 4–7) L: 25–40 / l: 8–12,5 / str. 11–13/10 µm . . . . **37.** ***N. praeterita*** vol. 2/1 p. 113
15a Aire centrale nettement plus élargie transapicalement qu'apicalement, presque toujours plus que la moitié de la largeur de la valve. La forme de base peut être décrite comme transversale-rectangulaire. (Toutes les formes avec aire centrale variable sur ce point sont présentées à nouveau à partir du numéro-clef 30) **16**
15b Aire centrale ronde, rhomboédrique, lancéolée ou très faiblement prononcée **30**
16a Valves plus fortement ondulées si bien que les stries apparaissent coudées sur les bords. Formes d'eaux marines et saumâtres **29**
16b Caractéristiques différentes **17**
17a Longueur des valves toujours < 20 µm sans extrémités étirées (vol. 2/1 Fig. 35: 14–20) L: 5,5–20 / l: 2–4 / str. 14–18/10 µm . . . . **35.** ***N. perminuta*** vol. 2/1 p. 112
17b Caractéristiques différentes **18**
18a Valves toujours < 30 µm. Aire centrale étroite. Stries très grossières par rapport à la petite taille des valves. Stries moins de 12/10 µm (vol. 2/1 Fig. 42: 16–17). Cf. aussi *N. costulata* (vol. 2/1 Fig. 42: 13–15) et *N. lesmonensis* (vol. 2/1 Fig. 41: 8–11) L: 12–22 / l: 2,4–4 / str. < 12/10 µm . . . . **60.** ***N. subcostulata*** vol. 2/1 p. 124

18b Caractéristiques différentes **19**

19a Formes de taille moyenne à grande. Fissures du raphé interne et externe parallèles sur presque toute la longueur jusqu'aux pores centraux très marqués. Stries lignées très grossières, 5–12/10 µm et 18–25/10 µm resp. **26**

19b Caractéristiques différentes **20**

20a Stries lignées régulièrement > 30/10 µm, difficiles à différencier **23**

20b Stries lignées env. 25/10 µm, encore relativement faciles à différencier . **21**

21a Valves aux extrémités rostrées. Str. moyennement radiales (vol. 2/1 Fig. 31: 6, 7) L: 30–45 / l: 6–9 / str. 12–15/10 µm . . . . **16 *N. subrhynchocephala*** vol. 2/1 p. 102

Stries plus fortement radiales (vol. 2/1 Fig. 31: 15) . . . . **18. *N. expecta*** vol. 2/1 p. 103

21b Extrémités pas ou très peu étirées **22**

22a Stries fortement radiales (vol. 2/1 Fig. 27: 12–17) L: 25–80 / l: 6–11 / str. (9) 12–16/10 µm . . . . **4. *N. cari*** vol. 2/1 p. 96

22b Stries faiblement radiales (vol. 2/1 Fig. 28: 17–19) L: 25–40 / l: 4,4–8 / str. 12–13,5/10 µm . . . . **8. *N. libonensis*** vol. 2/1 p. 99

23a Stries faiblement radiales **25**

23b Stries plus fortement radiales **24**

24a Valves larges-lancéolées, aux extrémités +/- arrondies-pointues (vol. 2/1 Fig. 32: 16, 17) L: 15–50 / l: 7,5–12 / str. 8–12/10 µm . . . . **22. *N. menisculus*** var. ***upsaliensis*** vol. 2/1 p. 105

24b Valves plus étroites lancéolées aux extrémités arrondies-tronquées (vol. 2/1 Fig. 27: 12–17) L: 25–80 / l. 6–11 / str. 9–12(16)/10 µm . . . . **4. *N. cari*** vol. 2/1 p. 96

Cf. aussi *N. tenelloides*, plus petites et avec aire centrale moins élargie (vol. 2/1 Fig. 38: 16–21), et *N. angusta*, avec aire centrale très variable (vol. 2/1 Fig. 28: 1–5).

25a Valves linéaires. Longueur > 30 µm (vol. 2/1 Fig. 27: 1–3). Cf. aussi *N. recens* (vol. 2/1 Fig. 27: 7–11) et *N. margalithii* (vol. 2/1 Fig. 27: 4–6) L: 30–70 / l: 6–10 (8–9) / str. 9–12 (11–10)/µm **1. *N. tripunctata*** vol. 2/1 p. 95

25b Valves lancéolées. Longueur < 30 µm (vol. 2/1 Fig. 32: 1–4). L: 13–30 / l: 5–6 / str. 13,5–15/10 µm . . . . **19. *N. veneta*** p 104

26a Valves aux extrémités pas ou peu étirées mais toujours arrondies-tronquées **28**

26b Valves aux extrémités plutôt arrondies-pointues et/ou +/- rostrées . . **27**

27a Extrémités des valves pas ou peu étirées (vol. 2/1 Fig.: 30: 2–4) L: 35–70 / l: 10–20 / str. 7–8,5/10 µm . . . . **13. *N. meniscus*** vol. 2/1 p. 101

27b Extrémités des valves régulièrement plus longuement étirées (vol. 2/1 Fig. 30: 5–8; Fig. 31: 1–2)L: 35–80 / l: 9–14 / str. 7–12/10 µm . . . . **14. *N. rhynchocephala*** vol. 2/1 p. 101

28a Str. 5–7/10 µm. Lineolae 18–25/10 µm (vol. 2/1 Fig. 30: 1) L: 40–180 / l: 10–30 / str. 5–7/10 µm . . . . **12. *N. peregrina*** vol. 2/1 p. 100

28b Str. 8–9/10 µm. Lineolae +/- 25/10 µm (vol. 2/1 Fig. 31: 3–5) L: 25–60 / l: 8–11 / str. 8–9/10 µm . . . . **15. *N. slesvicensis*** vol. 2/1 p. 102

29a Aire centrale en règle générale +/- asymétrique. Extrémités des valves très variables (vol. 2/1 Fig. 39: 3) L: 30–100 / l: 6–16 / str. 5–11/10 µm . . . . **47. *N. cancellata*** vol. 2/1 p. 118

29b Aire centrale symétrique. Extrémités des valves cunéiformes, arrondies-pointues (vol. 2/1 Fig. 39: 2) L: 30–80 / l: 8–11: str. 9–10/10 µm . . . . **46. *N. arenaria*** var. ***rostellata*** vol. 2/1 p. 118

30a Aire centrale petite ou peu différenciée de l'aire axiale formant avec elle une unité lancéolée **55**

30b Aire centrale plus grande, de forme variable, ronde ou rhomboédrique . **31**
31a Stries autour du nodule central +/- régulièrement plus courtes et plus longues **32**
31b Stries autour du nodule central raccourcies autrement **41**
32a Extrémités étirées rostrées **36**
32b Extrémités tronquées, plus rarement arrondies-pointues cependant pas visiblement étirées **33**
33a Extrémités tronquées à largement arrondies **34**
33b Extrémités arrondies-pointues (vol. 2/1 Fig. 32: 17; 34: 5) **37**
34a Grande aire centrale ronde, stries 13–16/10 µm (vol. 2/1 Fig. 39: 12, 13) L: 40–75 / l: 10–15 / str. 13–16/10 µm . . . . **50.** ***N. bottnica*** vol. 2/1 p. 119
34b Aire centrale différente. Str. le plus souvent nettement < 13/10 µm. Chez *N. digitoradiata* var. *minima* stries plus rapprochées, aire centrale jamais grande ni ronde **35**
35a Stries lignées grossières, 20–25/10 µm **38**
35b Stries lignées pas nettes, plus de 30/10 µm (vol. 2/1 Fig. 34: 1–9) L: 25–80 / l: 7–28 / str. 7–14/10 µm . . . . **27.** ***N. digitoradiata*** vol. 2/1 p. 108
36a Grande aire centrale, ronde (vol. 2/1 Fig. 35: 5–8). Cf. aussi *N. digitoradiata* var. *rostrata* sensu Hustedt (vol. 2/1 Fig. 39: 6) L: 20–40 / l: 8–12/ str. 13–17/10 µm . . . . **31.** ***N. salinarum*** vol. 2/1 p. 110
36b Aire centrale petite, irrégulièrement délimitée (vol. 2/1 Fig. 32: 12–15) L: 25–45 /l: 7–10 / str. 11–14/10 µm **21.** ***N. capitatoradiata*** vol. 2/1 p. 105
37a Aire centrale élargie transversalement (vol. 2/1 Fig. 32: 17) L: 15–50 / l: 7,5–12 / str 8–12/10 µm . . . . **22.** ***N. menisculus*** var. ***upsaliensis*** part. vol. 2/1 p. 105
37b Aire centrale plus petite, irrégulièrement délimitée, variable (cf. Fig. 34: 5 et quelques formes de la planche 33). Formes supposées des espèces: ***N. digitoradiata*****,** ***N. stankovicii*****,** ***N. cryptotenella*** **et** ***N. species 1*** et **2**
38a Branches du raphé approximativement droites, soit avec des pores centraux fortement marqués et des stries très grossières (7–9/10 µm), soit avec des stries plus rapprochées (12–16/10 µm), et des extrémités faiblement étirées **40**
38b Branches du raphé arquées ou croisement frappant des fissures internes et externes du raphé près du nodule central **39**
39a Lineolae très grossières +/- 20/10 µm (vol. 2/1 Fig. 40: 3–7) L: 50–70 / l: 10–14 / str. 7–9/10 µm . . . . **52.** ***N. striolata*** vol. 2/1 p. 120
39b Lineolae moins grossières (+/- 24/10 µm). Dans l'ensemble, valves un peu plus petites avec des extrémités moins tronquées (vol. 2/1 Fig. 40: 8, 9) L: 30–60 / l: 8,5–12 / str. 7–12/10 µm . . . **53.** ***N. oppugnata*** vol. 2/1 p. 121
40a Str. 7–9/10 µm (vol. 2/1 Fig. 40: 1, 2) L: 35–70 / l: 11–18 / str. 7–9/10 µm . . . . **51.** ***N. reinhardtii*** vol. 2/1 p. 120
40b Str. 12–16/10 µm (vol. 2/1 Fig. 36: 1–3) L: 30–46 / l: 7–9 / str. 12–16/10 µm . . . . **36.** ***N. splendicula*** vol. 2/1 p. 112
41a Grande aire centrale, unilatéralement ou bilatéralement atteignant au moins la moitié de la largeur valvaire **42**
41b Aire centrale plus petite ou irrégulièrement délimitée par quelques stries allongées ou éloignées **54**
42a Lineolae nettes, jusqu'à env. 25/10 µm **43**
42b Lineolae moins nettement marquées, > 25/10 µm **45**
43a Valves longues (L. 50–140 µm). Branches du raphé +/- filiformes. Nodule central nettement renforcé d'un seul côté. Aire centrale grande, ronde à rhomboédrique. Formes d'eau douce oligosaprobe (vol. 2/1 Fig. 41: 1) L: 50–140 / l: 10–20 / str. 8–11/10 µm . . . . **54.** ***N. vulpina*** vol. 2/1 p. 121
43b Caractéristiques différentes **44**

44a Formes moyennement grandes à extrémités rostrées. Str. > 12/10 µm (vol. 2/1 Fig. 31: 6, 7) L: 30–45 / l: 6–9 / str. 12–15/10 µm . . . . **16. *N. subrhynchocephala*** vol. 2/1 p. 102

44b Formes plus grandes avec stries plus grossières, < 12/10 µm (cf. aussi *N. expecta*, Fig. 31: 15) **26–28**

45a Valves fortement ondulées d'où stries apparaissant coudées sur les bords. Formes d'eaux douces et marines **29**

45b Caractéristiques différentes **46**

46a Stries faiblement à moyennement radiales **47**

46b Stries radiales prononcées **51**

47a Valves linéaires (vol. 2/1 Fig. 27: 1–3) L: 30–70 / l: 6–10 svt 8–9 / str. 9–12 svt 10–11/10 µm . . . . **1. *N. tripunctata*** vol. 2/1 p. 95

47b Valves lancéolées à linéaires-lancéolées (mais cf. aussi quelques formes de *N. halophila*, Fig. 44: 18) **48**

48a Valves petites, largeur < 6,5 µm **49**

48b Valves plus grandes, nettement plus larges **50**

49a Valves +/- elliptiques (vol. 2/1 Fig. 28: 13–15) L: 13–45(55) / l: 5–8 / str. 8–12(17)/10 µm . . . . . **7. *N. cincta* petites formes limite** vol. 2/1 p. 98

49b Valves lancéolées (vol. 2/1 Fig. 32: 1–4) L: 13–30 / l: 5–5 / str. 13,5–15/10 µm . . . . **19. *N. veneta*** vol. 2/1 p. 104

50a Str. 9–10/10 µm, lineolae +/- 30/10 µm (vol. 2/1 Fig. 27: 4–6) . . . . **2. *N. margalithii*** vol. 2/1 p. 95

50b Str. 11–14/10 µm, lineolae +/- 35/10 µm (vol. 2/1 Fig. 27: 7–11) L: 16–40 / l: 6,5–9 / str. 11–14/10 µm . . . . **3. *N. recens*** vol. 2/1 p. 95

51a Valves nettement linéaires (sauf les plus petites formes) (vol. 2/1 Fig. 28: 1–5) L: 30–72 / l: 5–8 / str. 11–12/10 µm . . . . **5. *N. angusta*** p 97

51b Valves lancéolées, tout au plus linéaires-lancéolées **52**

52a Aire centrale bien délimitée, +/- circulaire à rhomboédrique **53**

52b Aire centrale rhomboédrique (vol. 2/1 Fig. 29: 1–4) . . . . **9. *N. radiosa*** vol. 2/1 p. 99

53a Largeur valvaire < 8 µm (vol. 2/1 Fig. 31: 8–14) L: 20–40 / l: 5–7 / str. 14–17/10 µm . . . . **17. *N. cryptocephala*** vol. 2/1 p.102

53b Largeur valvaire > 8 µm (vol. 2/1 Fig. 29: 5–7) L: 28–70 /l: (8)9–12 / str. 10–13/10 µm . . . . **10. *N. lanceolata*** vol. 2/1 p. 100

54a Valves elliptiques à linéaires-lancéolées (vol. 2/1 Fig. 28:8–15) – cf. aussi *N. eidrigiana* (vol. 2/1 Fig. 28: 6, 7) L: 13–45(55) / l: 5–8 / str. (8)12–17/10 µm . . . . **7. *N. cincta*** vol. 2/1 p. 98

54b Au cas où les valves sont nettement lancéolées, cf. aussi *N. cryptocephala* (vol. 2/1 Fig. 31: 8–14) *N. cryptotenella* (vol. 2/1 Fig. 33: 9–11) *N. menisculus* (vol. 2/1 Fig. 32: 16–25) *N. heimansii* (vol. 2/1 Fig. 29: 8–11) *N. phyllepta* (vol. 2/1 Fig. 32: 9–11)

55a Lineolae apparaissant nettement écartées. 20–25/10 µm **56**

55b Lineolae plus rapprochées **59**

56a Stries plus délicates, au moins 16/10 µm (vol. 2/1 Fig. 41: 3, 4) L: 35–60 / l: 8–12(15) / str. 16–18/10 µm . . . . . . **56 *N. gottlandica*** vol. 2/1 p. 122

56b Stries plus grossières, au maximum 14/10 µm **57**

57a Extrémités des valves +/- arrondies-tronquées (vol. 2/1 Fig. 28: 6, 7) L: 20–60 / l: 6–8 / str. 10–12,5/10 µm . . . . . **6. *N. eidrigiana*** vol. 2/1 p. 97

57b Extrémités des valves +/- étirées, rostrées ou légèrement capitées ou très arrondies-pointues **58**

58a Formes d'eaux salées ou saumâtres. Extrémités arrondies-pointues (vol. 2/1 Fig. 34: 10–11) L: (14) 30–55 / l: (4,5) 7–9 / str. 9–11/10 µm . . . . **28. *N. flanatica*** vol. 2/1 p. 109

58b Formes d'eaux douces oligosaprobes. Extrémités le plus souvent rostrées à légèrement capitées (vol. 2/1 Fig. 36: 4–7) L: 25–40 /l: 5,5–7,5 / str. 12–14/10 µm . . . . **37.** ***N. praeterita*** vol. 2/1 p. 113

59a Stries presque parallèles à faiblement radiales **60**

59b Stries plus nettement radiales au moins au centre **62**

60a Uniquement de très petites formes d'une largeur maximale de 4 µm. Pores centraux rapprochés **61**

60b Formes plus grandes. Pores centraux +/- éloignés. Si formes vivant en eaux très alcalines ou dans des eaux où la pression osmotique varie dans les valeurs élevées (vol. 2/1 Fig. 43: 1–11) L: 7–140/l.: 4,5–18/str.: 15–24/10 . . . . **62.** ***N. halophila*** vol. 2/1 p. 126
Si formes vivant dans le marais +/- acides et avec poles sans stries, parfois difficilement observables (vol. 2/1 Fig. 44: 12, 13) L: 32–45/l. 7–9/str.: env. 20/10 . . . . **63.** ***N. riparia*** vol. 2/1 p. 127

61a Aucune aire centrale (vol. 2/1 Fig. 35: 9, 10) L: (7)17–20/l.: 2–3(4,5)/St. 17–20/10 µm . . . . **32.** ***N. salinicola*** vol. 2/1 p. 111

61b Aire centrale peu ébauchée (vol. 2/1 Fig. 35: 21–24) L: 7–20/l.: 2,5–4,5/St. 13–16/10 µm . . . . **34.** ***N. incertata*** vol. 2/1 p. 111

62a Très petites formes (L. max. 12 µm) légèrement ventrues au milieu et aux extrémités très arrondies. L'appartenance aux *Lineolatae* est douteuse (vol. 2/1 Fig. 35: 11–13) L: 9–12/l.: 2,5–3/St. 18–20/10 µm . . . . **33.** ***N. bremensis*** vol. 2/1 p. 111

62b Valves plus longues **63**

63a Largeur valvaire max. 4 µm (vol. 2/1 Fig. 38: 16–20) L: 14–21/l.: 2,5–4/St. 15–17/10 µm . . . . **45.** ***N. tenelloides*** vol. 2/1 p. 117

63b Largeur valvaire > 5 µm **64**

64a Côte médiane apparaissant renforcée. Pores centraux rapprochés. Formes d'eaux marines ou saumâtres, également (mais rarement) en eau douce très alcaline (vol. 2/1 Fig. 32: 5–11) L: 12–45/l.: 4–8/St. 14–20/10 µm . . . . **45.** ***N. phyllepta*** vol. 2/1 p. 104

64b Caractéristiques différentes **65**

65a Valves larges-lancéolées. Stries 8–12/10 µm (vol. 2/1 Fig. 32: 16–25) L: 15–50/l.: 7,5–12/St. 8–12/10 µm . . . . **22.** ***N. menisculus*** vol. 2/1 p. 105

65b Valves plus étroites et lancéolées. St. > 12/10 µm **66**

66a Raphé courbé dans la médiane, uniquement vers le centre de la valve, pas reconnaissable dans la plupart des cas car très proche du bord de l'aire axiale très étroite. Uniquement en eaux acides (vol. 2/1 Fig. 29: 8–11) L: 25–48/l.: 5–6/St. 15–18/10 µm . . . . . . **11.** ***N. heimansii*** vol. 2/1 p. 100

66b Caractéristiques différentes **67**

67a Lineolae env. 30/10 µm. Formes typiques linéaires-lancéolées (vol. 2/1 Fig. 33: 1–4) L: 25–46/l.: 6–8/St. 14–18/10 µm . . . . **23.** ***N. stankovicii*** vol. 2/1 p. 106

67b Lineolae plus rapprochées, presqu'irrésolvables. Formes toujours lancéolées L: 14–40/l.: 5–7/St. (12)14–16(18)/10 µm (vol. 2/1 Fig. 33: 9–11; 13–17) . . . . **24.** ***N. cryptotenella*** vol. 2/1 p. 106
L: 30–42/l.: 6,5–7,5/St. env. 12/10 µm (vol. 2/1 Fig. 33: 21, 22) . . . . **25.** *N.* species 1 vol. 2/1 p. 107
L: 12–22/l.: 5–6/St. 14–16/10 µm (vol. 2/1 Fig. 33: 23–25) . . . . **26.** *N.* species 2 vol. 2/1 p. 108

68a Stries continuellement radiales (vol. 2/1 Fig. 38: 1–4) L: 30–55/l.: 5–9/St. 12–16/10 µm . . . . **42.** ***N. schroeteri*** vol. 2/1 p. 115

68b Stries convergentes aux extrémités **69**

69a Extrémités non étirées, émoussées à plus rarement arrondies-pointues (vol. 2/1 Fig. 38: 5–9) L: 25–35/l. 5–7/St. 12–14/10 µm . . . . **43.** ***N. erifuga*** vol. 2/1 p. 116

69b Extrémités plus ou moins étirées 70
70a Stries 7–10/10 µm (vol. 2/1 Fig. 37: 1–9) L: 34–100/l.: 7–15/St. 7–14/10 µm . . . . **41. *N. viridula*** vol. 2/1 p. 114
70b Stries 13–22/10 µm (vol. 2/1 Fig. 38: 10–15) L: 13–42/l.: 5–10/St. 13–22/10 µm . . . . **44. *N. gregaria*** vol. 2/1 p. 116
71a Aire centrale peu ou pas allongée transapicalement (vol. 2/1 Fig. 42: 1–11) L: 10–47/l.: 4–10/St. 8–11/10 µm . . . . . . **58. *N. capitata*** vol. 2/1 p. 123
71b Aire centrale remarquablement allongée transapicalement 72
72a Aire centrale délimitée des deux côtés par une strie courte (vol. 2/1 Fig. 41: 8–11) L: 12–30/l.: 4–7/St. 11–16/10 µm . **57. *N. lesmonensis*** vol. 2/1 p. 122
72b Aire centrale étirée jusqu'aux bords des valves 73
73a Extrémités arrondies-pointues, stries 7–10/10 µm (vol. 2/1 Fig. 42: 13–15) 10/10 µm . . . . **59. *N. costulata*** vol. 2/1 p. 124
73b Extrémités arrondies-tronquées, stries +/- 12/10 µm (vol. 2/1 Fig. 42: 16, 17) L: 12–22/l.:2,5–4 . . . . **60. *N. subcostulata*** vol. 2/1 p. 124

Groupe-clef B "Orthostichae" (vol. 2/1 p. 124)

1a Valves étroites, linéaires-lancéolées (fusiformes) avec un nodule central stauroïde élargi (vol. 2/1 Fig. 52: 4–6) . . . . **voir "Fusiformes"** [*Haslea*] vol. 2/1 p. 132
1b Valves différentes 2
2a Valves sans côtes longitudinales nettement reconnaissables ou > 25 côtes/ 10 µm 3
2b Valves avec côtes longitudinales nettes, 8–20/10 µm (vol. 2/1 Fig. 43: 1–8). *N. perrotettii* manquant dans la région L: (30)120–250/l: 13–25(44). St. 11–19(24)/10 µm . . . . **61. *N. cuspidata*** vol. 2/1 p. 126
3a Côtes longitudinales 28/10 µm ou plus, encore reconnaissables (vol. 2/1 Fig. 44: 1, 2) L: 7–140/l. 4,5–18/St. 15–24/10 µm/Lineolae 28–55/10 µm . . . . **62. *N. halophila* (formes robusta)** vol. 2/1 p. 126
3b Côtes longitudinales non reconnaissables 4
4a Valves plus larges, elliptiques-lancéolées (vol. 2/1 Fig. 45: 13–20) L: (15)17–25(37)/l. (4,5)8–11,5/st. 17–25/10 µm au milieu . . . . **64. *N. accomoda*** vol. 2/1 p. 128
4b Valves plus étroites-lancéolées 5
5a Pores centraux rapprochés (vol. 2/1 Fig. 39: 6–11) L: 20–60/l.: 4–6/St. 14–15/10 µm . . . . **49. *N. duerrenbergiana*** vol. 2/1 p. 119
5b Pores centraux plus écartés 6
6a Longueur valvaire régulièrement > 20 µm 7
6b Longueur valvaire régulièrement < 20 µm . . . . **Groupe-clef C**
7a Formes d'eaux "acides". Pôles sans stries (difficile à discerner) (vol. 2/1 Fig. 44: 12, 13) L: 32–45/l.: 7–9/St. 20/10 µm **63. *N. riparia*** vol. 2/1 p. 127
7b Formes d'eaux alcalines. Stries jusque dans les pôles (vol. 2/1 Fig. 44: 1–11; 14–18) . . . . **62. *N. halophila*** vol. 2/1 p. 126

Groupe-clef C (vol. 2/1 p. 129)

1a Valves plus grandes d'une population, lancéolées aux extrémités +/- étirées 6
1b Les valves les plus grandes d'une population, +/- elliptiques aux . . . 2 extrémités tronquées à larges-arrondies. Cf. aussi *N. digitulus* (vol. 2/1 Fig. 77: 19–28), *N. subhamulata* (vol. 2/1 Fig. 66: 32–34)et *N. citrus* (vol. 2/1 Fig. 58: 6–8) si extrémités courtes et pointues
2a Largeur valvaire: 6–7 µm/St. 14–18/10 µm (vol. 2/1 Fig. 64: 26–28) . . . . **71. *N. kriegeri*** vol. 2/1 p. 132
2b Valves plus étroites et/ou stries plus rapprochées 3

3a Stries > 25/10 µm (vol. 2/1 Fig. 45: 24,25) L: 10–16/l. 4,5–5,5/St. 17–18/10 µm . . . . **69. *N. fluens*** vol. 2/1 p. 131
3b Stries en règle générale < 25/10 µm **4**
4a Aire centrale au moins faiblement élargie transversalement (vol. 2/1 Fig. 73:12–15) L:8–12/l. 4,5–5,5/St. 17–18/10 µm: . . . . **192. *N. submuralis*** vol. 2/1 p. 203
4b Aire centrale à peine remarquable ou très peu élargie **5**
5a Côte médiane et nodules terminaux très prononcés (vol. 2/1 Fig. 45: 31–33) L: 10–12/l. 3,5–5/St. 22–24/10 µm . . . . **70. *N. muraliformis*** vol. 2/1 p. 132
5b Caractéristiques différentes. Extrémités rarement tronquées-arrondies. Stries le plus souvent nettement radiales (vol. 2/1 Fig. 76: 21–26) L: 7–12,5/l. 3,5–6/St. 15–26/10 µm . **224. *N. subminuscula*** vol. 2/1 p. 223
6a Stries au centre régulièrement < 26/10 µm **7**
6b Stries au centre presque toujours plus denses, jusqu'à plus de 40/10 µm aux extrémités **8**
7a Valves étroites-lancéolées. Formes d'eaux plus acides (vol. 2/1 Fig. 45: 26–30) L: 13,5–17/l. 3,3–4,5/St. 19–24/10 µm . . . . **68. *N. submolesta*** vol. 2/1 p. 131
7b Valves plus larges-lancéolées. Formes dominant dans des eaux à contenu électrolytique moyen à élevé ou dans biotopes fluctuants (vol. 2/1 Fig. 45: 10–12) L: 15–16,5/l. 4,5–6/St. 24–26/10 µm milieu-28–30/10 µm pôles . . . . **65. *N. minusculoides*** vol. 2/1 p. 129
8a Répartition jusqu'ici: Neusiedler See et eaux saumâtres du Burgenland (vol. 2/1 Fig. 45: 21–23) L: 12–17/l. 3–4,5/St. 30/10 µm milieu-35–40/10 µm pôles . . . . **67. *N. halophilioides*** p 131
8b Cosmopolite, avec optimum écologique des eaux α-mésosaprobes à modérément polysaprobes (vol. 2/1 Fig. 45: 1–9) L: 9,5–22/l. 3–5/St. 23–36/10 µm milieu – 40 et + µm pôles **65. *N. molestiformis*** vol. 2/1 p. 130

Fusiformes (vol. 2/1 p. 133)
1a Stries environ 12/10 µm (vol. 2/1 Fig. 52: 4) . . . . **72. *N. crucigera*** vol. 2/1 p. 133
1b Stries > 25/10 µm (vol. 2/1 Fig. 52: 5,6) . . . **73. *N. spicula*** vol. 2/1 p.133

Groupe-clef D (vol. 2/1 p. 134)
Groupe d'espèces proches de *Navicula elginensis*
1a Formes à petites valves, lancéolées à linéaires-lancéolées, régulièrement < 20 µm de longueur avec structures fines non nettes des stries. Toujours sans stigma isolé dans l'aire centrale (cf. vol. 2/1 p. e. Fig. 35: 9–20) . . . . **Cf. Lineolatae ou autres groupes-clefs**
1b Taille moyenne des valves des populations plus grandes et/ou contour +/- elliptique, rhomboédrique, large-lancéolé, souvent avec extrémités étirées-tronquées ou capitées. Stigmata isolés dans l'aire centrale, qui n'est jamais stauroïde comme dans la section autour de *N. mutica* 2
2a Fissures terminales courbées dans des directions opposées 3
2b Fissures terminales courbées du même côté 10
3a Aire centrale avec stigmata distincts des stries 4
3b Aire centrale sans stigmata distincts des stries 7
4a Aire centrale avec 2 ou plus de stigmata 5
4b Aire centrale avec 1 seul stigma 6
5a Plusieurs stries courtes intercalées au milieu (vol. 2/1 Fig. 47: 1–9) L: 15–50/l. 7–15/St. 8–15/10 µm . . . . . . **78. *N. clementis*** vol. 2/1 p. 139

5b Sans stries ainsi raccourcies (vol. 2/1 Fig. 48: 3–8) L: 4–32/l. 7–12/St. 13–14/10 µm . . . . **79.** ***N. clementoides*** vol. 2/1 p. 140
6a Valves aux extrémités courtes et rostrées (vol. 2/1 Fig. 48: 10, 11). Cf. aussi *N. Latens* avec fissures terminales difficiles à différencier (vol. 2/1 Fig. 48: 15, 16) L: 25–44/l. 10–14/St. 10–14/10 µm **80.** ***N. constans*** vol. 2/1 p. 140
6b Valves larges-rhomboédriques-lancéolées aux extrémités à peine étirées (vol. 2/1 Fig. 47: 19–21) L: 11–22/l.:6–12/St. 13–16/10 µm . . . . **82.** ***N. porifera*** vol. 2/1 p. 141
7a Plusieurs stries courtes intercalées au milieu **8**
7b Sans de telles stries **9**
8a Valves avec extrémités courtes et rostrées (vol. 2/1 Fig. 48: 12–14). Cf. aussi *N. pseudanglica* (vol. 2/1 Fig. 46: 13, 14) L: 25–44/l. 10–14/St. 10–14/10 µm . . . . **80.** ***N. constans*** **var.** ***symmetrica*** vol. 2/1 p. 140
8b Valves larges-rhomboédriques-lancéolées avec extrémités à peine étirées (vol. 2/1 Fig. 47: 22–24) L: 11–22/l. 6–12/St. 13–16/10 µm . . . . **82.** ***N. porifera*** var. ***opportuna*** vol. 2/1 p. 141
9a Valves linéaires-lancéolées avec stries approx. parallèles au niveau des extrémités rostrées (vol. 2/1 Fig. 49: 1–3) L: 28–40/l. 9–12/St. 10–12/10 µm et 20/10 µm extr. . . . . **85.** ***N. explanata*** vol. 2/1 p. 143
9b Caractéristiques différentes (vol. 2/1 Fig. 50: 5–8). Cf. aussi *N. pseudanglica* (vol. 2/1 Fig. 46: 13–15) L: 30–70/l. 12–28/St. 6–9(12)/10 µm . . . . **89.** ***N. subplacentula*** vol. 2/1 p. 145
10a Dans chaque pôle, 2 structures en forme de points jusqu'en forme de lignes très proches du raphé **14**
10b Pôles sans structures semblables **11**
11a Aire centrale avec un stigma **12**
11b Aire centrale sans stigma **15**
12a Stries alternativement courtes et longues au centre (vol. 2/1 Fig. 47: 10–18) L: 15–27/l. 6–9/St. 14–18/10 µm . . . . . . **81.** ***N. decussis*** vol. 2/1 p. 141
12b Stries courtes régulières, 14–18/10 µm, ou une seule strie courte, mais pas alternativement courtes et longues **13**
13a Extrémités courtes et larges finissant de façon arrondies-plates (vol. 2/1 Fig. 49: 7–9) L: 20–60/l. 10–20/St. 8–13/10 µm . . . . **86.** ***N. gastrum*** var. ***signata*** vol. 2/1 p. 143
13b Extrémités le plus souvent +/- rostrées à capitées (vol. 2/1 Fig. 46: 15,18) L: 20–40(50)/l. 8–14(20)/St. 9–12/10 µm . . . . **75.** ***N. pseudanglica*** var. ***signata*** vol. 2/1 p. 137
L: 15–40/l. 7–15/St. 10–14/10 µm **76.** ***N. exigua*** **var. signata** vol. 2/1 p.138
14a Valves petites, longueur 10–20 µm (vol. 2/1 Fig. 41: 5–7) L: 10–20/l. 5–7/St. 5/10 µm . . . . **84.** ***N. similis*** vol. 2/1 p. 143
14b Valves plus grandes, longueur 24–40 µm (vol. 2/1 Fig. 48: 15, 16) L: 24–40/l. 9–15/St. 10–16/10 µm . . . . **83.** ***N. latens*** vol. 2/1 p. 142
15a Stries ponctuées relativement grossièrement au milieu. Généralement, formes d'eaux marines à saumâtres ou de biotopes soumis à de hautes pressions osmotiques variables (cf. aussi groupes-clefs G et H) **26**
15b Stries plus finement ponctuées **16**
16a Extrémités nettement rostrées à capitées **22**
16b Extrémités autrement formées, le plus souvent tronquées et larges-arrondies **17**
17a Aire centrale nettement élargie de manière oblique-rhomboédrique, au bord le plus souvent avec de nombreuses stries courtes insérées (vol. 2/1 Fig. 51: 3–5) L: 30–60/l. 14–22/St. 14–18/10 µm . . . . **93.** ***N. platystoma*** vol. 2/1 p. 146
17b Caractéristiques différentes **18**

18a Strie centrale remarquablement étirée des deux côtés, les stries parallèles, en revanche, sont courtes (vol. 2/1 Fig. 50: 9–13) L: 12–24/l. 5–8/St. 13–17/10 µm . . . . **87.** ***N. hambergii*** vol. 2/1 p. 146
18b Caractéristique différente **19**
19a Valves légèrement cymbelloïdes **20**
19b Valves toujours totalement symétriques **21**
20a Largeur valvaire régulièrement moins de 10 µm, stries au centre +/- de même longueur (vol. 2/1 Fig. 49: 10–13) L: 12–35/l. 4,5–10/St. 9–15(22)/10 µm . . . . **87.** ***N. diluviana*** vol. 2/1 p. 144
20b Largeur valvaire plus de 10 µm, au centre quelques stries courtes (vol. 2/1 Fig. 49: 4–9) L: 20–60/l. 10–20/St. 8–13/10 µm . . . . **86.** ***N. gastrum*** vol. 2/1 p. 143
21a Valves arrondies-tronquées, aire centrale grande (vol. 2/1 Fig. 46: 10–12) L: 20–40/l. 8–15/St. 8–12/10 µm **74.** ***N. elginensis*** **var.** ***cuneata*** vol. 2/1 p. 136
21b Caractéristiques différentes (vol. 2/1 Fig. 50: 1–4). Cf. aussi *N. diluviana* (vol. 2/1 Fig. 49: 10–13) et *N. gastrum* (vol. 2/1 Fig. 49: 4–9) L: 30–70/l. 12–28/St. 6–9(12)/10 µm . . . . **88.** ***N. placentula*** vol. 2/1 p. 145
22a Stries déjà nettement convergentes avant les extrémités L: 20–36/l. (5)7–10/St. 17–21/10 µm . . . . **173.** ***N. laterostrata*** vol. 2/1 p. 197
22b Stries radiales ou parallèles aux extrémités, dans tous les cas les extrêmement distales sont convergentes **23**
23a Stries apparaissant constituées de simples rangées de points **24**
23b Pas de rangées de points simples reconnaissables (points doubles sous micr. électronique, Fig. 8: 6) – formes plus larges-elliptiques (vol. 2/1 Fig. 50: 1–4) L: 30–70/l.: 12–28/St. 6–9(12)/10 µm **88.** ***N. placentula*** vol. 2/1 p. 145
24a Valves linéaires aux extrémités larges (env. 6 µm) abruptement terminées (vol. 2/1 Fig. 48: 1) L: 35–46/l. 9–12/St. 8–12/10 µm . . . . **77.** ***N. abiskoensis*** vol. 2/1 p. 139
24b Valves, si linéaires, à extrémités moins larges **25**
25a Aire centrale grande, élargie transversalement (vol. 2/1 Fig. 46: 1–12) L: 0–40/l. 8–15/St. 8–12/10 µm . . . . **74.** ***N. elginensis*** vol. 2/1 p. 136
25b Aire centrale petite, parfois irrégulièrement délimitée par des stries courtes isolées (vol. 2/1 Fig. 46: 13–18) L: 20–40(50)/l. 8–14(20/St. 9–12/10 µm . . . . **75.** ***N. pseudanglica*** vol. 2/1 p. 137
L: 15–40/l. 7–15/St. 10–14/10 µm . . . . **76.** ***N. exigua*** sensu Hustedt vol. 2/1 p. 138
26a Stries continuellement radiales (vol. 2/1 Fig. 51: 1) L: 30–80/l. 22–30/St. 5–9/10 µm . . . . **91.** ***N. amphibola*** vol. 2/1 p. 146
26b Stries aux extrémités parallèles à légèrement convergentes (vol. 2/1 Fig. 51: 2) L: 45–120/l. 20–30/St. 6–8/10 µm centre St. 12–14/10 µm pôles . . . . **92.** ***N. semen*** vol. 2/1 p. 146

Groupe-clef E (vol. 2/1 p. 147)
Espèces voisines de *N. mutica*

1a Points en bordure valvaire. Points reconnaissables en vue connective comme de petites épines (vol. 2/1 Fig. 63: 17–19) L: 9–15/l.: 7–9/St.15–18/10 µm . . . . **107.** ***N. spinifera*** vol. 2/1 p. 157
1b Valves toujours sans petites épines **2**
2a Bords valvaires présentant 2 ou plusieurs ondulations, ou partie centrale fortement rhomboédrique en relation avec des extrémités fortement étranglées **16**
2b Valves sans ondulation, +/- elliptiques ou avec partie centrale renflée et extrémités capitées **3**

3a Branches du raphé avec pores centraux ou fissures centrales nettement courbées dans le sens opposé au stigma, +/- clairement marquées . . . **4**
3b Extrémités proximales du raphé sans pores centraux à contours nets, en partie légèrement arquées depuis le stigma. Un changement de la direction des fissures centrales extérieures peut souvent être constaté après ajustement du microscope (ouverture maxi.) **13**
4a Stigma en point et en position centrale ou +/- proche du nodule central . **5**
4b Caractéristiques différentes **9**
5a Extrémités des valves larges à plates-arrondies **6**
5b Valves plus fortement rétrécies aux extrémités ou +/– capitées **7**
6a Fissures centrales et terminales nettement courbées en crochet (vol. 2/1 Fig. 60: 20) . . . . ***N. plausibilis***
6b Caractéristique différente (vol. 2/1 Fig. 63: 1–3) – cf. aussi *N. suecorum* (vol. 2/1 Fig. 637–12) L: 10–30/l. 6–12/str. 15–20/10 µm . . . . **96. *N. cohnii*** vol. 2/1 p. 152
7a Extrémités largement étirées et capitées **8**
7b Extrémités étroites, +/- rostrées ou seulement tronquées à arrondies pointues (vol. 2/1 Fig. 62: 14–18). Cf. aussi les formes *simplex* de *N. charlatii* . . . . ***N. obligata* et nouvelles combinaisons de formes de *N. mutica***
8a Extrémités capitées, presqu'aussi larges que la partie centrale de la valve . . . . ***N. palaearctica***
8b Extrémités capitées nettement plus étroites (vol. 2/1 Fig. 61: 12–15). Cf. aussi *N. pseudonivalis* (vol. 2/1 Fig. 61: 23–26•) L: 10–25/l. 6–10/str. 15–18/10 µm . . . . **97. *N. muticopsis*** vol. 2/1 p. 153
9a Canal des stigma en forme de rayures, proche du bord valvaire **10**
9b Canal des stigma en position plus ou moins centrale, le plus souvent en forme de rayure, à focalisation ajustée souvent entourée d'une structure en forme d'anneau (vol. 2/1 Fig. 62: 1–12). *N. mitigata* à extrémités rostrées (vol. 2/1 Fig. 62: 13) n'est présentée que pour la comparaison L: 10–65(144)/ l. 6–15(36)/ str. 18–24/10 µm . . . . **95. *N. goeppertiana*** vol. 2/1 p. 150
10a Valves +/- rhomboédriques avec extrémités arrondies-plates (vol. 2/1 Fig. 63: 13) . . . . ***N. mutica* f. *intermedia***
10b Valves +/- elliptiques à extrémités tronquées à larges-arrondies **11**
11a Str. 28–30/10 µm – formes tropicales (vol. 2/1 Fig. 60: 21, 22) . . . . ***N. muticoides***
11b Stries plus grossières **12**
12a Valves elliptiques-lancéolées, à extrémités +/- faiblement étirées. Stigma difficile à différencier (vol. 2/1 Fig. 60: 16–19). L: 12–25/l. 5–10/str. 19–23/10 µm . . . . **102. *N. pseudokotschyi*** vol. 2/1 p. 155
12b Extrémités plus largement arrondies. Stigma bien différenciable, un peu plus éloigné du bord (vol. 2/1 Fig. 63: 4–6) L: 8–32/l. 6–10/str. 21–23/10 µm . . . . **103. *N. saxophila*** vol. 2/1 p. 155
13a Extrémités capitées **14**
13b Extrémités toujours +/- étirées **15**
14a Stigma ponctiforme en position centrale (vol. 2/1 Fig. 61: 23–26) L: 14–20/l. 5,5–6,5/str. 19–21/10 µm . **100. *N. pseudonivalis*** vol. 2/1 p. 154
14b Stigma proche du bord (vol. 2/1 Fig. 61: 9–11) L: 6–30(40)/l. 4–9(12)/str. 14–20(25)/10 µm . . . . **94. *N. mutica* var. *ventricosa*** vol. 2/1 p. 149
15a Extrémités +/- étirées (vol. 2/1 Fig. 61: 27–31) L: 10–22/l. 4–6/str. 17–22/10 µm . . . . **101. *N. paramutica*** vol. 2/1 p. 155
15b Extrémités pas nettement étirées (vol. 2/1 Fig. 61: 1–8) L: 6–30(40)/l. 4–9(12)/str. 14–20/10 µm . . . . **94. *N. mutica*** vol. 2/1 p. 149

16a Valves fortement rhomboédriques au centre, extrémités très capitées-étranglées (vol. 2/1 Fig. 63: 13–15) L: 20–35/l. 8–12/str. 15–16/10 µm . . . . 105. *N. heufleriana* vol. 2/1 p. 156
16b Caractéristique différente 17
17a Partie centrale valvaire présentant 2 ondulations (vol. 2/1 Fig. 61: 30, 31). Cf. aussi *N. incoacta* Hustedt (du Pérou) et *N. mutica* var. *binodis* Hustedt Probablement pas co-spécifique . . . . . 101. ***N. paramutica*** var. ***binodis***
17b Partie centrale présentant plusieurs ondulations 18
18a Valves rhomboédriques à elliptiques-lancéolées 19
18b Valves linéaires 21
19a Valves aux extrémités plus fortement rostrées à capitées 20
19b Extrémités pas nettement marquées, bords faiblement ondulés (vol. 2/1 Fig. 63: 7–12) L: 15–50/l. 6–18/str. 14–21/10 µm . . . . 104. ***N. suecorum*** vol. 2/1 p. 156
20a Grandes valves (vol. 2/1 Fig. 63: 20, 21) L: 30–45/l. 10–16/str. 12–15/10 µm . . . . 106. ***N. charlatii*** vol. 2/1 p. 157
20b Valves beaucoup plus petites (vol. 2/1 Fig. 61: 23–26•) L: 14–20/l. 5,5–6,5/str; 19–21/10 µm . . . . . . 100. ***N. pseudonivalis*** vol. 2/1 p. 154
21a Stries généralement constituées de trois points seulement. Les rangées longitudinales sont démarquées des autres près de l'aire axiale (vol. 2/1 Fig. 61: 21, 22) L: 8–25/l. 6–9/str. 15–20/10 µm . . . . 99. ***N. nivaloides*** vol. 2/1 p. 154
21b Stries autrement ponctuées 22
22a Pores centraux courbés en direction opposée à celle du stigma et fortement marqués (vol. 2/1 Fig. 61: 17–20)L: 12–42/l. 5,5–13/str. 17–20(24)/10 µm . . . . 98. ***N. nivalis*** vol. 2/1 p. 153
22b Pas de pores centraux nettement marqués (vol. 2/1 Fig. 61: 8). Cf. aussi *N. pseudonivalis* (vol. 2/1 Fig. 61: 23–26) L: 6–30(40)/l. 4–9/str. 14–20(25) . . . . 94. ***N. mutica*** vol. 2/1 p. 149

Groupe-clef F (vol. 2/1 p. 158)
"Punctatae" elliptiques plus ou moins largement arrondies

1a Valves strictement elliptiques à presque circulaires, avec extrémités +/- largement arrondies 2
1b Contour valvaire différent 8
2a Raphé dans un sillon médian très marqué, bordé de rangées de points saillants sans stries courtes intercalées (vol. 2/1 Fig. 67: 14, 15) L: 9–20/l. 6–8/str. 16–20/10 µm . . . . 159. ***N. aboensis*** vol. 2/1 p. 191
2b Caractéristiques différentes 3
3a Stries très grossièrement ponctuées, 10–16/10 µm (vol. 2/1 Fig. 59: 16–19) . . . . 112. ***N. scutelloides*** vol. 2/1 p. 160
3b Points plus rapprochés 4
4a Stries faiblement radiales 7
4b Stries fortement radiales 5
5a Aire centrale assez grande, plus ou moins ronde (vol. 2/1 Fig. 59: 10, 11) L: 24–48/l. 17–25/str. (18)22–30/10 µm . 110. ***N. scutiformis*** vol. 2/1 p. 159
5b Aire centrale petite ou absente 6
6a Valves presque circulaires-elliptiques (vol. 2/1 Fig. 59: 12–15) L: 3,5–25/l. 3–17/str. 20–26/10 µm . . . . . 11. ***N. pseudoscutiformis*** vol. 2/1 p. 159
6b Valves +/- rhomboédriques-lancéolées (vol. 2/1 Fig. 59: 2–5) L: 12–40/l. 7–15/str. 24–36/10 µm . . . . 108. ***N. cocconeiformis*** vol. 2/1 p. 158
7a Stries centrales 9–12/10 µm (vol. 2/1 Fig. 60: 1, 2) L: 16–40/l. 8–20/str. 10/10 µm . . . . 113. ***N. jentzschii*** vol. 2/1 p. 160

7b Stries 25–36/10 µm (vol. 2/1 Fig. 59: 6–9) L: 8–20/l. 6–11/str. 25–36/10 µ . . . . **109.** ***N. jaernefeltii*** vol. 2/1 p. 159
8a Aire centrale à peine marquée (vol. 2/1 Fig. 59: 2–5) L: 12–40/l. 7–15/str. 24–36/10 µ . . . . **108.** ***N. cocconeiformis*** vol. 2/1 p. 158
8b Aire centrale +/- grande, presque elliptique (vol. 2/1 Fig. 75: 29–31) L: 9–28/l. 4–10/str. (15)18–26/10 µ . . . **223.** ***N. confervacea*** vol. 2/1 p. 221

Groupe-clef G (vol. 2/1 p. 160)
Microstigmaticae

1a Longueur valvaire +/- 200 µm (vol. 2/1 Fig. 52: 3) L: +/- 200/l. +/- 30/str. 11–12/10 µm . . . . **114.** ***N. bergenensis*** vol. 2/1 p. 161
1b Valves plus courtes **2**
2a Structure particulière de l'aire centrale selon Fig. 55: 11, L: 45–110/l. 18–30/str. 14–16/10 µm ext. . . . . **119.** ***N. zeta*** vol. 2/1 p. 163
2b Caractéristique différente **3**
3a Valves étroites-lancéolées. Largeur toujours < 7,5 µm avec (souvent difficilement reconnaissable) un point isolé dans la petite aire centrale **7**
3b Caractéristiques différentes **4**
4a Valves aux extrémités pointues et pseudoseptae (vol. 2/1 Fig. 55: 1–3) L: 25–45/l. 8–10/str. 17–23/10 µm . . . . **116.** ***N. integra*** vol. 2/1 p. 162
4b Caractéristique différente **5**
5a Contour valvaire lancéolé, progressivement effilé depuis le milieu, avec extrémités +/- étirées (vol. 2/1 Fig. 54: 1–13) L: 20(35)-100/l. 8–23/str. 14–18/10 µm . . . . **115.** ***N. crucicula*** vol. 2/1 p. 161
5b Valves plus elliptiques à linéaires, aux extrémités pas ou largement étirées à arrondies-tronquées **6**
6a Stries très faiblement radiales. Celles du centre à peine plus espacées que les autres (vol. 2/1 Fig. 55: 4) L: 50–90/l. 5–10/str. 17–20/10 µm . . . . **117.** ***N. plicata*** vol. 2/1 p. 163
6b Stries nettement plus radiales. Celles du centre nettement plus espacées que les autres (vol. 2/1 Fig. 55: 5–10) L. 17–60/l. 5–10/str. 14–20/10 µm . . . . **118.** ***N. protracta*** vol. 2/1 p. 163
7a Valves rostrées (vol. 2/1 Fig. 56: 6–9) L: 15–30/l. 2–4/str. 30(40)/10 µm . . . . **21.** ***N. bulnheimii*** vol. 2/1 p. 164
7b Valves aux extrémités pointues à tronquées, non effilées **8**
8a Extrémités tronquées (vol. 2/1 Fig. 56: 10) L. 15–30/l. 2–4/str. 20/10 µm . . . . **121.** ***N. bulnheimii*** **var.** ***belgica*** vol. 2/1 p. 165
8b Extrémités pointues (vol. 2/1 Fig. 56: 1–5) L: 25–75/l. 4–7,5/str. 20/10 µm . . . . **120.** ***N. complanata*** vol. 2/1 p. 164

Groupe-clef H (vol. 2/1 p. 165)
Restes des punctatae hétérogènes

1a Valves aux extrémités +/- brusquement effilées, rostrées à capitées . . **2**
1b Valves aux extrémités différemment effilées **10**
2a Valves linéaires **3**
2b Valves à bords +/- nettement convexes **6**
3a Valves larges-linéaires aux extrémités plus étroites-étirées **5**
3b Valves linéaires aux extrémités larges **4**
4a Valves max. 30 µm de long (vol. 2/1 Fig. 57: 5). Cf. aussi la variété *capitata* inconnue dans la région (vol. 2/1 Fig. 57: 6) L: 28–30/l. 7,5–8,5/str. 15–18/10 µm . . . . **124.** ***N. pusilla*** **var.** ***incognita*** vol. 2/1 p. 167
4b Longueur valvaire au moins 40 µm L: 35–46/l. 9–12/str. 8–12/10 µm . . . . **77.** ***N. abiskoensis*** vol. 2/1 p. 139

5a Stries fortement radiales avec de nombreuses stries courtes intercalées au centre (vol. 2/1 Fig. 58: 1) L: 33–100/l. 20–42/str. 9–12/10 µm . . . . **125.** ***N. humerosa*** vol. 2/1 p. 168
5b Stries faiblement radiales (vol. 2/1 Fig. 58: 4–5) L: 35–72/l. 16–25/str. 10–15/10 µm . . . . **127.** ***N. maculosa*** vol. 2/1 p. 168
6a Aire centrale élargie +/- transversalement **7**
6b Aire centrale +/- isodiamétrique **8**
7a Extrémités rostrées très pointues. Stries presque parallèles. Points env. 20/10 µm (vol. 2/1 Fig. 58: 6–8) L: 18–22/l. 6–7/str. 18–20/10 µm . . . . **128.** ***N. citrus*** vol. 2/1 p. 169
7b Extrémités plus arrondies-tronquées. Stries radiales. Points env. 24/10 µm (vol. 2/1 Fig. 60: 10–15). Cf. aussi *N. pseudotuscula* (vol. 2/1 Fig. 82: 1–4) L: 15–16/l. 5–7/str. 20–26/10 µm . . . . . **131.** ***N. kotschyi*** vol. 2/1 p. 169
8a Largeur valvaire < 10 µm. Stries centrales > 20/10 µm (vol. 2/1 Fig. 57: 1–4) L: 26–35/l. 7–10/str. 27–30/10 µm . . . **123.** ***N. ordinaria*** vol. 2/1 p. 166
8b Largeur valvaire > 10 µm **9**
9a Aire centrale ronde (vol. 2/1 Fig. 57: 7–9) L: 24–70/l. 7,5–26/str. 15–18/10 µm . . . . **124.** ***N. pusilla*** vol. 2/1 p. 167
9b Aire centrale prolongée transversalement (vol. 2/1 Fig. 51:1) L: 30–80/l. 23–30/str. 5–9/10 µm . . . . **91.** ***N. amphibola*** vol. 2/1 p. 146
10a Valves linéaires à bords +/- triondulaires (vol. 2/1 Fig. 60: 3–5). Cf. aussi *N. levanderi* (vol. 2/1 Fig. 65: 16–17) L: 25–50/l. 5–10/str. 18–22/10 µm . . . . **130.** ***N. pseudosilicula*** vol. 2/1 p. 169
10b Caractéristique différente **11**
11a Longueur valvaire jusqu'à +/- 10 µm, rhomboédrique-elliptique aux extrémités arrondies-pointues (vol. 2/1 Fig. 60: 6–9) L: 7,5–9/l. 3–4/str. +/– 21/10 µm . . . . **129.** ***N. ingenua*** vol. 2/1 p. 169
11b Caractéristiques différentes **12**
12a Stries nettement convergentes aux extrémités **13**
12b Stries continuellement radiales, mais les stries extrêmement distales sont convergentes **14**
13a Stries du milieu 6–8/10 µm (vol. 2/1 Fig. 51: 2) L: 45–120/l. 20–30/str. 6–8/10 µm centr. 12–14/10 µm pôles . . . . **92.** ***N. semen*** vol. 2/1 p. 146
13b Stries considérablement plus rapprochées (vol. 2/1 Fig. 52: 1–2) L: 35–75/l. 9–18/str. 18–20/10 µm . . . . **122.** ***N. brasiliana*** vol. 2/1 p. 166
14a Aire centrale élargie transversalement **15**
14b Aire centrale ronde (vol. 2/1 Fig. 58: 2–3). Cf. aussi *N. pusilla* (vol. 2/1 Fig. 57: 9) L: 30–90/l. 12–22/str. (8)11–16/10 µm . . . . **126.** ***N. lacustris*** vol. 2/1 p. 168
15a Largeur valvaire > 20 µm. Str. < 10/10 µm (vol. 2/1 Fig. 51: 1) L: 30–80/l. 23–30/str. 5–9/10 µm . . . . **91.** ***N. amphibola*** vol. 2/1 p. 146
15b Formes plus petites, stries plus nombreuses (vol. 2/1 Fig. 60: 10–15) L: 15–26/l. 5–7/str. 20–26/10 µm . . . . . . **131.** ***N. kotschyi*** vol. 2/1 p. 169

Groupe-clef I (vol. 2/1 p. 171)
Lyratae

1a Stries grossières (13–16, rarement jusqu'à 22/10 µm). Côte médiane élargie en forme de vésicule autour des pores centraux. Aires centrale et latérale reliées en parenthèse (vol. 2/1 Fig. 65: 12–13) L: 20–80/l. 10–24/str. 13–16/10 µm rar. 22/10 µm . . . . **135.** ***N. forcipata*** vol. 2/1 p. 172
1b Caractéristiques différentes **2**
2a Pores centraux très espacés **3**
2b Caractéristique différente **4**

3a Aires latérales courbées en forme de parenthèse (vol. 2/1 Fig. 65: 7–9) L: 8–15/l. 5–6/str. 24–25/10 µm . . . . . . **133. *N. cryptolyra*** vol. 2/1 p. 172
3b Aires latérales +/- sous forme de lignes droites (vol. 2/1 Fig. 65: 11) L: 6–8/l. 3,5–4/str. 33–35/10 µm . . . . **136. *N. muralibionta*** vol. 2/1 p. 173
4a Aires centrale et latérale nettement liées. Pores centraux fortement marqués (vol. 2/1 Fig. 65: 1–6') . . . . **132. *N. pygmaea*** vol. 2/1 p. 171
4b Caractéristiques différentes (vol. 2/1 Fig. 65: 10) L: 9–10/l. 5–6/str. 22–24/10 µm . . . . **134. *N. pseudoforcipata*** vol. 2/1 p. 172

Groupe-clef J (vol. 2/1 p. 173)
1a Aire centrale formant, d'un ou des 2 côtés, une large bande transversale atteignant les bords valvaires **2**
1b Caractéristique différente **3**
2a Valves transapicalement élargies au milieu et aux extrémités (vol. 2/1 Fig. 66: 5–8) L: 10–19/l. +/- 4/str. 21–24/10 µm . . . . **140. *N. occulta*** vol. 2/1 p. 176
2b Bords valvaires tout au plus faiblement convexes (vol. 2/1 Fig. 66: 9–11) L: 7,5–16/l. 3–4/str. 18–22/10 µm . . . . . **141. *N. lucinensis*** vol. 2/1 p. 176
3a Une rangée apicale de points nettement marquée, reposant d'un côté (rarement des 2 côtés et de façon fragmentaire) dans la côte médiane, et interrompue par l'aire centrale. En outre, une ligne hyaline parallèle au bord valvaire sépare les stries en une rangée de points périphérique et une rangée de points centrale (vol. 2/1 Fig. 66: 19–23) L: 9–27/l. 4–9/str. 13–22/10 µm . . . . **139. *N. tenera*** vol. 2/1 p. 175
3b Au cas où il y aurait une rangée isolée de points dans une telle combinaison de caractéristiques, elle se situe non pas dans, mais à côté de la côte médiane (contrastante) **4**
4a Valves linéaires avec extrémités largement arrondies (en partie étirées) **5**
4b Valves aux bords nettement convexes (ou concaves), elliptiques à rhomboédriques **8**
5a Valves relativement grandes (28–45 µm), linéaires, le plus souvent à 3 ondulations, str. nettement ponctuées (vol. 2/1 Fig. 65: 16, 17). Si extrémités capitées à rostrées, cf. *N. hoefleri* (vol. 2/1 Fig. 79: 28) L: 28–45/l. 4,5–7/str. 20–24/10 µm . . . . **142. *N. levanderi*** vol. 2/1 p. 177
5b Caractéristiques différentes **6**
6a Pores centraux du raphé espacés. Aire centrale assez grande, ronde (vol. 2/1 Fig. 65: 11) L: 6–8/l. 3,5–4/str. 33–35/10 µm . . . . **136. *N. muralibionta*** vol. 2/1 p. 173
6b Caractéristiques différentes **7**
7a Une seule fine rangée de points sépare chaque ligne longitudinale hyaline des stries (vol. 2/1 Fig. 66: 24–27) . . . . **160. *N. helensis*** vol. 2/1 p. 192
7b Vers le milieu de chaque côté valvaire, les lignes longitudinales délimitent une zone faiblement hyaline parallèle à la côte médiane. Fissures terminales longues, en forme de crochets (vol. 2/1 Fig. 66: 32–34) L: 12–25/l. 4–7/str. 26–30/10 µm . . . . **162. *N. subhamulata*** vol. 2/1 p. 192
8a Bords valvaires concaves (vol. 2/1 Fig. 75: 26–28) L: 8–12/l. 2,5–3,5/str. 26–28/10 µm . . . . **230. *N. aerophila*** vol. 2/1 p. 228
8b Caractéristique différente **9**
9a Extrémités valvaires plus ou moins étirées ou arrondies-pointues . . . . **143. *N. krasskei*** vol. 2/1 p. 177 (vol. 2/1 Fig. 72: 21–24) L: 5,8–17/l. 3,4–5,5/str. 45–55/10 µm cf. aussi groupe-clef M, numéros 36–38
9b Extrémités tronquées ou arrondies, non étirées **10**
10a Les aires axiale et centrale forment un espace hyalin +/- large, en plus

d'une ligne hyaline près du bord valvaire (vol. 2/1 Fig. 66: 1–4) L: 7–22/l. 4,5–7/str. 20–25/10 µm . . . . **138. *N. insociabilis*** vol. 2/1 p. 175

10b Un espace hyalin est présent, même si peu perceptible. La ligne hyaline s'étend en position +/- centrale (vol. 2/1 Fig. 66: 12–18) L: 8–22/l. 3–6,5/str. 20–30/10 µm. . . . . **137. *N. monoculata*** vol. 2/1 p. 174

Groupe-clef K (vol. 2/1 p. 178)
Annulatae

1a Contour des valves ondulé (vol. 2/1 Fig. 64: 12–15) L: 17–25/l. 4–6/str. 13–18/10 µm . . . . **146. *N. ignota*** var. ***ignota*** vol. 2/1 p. 179

1b Contour des valves non ondulé **2**

2a Stries plus fortement radiales, surtout celles du milieu **4**

2b Stries centrales faiblement radiales. Le plus souvent, un stigma isolé proche du nodule central **3**

3a Longueur valvaire 16–24 µm (vol. 2/1 Fig. 64: 16–19) L: 16–24/l. 4–6/str. 13–18/10 µm . . . . **146. *N. ignota*** var. ***palustris*** vol. 2/1 p. 180

3b Longueur valvaire 6–14 µm (vol. 2/1 Fig. 64: 20–24) L: 6–14/l. 4–6/str. 13–18/10 µm . . . . **146. *N. ignota*** var. ***acceptata*** vol. 2/1 p.180

4a Aire centrale grande, atteignant presque les bords (vol. 2/1 Fig. 64: 29–32) L: 11–20/l. 4–5/str. 16–19/10 µm . . . **145. *N. dolomitica*** vol. 2/1 p. 179

4b Aire centrale plus petite, irrégulièrement délimitée (vol. 2/1 Fig. 64: 1–11) L: 6–28/l. 4–9/str. 12–20/10 µm . . . **144. *N. schoenfeldii*** vol. 2/1 p. 178

Groupe-clef L (vol. 2/1 p. 181)
Espèces voisines de *N. bryophila* et *N. subtilissima*

1a Stries croisées par 2 lignes hyalines relativement larges et de forme lancéolée (vol. 2/1 Fig. 79: 28). Cf. aussi certaines formes de *N. subtilissima* L: 25–48/l. 5–10/str. 40–42/10 µm . . . . **151. *N. hoefleri*** vol. 2/1 p. 183

1b Caractéristique différente **2**

2a Aire centrale relativement grande, rectangulaire, elliptique ou rhomboédrique **3**

2b Aire centrale petite ou faiblement reconnaissable **4**

3a Stries > 30/10 µm. Extrémités peu étirées (vol. 2/1 Fig. 79: 18–21) L: 20–45/l. 4,5–6/str. 30–36/10 µm . . . . . . **153. *N. jaagii*** vol. 2/1 p. 184

3b Stries 24–27/10 µm, extrémités plus fortement étirées capitées (vol. 2/1 Fig. 79: 16, 17) L: 17–26/l. 4,5–6/str. 24–27/10 µm . . . . **152. *N. brockmannii*** vol. 2/1 p. 183

4a Stries à la limite de la résolution, +/- 42/10 µm (vol. 2/1 Fig. 79: 22–26) L: 18–38/l. 3,5–6/str. 40–42/10 µm . . . **150. *N. subtilissima*** vol. 2/1 p. 182

4b Stries plus grossières, nettement < 40/10 µm **5**

5a Valves étroites-linéaires à extrémités progressivement étirées (vol. 2/1 Fig. 79: 9–12) L: 11–20/l. 2–3,5/str. 24–30/10 µm . . . . **148. *N. suchlandtii*** vol. 2/1 p. 181

5b Valves plus largement linéaires aux extrémités plus marquées, le plus souvent rostrées à capitées **6**

6a Largeur 2,5–4 µm (vol. 2/1 Fig. 79: 1–8') L: 10–25/l. 2,5–4/str. 24–26(38)/10 µm . . . . **147. *N. bryophila*** vol. 2/1 p. 189

6b Largeur 4,5–5,5 µm (vol. 2/1 Fig. 79: 13–15) L: 18–26/l. 4,5–5,5/str. (26)32–36(40)/10 µm . . . . . . **149. *N. pseudobryophila*** vol. 2/1 p. 182

Groupe-clef M (vol. 2/1 p. 185)
Bacillares

1a Nodules terminaux élargis transapicalement **2**

1b Caractéristique différente **3**

2a Extrémités non étirées. Sillons longitudinaux (longeant la côte médiane) nettement marqués. Conopeum reconnaissable sous microscope électronique (vol. 2/1 Fig. 67: 2–4) L: (25)30–90/l. 10–20/str. 12–14(16)/10 µm prox. à 22(24)/10 µm dist. . . . . **154.** ***N. bacillum*** vol. 2/1 p. 187

2b Extrémités le plus souvent +/- étirées. Sillons longitudinaux uniquement en lignes ou reconnaissable au fait que les extrémités des stries semblent plus contrastées près de l'aire axiale (vol. 2/1 Fig. 68: 1–21) . . . . **58.** ***N. pupula*** vol. 2/1 p. 189

3a Extrémités valvaires étirées rostrées à +/- capitées **27**

3b Extrémités tout au plus faiblement étirées **4**

4a Valves linéaires à extrémités largement arrondies, plus rarement faiblement concaves ou convexes au milieu **5**

4b Valves rhomboédriques, elliptiques ou lancéolées **14**

5a Valves relativement grandes. Largeur > 10 µm. Sillons très larges. Nodules centraux élargis elliptiquement et à contours nets. Raphé onduleux (vol. 2/1 Fig. 67: 1). Si sillons moins larges cf. aussi *N. bacillum* (vol. 2/1 Fig. 67: 2–4) L: 30–140/l. 10–30/str. 13–18/10 µm . . . . **156.** ***N. americana*** vol. 2/1 p. 188

5b Caractéristiques différentes **6**

6a Stries croisées par des aires latérales ou par de nettes structures lignées . . . . **cf. Groupe-clef J**

6b Pas de telles structures **7**

7a Aire centrale élargie +/- transversalement **8**

7b Aire centrale différente **10**

8a Aire centrale formant une bande transversale atteignant les bords (vol. 2/1 Fig. 77: 2–3) L: 7–13/l. +/- 3,5/str. 24–28/10 µm . . . . **242.** ***N. muraloides*** vol. 2/1 p. 232

8b Caractéristique différente **9**

9a Largeur valvaire toujours > 5 µm. Raphé onduleux (vol. 2/1 Fig. 67: 6–13) L: 20–70/l. 6–11/tr. 15–22/10 µm . . . **157.** ***N. laevissima*** vol. 2/1 p. 189

9b Largeur valvaire < 5 µm. Petites formes dans l'ensemble (vol. 2/1 Fig. 69: 1–10). Cf. aussi *N. digitulus* (vol. 2/1 Fig. 77: 27) L: 7,8–24/l. 3,2–5,1/str. 23–28/10 µm . . . . **165.** ***N. stroemii*** vol. 2/1 p. 194

10a Sillons largement aplatis, sur 1/3–1/2 de la largeur valvaire. Fissures terminales très longues, en forme de crochets (vol. 2/1 Fig. 66: 32–34) L: 12–25/l. 4–7/str. 26–30/10 µm . . . **162.** ***N. subhamulata*** vol. 2/1 p. 192

10b Caractéristiques différentes **11**

11a Largeur valvaire < 6 µm. Stries > 30/10 µm (vol. 2/1 Fig. 66: 35–39) L: 8–17/l. 3–5/str. 32–40/10 µm . . . . **163.** ***N. lenzii*** vol. 2/1 p. 193

11b Caractéristiques différentes **12**

12a Stries 22–26/10 µm (vol. 2/1 Fig. 66: 24–27) L: 13–30/l. 4,5–8/str. 21–26/10 µm . . . . **160.** ***N. helensis*** vol. 2/1 p. 192

12b Stries 26–32/10 µm **13**

13a Longueur des valves > 14 µm (vol. 2/1 Fig. 66: 31) L: 14–21/l. 6–7/str. 26–32/10 µm . . . . **161.** ***N. fracta*** vol. 2/1 p. 192

13b Longueur des valves < 10 µm (vol. 2/1 Fig. 66: 40–42) L: +/- 9,5/l. 4,5/str. +/- 30/10 µm . . . . **164.** ***N. sublucidula*** vol. 2/1 p. 193

14a Valves +/- elliptiques aux extrémités largement arrondies **15**

14b Valves rhomboédriques, lancéolées, extrémités plus fortement étirées . **22**

15a Stries +/- plus grossièrement ponctuées, environ 25 points/10 µm (vol. 2/1 Fig. 67: 14–15) L: 9–20/l. 6–8/str. 16–20/10 µm . . . . **159.** ***N. aboensis*** vol. 2/1 p. 191

15b Ponctuation plus fine ou non reconnaissable **16**

16a Stries encadrant le nodule central alternativement plus courtes et plus longues. Extrémités larges-arrondies (vol. 2/1 Fig. 69: 11–13) L:5–22/l. 3,5–10/str. 18–22/10 µm . . . . **166. *N. rotunda*** vol. 2/1 p. 194
16b Caractéristiques différentes **17**
17a Aire centrale grande, +/- circulaire. Stries faiblement radiales, courtes au centre (vol. 2/1 Fig. 67: 5) . . . . **155. *N. bacilloides*** vol. 2/1 p. 188
17b Caractéristiques différentes **18**
18a Aire centrale formant une bande transversale allant au-delà de la moitié de la largeur de la valve (vol. 2/1 Fig. 73: 4–7) L: 12–24/l. 7–9/str. (24)26–30/10 µm . . . . **190. *N. lapidosa*** vol. 2/1 p. 203
18b Caractéristique différente **19**
19a Valves linéaires-elliptiques à linéaires avec une aire centrale assez grande, +/- elliptique-transversale (vol. 2/1 Fig. 73: 8–11) . . . . **191. *N. variostriata*** vol. 2/1 p. 203
19b Caractéristiques différentes, valves elliptiques **20**
20a Côte médiane avec nodules centraux et terminaux très saillants, cependant sans indication de sillons parallèles (vol. 2/1 Fig. 74: 1–38) . . . . **Groupe-clef Nc autour de** *N. atomus*
20b Côte médiane avec 3 nodules pas aussi saillants **21**
21a Aire centrale au moins de taille moyenne (vol. 2/1 Fig. 73: 1–2) L: 19–21,5/l. 7,5–9/str. 22–26/10 µm . . **189. *N. weinzierlii*** vol. 2/1 p. 202
21b Vu que quelques stries centrales se prolongent jusqu'à l'aire axiale, pas d'aire centrale prononcée (vol. 2/1 Fig. 59: 2–9) L: 12–40/l. 7–15/str. 24–36/10 µm . . . . **108. *N. cocconeiformis*** vol. 2/1 p. 158
L: 8–20/l. 6–11/str. 25–36/10 µm . . . **109. *N. jaernefeltii*** vol. 2/1 p. 159
22a Aire centrale nettement prononcée **23**
22b Aire centrale très petite ou absente **25**
23a Aire centrale +/- circulaire **24**
23b Aire centrale formant une bande transversale +/- longue (vol. 2/1 Fig. 73: 4–7) L: 12–24/l. 7–9/str. (24)26–30/10 µm . **190. *N. lapidosa*** vol. 2/1 p. 203
24a Largeur valvaire 8–12 µm (vol. 2/1 Fig. 67: 5) L: 17–30/l. 8–12/str. (21)26–30 . . . . **155. *N. bacilloides*** vol. 2/1 p. 188
24b Largeur valvaire < 6 µm (vol. 2/1 Fig. 78: 21–25) L: (6)10–15,5/l. 4–5,5/str. 22–25/10 µm . . . . **187. *N. subadnata*** vol. 2/1 p. 202
25a Stries > 30/10 µm (vol. 2/1 Fig. 72: 29–33) L: 15–17/l. 3,5–5/str. 30–35/10 µm . . . . **188. *N. egregia*** vol. 2/1 p. 202
25b Stries plus grossières **26**
26a Longueur valvaire > 12 µm, largeur 6–7 µm (vol. 2/1 Fig. 64: 26–28) L: 12–21/l. 6–7/str. 14–18/10 µm . . . . **71. *N. kriegeri*** vol. 2/1 p. 132
26b Longueur valvaire < 12 µm, largeur < 6 µm (vol. 2/1 Fig. 76: 21–26) L: 7–12,5/l. 3,5–6/str; 15–26(34)/10 µm **224. *N. subminuscula*** vol. 2/1 p. 223
27a Extrémités +/- étirées-rostrées mais non capitées **28**
27b Extrémités +/- capitées **39**
28a Sillons très larges constituant au moins des zones n'étant pas dans le même plan, (vis micr.) **36**
28b Sillons plus étroits ou indiqués seulement par des lignes ou à peine visibles **29**
29a Stries +/- 20/10 µm **30**
29b Stries plus étroites, > 22/10 µm **32**
30a Largeur valvaire 7–10 µm, stries nettement ponctuées (vol. 2/1 Fig. 70: 22–24) L: 20–36/l. (5)7–10/str. 17–21/10 µm . . . . **173. *N. laterostrata*** vol. 2/1 p. 197
30b Valves plus fines et/ou non nettement ponctuées **31**
31a Extrémités largement étirées et arrondies. Partie centrale plus fortement

convexe. Aire centrale relativement grande (vol. 2/1 Fig. 71: 3–8) . . . . **176.** ***N. pseudoventralis*** vol. 2/1 p. 198

31b Caractéristiques pas ou beaucoup plus faiblement exprimées (vol. 2/1 Fig. 71: 9–13) L: 10–16/l. 5–6/str. +/- 18/10 µm . . . . **177.** ***N. modica*** vol. 2/1 p. 198

32a Plusieurs stries convergentes avant les extrémités, plus rarement parallèles (vol. 2/1 Fig. 71: 15–21) . . . . **178.** ***N. absoluta*** p 198

32b Stries continues radiales, pour autant que reconnaissables **33**

33a Extrémités étirées-étroites presque arrondies-pointues (vol. 2/1 Fig. 72: 14–16) Str. 27–28/10 µm . . . . **185.** ***N. ingrata*** vol. 2/1 p. 201

33b Extrémités plus tronquées à largement arrondies **34**

34a Aire centrale irrégulièrement élargie transversalement (vol. 2/1 Fig. 71: 25–31') L: 8–22/l. 3,5–7/str. 22–26/10 µm . . . . **180.** ***N. vitabunda*** vol. 2/1 p. 199

34b Aire centrale toujours faiblement et elliptiquement élargie **35**

35a Stries nettes alternativement courtes et longues au centre (vol. 2/1 Fig. 70: 19–21) L: 15–30/l. 6–13/str. 30–36/10 µm . . **168.** ***N. pusio*** vol. 2/1 p. 195

35b Stries à peu près de la même longueur (vol. 2/1 Fig. 69: 14–17) L: 15–18/l. 4,5–6/str. 28–33/10 µm . . . . **167.** ***N. detenta*** vol. 2/1 p. 195

36a Stries non ou à peine résolvables ; +/- 50/10 µm (vol. 2/1 Fig. 72: 21–24) chez les formes typiques. Cf. aussi *N. kuelbsii à* extrémités non étirées (vol. 2/1 Fig. 80: 33–35) . . . . **143.** ***N. krasskei*** vol. 2/1 p. 177

36b Stries résolvables au moins sur les bords, <= 30/10 µm **37**

37a Conopeum indiqué par des lignes longitudinales +/- nettes **38**

37b Lignes longitudinales absentes ou peu nettes (vol. 2/1 Fig. 72: 17–20) L: 13–20/l. 4–5,5/str. 28–30(36)/10 µm . . . **186.** ***N. maceria*** vol. 2/1 p. 201

38a Conopeum (dépression) élargi de façon lancéolée (convexe) au centre (vol. 2/1 Fig. 72: 8–10) Str. 24–27/10 µm . **184.** ***N. naumannii*** vol. 2/1 p. 201

38b Conopeum sans élargissement convexe (vol. 2/1 Fig. 72: 1–7) L: 11–33/l. 4–7,5/str. (22)24–30(36)/10 µm . . . . **183.** ***N. festiva*** vol. 2/1 p. 200

39a Stries non résolvables en microscopie optique (vol. 2/1 Fig. 70: 14–15) L: 14–16/l. +/- 4 µm . . . . **197.** ***N. impexa*** vol. 2/1 p. 207

39b Stries reconnaissables **40**

40a Aire centrale au moins de taille moyenne, le plus souvent élargie transversalement **41**

40b Aire centrale petite, pas clairement distincte de l'aire axiale **49**

41a Largeur des valves 5–10 µm **42**

41b Valves plus étroites, largeur 2–5 µm **43**

42a Au moins les stries du centre nettement ponctuées (vol. 2/1 Fig. 70: 22–24) L: 20–36/l. (5)7–10/str. 17–21/10 µm **173.** ***N. laterostrata*** vol. 2/1 p. 197

42b Stries non nettement ponctuées (vol. 2/1 Fig. 71: 1–2) L: 11,5–25/l. 4,5–6,5/str. 24–29/10 µm . . . . **175.** ***N. ventralis*** vol. 2/1 p. 197

43a Stries +/- 30/10 µm **44**

43b Stries plus éloignées, < 30/10 µm **45**

44a Longueur valvaire maximale 10 µm (vol. 2/1 Fig. 70: 8–13) L: 6–10/l. 2,5–3/str. +/- 30/10 µm . . . . . . **170.** ***N. schmassmannii*** vol. 2/1 p. 196

44b Longueur valvaire > 10 µm (vol. 2/1 Fig. 70: 1–7) L. 10–16/l. 3,5–4,5/str. +/- 30/10 µm . . . . **169.** ***N. medioconvexa*** vol. 2/1 p. 195

45a Valves aux bords fortement convexes. Microcéphales terminales beaucoup plus étroites que le milieu de la valve **46**

45b Différence de largeur entre microcéphales et milieu de la valve moins nette **48**

46a Valves étroites elliptiques-lancéolées (vol. 2/1 Fig. 71: 22–24) L: 12–17/l. 4–5/str. 24–28/10 µm . . . . **179.** ***N. hustedtii*** vol. 2/1 p. 199

46b Valves avec partie centrale fortement bombée **47**
47a Stries environ 25/10 µm (vol. 2/1 Fig. 71: 32–38) L: 8–17/l. 4,5–8/str. 22–26/10 µm . . . . **181.** ***N. schadei*** vol. 2/1 p. 199
47b Stries >= 30/10 µm (vol. 2/1 Fig. 71: 39) L: 8–17/l. 4,5–8/str. 30–32/10 µm . . . . **182.** ***N. glomus*** vol. 2/1 p. 200
48a Extrémités fortement étranglées, aire centrale +/- ronde (vol. 2/1 Fig. 70: 18) L: 20–22/l. 4,5/str. 24/10 µm . . . . . **172.** ***N. laticeps*** vol. 2/1 p. 196
48b Extrémités faiblement étranglées. Aire centrale en bande transversale atteignant presque les bords valvaires (vol. 2/1 Fig. 70: 16, 17) L: 21–28/l. 4,5/str. +/- 25/10 µm . . . . **171.** ***N. disjuncta*** vol. 2/1 p. 196
49a Valves aux extrémités étroites, étranglées-capitées. Un stigma isolé dans l'aire centrale (vol. 2/1 Fig. 70: 25–27) L: 19–22/l. 6–7/str. 16–26/10 µm . . . . **174.** ***N. declivis*** vol. 2/1 p. 197
49b Caractéristiques différentes **50**
50a Stries alternativement courtes et longues au centre (vol. 2/1 Fig. 70: 19–21) L: 15–30/l. 6–13/str. 30–36/10µm . . . . . . **168.** ***N. pusio*** vol. 2/1 p. 195
50b Stries à peu près de la même longueur (vol. 2/1 Fig. 69: 14–17) L: 15–18/l. 4,5–6/str. 28–33/10 µm . . . . **167.** ***N. detenta*** vol. 2/1 p. 195

Groupe-clef N Minusculae (vol. 2/1 p. 205)

Clef des sous-groupes (artificiels)

1a Stries non ou difficilement résolvables en microscopie optique . . . . **Sous-groupe Na**) (p. 247)
1b Stries bien résolvables au moins dans la partie centrale de la valve . . . **2**
2a Extrémités étirées rostrées à +/- capitées . . . **Sous-groupe Nd**) (p. 251)
2b Extrémités au plus faiblement étirées **3**
3a Valves +/- elliptiques. Côte médiane très contrastée avec nodules centraux et terminaux en forme de points mais sans sillons longitudinaux . . . . **Sous-groupe Nb; groupe d'espèces autour de *N. atomus*)** (p. 249)
3b Caractéristiques différentes **4**
4a En vue connective, frustules enchaînés en bandes (chaînes souvent détruites par traitement acide, non par la technique de grillage) . . . . **Sous-groupe Nc**) (p. 250)
4b Caractéristique différente **5**
5a Formes présentant une aire centrale très petite ou peu nettement délimitée, ou aire centrale absente . . . . **Sous-groupe Ne**) (p. 252)
5b Formes présentant une aire centrale +/- bien formée et sans les caractéristiques des groupes-clefs et sous-groupes restants **Sous-groupe Nf**) (p. 252)

Sous-groupe Na (vol. 2/1 p. 205)
(Formes pour lesquelles les stries sont pas ou à peine résolvables au microscope optique)

1a Valves strictement elliptiques aux extrémités larges-arrondies **2**
1b Valves plutôt linéaires, elliptiques-lancéolées, rhomboédriques ou avec extrémités +/- étirées **6**
2a Stries partiellement résolvables avec des moyens optiques particuliers . **5**
2b Stries et autres structures délicates, résolvables uniquement en microscopie électronique **3**
3a Valves larges-elliptiques. Nodules terminaux proches des pôles **4**
3b Nodules terminaux déplacés proximalement (vol. 2/1 Fig. 75: 34–37) L: 7–8,5/l. 3,5–4/str. n.r. . . . . **201.** ***N. lacunolaciniata*** vol. 2/1 p. 209
4a Longueur des valves 9–12,5 µm. Largeur 4–6,2 µm (vol. 2/1 Fig. 74: 37, 38) . . . . **199.** ***N. pelliculosa*** vol. 2/1 p. 208

4b Longueur des valves 3,8–7,6 µm. Largeur 2–4 µm (vol. 2/1 Fig. 74: 35, 36) . . . . **198.** ***N. saprophila*** vol. 2/1 p. 207
5a Nodules terminaux très éloignés proximalement des pôles. Stries radiales, env. 35/10 µm (vol. 2/1 Fig. 75: 32, 33) L: 7,5–9/l. 3,5–5/str. env. 35/10 µm. . . . . **200.** ***N. nolensoides*** vol. 2/1 p. 209
5b Nodules terminaux peu éloignés proximalement des pôles (vol. 2/1 Fig. 74: 35, 36) L: 6–9/l. 3–6,5/str. (25)30–36/10 µm . . . . **216.** ***N. atomus*** var. ***permitis*** vol. 2/1 p. 216
6a Extrémités valvaires peu ou à peine étirées **7**
6b Extrémités +/- étirées, bords parfois ondulés **11**
7a Aire centrale encore reconnaissable **8**
7b Pas d'aire centrale reconnaissable **9**
8a Aire centrale unilatéralement plus marquée (vol. 2/1 Fig. 80: 31) L: 8–17/l. 3–4/str. 35–44/10 µm . . . . **210.** ***N. vitiosa*** vol. 2/1 p. 213
8b Aire centrale +/- isodiamétrique (vol. 2/1 Fig. 76: 48 et 77: 10–12) L: 10–15/l. 3–3,5/str. 36–39/10 µm . . . **232.** ***N. semihyalina*** vol. 2/1 p. 229
L: 12–19/l. 3,5–4,5/str. +/– 35(40)/10 µm . **243.** ***N. tenerrima*** vol. 2/1 p. 233
9a Valves aux extrémités arrondies-tronquées, lg. 5–6 µm, l. 2–2,5 µm (vol. 2/1 Fig. 78: 28) stries +/- 35/10 µm . . **202.** ***N. diabolica*** vol. 2/1 p. 210
9b Valves le plus souvent plus larges, extrémités plus largement arrondies . **10**
10a Raphé enfermé dans une côte médiane très renforcée (vol. 2/1 Fig. 80: 32–34) L: 4,5–13/l. 2,5–4/str. 35/10 µm . **213.** ***N. kuelbsii*** vol. 2/1 p. 213
10b Caractéristique différente (vol. 2/1 Fig. 80: 26, 27) L: 8–10/l. +/- 3/stries n.r . . . . **209.** ***N. difficilimoides*** vol. 2/1 p. 213
11a Contour des valves présentant 3 ondulations (vol. 2/1 Fig. 80: 1–3) L: 11–19/l. 3,5–4/str. n. r. . . . . **203.** ***N. tridentula*** vol. 2/1 p. 210
11b Contour valvaire pas ou rarement ondulé **12**
12a Aire centrale encore reconnaissable **13**
12b Plus d'aire centrale différenciable avec certitude **14**
13a Valves aux extrémités +/- fortement étirées. Sous microscope électronique, stries 48–55/10 µm (vol. 2/1 Fig. 80: 18–20). Si stries situées au centre encore différenciables en partie, env. 40/10 µm (vol. 2/1 Fig. 80: 21–22) L: 10–21/l. 3,4–4,5/str. 48–55/10 µm . . . . **207.** ***N. gerloffii*** vol. 2/1 p. 212
. . . . **205.** ***N. arvensis*** **var.** ***major***
13b Extrémités toujours très faiblement étirées. Stries, si reconnaissables, 36–39/10 µm (vol. 2/1 Fig. 76:48) L: 5–17/l. 3–3,5/str. 36–39/10 µm . . . . **232.** ***N. semihyalina*** vol. 2/1 p. 229
14a Valves aux calottes terminales très siliceuses, contour sombre (lors de focalisation profonde) (vol. 2/1 Fig. 80: 7, 8) L: 8–15/l. 3–4/str. 55–65/10 µm . . . . **204.** ***N. difficillima*** vol. 2/1 p. 210
14b Caractéristique différente **15**
15a Stries +/- 40/10 µm, à la limite de la résolution **16**
15b Stries nettement > 40/10 µm, ne sont plus résolvables **17**
16a Valves le plus souvent plus grandes. Largeur 3,2–5,5 µm (vol. 2/1 Fig. 69: 18–27) L: 8–16/l. 3,2–5,5/str. 30–45/10 µm . . . . **196.** ***N. minuscula*** vol. 2/1 p. 207
16b Valves le plus souvent plus petites, largeur 2,5–3,5 µm avec extrémités fortement marquées, rostrées (vol. 2/1 Fig. 80: 10–12) · Cf. aussi *N. subarvensis* (vol. 2/1 Fig. 80: 13) L: 5–9/l. 2,5–3,5/str. 34–40/10 µm . . . . **205.** ***N. arvensis*** var ***arvensis*** vol. 2/1 p. 211
17a Valves larges-linéaires aux extrémités brusquement distinctes. Longueur max. 10 µm (vol. 2/1 Fig. 80: 17) L: 8–10/l. 3–3,5/str. n. r. . . . . **206.** ***N. pseudoarvensis*** vol. 2/1 p. 212
17b Caractéristiques différentes **18**

18a Valves elliptiques-lancéolées aux extrémités étirées +/- capitées **20**
18b Extrémités non capitées **19**
19a Largeur valvaire 2,5–3 µm (vol. 2/1 Fig. 80: 28–30) L: 6–8/l. 3,2,5–3/str. n. r. . . . . **211.** ***N. indifferens*** vol. 2/1 p. 213
19b Largeur valvaire 3,4–5,5 µm (vol. 2/1 Fig. 72: 21–25) L: 5,8–17/l. 3,4–5,5/str. 45–55/10 µm . . . . **143.** ***N. krasskei*** vol. 2/1 p. 177
20a Largeur valvaire 3–4 µm (vol. 2/1 Fig. 80: 25; 80: 7, 8) L: 8–15/l. 3–4/str. 55–65/10 µm . . . . **204.** ***N. difficillima*** vol. 2/1 p. 210
L: 14–16/l. env. 4/str. n.r. . . . . **197.** ***N. impexa*** vol. 2/1 p. 207
20b Largeur valvaire 2–2,5 µm (vol. 2/1 Fig. 80: 23, 24) L: 6–15/l. 2–2,5/str. env. 50/10 µm . . . . **208.** ***N. enigmatica*** vol. 2/1 p. 212

Sous-groupe Nb (vol. 2/1 p. 214)
Formes autour de *N. atomus*
Valves +/- elliptiques, côte médiane avec nodules centraux et terminaux très saillants, toutefois sans sillons longitudinaux parallèles

1a Stries non résolvables en microscopie optique, même avec des moyens particuliers **2**
1b Caractéristique différente **5**
2a Nodules terminaux très proches des pôles **3**
2b Nodules terminaux quelque peu éloignés des pôles **4**
3a Lg. valvaire: 9–12,5 µm/l. 4–6,2 µm/stries 45–55/10 µm (vol. 2/1 Fig. 74: 37–38). Cf. aussi *N. minuscula* var. *muralis* (vol. 2/1 Fig. 69: 25–27) . . . . **199.** ***N. pelliculosa*** vol. 2/1 p. 208
3b Lg. valvaire: 4–7,6 µm/l. 2–4 µm/stries 48–81/10 µm (vol. 2/1 Fig. 74: 35, 36) . . . . **198.** ***N. saprophila*** vol. 2/1 p. 207
4a Nodules terminaux moyennement éloignés des pôles. Stries non visibles même sous microscope électronique (vol. 2/1 Fig. 75: 34–37) L: 7–8,5 µm/l. 3,5–4 µm/stries n.r. . . . . **201.** ***N. lacunolaciniata*** vol. 2/1 p. 209
4b Nodules terminaux plus éloignés proximalement des pôles. Stries encore reconnaissables sous conditions optimales en microscopie optique, +/- 35/10 µm (vol. 2/1 Fig. 75: 32, 33). D'autres taxons avec caractéristiques semblables ne se retrouvent qu'aux bords des mers et/ou en dehors de la région . . . . **200.** ***N. nolensoides*** vol. 2/1 p. 209
5a Nodules terminaux très éloignés proximalement des pôles **6**
5b Caractéristique différente **7**
6a Str. +/- 35/10 µm, reconnaissables seulement en conditions optiques optimales (vol. 2/1 Fig. 75: 32, 33) L:7,5–9 µm/l. 3,5–5 µm/str. 35/10 µm . . . . **200.** ***N. nolensoides*** vol. 2/1 p. 209
6b Stries grossièrement ponctuées (vol. 2/1 Fig. 74: 34) L: 12–16 µm/l. 5–6 µm/str. 15–20/10 µm . . . . **218.** ***N. asellus*** vol. 2/1 p. 218
7a Nodules terminaux très proches des pôles souvent +/- étirés. Stries 30–45/10 µm, radiales au centre, brusquement convergentes avant les extrémités (vol. 2/1 Fig. 69: 18–27) L: 8–16 µm/l. 3,2–5,5 µm/stries 30–45/10 µm . . . . **196.** ***N. minuscula*** vol. 2/1 p. 207
7b Caractéristiques différentes **8**
8a Stries relativement grossières, apparaissant parfois ponctuées, +/- 20/10 µm **10**
8b Stries plus rapprochées, parfois à peine résolvables **9**
9a Valves le plus souvent étroites-elliptiques-lancéolées avec des extrémités moyennement tronquées à presque arrondies-pointues (vol. 2/1 Fig. 74: 1–7) L: 8–11,5 µm/l. 2,5–3,5 µm/stries 24–28/10 µm . . . . **214.** ***N. agrestis*** vol. 2/1 p. 215

9b Valves plus largement elliptiques aux extrémités plus largement arrondies (vol. 2/1 Fig. 74: 14–17) L:6–9 µm/l. 3–6,5 µm/stries (25)30–36/10 µm . . . . **216.** ***N. atomus*** var. ***permitis*** vol. 2/1 p. 216
10a Stries continues radiales **13**
10b Stries continues parallèles ou parallèles aux extrémités ou légèrement convergentes **11**
11a Stries continues parallèles (vol. 2/1 Fig. 45: 31–33) L:10–12 µm/l. 3,5–5 µm/Stries 22–24/10 µm . . . . . **70.** ***N. muraliformis*** vol. 2/1 p. 132
11b Caractéristique différente **12**
12a Aire centrale elliptique-transversale (vol. 2/1 Fig. 74: 9) L: 12,5 µm/l. 5 µm/Stries 21/10 µm . . . . **219.** ***N. fossaloides*** vol. 2/1 p. 218
12b Aire centrale petite, ronde. Stries +/- 25/10 µm (vol. 2/1 Fig. 74: 8) L: 11,5 µm/l. 4,5 µm/stries 25/10 µm . . . . **215.** ***N. destricta*** vol. 2/1 p. 215
13a Aire centrale non marquée **14**
13b Aire centrale distincte de l'aire axiale ou formant avec elle une aire hyaline +/- grande **15**
14a Valves L: 12–16 µm/stries +/- 17/10 µm (vol. 2/1 Fig. 74: 11–13) . . . . **216.** ***N. atomus*** var. ***excelsa*** vol. 2/1 p. 217
14b Valves le plus souvent un peu plus petites. Stries +/- 20/10 µm. Côte médiane apparaissant le plus souvent moins large (vol. 2/1 Fig. 74: 10, voir aussi 18–26) L: 8,5–13/l. 4–5,5/stries +/- 20/10 µm . . . . **216.** ***N. atomus*** var. ***atomus*** vol. 2/1 p. 215
15a Aires centrale et axiale formant une aire hyaline +/- large **16**
15b Aire centrale distincte de l'aire axiale et élargie transversalement **17**
16a Valves L: 8–9 µm/stries +/- 20/10 µm (vol. 2/1 Fig. 74: 27, 28) . . . . **216.** ***N. atomus*** var. ***recondita*** vol. 2/1 p. 217
16b Valves L: 10–13 µm/stries 16–20/10 µm, apparaissant ponctuées (vol. 2/1 Fig. 74: 29–31) . . . . **217.** ***N. fossalis*** var. ***obsidialis*** vol. 2/1 p. 218
17a Valves L: 9–12 µm (vol. 2/1 Fig. 74: 32, 33) . . . . **217.** ***N. fossalis*** var. ***fossalis*** vol. 2/1 p. 218
17b Valves L: 12–16 µm. Nodules terminaux plus éloignés des pôles (vol. 2/1 Fig. 74: 34). Cf. aussi *N. parsura* (vol. 2/1 Fig. 76: 8–10) . . . . **218.** ***N. asellus*** vol. 2/1 p. 218

Sous-groupe Nc (vol. 2/1 p. 219)
Frustules formant des chaînes en vue connective

1a Valves linéaires, non ponctuées en bordure. Nodules terminaux éloignés proximalement des pôles (vol. 2/1 Fig. 75: 23–25) L:10–55 µm/l. 2,5–5 µm/stries 24–36/10 µm . . . . **222.** ***N. brekkaensis*** vol. 2/1 p. 220
1b Caractéristiques différentes **2**
2a Valves étroites-linéaires-lancéolées. Longueur 10–36 µm (vol. 2/1 Fig. 75: 18–22) L: 6–45 µm/l. 2→5 µm/stries 26–40/10 µm . . . . **221.** ***N. gallica*** var. vol. 2/1 p. 220
2b Caractéristiques différentes **3**
3a Stries grossières, nettement ponctuées (vol. 2/1 Fig. 75: 29–31) L: 9–28/l. 4–10/str. (15)18–26/10 µm . . . . . . **223.** ***N. confervacea*** vol. 2/1 p. 221
3b Stries plus délicates, ponctuation non reconnaissable **4**
4a Valves linéaires avec tout au plus des bords faiblement concaves. Aire centrale le plus souvent distincte de l'aire axiale. Stries parallèles, sans points en bordure, raphé toujours présent (vol. 2/1 Fig. 75: 1–5) . . . . **220.** ***N. contenta*** vol. 2/1 p. 219
4b Caractéristiques différentes (vol. 2/1 Fig. 75: 6–17') L: 6–45 µm/l. 2–5 µm/stries 26–40/10 µm vol. 2/1 p. 220 . . . . **221.** ***N. gallica***

Sous-groupe Nd (vol. 2/1 p. 222)

Aire centrale très petite ou irrégulièrement délimitée ou totalement absente

1a Stries parallèles ou très faiblement radiales **2**

1b Stries nettement radiales au moins au centre **12**

2a Valves linéaires à linéaires-elliptiques **3**

2b Valves sans tendance linéaire **10**

3a Aire axiale nettement délimitée au milieu et non élargie (vol. 2/1 Fig. 45: 31–33) L: 10–12 µm/l. 3,5–5 µm/stries 22–24/10 µm . . . . **70.** ***N. muraliformis*** vol. 2/1 p. 122

3b Caractéristique différente **4**

4a Valves étroites-linéaires. Stries relativement éloignées (env. 20/10 µm). Aire centrale élargie transversalement et souvent peu nettement délimitée à cause de cet élargissement ou à cause des stries faiblement marquées . **7**

4b Stries plus rapprochées **5**

5a Extrémités largement arrondies **6**

5b Extrémités +/- cunéiformes (vol. 2/1 Fig. 74: 8) L: 11,5 µm/l. 4,5 µm/stries 25/10 µm . . . . **215.** ***N. destricta*** vol. 2/1 p. 215

6a Branches du raphé arquées. Pores centraux fortement marqués (vol. 2/1 Fig. 66: 40–42) L: 9,5 µm/l. 4,5 µm/stries 30/10 µm . . . . **164.** ***N. sublucidula*** vol. 2/1 p. 193

6b Caractéristiques différentes (vol. 2/1 Fig. 77: 19–24, éventuellement 25–28) L: 14–22 µm/l. 3–5,5 µm/stries 28–40/10 µm . . . . **195.** ***N. digitulus*** vol. 2/1 p. 204

7a Valves avec un léger mais net renflement au centre (vol. 2/1 Fig. 78: 14–16) L: 9–16 µm/l. 2–3 µm/Stries 20–23/10 µm . . . . **227.** ***N. mediocris*** vol. 2/1 p. 225

7b Bords valvaires strictement linéaires ou progressivement légèrement renflés vers le centre ou ondulés **8**

8a Bords strictement linéaires ou faiblement concaves (vol. 2/1 Fig. 78: 17–20) L:10–16 µm/l. 2–3,6 µm/stries 17–19/10 µm . **228.** ***N. begeri*** vol. 2/1 p. 225

8b Bords ondulés ou un peu élargis au milieu **9**

9a Stries continues +/- parallèles, jamais convergentes aux extrémités (vol. 2/1 Fig. 78: 1–13) L: 9–16 µm/l. 2–3,5 µm/stries 17–24/10 µm . . . . **226.** ***N. soehrensis*** vol. 2/1 p. 224

9b Stries faiblement radiales au centre, convergentes aux extrémités (vol. 2/1 Fig. 35: 11–13) L: 9–12 µm/l. 2,5–3 µm/stries 18–20/10 µm . . . . **33.** ***N. bremensis*** vol. 2/1 p. 111

10a Stries le plus souvent +/- 30/10 µm (vol. 2/1 Fig. 45: 24–25) L: 10–16 µm/l. 4–5 µm/stries 25–30/10 µm . . . . **69.** ***N. fluens*** vol. 2/1 p. 131

10b Stries le plus souvent nettement < 30/10 µm **11**

11a Largeur valvaire 6–7 µm (vol. 2/1 Fig. 64: 26–28) L: 12–21 µm/l. 6–7 µm/stries 14–18/10 µm . . . . **71.** ***N. kriegeri*** vol. 2/1 p. 132

11b Largeur valvaire < 6 µm (vol. 2/1 Fig. 76: 21–26) L: 7–12 µm/l. 3,5–6 µm/stries 30/10 µm (15–26(34)/10 µm) . **224.** ***N. subminuscula*** vol. 2/1 p. 223

12a Un stigma isolé au bord du nodule central. Forme d'eaux saumâtres et marines (vol. 2/1 Fig. 77: 29, 30). Cf. aussi *N. porifera* (vol. 2/1 Fig. 47: 19–24) pour formes d'eaux douces . . **225.** ***N. bahusiensis*** vol. 2/1 p. 223

12b Caractéristique différente **13**

13a Longueur valvaire seulement 2,7–4,5 µm. Probablement des formes chétives de *Achnanthes minutissima* (vol. 2/1 Fig. 78: 26, 27) L: 2,7–4,5 µm/l. 2–2,7 µm/stries 26–30/10 µm . . . . . . **229.** ***N. strenzkei*** vol. 2/1 p. 226

13b Valves plus grandes. Raphé toujours bien formé **14**

14a Stries fortement radiales **15**

14b Stries faiblement à moyennement radiales **17**

15a Valves étroites-elliptiques, souvent à extrémités étirées. Stries à la limite de la résolution (vol. 2/1 Fig. 69: 18–27) L: 8–16 µm/l. 3,2–5,5 µm/stries 30–45/10 µm . . . . **196. *N. minuscula*** vol. 2/1 p. 207
15b Valves larges-elliptiques **16**
16a Stries 24–36/10 µm (vol. 2/1 Fig. 73: 25, 26) L: 8–12 µm/l. 4,5–6 µm/stries 24–36/10 µm . . . . **194. *N. utermoehlii*** vol. 2/1 p. 204
16b Stries presque toujours < 25/10 µm (vol. 2/1 Fig. 73: 16–20). Cf. aussi *N. porifera* (vol. 2/1 Fig. 47: 19–24)vol. 2/1 p. 204 L: 7–13 µm/l. 4–5,5 µm/stries 21–25(30?)/10 µm . . . . **193. *N. subrotundata***
17a Valves étroites (2,5–3,5 µm), elliptiques-lancéolées (vol. 2/1 Fig. 74: 1–7) L: 8–11,5 µm/l. 2,5–3,5 µm/stries 24–28/10 µm . . . . **214. *N. agrestis*** vol. 2/1 p. 215
17b Caractéristique différente . . . . **clef n.° 2–11**

Sous-groupe Ne (vol. 2/1 p. 226)
Formes à extrémités étirées, +/- rostrées ou capitées
1a Nodules terminaux nettement proximalement éloignés des pôles, ce qui donne un raccourcissement +/- prononcé des branches du raphé . . . **2**
1b Caractéristique différente **3**
2a Bords valvaires apparaissant ponctués (vol. 2/1 Fig. 75: 18–22) . . . . **221. *N. gallica*** var. ***laevissima*** vol. 2/1 p. 220
2b Bords valvaires sans points (vol. 2/1 Fig. 75:23–25) . . . . **222 *N. brekkaensis*** vol. 2/1 p. 220
3a Aires axiale et centrale reliées en un espace hyalin lancéolé (vol. 2/1 Fig. 75: 29–31) . . . . **223. *N. confervacea*** (**f. *rostrata***) vol. 2/1 p. 221
3b Caractéristique différente **4**
4a Stries presque parallèles, 25–36/10 µm. Frustules souvent reconnaissables comme colonies constituant des bandes (vol. 2/1 Fig. 75: 1–5)- . . . . **220. *N. contenta*** vol. 2/1 p. 219
4b Caractéristiques différentes **5**
5a Valves d'une largeur max. de 3,5 µm, linéaires, à bords triondulaires ou centre bombé. Raphé à pores centraux espacés. Stries relativement grossières vu la taille réduite des valves, environ 18/10 µm au milieu (vol. 2/1 Fig. 78: 1–9) . . . . **226. *N. soehrensis*** vol. 2/1 p. 224
5b Caractéristiques différentes (cf. aussi *N. bulnheimii*, Fig. 56: 6–10) . . . . **cf. groupe-clef M n° 26–48**

Sous-groupe Nf (vol. 2/1 p. 226)
Espèces restantes avec aire centrale +/- bien formée, sans les caractéristiques des autres groupes et sous-groupes
1a Aire centrale nettement élargie transversalement **2**
1b Forme de base de l'aire centrale ronde ou elliptique **25**
2a Valves rhomboédriques-elliptiques, longueur max. 9 µm. Vu cette taille, stries très grossièrement ponctuées, env. 24 points/10 µm (vol. 2/1 Fig. 60: 6–9) L: 7,5–9 µm/l. 3–4 µm/stries +/- 21/10 µm . . . . **129. *N. ingenua*** vol. 2/1 p. 169
2b Caractéristiques différentes **3**
3a Stries relativement rapprochées, > 24/10 µm **4**
3b Stries plus grossières, < 24/10 µm **12**
4a Largeur valvaire > 4 µm **5**
4b Largeur valvaire < 4 µm **8**
5a Aire centrale très étroite mais fortement élargie transapicalement. Stries presque parallèles (vol. 2/1 Fig. 77: 13–18) . . . . ***Neidium alpinum*** vol. 2/1 p. 273

5b Caractéristiques différentes **6**
6a Contraste important entre les stries courtes de l'aire centrale et les autres stries (vol. 2/1 Fig. 73: 4–7) L: 12–24 µm/l. 7–9 µm/stries (24)26–30/10 µm . . . . **190.** ***N. lapidosa*** vol. 2/1 p. 203
6b Stries centrales irrégulièrement courtes avec des passages plutôt continus **7**
7a Valves linéaires à linéaires-elliptiques (vol. 2/1 Fig. 73: 8–11) L: 15–44 µm/l. 5–10 µm/stries 27–30/10 µm . . . . **191.** ***N. variostriata*** vol. 2/1 p. 203
7b Valves à bords fortement convexes (vol. 2/1 Fig. 73: 1–3) L: 19–21,5 µm/l. 7,5–9 µm/stries 22–26/10 µm . . . . . **189.** ***N. weinzierlii*** vol. 2/1 p. 202
8a Stries très faiblement radiales (vol. 2/1 Fig. 77: 2, 3) L: 7–13 µm/l. +/- 3,5 µm/stries 24–28/10 µm . . . . . . **240.** ***N. muraloides*** vol. 2/1 p. 232
8b Stries plus nettement radiales **9**
9a Stries 35–44/10 µm, résolvables seulement au milieu de la valve avec des effets optiques spéciaux **10**
9b Stries au maximum environ 30/10 µm **11**
10a Aire centrale nette des deux côtés (vol. 2/1 Fig. 76: 48) L: 10–15 µm/l. 3–3,5 µm/stries 36–39/1 µm . . . . . **232.** ***N. semihyalina*** vol. 2/1 p. 229
10b Aire centrale élargie jusqu'au bord d'un seul côté (vol. 2/1 Fig. 80: 31, 32) L: 8–17 µm/l. 3–4 µm/stries 35–44/10 µm . **210.** ***N. vitiosa*** vol. 2/1 p. 213
11a Extrémités des valves larges-arrondies. Aire axiale toujours étroite, linéaire (vol. 2/1 Fig. 76: 39–47) L: 5–18 µm/l. 2–4,5 µm/stries 25–30/10 µm . . . . **231.** ***N. minima*** vol. 2/1 p. 229
11b Extrémités valvaires plus variables, le plus souvent légèrement cunéiformes, arrondies tronquées à presque arrondies-pointues. Dans certains cas, aire axiale élargie lancéolée (vol. 2/1 Fig. 76: 1–7) L: 9–15 µm/l. 3–4 µm/stries 23–27/10 µm . . . . **233.** ***N. harderi*** vol. 2/1 p. 229
12a Aire centrale constituant, au moins d'un côté, une bande transversale atteignant les bords **13**
12b Régulièrement, il y a encore des stries courtes entre le bord valvaire et le nodule central **16**
13a Aire centrale large. Valves élargies transapicalement au milieu et aux extrémités (vol. 2/1 Fig. 66: 5–8) L: 10–19 µm/l. +/- 4 µm/stries 21–24/10 µm . . . . **140.** ***N. occulta*** vol. 2/1 p. 176
13b Caractéristique différente **14**
14a Stries apparaissant interrompues par une ligne +/- nette (vol. 2/1 Fig. 66: 9–11) L: 7,5–16 µm/l. 3–4 µm/stries 18–22/10 µm . . . . **141.** ***N. lucinensis*** vol. 2/1 p. 176
14b Caractéristique différente **15**
15a Au milieu des bords valvaires petit renflement abrupt (vol. 2/1 Fig. 78: 14–16) L: 9–16 µm/l. 2–3 µm/stries 20–23/10 µm . . . . **227.** ***N. mediocris*** vol. 2/1 p. 225
15b Caractéristique différente. Nodule central épaissi en stauros (vol. 2/1 Fig. 77: 6–9) L: 15–36 µm/l. 5–7 µm/stries 16–24/10 µm . . . . **242.** ***N. soodensis*** vol. 2/1 p. 233
16a Bords valvaires avec petit renflement abrupt au milieu (vol. 2/1 Fig. 78: 14–16) . . . . **227.** ***N. mediocris***
16b Caractéristique différente **17**
17a Valves relativement larges (5–7 µm). Stries pour la plupart très faiblement radiales. L'aire centrale étroite avec nodule central montre un épaississement stauroïde (vol. 2/1 Fig. 77: 6–9) L: 15–36 µm/l. 5–7 µm/stries 16–24/10 µm . . . . **242.** ***N. soodensis*** vol. 2/1 p. 233
17b Caractéristiques différentes **18**
18a Aire centrale rhomboédrique-elliptique. Fortement élargie surtout en sens

apical. Nodules terminaux très marqués (vol. 2/1 Fig. 76: 8–10) L: 11–19 µm/l. 3,5–4 µm/stries 21–23/10 µm **238.** ***N. parsura*** vol. 2/1 p. 232

18b Caractéristique différente **19**

19a Extrémités valvaires larges et arrondies, linéaires ou linéaires-elliptiques **20**

19b Extrémités valvaires beaucoup plus étroites **23**

20a Bords valvaires strictement linéaires ou légèrement concaves (vol. 2/1 Fig. 78: 17–20) L: 10–16 µm/l. 2–3,6 µm/stries 17–19/10 µm . . . . **228.** ***N. begeri*** vol. 2/1 p. 225

20b Bords valvaires plus ou moins convexes **21**

21a Extrémités légèrement étirées. Stries plus fortement radiales (vol. 2/1 Fig. 76: 37, 38) L: (7)11,5–15 µm/l. 3,5–4,5 µm/stries 19–21/10 µm . . . . **235.** ***N. joubaudii*** vol. 2/1 p. 231

21b Extrémités non étirées. Str. faiblement radiales **22**

22a Stries 17–18/10 µm. Aire centrale très étroite (vol. 2/1 Fig. 73: 12–15) L: 8–12 µm/l. 4,5–5,5 µm/stries 17–17/10 µm . . . . **192.** ***N. submuralis*** vol. 2/1 p. 203

22b Stries 18–22/10 µm. Aire centrale plus large (vol. 2/1 Fig. 76: 30–36) L: 3–21 µm/l. 2–5 µm/stries 18–22/10 µm . **234.** ***N. seminulum*** vol. 2/1 p. 230

23a Extrémités +/- faiblement étirées. Stries fortement radiales (vol. 2/1 Fig. 64: 29–32) L: 11–20 µm/l. 4–5 µm/stries 16–19/10 µm . . . . **145.** ***N. dolomitica*** vol. 2/1 p. 179

23b Caractéristiques différentes **24**

24a Largeur valvaire 3,5–4,5 µm (vol. 2/1 Fig. 76: 27–29) L: (7)11,5–15 µm/l. 3,5–4,5 µm/stries 19–21/10 µm . . . . **236.** ***N. vaucheriae*** vol. 2/1 p. 231

24b Largeur valvaire 2–2,5 µm (vol. 2/1 Fig. 76: 17–20) L: 8–11 µm/l. 2–2,5 µm/stries 20–24/10 µm . . . . **237.** ***N. obsoleta*** vol. 2/1 p. 231

25a Aires centrale et axiale reliées en une aire hyaline +/- grande **26**

25b Caractéristique différente **27**

26a Bords valvaires concaves au milieu (vol. 2/1 Fig. 75: 26–28). Cf. aussi *N. contenta* (vol. 2/1 Fig. 75: 1–5) L: 8–12 µm/l. 2,5–3,5 µm/stries 26–28/10 µm . . . . **230.** ***N. aerophila*** vol. 2/1 p. 228

26b Bords valvaires non concaves au milieu (vol. 2/1 Fig. 75: 29–31). Cf. aussi *N. gallica* et *N. contenta* (vol. 2/1 Fig. 75: 1–22) . . . . **223.** ***N. confervacea*** vol. 2/1 p. 221

27a Bords valvaires strictement linéaires ou faiblement concaves, aux extrémités largement arrondies (vol. 2/1 Fig. 78: 17–20) L: 10–16 µm/l. 2–3,6 µm/stries 17–19/10 µm . . . . **228.** ***N. begeri*** vol. 2/1 p. 225

27b Caractéristiques différentes **28**

28a Stries très denses, env. 35/10 µm ; aussi, l'aire centrale pourtant présente est souvent inaperçue (vol. 2/1 Fig. 77: 10–12) L: 12–19 µm/l. 3,5–4,5 µm/stries+/- 35 (40?)/10 µm . . . . **241.** ***N. tenerrima*** vol. 2/1 p. 233

28b Stries plus grossières **29**

29a Valves linéaires-elliptiques. Aire centrale grande, rhomboédrique-elliptique. Nodules terminaux très marqués (vol. 2/1 Fig. 76: 8–10) L: 6–10 µm/l. 3,5–4 µm/stries 21–23/10 µm . . . . . . **238.** ***N. parsura*** vol. 2/1 p. 232

29b Caractéristiques différentes **30**

30a Raphé débouchant dans une côte médiane plus contrastée. Pores centraux très marqués. Aire centrale grande. Stries fortement radiales (vol. 2/1 Fig. 78: 21–25) L: (6)10–15,5 µm/l. 4–5,5 µm/stries 22–25/10 µm. . . . . **187.** ***N. subadnata*** vol. 2/1 p. 202

30b Caractéristiques différentes **31**

31a Valves très petites. Longueur 6–10 µm, largeur 2,5–3 µm, stries 24–28/10 µm (vol. 2/1 Fig. 76: 11–16) . . **239.** ***N. evanida*** vol. 2/1 p. 232

31b Valves plus grandes (vol. 2/1 Fig. 77: 19–28) L: 14–22 µm/l. 3–5,5 µm/stries 28–40/10 µm . . . . **195. *N. digitulus*** p 204

2. Genre *Stauroneis* Ehrenberg 1843 (vol. 2/1 p. 236)

Typus generis: *Stauroneis phoenicentreon* (Nitzsch) Ehrenberg; [*Bacillaria phoenicentreon* Nitzsch 1817] (Boyer 1927)

Clef pour les espèces

1a Stries très délicatement ponctuées, le plus souvent non résolvables au microscope optique, > 30/10 µm **2**
1b Points sur les stries reconnaissables à plus grande ouverture **13**
2a Présentes exclusivement en eaux très alcalines **3**
2b Normalement présentes en eaux douces, plus rarement dans des eaux côtières **4**
3a Longueur < 40 µm (vol. 2/1 Fig. 91: 16, 17) L: 20–28 µm/l. 5–9 µm/stries 22–24/10 µm milieu jusqu'à 28/10 µm extr. . . . . **28. *St. wislouchii*** vol. 2/1 p. 250
3b Longueur > 50 µm (vol. 2/1 Fig. 91: 14, 15) L: 50–110 µm/l. 11–20 µm/stries 18–25/10 µm . . . . **27. *St. salina*** vol. 2/1 p. 250
4a Stries difficilement résolvables au microscope optique, > 33/10 µm . . **5**
4b Stries fines mais nettes au microscope optique **8**
5a Longueur valvaire < 12 µm (vol. 2/1 Fig. 91: 8, 9) L: 9–12 µm/l. 3–4 µm/stries >= 33/10 µm . . . . **24. *St. nana*** vol. 2/1 p. 249
5b Longueur valvaire > 13 µm **6**
6a Largeur valvaire > 6 µm (vol. 2/1 Fig. 90: 15) L: 17–24 µm/l. 6–7,5 µm/stries 34–38/10 µm . . . . **19. *St. recondita*** vol. 2/1 p. 247
6b Largeur valvaire < 6 µm **7**
7a Extrémités capitées (vol. 2/1 Fig. 90: 28–30) L.: 12–21 µm/l. 3–5 µm/stries n. r. **21. *St. gracillima*** vol. 2/1 p. 248
7b Extrémités tout au plus légèrement espacées (vol. 2/1 Fig. 91: 18, 19) L: 17–20 µm/l. 4–4,5 µm/stries env. 35–40/10 µm **29. *St. alpina*** vol. 2/1 p. 251
8a Extrémités nettement étirées ou capitées **9**
8b Extrémités faiblement ou non marquées et arrondies-tronquées **11**
9a Stries relativement grossières, 20–24/10 µm (vol. 2/1 Fig. 90: 31–34) L: 8–17 µm/l. 3–5 µm/stries 20–24/10 µm centre . . . . **22. *St. thermicola*** vol. 2/1 p. 248
9b Stries plus fines, 26 et +/10 µm **10**
10a Aire centrale un fascia large (vol. 2/1 Fig. 90: 21–22) L: 19–28/l. 4–6/str. 27–34/10 µm . . . . **18. *St. agrestis*** vol. 2/1 p. 247
10b Aire centrale un fascia étroit (vol. 2/1 Fig. 90: 23–27) L: 17–24 µm/l. 4–6 µm/stries 26–30/10 µm . . . . **20. *St. kriegeri*** vol. 2/1 p. 248
11a Valves finement rayées, 29–32/10 µm. Extrémités non marquées et largement arrondies (vol. 2/1 Fig. 91: 10, 11) L: 6,6–9,7 µm/l. +/- 3,1 µm/stries 29–32/10 µm . . . . **25. *St. lapidicola*** vol. 2/1 p. 249
11b Valves plus grossièrement rayées. Extrémités faiblement marquées . . **12**
12a Valves relativement larges. L/l < 3,6 µm. Stries 24–26/10 µm jusqu'à 32/10 µm aux pôles (vol. 2/1 Fig. 91: 12, 13) **26. *St. tackei*** vol. 2/1 p. 249
12b Valves étroites. L/l > 3,7. Str. 20–21/10 µm (vol. 2/1 Fig. 91: 1–7) jusqu'à 23/10 µm aux extrémités . . . **23. *St. pseudosubobtusoides*** vol. 2/1 p. 249
13a Stries < 20/10 µm **14**
13b Stries > 20/10 µm **23**
14a Pseudosepta absents ou difficilement reconnaissables **15**
14b Pseudosepta bien visibles, du moins en vue connective **19**

15a Formes petites, < 30 µm de long **16**
15b Formes plus grandes **17**
16a Aire centrale un fascia moyennement large (vol. 2/1 Fig. 90: 10–12) L: 10–24 µm/l. 3–5 µm/Stries 18–23/10 µm *14. St. borrichii* vol. 2/1 p. 245
16b Aire centrale un fascia large (vol. 2/1 Fig. 90: 13) L: +/- 25 µm/l. 8 µm/ stries 16–20/10 µm milieu jusqu'à 24/10 µm aux extr. . . . . *15. St. laterostrata* vol. 2/1 p. 246
17a Ponctuation de la zone de bordure différente de celle du milieu (vol. 2/1 Fig. 87: 1, 2) L: 100–185µm/l. 23–38 µm/stries 14–17/10 µm . . . . *5. St. Nobilis* vol. 2/1 p. 242
17b Ponctuation identique sur toute la valve **18**
18a Côtés presque parallèles. Extrémités marquées cunéiformément (vol. 2/1 Fig. 88: 5) L: 56–80 µm/l. 15–22 µm/stries 16–18/10 µm au milieu jusqu'à 22/10 µm aux extr . . . . **6. *St. dilatata*** vol. 2/1 p. 242
18b Valves lancéolées ou elliptiques-lancéolées (vol. 2/1 Fig. 84: 1–3; 85: 1–6) L: 70–360 µm/l. 16–53 µm/stries 12–20/10 µm . . . . **1. *St. phoenicenteron*** vol. 2/1 p. 239
19a Longueur < 30 µm (vol. 2/1 Fig. 90: 13) L: +/- 25 µm/l. 8 µm/stries 16–20/10 µm au milieu jusqu'à 24/10 µm aux extr. . . . . **15. *St. laterostrata*** vol. 2/1 p. 246
19b Longueur > 30 µm **20**
20a Raphé proximalement nettement revers-latéral (vol. 2/1 Fig. 86: 1–6) L: 50–260 µm/l. 12–44 µm/stries 11–18/10 µm . **3. *St. javanica*** vol. 2/1 p. 241
20b Fissure extérieure du raphé proximalement droite **21**
21a Extrémités étroites et longuement étirées (vol. 2/1 Fig. 85: 7–9) L: 30–48 µm/l. 9–11 µm/stries 18–22/10 µm . . . . **4. *St. lauenburgiana*** vol. 2/1 p. 241
21b Extrémités tronquées-arrondies et à peine étirées **22**
22a Valves rhomboédriques-lancéolées, au milieu +/- élargies (vol. 2/1 Fig. 88: 6–9) L: 80–180 µm/l. 15–35 µm/stries 11–16/10 µm au milieu 13–17/10 µm aux extrémités . . . . **7. *St. acuta*** vol. 2/1 p. 242
22b Valves linéaires ou linéaires-elliptiques à côtés réguliers convexes (vol. 2/1 Fig. 90: 1–6) L: 20–120 µm/l. 4–13 µm/stries 18–24/10 µm . . . . **13. *St. obtusa*** vol. 2/1 p. 244
23a Pseudosepta flous ou absents **24**
23b Pseudosepta nets **28**
24a Valves étroites, linéaires, L/l > 5 µm (vol. 2/1 Fig. 89: 11) . . . . **11. *St. lundii*** vol. 2/1 p. 244
24b Valves plus largement lancéolées à elliptiques-lancéolées L/l < 5 µm . . **25**
25a Stries au centre peu radiales, donc aire centrale en fascia presque linéaire (vol. 2/1 Fig. 87: 3–9; 88: 1–4) L: 20–130 µm/l. 6–18 µm/stries 20–33/10 µm . . . . **2. *St. anceps*** vol. 2/1 p. 240
25b Stries au centre plus fortement radiales, donc fascia +/- élargie vers l'extérieur **26**
26a Côtés des valves présentant le plus souvent 3 ondulations (vol. 2/1 Fig. 89: 1–10) L: 20–30 µm/l. 5–7 µm/stries 21–23/10 µm . . . . **10. *St. undata*** vol. 2/1 p. 244
26b Côtés des valves non triondulés **27**
27a Stries courbées au moins au milieu de la valve (vol. 2/1 Fig. 90: 10–12) L: 10–24 µm/l. 3–5 µm/Stries 18–23/10 µm **14. *St. borrichii*** vol. 2/1 p. 245
27b Stries droites au milieu de la valve (vol. 2/1 Fig. 90: 7–9) . . . . **16. *St. schimanskii*** vol. 2/1 p. 246
28a Extrémités des valves arrondies-tronquées, tout au plus faiblement mar-

quées (vol. 2/1 Fig. 90: 1–6) L: 20–120 µm/l. 4–13 µm/stries 18–24/10 µm . . . . **13. *St. obtusa*** vol. 2/1 p. 244

28b Extrémités des valves nettement marquées et étirées à capitées **29**

29a Extrémités étirées en pointe. Pseudosepta allant jusqu'aux pieds de la partie étirée **30**

29b Extrémités étirées plus en largeur à capitées · Pseudosepta ne remplissant que la fin de la partie étirée **31**

30a Valves étroites, linéaires. Bords parallèles à triondulés, onde centrale env. de la même largeur que les autres (vol. 2/1 Fig. 90: 16–20) L: 18–42 µm/l. 3–8 µm/stries 20–24/10 µm . . . . **17. *St. prominula*** vol. 2/1 p. 247

30b Valves plus largement lancéolées ou triondulées ; onde centrale nettement plus large que les autres (vol. 2/1 Fig. 89: 16–23) . . . . **12. *St. smithii*** vol. 2/1 p. 244

31a Côtés des valves fortement triondulés (vol. 2/1 Fig. 89: 12–15) L: 16–45 µm/l. 4–10 µm/stries 24–29/10 µm . **9. *St. legumen*** vol. 2/1 p. 243

31b Côtés des valves convexes **32**

32a Stries grossièrement ponctuées. Points irrégulièrement placés sur les stries (vol. 2/1 Fig. 85: 7–9) L: 30–48 µm/l. 9–11 µm/stries 18–22/10 µm . . . . **4. *St. lauenburgiana*** vol. 2/1 p. 241

32b Stries finement ponctuées. 28–33 points/10 µm (vol. 2/1 Fig. 89: 1–7) L: 30–50 µm/l. 8–11 µm/stries 22–28/10 µm . **8. *St. producta*** vol. 2/1 p. 243

3. Genre *Anomoeoneis* Pfitzer 1871 (vol. 2/1 p. 251)

Typus generis: *Anomoeoneis sphaerophora* (Ehrenberg) Pfitzer; [*Navicula sphaerophora* Ehrenberg 1841]

Clef pour les espèces

1a Valves fortement renflées au milieu. Extrémités longuement étirées et souvent légèrement capitées (vol. 2/1 Fig. 93: 4) L: 20–40 µm/l. 12–17 µm/stries 23–26/10 µm au milieu . . . . **2. *A. follis*** vol. 2/1 p. 253

1b Milieu des valves non renflé. Côtés convexes **2**

2a Formes plus grandes avec le plus souvent < 18 str/10 µm (vol. 2/1 Fig. 92: str.1–6; 93: 1–3) L: 25–200 µm/l. 12–60 µm/stries 13–20/10 µm . . . . **1. *A. sphaerophora*** vol. 2/1 p. 252

2b Structure plus délicate. >= 19 str./10 µm **3**

3a Nodules centraux le plus souvent prolongés (vol. 2/1 Fig. 94: 15–20) L: 20–44 µm/l. 5–8 µm/stries 24–27/10 µm . . . **5. *A. styriaca*** vol. 2/1 p. 255

3b Nodules centraux non prolongés **4**

4a > 30 str./10 µm **5**

4b < 30 str./10 µm **7**

5a Côte longitudinale nette dans la face valvaire. Extrémités à peine marquées (vol. 2/1 Fig. 93: 8, 9) L: 14–35 µm/l. 4–5,5 µm/stries 35–42/10 µm . . . . **7. *Brachysira aponina*** vol. 2/1 p. 257

5b Côtes longitudinales absentes dans la face valvaire **6**

6a Stries >= 36/10 µm. Extrémités non marquées (vol. 2/1 Fig. 103a: 10–13) L: 14–42 µm/l. 4–6 µm/stries 30–37/10 µm . . **7. *A. garrensis*** vol. 2/1 p. 257

6b Stries <= 36/10 µm. Extrémités presque toujours marquées ou capitées (vol. 2/1 Fig. 94: 21–28; 103a: 14) . . . . **6. *A. vitrea*** vol. 2/1 p. 256

7a Le plus souvent des formes plus grandes à extrémités arrondies-pointues. Largeur valvaire > 10 µm. Stries < 23/10 µm (vol. 2/1 Fig. 93: 5–7) L: 40–115 µm/l. 9–21 µm/stries 19–23/10 µm . **3. *A. serians*** vol. 2/1 p. 254

7b Formes plus petites. Extrémités arrondies-tronquées, ou marquées, cunéiformes ou tronquées arrondies-pointues. Stries > 23–10 µm (vol. 2/1 Fig. 94: 1–14) L: 14–47 µm/l. 4–10 µm/stries 26–30/10 µm . . . . 4. *A. brachysira* vol. 2/1 p. 254

4. Genre *Frustulia* Rabenhorst 1853 nom. cons. (vol. 2/1 p. 258)

Typus generis: *Frustulia saxonica* Rabenhorst 1853 (typ. cons.)

Clef pour les espèces

1a Branches du raphé avec des fissures centrales courbées dans le même sens 2
1b Branches du raphé sans fissures centrales visibles sous microscope optique 3
2a Branches du raphé accompagnées par des côtes axiales nettes. Stries ayant environ 40 points/10 µm, très délicatement ponctuées (vol. 2/1 Fig. 97: 12–14) L.: 32–60 µm/l.: 6,5–10 µm/stries env. 30–34/10 µm jusqu'à 40/10 µm aux extrémités . . . . **5. *F. weinholdii*** vol. 2/1 p. 262
2b Côtes axiales absentes. Points 28–30/10 µm (vol. 2/1 Fig. 97: 10, 11) L.: 30–34 µm/l.: 6–8 µm/stries 24–30/10 µm au centre, points 28–30/10 µm . . . . **4. *F. creuzburgensis*** vol. 2/1 p. 261)
3a Nodule central très étroit, linéaire. Points relativement grossiers, irrégulièrement répartis sur les stries (vol. 2/1 Fig. 97: 7–9) L.: 27–52 µm/l.: 5–8,5 µm/stries 26–30/10 µm, points 23–30/10 µm . . . . **3. *F. spicula*** vol. 2/1 p. 260
3b Nodule central autrement structuré, points réguliers **4**
4a Côtés du nodule central concaves, valves linéaires-lancéolées (vol. 2/1 Fig. 95: 1–7; 96: 4, 5) L.: 40–160 µm/l.: 12–30 µm/stries 20–40/10 µm, points 2 env. 20–40/10 µm . . . . **1. *F. rhomboides*** vol. 2/1 p. 258
4b Côtés du nodule central convexes **5**
5a Formes plus grandes > 80 µm, linéaires-lancéolées (vol. 2/1 Fig. 96: 1–3) L.: 80–110 µm/l.:13–16 µm/stries et points 25–30/10 µm . . . . **1. *F. rhomboides* var. *viridula*** vol. 2/1 p. 259
5b Formes plus petites < 70 µm, linéaires-elliptiques (vol. 2/1 Fig. 97: 1–6) . . . . **2. *F. vulgaris*** vol. 2/1 p. 253

5. Genre *Amphipleura* Kützing 1844 (vol. 2/1 p. 262)

Typus generis: *Amphipleura pellucida* (Kützing) Kützing [*Frustulia pellucida* Kützing 1833] (Boyer 1927)

Clef pour les espèces

1a Largeur des valves d'env. 2 µm seulement (vol. 2/1 Fig. 98: 7, 8) L.: 15–35 µm/l.: env. 2 µm/stries env. 25/10 µm . . . . **4. *A. kriegeriana*** vol. 2/1 p. 264
1b Valves considérablement plus larges **2**
2a Stries très délicates, résolvables seulement à l'aide de matériel pointu (vol. 2/1 Fig. 98: 4–6) L.: 80–140 µm/l.: 7–9 µm/stries 37–40/10 µm . . . . **1. *A. pellucida*** vol. 2/1 p. 263
2b Stries plus grossières, < 30/10 µm **3**
3a L > 120 µm, forme d'eau douce (vol. 2/1 Fig. 98: 1–3) L.: 120–330 µm/l.: 23–27 µm/stries 26–28/10 µm . . . . . **2. *A. lindheimeri*** vol. 2/1 p. 263
3b L < 35 µm, forme d'eau saumâtre (vol. 2/1 Fig. 98: 9–11) L.: 15–35 µm/l.: 4–6 µm/stries 24–28/10 µm . . . . **3. *A. rutilans*** vol. 2/1 p. 264

6. Genre *Neidium* Pfitzer 1871 (vol. 2/1 p. 265)

Typus generis: *Neidium affine* (Ehrenberg) Pfitzer [*Navicula affinis* Ehrenberg 1841] (Boyer 1927)

Clef pour les espèces

1a Extrémités proximales du raphé à pores centraux ou très brièvement courbées latéralement (cf. aussi 14. *N. alpinum*) 2

1b Extrémités proximales du raphé à fissures centrales nettes, courbées vers différents côtés 10

2a Str. < 24/10 µm 3

2b Str. > 24/10 µm 7

3a Branches du raphé à pores centraux 4

3b Pas de pores centraux, extrémités proximales du raphé très peu courbées latéralement dans différentes directions 6

4a Aire centrale constituant une large bande transversale atteignant presque le bord (vol. 2/1 Fig. 100: 1–2) L.: 19–44 µm/l.: 10–17 µm/stries 14–20/10 µm . . . . **5. *N. ladogensis*** vol. 2/1 p. 269

4b Aire centrale de structure différente 5

5a Dans le centre de la valve, stries plus fortement radiales, courbées, alternativement courtes et longues dans la zone du nodule central; valves larges-lancéolées (vol. 2/1 Fig. 99: 8, 9) L.: 28–76 µm/l.: 16–27 µm/stries 16–19/10 µm . . . . **2. *N. opulentum*** vol. 2/1 p. 268

5b Stries peu radiales dans le centre valvaire, non courbées, valves linéaires à largement linéaires-elliptiques (vol. 2/1 Fig. 99: 1–7) L.: 30–58 µm/l.: 10–16 µm/stries 16–24/10 µm . . . . **1. *N. dubium*** vol. 2/1 p. 267

6a Valves étroites, linéaires-elliptiques, extrémités arrondies tronquées (vol. 2/1 Fig. 99: 11) L.: env. 70 µm/l.: 14 µm./stries 19–20/10 µm . . . . **4. *N. juba*** vol. 2/1 p. 269

6b Valves larges-elliptiques, extrémités cunéiformes 7

7a Valves bi-ondulées, rétrécies au centre 8

7b Valves à côtés convexes 9

8a Aire centrale petite, ronde ou absente. Ponctuation très délicate (vol. 2/1 Fig. 100: 3–5) L.: 15–31 µm/l.: 5–9,5 µm/stries 24–30/10 µm . . . . **8. *N. binodis*** vol. 2/1 p. 270

8b Aire centrale grande, rectangulaire ou transversale-elliptique. Str. plus grossièrement ponctuées (vol. 2/1 Fig. 100: 6–8) L.: 16–33 µm/l.: 5–8 µm/stries 24–28/10 µm . . . . **9. *N. binodeformis*** vol. 2/1 p. 271

9a Valves linéaires-elliptiques, extrémités arrondies tronquées, ponctuation env. 20/10 µm (vol. 2/1 Fig. 99: 10) L.: 34 µm/l.: 9 µm/ponctuation env. 20/10 µm . . . . **3. *N. testa*** vol. 2/1 p. 268

9b Valves larges-lancéolées, extrémités cunéiformes, ponctuation 24–29/10 µm (vol. 2/1 Fig. 100: 10–13) L.: 19–46 µm/l.: 12–17 µm/stries 26–28/10 µm . . . . **6. *N. densestriatum*** vol. 2/1 p. 269

10a Fissures centrales très longues, atteignant presque les canaux longitudinaux des bords 11

10b Fissures centrales plus courtes 13

11a Largeur < 7 µm 12

11b Largeur > 7 µm (vol. 2/1 Fig. 101: 2–4) L.: 29–58 µm/stries 16–18/10 µm . . . . **11. *N. calvum*** vol. 2/1 p. 272

12a Stries nettement ponctuées, ponctuation 21–25/10 µm (vol. 2/1 Fig. 101: 5) L.: 23–29 µm/l.: 5–6 µm/stries 19–25/10 µm . . . . **12. *N. minutissimum*** vol. 2/1 p. 272

12b Stries très délicatement ponctuées, presque non reconnaissables, ponctuation 27–32/10 µm (vol. 2/1 Fig. 101: 6, 7) L.: 16–32 µm/l.: 4,5–6 µm/stries 28–34/10 µm . . . . **13. *N. javanicum*** vol. 2/1 p. 272

13a Stries > 30/10 µm **14**

13b Stries < 30/10 µm **16**

14a Côtés convexes, non-ondulés, extrémités arrondies tronquées (vol. 2/1 Fig. 101: 13–17) L.: 13–37 µm/l.: 4–6 µm/stries 36–42/10 µm . . . . **14. *N. alpinum*** vol. 2/1 p. 273

14b Extrémités nettement marquées **15**

15a Valves linéaires, côtés le plus souvent bi- à tri-ondulées (vol. 2/1 Fig. 101: 8–12) L.: 20–40 µm/l.: 5–6,7 µm/stries 30–35/10 µm . . . . **15. *N. septentrionale*** vol. 2/1 p. 273

15b Valves linéaires, côtés non-ondulées (vol. 2/1 Fig. 103a: 4, 5) L.: 25–38 µm/ l.: 4,5–6,4 µm/stries 30–33/10 µm . . . . **27. *N. affine* var. *longiceps*** vol. 2/1 p. 281

16a Stries et fissures centrales très obliques (vol. 2/1 Fig. 103a: 1, 2) L.: 39–65 µm/l.: 10,5–12 µm/stries 22–26/10 µm **24. *N. carteri*** vol. 2/1 p. 278

16b Stries moins ou pas du tout obliques, jamais de fissures centrales obliques **17**

17a Stries > 26/10 µm **18**

17b Stries < 24/10 µm **20**

18a Extrémités nettes, largement étirées (vol. 2/1 Fig. 106: 8–10) L.: 20–80 µm/ l.: 6–17 µm/stries 20–33/10 µm . . . . **27. *N. affine*** part.

18b Extrémités arrondies tronquées **19**

19a Fissures centrales longues. 1–2 canaux longitudinaux nets, distincts du bord (vol. 2/1 Fig. 103: 1–10) L.: 28–82 µm/l.: 7–12 µm/stries 26–30/120 µm . . . . **22. *N. bisulcatum*** vol. 2/1 p. 277

19b Fissures centrales courtes. Canaux longitudinaux presque confondus avec le bord et souvent non reconnaissables en vue valvaire (vol. 2/1 Fig. 103: 11–16) L.: 17–80 µm/l.: 5–13 µm/stries 26–30/10 µm . . . . **23. *N. hercynicum*** vol. 2/1 p. 277

20a Formes linéaires-elliptiques. L/l le plus souvent considérablement < 3/1 µm. Fissures centrales très courtes **21**

20b Caractéristiques différentes **22**

21a Stries ponctuées grossièrement, 16–18 points/10 µm (vol. 2/1 Fig. 101: 1) L.: 45–60 µm/l.: 20–28 µm/stries env. 16/10 µm . . . . **10. *N. dilatatum*** vol. 2/1 p. 271

21b Stries ponctuées plus délicatement, 24–26 points/10 µm (vol. 2/1 Fig. 100: 9) L.: 30–48 µm/l.: 15–20 µm/stries 19–22/10 µm . . . . **7. *N. apiculatum*** vol. 2/1 p. 270

22a Longueur valvaire < 36 µm **23**

22b Longueur valvaire > 36 µm **25**

23a Ponctuation non en rangées longitudinales ondulées (vol. 2/1 Fig. 102: 1, 2) cf. aussi sous 27. *N. affine* L.: 18–36 µm/l.: env. 6 µm/stries env. 20/10 µm . . . . **16. *N. hustedtii*** vol. 2/1 p. 274

23b Ponctuation en rangées ondulées au moins vers les bords **24**

24a Ponctuation fine, env. 30/10 µm, une bande longitudinale courbée sur le bord (vol. 2/1 Fig. 102: 7–9) L.: 20–26 µm/l.: 4,8–5,4 µm/stries env. 22/10 µm . . . . **18. *N. perforatum*** vol. 2/1 p. 274

24b Ponctuation plus grossière, plusieurs bandes longitudinales courbées dans la face valvaire (vol. 2/1 Fig. 102: 3–6) L.: 24–36 µm/l.: 5,5–9 µm/stries 21–24/10 µm . . . . **17. *N. bergii*** vol. 2/1 p. 274

25a Points en perles, étonnamment grands **26**

25b Points ayant une autre structure **27**

26a Pas d'aire centrale (vol. 2/1 Fig. 102: 10, 11) L.: 35–70 µm/l. 12–21 µm/stries 10–12/10 µm . . . . **20.** ***N. distinctepunctatum*** vol. 2/1 p.275
26b Aire centrale ronde ou un fascia oblique (vol. 2/1 Fig. 102: 12–15) L.: 36–78 µm/l.: 9–24 µm/stries 10–18/10 µm **21.** ***N. kozlowii*** vol. 2/1 p. 276
27a Ponctuation 20/10 µm et plus (vol. 2/1 Fig. 106: 8–10) L.: 20–80 µm/l.: 6–17 µm/stries 20–30/10 µm . . . . **27.** ***N. affine*** vol. 2/1 p. 280
27b Ponctuation < 20/10 µm **28**
28a Côtés tri-ondulées (vol. 2/1 Fig. 107: 3) L.: 40–100 µm/l.: 10–19 µm/stries 18–22/10 µm . . . . **29.** ***N. hitchcockii*** vol. 2/1 p. 282
28b Côtés de structure différente **29**
29a Extrémités étirées étroitement, le plus souvent légèrement capitées (vol. 2/1 Fig. 107: 4–6) L.: 40–100 µm/l.: 12–32 µm/stries 18–22/10 µm . . . . **28.** ***N. productum*** vol. 2/1 p. 281
29b Extrémités de structure différente **30**
30a Sur chaque côté valvaire, la ponctuation est faite de 3 à 4 rangées longitudinales (vol. 2/1 Fig. 103a: 3) L.: 40–100 µm/l.: 12–23 µm/stries 13–19/10 µm . . . . **19.** ***N. decoratum*** vol. 2/1 p. 275
30b Rangées longitudinales des points seulement dans la zone des canaux longitudinaux **31**
31a Fissures centrales très petites. Extrémités cunéiformes, formes très grandes (vol. 2/1 Fig. 104: 1–4; 105: 1) L.: 37-env. 300 µm/l.: 15–40 µm/stries 12–18/10 µm . . . . **25.** ***N. iridis*** vol. 2/1 p. 279
31b Fissures centrales plus longues. Extrémités arrondies tronquées à étirées capitées (vol. 2/1 Fig. 105: 2–6; 106: 1–7; 107: 1, 2) L.: 40–100 µm/l.: 14–24 µm/stries 16–20/10 µm . . . . . **26.** ***N. ampliatum*** vol. 2/1 p. 279

8. Genre *Diploneis* Ehrenberg 1844 (vol. 2/1 p. 283)

Typus generis: *Diploneis didyma* (Ehrenberg) Ehrenberg [*Navicula* (*Pinnularia*) *didyma* Ehrenberg] (Boyer 1927)

Clef pour les espèces

1a Petites formes avec plus de 18 stries/10 µm **2**
1b Grandes formes avec moins de 18 stries/10 µm **8**
2a Stries avec doubles points (vol. 2/1 Fig. 109: 15, 16) . . . . **8.** ***D. puella*** vol. 2/1 p. 289
2b Stries avec simples points **3**
3a Stries très fines, à peine résolvables au microscope optique, 35–39/10 µm (vol. 2/1 Fig. 110: 6–8) . . . . **17.** ***D. minuta*** vol. 2/1 p. 292
3b Stries bien résolvables au microscope optique **4**
4a Canaux longitudinaux étroits **5**
4b Canaux longitudinaux larges **7**
5a Bords des canaux longitudinaux concaves au niveau du nodule central (vol. 2/1 Fig. 108: 7–10) . . . . **3.** ***D. oblongella*** vol. 2/1 p. 287
5b Bords externes des canaux longitudinaux non concaves au niveau du nodule central **6**
6a Stries fortement contrastées, 18–20/10 µm (vol. 2/1 Fig. 110: 9–12) . . . . **18.** ***D. modica*** **vol. 2/1 p. 293**
6b Stries faiblement contrastées, 20–24/10 µm (vol. 2/1 Fig. 110: 13–15) . . . . **19.** ***D. oculata*** **vol. 2/1 p. 293**
7a Bords externes des canaux longitudinaux fortement concaves, les 2 renfermant une large aire lancéolée (vol. 2/1 Fig. 110: 16, 17) . . . . **20.** ***D. petersenii*** vol. 2/1 p. 293

7b Bords externes des 2 canaux longitudinaux presque parallèles, les 2 renfermant une aire linéaire-elliptique (vol. 2/1 Fig. 110: 3–5) . . . . **16.** ***D. marginestriata*** vol. 2/1 p. 292
8a Côtés des valves concaves au milieu **9**
8b Côtés des valves droits ou convexes **11**
9a Forme d'eau douce dans des eaux à teneur moyenne en électrolytes (vol. 2/1 Fig. 111: 4) . . . . **11.** ***D. alpina*** **part. vol. 2/1 p. 290**
9b Formes d'eau saline ou d'eaux continentales à haute teneur en électrolytes 10
10a Valves à côtes longitudinales, côtés faiblement rentrés au milieu (vol. 2/1 Fig. 112: 7) . . . . **15.** ***D. didyma*** vol. 2/1 p. 292
10b Côtes longitudinales absentes, côtés fortement rentrés au milieu (vol. 2/1 Fig. 112: 5, 6) . . . . **14.** ***D. interrupta*** vol. 2/1 p. 292
11a Stries à simples points **12**
11b Stries à doubles points, du moins dans la zone des bords **18**
12a Canaux longitudinaux ou leurs côtés externes non ondulés au niveau du nodule central **13**
12b Canaux longitudinaux ou leurs côtés externes ondulés au niveau du nodule central **14**
13a Canaux longitudinaux très larges (vol. 2/1 Fig. 110: 1, 2) . . . . 10. ***D. finnica*** **vol. 2/1 p. 290**
13b Canaux longitudinaux étroits (vol. 2/1 Fig. 111: 5, 6) . . . . **12.** ***D. domblittensis*** **vol. 2/1 p. 290**
14a Valves linéaires-elliptiques aux côtés presque parallèles **15**
14b Valves larges-elliptiques aux côtés convexes **16**
15a 14–15 stries/10 µm (vol. 2/1 Fig. 109: 10, 11) . . . . **7.** ***D. boldtiana*** vol. 2/1 p. 288
15b 7–11 stries/10 µm (vol. 2/1 Fig. 111: 1–4) . . **11.** ***D. alpina*** **vol. 2/1 p. 290**
16a Petites formes, généralement plus étroites que 12 µm (vol. 2/1 Fig. 109: 10–11) . . . . **7.** ***D. boldtiana*** **vol. 2/1 p. 288**
16b Valves généralement beaucoup plus larges que 12 µm **17**
17a Structure grossière, canaux longitudinaux déviant nettement des stries, aréoles avec aréoles foramina articulées (vol. 2/1 Fig. 108: 6) . . . . **1.** ***D. elliptica*** **vol. 2/1 p. 285**
17b Structure plus délicate, canaux longitudinaux structurés comme les stries, aréoles avec un foramen en forme de point (vol. 2/1 Fig. 108: 16) . . . . **2.** ***D. ovalis*** vol. 2/1 p. 286
18a Doubles points sur stries en quinconce **19**
18b Doubles points sur stries en vis-à-vis **23**
19a Canaux longitudinaux larges **20**
19b Canaux longitudinaux étroits **21**
20a Canaux longitudinaux finement ponctués (vol. 2/1 Fig. 112: 2–4) . . . . **13.** ***D. smithii*** vol. 2/1 p. 291
20b Canaux longitudinaux à taches grossières (vol. 2/1 Fig. 109: 12–14) . . . . **9.** ***D. mauleri*** vol. 2/1 p. 289
21a Bords externes des canaux longitudinaux non concaves au niveau du nodule central (vol. 2/1 Fig. 112: 2) . . . . **13.** ***D. smithii*** **var.** ***pumila*** **vol. 2/1 p. 291**
21b Bords externes des canaux longitudinaux concaves au niveau du nodule central **22**
22a Nodule central petit, 14–17 stries/10 µm (vol. 2/1 Fig. 109: 1–7) . . . . **5.** ***D. parma*** vol. 2/1 p. 287
22b Nodule central grand, 10–12 stries/10 µm (vol. 2/1 Fig. 109: 8–9) . . . . **6.** ***D. subovalis*** vol. 2/1 p. 288

23a Stries très finement ponctuées, 25–30/10 µm (vol. 2/1 Fig. 109: 10, 11) . . . . **7.** ***D. boldtiana*** vol. 2/1 p. 288
23b Stries très grossièrement ponctuées **24**
24a Grandes formes, longueur > 35 µm (vol. 2/1 Fig. 112: 1) . . . . **10.** ***D. finnica*** vol. 2/1 p. 290
24b Petites formes, longueur < 32 µm **25**
25a Petit nodule central, milieu des canaux longitudinaux non concave (vol. 2/1 Fig. 109: 15, 16) . . . . **8.** ***D. puella*** vol. 2/1 p. 289
25b Grand nodule central, milieu des canaux longitudinaux concave (vol. 2/1 Fig. 108: 11–13) . . . . **4.** ***D. pseudovalis*** vol. 2/1 p. 287

9. Genre *Pleurosigma* W. Smith 1852 nom. cons. (vol. 2/1 p. 294)

Typus generis: *Pleurosigma angulatum* (Quekett) W. Smith; [*Navicula angulata* Quekett 1848]

Clef pour les espèces

1a Bandes transversales et obliques à égale distance l'une de l'autre et de structure identique (vol. 2/1 Fig. 113: 1, 2; 114: 1, 2) . . . . **1.** ***vol. 2/1 P. angulatum*** vol. 2/1 p. 294
1b Bandes transversales plus grossières ou plus fines que les bandes obliques **2**
2a Bandes transversales plus grossières que les bandes obliques (vol. 2/1 Fig. 113: 3) . . . . **2.** ***vol. 2/1 P. salinarum*** **vol. 2/1 p. 294**
2b Bandes transversales plus fines que les bandes obliques (vol. 2/1 Fig. 113: 4; 114: 3) . . . . **3.** ***vol. 2/1 P. elongatum*** **vol. 2/1 p. 295**

10. Genre *Gyrosigma* Hassal 1843 nom. cons. (vol. 2/1 p. 295)

Typus generis: *Gyrosigma hippocampus* (Ehrenberg) Hassall 1845; [*Navicula hippocampus* Ehrenberg 1838]

Clef pour les espèces

1a Extrémités longuement étirées **2**
1b Extrémités arrondies tronquées ou arrondies pointues **4**
2a Grandes formes, longueur > 200 µm (vol. 2/1 Fig. 116: 5) . . . . **11.** ***G. macrum*** vol. 2/1 p. 300
2b Longueur < 150 µm **3**
3a Extrémités relativement largement étirées (vol. 2/1 Fig. 116: 4) . . . . **10.** ***G. parkeri*** vol. 2/1 p. 299
3b Extrémités très finement étirées (vol. 2/1 Fig. 116: 6) . . . . **12.** ***G. fasciola*** vol. 2/1 p. 300
4a Bandes longitudinales et transversales éloignées à égale distance **5**
4b Bandes transversales plus proches ou plus éloignées que les bandes longitudinales **6**
5a Valves < 200 µm de long, 16–23 stries/10 µm, souvent formes d'eau douce (vol. 2/1 Fig. 114: 4, 8) . . . . **1.** ***G. acuminatum*** vol. 2/1 p. 296
5b Valves > 200 µm de long, 11–16 stries/10 µm, souvent formes d'eau saumâtre (vol. 2/1 Fig. 115: 5) . . . . **7.** ***G. balticum*** vol. 2/1 p. 299
6a Bandes transversales plus délicates que bandes longitudinales (vol. 2/1 Fig. 114: 5, 7, 9) . . . . **2.** ***G. attenuatum*** vol. 2/1 p. 297
6b Bandes transversales plus grossières que bandes longitudinales **7**

7a Structure grossière, 11–14 stries/10 µm (vol. 2/1 Fig. 114: 6) . . . . **3.** ***G. strigilis*** vol. 2/1 p. 297
7b Structure plus délicate **8**
8a Raphé en région distale fortement rapproché du côté convexe de la valve (vol. 2/1 Fig. 115: 3) . . . . **6.** ***G. peisonis*** vol. 2/1 p. 298
8b Raphé distal dans la ligne du milieu de la valve **9**
9a Stries difficilement résolvables au microscope optique (+/- 40/10 µm) (vol. 2/1 Fig. 116: 1, 2) . . . . **8.** ***G. obscurum*** vol. 2/1 p. 299
9b Stries bien résolvables au microscope optique, < 30/10 µm **10**
10a Plus de 28 lignes longitudinales/10 µm, extrémités généralement arrondies en biais (vol. 2/1 Fig. 116: 3) . . . . **9.** ***G. scalproides*** vol. 2/1 p. 299
10b 26 et moins de lignes longitudinales/10 µm **11**
11a Aire centrale non diagonale, formes d'eau de mer et d'eau saumâtre (vol. 2/1 Fig. 115: 2) . . . . **5.** ***G. spenceri*** vol. 2/1 p. 298
11b Aire centrale généralement oblique, formes d'eaux continentales de moyenne à haute teneur en électrolytes (vol. 2/1 Fig. 115: 1) . . . . **4.** ***G. nodiferum*** vol. 2/1 p. 297

11. Genre *Cymbella* Agardh 1830 (vol. 2/1 p. 300)

Typus generis: *Cymbella cymbiformis* Agardh 1830

Clef pour les sous-genres

1a Fissures terminales courbées ventralement, point isolé dorsal au niveau du nodule central ou absent **2**
1a Fissures terminales courbées dorsalement, stigma ventraux au niveau du nodule central **3**
2a Fissure interne du raphé toujours avec intermissio, généralement uniquement visible au microscope électronique . . . . **1.** ***Encyonema*** (vol. 2/1 p. 302) (p. 264)
2b Intermissio absent **3.** ***Cymbopleura*** n. 37–43, 46 – (vol. 2/1 p. 325) (p. 267)
3a Généralement visiblement dorsi-ventral, champs poreux apicaux toujours présents . . . . **2.** ***Cymbella*** (vol. 2/1 p. 312) (p. 266)
3b En général naviculoïde, peu de formes dorsi-ventrales, généralement absence des champs poreux apicaux et des stigma . . . . **3.** ***Cymbopleura*** part. (vol. 2/1 p. 321) (p. 267)

1. Sous-Genre *Encyonema* (Kützing 1833) Cleve-Euler 1955 (vol. 2/1 p. 302)

Typus subgeneris: *Monema prostratum* Berkeley 1832

Clef pour les sous-espèces

1a Points ou lineolae reconnaissables avec ouvertures > 1,0, en général < 30 points/10 µm **2**
1b Points ou lineolae peu ou pas reconnaissables avec ouvertures > 1,0 . . **16**
2a Aréoles apparaissant comme points à focalisation profonde **3**
2b Aréoles apparaissant comme lineolae à focalisation profonde **11**
3a Stries grossièrement ponctuées (< 24 points/10 µm) **4**
3b Stries plus finement ponctuées **5**
4a Extrémités valvaires arrondies tronquées, L/l. > 4 µm (vol. 2/1 Fig. 118: 1–8) . . . . ***2.** ***C. mesiana*** vol. 2/1 p. 304
4b Extrémités valvaires arrondies pointues, L/l < 4 µm (vol. 2/1 Fig. 122: 6–9) . . . . ***15.** ***C. elginensis*** vol. 2/1 p. 310

5a Valves étroites (L/l généralement 5–8 µm) (vol. 2/1 Fig. 120: 1–16) . . . . *11. *C. gracilis* vol. 2/1 p. 308
5b Rapport L/l plus petit 6
6a Valves (à part les cellules initiales) en forme de croissant de lune, côté ventral faiblement convexe, raphé fortement déplacé vers position ventrale 7
6b Côté ventral convexe, raphé faiblement déporté vers la position ventrale **8**
7a Dorsal 10–15 stries/10 µm (vol. 2/1 Fig. 117: 1–24) . . . . **1. *C. silesiaca*** vol. 2/1 p. 304
7b Dorsal 7–10 stries/10 µm (vol. 2/1 Fig. 199: 14–16) . . . . **4. *C. paucistriata*** vol. 2/1 p. 305
8a Extrémités non marquées (vol. 2/1 Fig. 119: 29–31) . . . . ***8. *C. brehmii*** vol. 2/1 p. 307
8b Extrémités marquées **9**
9a Extrémités étirées-capitées (vol. 2/1 Fig. 119: 21) . . . . **6. *C. latens*** vol. 2/1 p. 306
9b Extrémités finement étirées **10**
10a L/l < 3,8 µm (vol. 2/1 Fig. 119: 17–20) . . ***5. *C. obscura*** vol. 2/1 p. 306
10b L/l > 4 µm (vol. 2/1 Fig. 121: 10–11) . . ***14. *C. hillardii*** vol. 2/1 p. 309
11a Stries finement lignées, valves elliptiques (vol. 2/1 Fig. 123: 1–6) . . . . ***18. *C. alpina*** vol. 2/1 p. 311
11b Stries grossièrement lignées **12**
12a Valves naviculoïdes, à peine dorsi-ventrales (vol. 2/1 Fig. 124: 1–8) . . . . 19. *C. lacustris* **vol. 2/1 p. 312**
12b Valves fortement dorsi-ventrales **13**
13a Point isolé visible dorsalement au niveau du nodule central (vol. 2/1 Fig. 118: 1–8) . . . . ***2. *C. mesiana*** vol. 2/1 p. 304
13b Points isolés absents **14**
14a Surtout formes tropicales (vol. 2/1 Fig. 122: 10–15) . . . . ***17. *C. muelleri*** vol. 2/1 p. 311
14b Formes fréquentes dans la région **15**
15a Fissures terminales situées dans la face valvaire (vol. 2/1 Fig. 123: 7–10) . . . . **20. *C. prostrata*** vol. 2/1 p. 312
15b Fissures terminales situées en bordure du manteau (vol. 2/1 Fig. 121: 12–16; 122: 1–5) . . . . ***16. *C. caespitosa*** vol. 2/1 p. 310
16a Valves fortement dorsi-ventrales, côté ventral droit ou faiblement convexe, raphé déporté fortement vers la position ventrale **17**
16b Valves relativement ou peu dorsi-ventrales, raphé faiblement déporté vers la position ventrale **19**
17a Plus de 17 stries/10 µm (vol. 2/1 Fig. 119: 32–36) . . . . **9. *C. reichardtii*** vol. 2/1 p. 307
17b Moins de 16 stries/10 µm **18**
18a Extrémités pas ou peu marquées, mais en partie courbées vers le côté ventral (vol. 2/1 Fig. 119: 1–13) . . . . . . ***3. *C. minuta*** vol. 2/1 p. 305
18b Extrémités étranglées-capitées (vol. 2/1 Fig. 119: 21) . . . . **6. *C. latens*** vol. 2/1 p. 306
19a Petites formes avec stries délicates, généralement < 25 µm **20**
19b Formes > 25 µm, stries grossières **22**
20a 10–13 stries/10 µm, extrémités en grande partie non étirées (vol. 2/1 Fig. 119: 22–27) . . . . ***7. *C. perpusilla*** vol. 2/1 p. 306
20b 14–18 stries/10 µm, extrémités étirées ou capitées **21**
21a Extrémités généralement capitées, stries dorsales du milieu radiales (vol. 2/1 Fig. 119: 37–43) . . . . **10. *C. gaeumannii*** vol. 2/1 p. 308

21b Extrémités seulement étirées, stries presque parallèles au milieu (vol. 2/1 Fig. 119: 28) . . . . *7. ***C. perpusilla*** var. vol. 2/1 p. 306
22a Valves linéaires-elliptiques, extrémités arrondies tronquées (vol. 2/1 Fig. 120: 17–24) . . . . *12. ***C. norvegica*** vol. 2/1 p. 309
22b Valves linéaires-lancéolées, extrémités arrondies pointues (vol. 2/1 Fig. 112: 1–9) . . . . 13. ***C. hebridica*** vol. 2/1 p. 309

2. Sous-Genre *Cymbella* (vol. 2/1 p. 312)

Typus subgeneris: *Cymbella cymbiformis* C. Agardh 1830

Clef pour les sous-espèces

1a Plus de 20 points ou lineolae/10 µm **2**
1b Moins de 20 points ou lineolae/10 µm **8**
2a Valves cymbelloïdes avec côté ventral concave ou droit **3**
2b Valves peu cymbelloïdes avec côté ventral convexe ou naviculoïde . . **6**
3a En général avec un stigma placé en position ventrale du nodule central **4**
3b En général avec plusieurs stigma placés en position ventrale du nodule central **5**
4a 26–28 points ou lineolae/10 µm, stries délicates, aire axiale étroite (vol. 2/1 Fig. 125: 1–22) . . . . *21. ***C. affinis* (complexe)** vol. 2/1 p. 314
4b 22–24 lineolae/10 µm, stries grossières, aire axiale avec aire centrale large lancéolée (vol. 2/1 Fig. 126: 1–3) . . . . . 22. ***C. hungarica*** vol. 2/1 p. 314
5a 24–25 lineolae/10 µm, valves fortement cymbelloïdes (vol. 2/1 Fig. 128: 7–8) . . . . 26. ***C. arctica*** vol. 2/1 p. 316
5b 18–21 points ou lineolae/10 µm (vol. 2/1 Fig. 127: 8–11; 128: 1–6) . . . . 25. ***C. cistula*** vol. 2/1 p. 316
6a En général, un stigma . . . . *21. ***C. affinis*** vol. 2/1 p. 314
6b En général, deux ou plusieurs stigma **7**
7a Stries finement ponctuées, 28–32 points/10 µm (vol. 2/1 Fig. 126: 8–19) . . . . 24. ***C. tumidula*** vol. 2/1 p. 315
7b Stries grossièrement ponctuées, 20–25 points/10 µm (vol. 2/1 Fig. 126: 4–7) . . . . 23. ***C. turgidula*** vol. 2/1 p. 314
8a 6 ou plus de stigma (souvent difficilement visibles au microscope optique) en position ventrale du nodule central, grandes formes, généralement longueur > 100 µm **9**
8b Moins de 6 stigma, toujours distinctement visibles **12**
9a Raphé visiblement revers-latéral, stigma très grands . . . . 34. ***C. schimanskii*** vol. 2/1 p. 320
9b Raphé non revers-latéral **10**
10a Extrémité proximale du raphé en forme de crosse ou de crochet **11**
10b Extrémité proximale du raphé avec grande pore centrale et ronde (vol. 2/1 Fig. 131: 1) . . . . 31. ***C. aspera*** vol. 2/1 p. 319
11a Extrémité proximale du raphé en forme de crosse (vol. 2/1 Fig. 131: 2b) . . . . 32. ***C. lanceolata*** vol. 2/1 p. 319
11b Extrémité proximale du raphé en forme de crochet (vol. 2/1 Fig. 131: 3b) . . . . 33. ***C. helmckei*** vol. 2/1 p. 320
12a Section proximale du raphé revers-latéral **13**
12b Section proximale du raphé non revers-latéral **15**
13a Valves peu dorsi-ventrales, 14–18 lineolae/10 µm, foramen du stigma sous microscope électronique en forme de fissure (vol. 2/1 Fig. 129: 3) . . . . 29. ***C. simonsenii*** vol. 2/1 p. 318
13b Valves fortement dorsi-ventrales, 18–20 lineolae/10 µm, foramen du stigma rond au microscope électronique (vol. 2/1 Fig. 129: 9) **14**

14a Généralement, un stigma en position ventrale au niveau du nodule central, aire centrale absente sur le côté dorsal (vol. 2/1 Fig. 129: 2–9) . . . . **28.** *C. cymbiformis* vol. 2/1 p. 317
14b Généralement, plus de 3 stigma en position ventrale au niveau du nodule central, aire centrale présente en forme demi-circulaire sur le côté dorsal (vol. 2/1 Fig. 127: 8–11; 128: 1–6) . . . . . . **25.** *C. cistula* vol. 2/1 p. 316
15a Un grand stigma en position ventrale au niveau du nodule central. Canal du stigma traversant le nodule central très obliquement (vol. 2/1 Fig. 130: 4–6) . . . . **30.** *C. tumida* vol. 2/1 p. 318
15b Plus de 3 stigma avec canaux moins obliques (vol. 2/1 Fig. 128: 9: 129: 1) . . . . **27.** *C. proxima* vol. 2/1 p. 317

3. Sous-Genre *Cymbopleura* Krammer 1982 (vol. 2/1 p. 321)

Typus subgeneris: *Cymbella subaequalis* Grunow 1880

Clef pour les sous-espèces

1a Fissures terminales dirigées ventralement 2
1b Fissures terminales dirigées dorsalement 11
2a Contour linéaire-lancéolé ou rhomboédrique-lancéolé avec extrémités étirées 3
2b Contour linéaire-elliptique 8
3a Plus courtes que 30 µm, 18–22 stries/10 µm 4
3b Plus longues que 30 µm 6
4a Fissures terminales très longues, allant dans le sens du raphé (vol. 2/1 Fig. 134: 4–13) . . . . ***38.** *C. cesatii* vol. 2/1 p. 325
4b Fissures terminales courtes 5
5a Extrémités légèrement étirées, fissures terminales en forme de virgule (vol. 2/1 Fig. 134: 9, 11, 12) . . . . ***39.** *C. falaisensis* vol. 2/1 p. 326
5b Extrémités capitées, fissures terminales demi-circulaires (vol. 2/1 Fig. 135: 1–5) . . . . ***41.** *C. descripta* vol. 2/1 p. 327
6a Fissures terminales très longues, allant dans le sens du raphé (vol. 2/1 Fig. 134: 4–8, 10) . . . . ***38.** *C. cesatii* vol. 2/1 p. 325
6b Fissures terminales formées autrement 7
7a Valves grossièrement rayées, 11–14 stries/10 µm, stigma manquants, fissure externe du raphé peu courbée (vol. 2/1 Fig. 134: 1–3) . . . . ***37.** *C. aequalis* vol. 2/1 p. 325
7b Valves plus finement rayées, 14–17 stries/10 µm, un point isolé en position dorsale du nodule central, fissure externe du raphé ondulée (vol. 2/1 Fig. 135: 11–13) . . . . ***42.** *C. amphioxys* vol. 2/1 p. 328
8a Grandes formes (longueur > 40 µm), extrémités non étirées (vol. 2/1 Fig. 136: 13–15) . . . . **46.** *C. naviculacea* vol. 2/1 p. 330
8b Petites formes (longueur < 35 µm), extrémités capitées 9
9a 10–12 stries/10 µm (vol. 2/1 Fig. 135: 6–10) **43.** *C. similis* vol. 2/1 p. 328
9b Rayé plus finement 10
10a 18–21 stries/10 µm (vol. 2/1 Fig. 135: 1–5) ***41.** *C. descripta* vol. 2/1 p. 326
10b 22–25 stries/10 µm (vol. 2/1 Fig. 134: 23–32) . . . . *40. *C. microcephala* vol. 2/1 p. 327
11a Formes fortement dorsi-ventrales, côté dorsal plus convexe que côté ventral 12
11b Formes plus naviculoïdes, côté dorsal moins convexe que côté ventral **19**
12a Lineolae relativement grossiers et bien visibles 13
12b Lineolae ou points plus fins ou non visibles au microscope optique . . **16**

13a Stries du milieu et des extrémités placées à la même distance, nodule terminal subterminal, fissures terminales longues **14**
13b Stries nettement plus étroites aux extrémités qu'au centre **15**
14a Structure très grossière (7–8 stries/10 µm, 15–16 lineolae/10 µm). Valves très convexes sur côté dorsal (vol. 2/1 Fig. 133: 9) . . . . **36.** ***C. balatonis*** vol. 2/1 p. 325
14b Structure un peu plus fine (8–12 stries/10 µm, 16–23 lineolae/10 µm). Valves relativement convexes sur côté dorsal (vol. 2/1 Fig. 133: 1–8) . . . . **35.** ***C. helvetica*** vol. 2/1 p. 324
15a Généralement longueur > 40 µm, stries finement ponctuées à focalisation haute et profonde, fissure externe du raphé fortement ondulée (vol. 2/1 Fig. 138: 1–6) . . . . **50.** ***C. austriaca*** vol. 2/1 p. 331
15b Généralement longueur < 50 µm, stries très grossièrement lignées à focalisation profonde, fissure externe du raphé courbée, mais non ondulée (vol. 2/1 Fig. 143: 1–13) . . . . **61.** ***C. leptoceros*** vol. 2/1 p. 336
16a Fissures terminales et extrémités proximales du raphé en direction dorsale (vol. 2/1 Fig. 148: 1–9) . . . . ***72.** ***C. pusilla*** vol. 2/1 p. 340
16b Fissures terminales vers direction dorsale, extrémités proximales du raphé vers direction ventrale **17**
17a Valves étroites, stries très fines (16–22/10 µm) (vol. 2/1 Fig. 137: 1–11) . . . . **47.** ***C. delicatula*** vol. 2/1 p. 330
17b Valves plus larges, stries grossières (11–14/10 µm) **18**
18a Formes moyennement grandes, stries délicatement ponctuées, points très difficilement résolvables sous microscope optique (vol. 2/1 Fig. 139: 4–18) . . . . **52.** ***C. laevis*** vol. 2/1 p. 332
18b Formes petites, points au microscope optique généralement nets (vol. 2/1 Fig. 140: 9–17) . . . . **56.** ***C. hustedtii*** vol. 2/1 p. 333
19a Valves étroites ou larges-lancéolées **20**
19b Valves linéaires, étroites-elliptiques, larges-elliptiques ou linéaires-elliptiques **36**
20a Valves larges-lancéolées **21**
20b Valves linéaires-lancéolées **27**
21a Moins de 22 points/10 µm **22**
21b Plus de 25 points/10 µm **24**
22a Plus de 14 stries/10 µm (vol. 2/1 Fig. 139: 1–3) . . . . **51.** ***C. hauckii*** vol. 2/1 p. 332
22b Moins de 11 stries/10 µm **23**
23a Fissures centrales en forme de crochet ou de crosse (vol. 2/1 Fig. 147: 3) . . . . **71.** ***C. heteropleura*** vol. 2/1 p. 340
23b Fissures centrales manquantes, extrémité proximale du raphé avec grandes pores centrales (vol. 2/1 Fig. 144: 1–6) **64.** ***C. ehrenbergii*** vol. 2/1 p. 337
24a Valves linéaires, elliptiques-lancéolées, extrémités bien marquées et larges-arrondies, grande aire centrale, rhomboédrique (vol. 2/1 Fig. 147: 1, 2) . . . . **70.** ***C. tynnii*** vol. 2/1 p. 339
24b Valves larges-lancéolées, extrémités à peine marquées, petite aire centrale **25**
25a Stries plus resserrées aux extrémités qu'au milieu, très délicatement ponctuées (+/- 30/10 µm) (vol. 2/1 Fig. 141: 1–3) . . . . **57.** ***C. reinhardtii*** vol. 2/1 p. 334
25b Différence de la densité des stries entre le milieu et l'extrémité moindre; stries nettement ponctuées **26**
26a Raphé relativement revers-latéral (vol. 2/1 Fig. 146: 5) . . . . **68.** ***C. budayana*** vol. 2/1 p. 339
26b Raphé peu latéral (vol. 2/1 Fig. 143: 17, 18) . **63.** ***C. lata*** vol. 2/1 p. 337

27a 14 et moins de stries/10 µm **28**
27b 15 et plus de stries/10 µm **33**
28a Aire centrale nettement marquée, large-rhomboédrique ou transversale-elliptique **29**
28b Aire centrale formée autrement **30**
29a Extrémités arrondies pointues, peu étirées, +/- 30 points/ 10 µm (vol. 2/1 Fig. 145: 4, 5) . . . . **65.** ***C. hybrida*** var. vol. 2/1 p. 338
29b Extrémités arrondies-tronquées, 23–25 points/10 µm (vol. 2/1 Fig. 140: 1) . . . . **53.** ***C. moelleriana*** vol. 2/1 p. 332
30a Axe apical hétéropolaire (vol. 2/1 Fig. 148: 18–20) . . . . **74.** ***C. ancyli*** vol. 2/1 p. 341
30b Axe apical isopolaire **31**
31a Stries très larges, 7–10/10 µm (vol. 2/1 Fig. 137: 18) . . . . **49.** ***C. borealis*** vol. 2/1 p. 331
31b Stries plus étroites, > 12/10 µm **32**
32a Extrémités étirées pointues (vol. 2/1 Fig. 143: 14–16) . . . . **62.** ***C. designata*** vol. 2/1 p. 336
32b Extrémités arrondies tronquées (vol. 2/1 Fig. 140: 2–6) . . . . **54.** ***C. rupicola*** vol. 2/1 p. 333
33a Aire centrale, un fascia (vol. 2/1 Fig. 140: 7–8) . . . . **55.** ***C. stauroneiformis*** vol. 2/1 p. 333
33b Aire centrale formée autrement **34**
34a Petites formes étroites, ponctuation à peine reconnaissable au microscope optique (vol. 2/1 Fig. 137: 1–11) . . . . . **47.** ***C. delicatula*** vol. 2/1 p. 330
34b Formes plus grandes, ponctuation délicate, mais résolvable **35**
35a Côté dorsal souvent triondulé, extrémités généralement capitées, contour linéaire à elliptique-lancéolé (vol. 2/1 Fig. 135: 15–18) . . . . **44.** ***C. angustata*** vol. 2/1 p. 328
35b Presque naviculoïde, rhomboédrique-lancéolé, extrémités pointues et courtes (vol. 2/1 Fig. 137: 12–17) . . . . **48.** ***C. lapponica*** vol. 2/1 p. 330
36a Valves linéaires à linéaires-elliptiques **37**
36b Valves larges-elliptiques **42**
37a Extrémités capitées (cf. aussi 73 *C. sinuata*, Fig. 148: 13) **38**
37b Extrémités non capitées **39**
38a 9–13 stries/10 µm (vol. 2/1 Fig. 145: 1–3) . **65.** ***C. hybrida*** vol. 2/1 p. 337
38b 16–20 stries/10 µm (vol. 2/1 Fig. 135: 15–18) . . . . **44.** ***C. angustata*** vol. 2/1 p. 328
39a Stigma entre extrémités proximales du raphé (vol. 2/1 Fig. 148: 10–17) . . . . **73.** ***C. sinuata*** vol. 2/1 p. 341
39b Sans stigma **40**
40a Stries fines, mais nettement ponctuées **41**
40b Stries pas nettement ponctuées (vol. 2/1 Fig. 141: 4–19) . . . . **58.** ***C. subaequalis*** vol. 2/1 p. 334
41a Raphé étroit-latéral (vol. 2/1 Fig. 136: 1–12) . **45.** ***C. incerta*** vol. 2/1 p. 329
41b Raphé large-latéral (vol. 2/1 Fig. 142: 1, 2) . **59.** ***C. bernensis*** vol. 2/1 p. 335
42a Stries grossièrement ponctuées **43**
42b Stries finement ponctuées **44**
43a Raphé proximal avec pores centraux, fissures centrales manquantes (vol. 2/1 Fig. 146: 1–4) . . . . **67.** ***C. cuspidata*** vol. 2/1 p. 338
43b Raphé proximal avec fissures centrales en forme de crosse (vol. 2/1 Fig. 146: 6, 7) . . . . **69.** ***C. subcuspidata*** vol. 2/1 p. 339
44a Points non résolvables au microscope optique (vol. 2/1 Fig. 142: 3–21) . . . . **60.** ***C. amphicephala*** vol. 2/1 p. 335
44b Points fins, mais résolvables au microscope optique **45**

45a Raphé fortement latéral, fissure externe ondulée, extrémités arrondies tronquées (vol. 2/1 Fig. 138: 1–7) . . . . **50.** ***C. austriaca*** vol. 2/1 p. 331
45b Raphé relativement latéral, fissure externe courbée, mais pas ondulée, extrémités étirées ou capitées (vol. 2/1 Fig. 145: 6–11) . . . . **66.** ***C. naviculiformis*** vol. 2/1 p. 338

12. Genre *Amphora* Ehrenberg in Kützing 1844 (vol. 2/1 1 p. 342)

Typus generis: *Navicula amphora* Ehrenberg 1831 (Boyer 1927)

1a Cellules sans bandes intercalaires **2**
1b Cellules avec bandes intercalaires **8**
2a Branches du raphé fortement courbées, tournant d'une position proximale à une position dorsale **3**
2b Branches du raphé légèrement courbées ou droites, pas dans une position proximale à dorsale **6**
3a Formes plus petites, côté dorsal (residuum), visiblement cannelé au microscope optique . . . . **6.** ***A. aequalis*** vol. 2/1 p. 347
3b Formes plus grandes, côté dorsal non cannelé **4**
4a Valves à aire centrale plus visiblement dorsale (vol. 2/1 Fig. 149: 3–11) . . . . **2.** ***A. libyca*** vol. 2/1 p. 345
4b Aire centrale dorsale manquante **5**
5a Côté dorsal bi-ondulé avec chapeau siliceux en forme de lanterne (vol. 2/1 Fig. 153: 8, 9) . . . . **19.** ***A. calumetica*** vol. 2/1 p. 352
5b Côté dorsal convexe (vol. 2/1 Fig. 149: 1, 2) . **1.** ***A. ovalis*** vol. 2/1 p. 344
6a Aire dorsale manquante (vol. 2/1 Fig. 150: 14–17) . . . . **5.** ***A. fogediana*** vol. 2/1 p. 346
6b Aire dorsale présente **7**
7a Extrémités non marquées, 18–25 stries/10 µm (vol. 2/1 Fig. 150: 1–13) . . . . **4.** ***A. pediculus*** vol. 2/1 p. 346
7b Extrémités marquées et légèrement étirées ventralement, 15–17 stries/10 µm (vol. 2/1 Fig. 150: 1–6) . . . . **3.** ***A. inariensis*** vol. 2/1 p. 345
8a Plus de 20 stries/10 µm **9**
8b Moins de 20 stries/10 µm **14**
9a Forme plus grande avec beaucoup de bandes intercalaires visiblement ornementées (vol. 2/1 Fig. 150: 23, 24) . . **7.** ***A. lineolata*** vol. 2/1 p. 347
9b Formes < 25 µm **10**
10a Valves fortement convexes dorsalement, petite forme cymbelloïde (vol. 2/1 Fig. 152: 15–18) . . . . **15.** ***A. thumensis*** vol. 2/1 p. 351
10b Bord dorsal seulement relativement convexe **11**
11a Aire centrale du côté dorsal croisée par 2 stries puissantes (vol. 2/1 Fig. 151: 18–27) . . . . **10.** ***A. montana*** vol. 2/1 p. 349
11b Aire centrale manquante **12**
12a Côté dorsal généralement triondulé, extrémités ventrales étirées capitées (vol. 2/1 Fig. 152: 7, 8) . . . . **12.** ***A. dusenii*** vol. 2/1 p. 350
12b Côté dorsal pas ondulé, extrémités tout au plus faiblement étirées . . . **13**
13a < 26 stries/10 µm (vol. 2/1 Fig. 152: 12–14) . . . . **14.** ***A. spitzbergensis*** vol. 2/1 p. 350
13b +/- 30 stries/10 µm (vol. 2/1 Fig. 152: 19–23) . . . . ***16.** ***A. delicatissima*** vol. 2/1 p. 351
14a Structure grossière, en forme de côtes (vol. 2/1 Fig. 152: 9–11) . . . . **13.** ***A. commutata*** vol. 2/1 p. 350
14b Stries à structure déviante **15**

15a Stries très finement ponctuées (vol. 2/1 Fig. 151: 1–6) . . . . **8.** ***A. coffeaeformis*** **var.** ***coffeaeformis*** vol. 2/1 p. 347
15b Stries nettement ponctuées **16**
16a Rayures dorsales moyennes plus éloignées **17**
16b Rayures dorsales moyennes pas plus éloignées ou aire centrale dorsale présente **18**
17a Valves plus étroites que 18 µm, généralement plus de 20 stries/10 µm (vol. 2/1 Fig. 151: 7–17) . . . . **9.** ***A. veneta*** vol. 2/1 p. 348
17b Valves plus larges que 20 µm, moins de 20 stries/10 µm (vol. 2/1 Fig. 153: 1–3) . . . . **17.** ***A. subcapitata*** vol. 2/1 p. 351
18a Pores centraux proches l'un de l'autre (vol. 2/1 Fig. 151: 6') . . . . **8.** ***A. coffeaeformis*** **var.** ***acutiuscula*** vol. 2/1 p. 348
18b Pores centraux plus éloignés **19**
19a Extrémités proximales des branches du raphé fortement courbées en équerre vers une position dorsale, nodule central très net (vol. 2/1 Fig. 153: 4–7) . . . . **18.** ***A. normanii*** vol. 2/1 p. 352
19b Extrémités proximales du raphé seulement faiblement courbées dorsalement, nodule central flou (vol. 2/1 Fig. 152: 1–6) . . . . ***11.** ***A. holsatica*** vol. 2/1 p. 349

13. Genre *Gomphonema* Ehrenberg 1832 nom. cons. (vol. 2/1 p. 352)

Typus generis: *Gomphonema acuminatum* Ehrenberg 1832 (ty vol. 2/1 p. cons.)

1a Grandes valves avec milieu fortement enflé et pôles larges, étranglés-capités, 2–5 stigma apicalement placé d'un côté de l'aire centrale, fissures terminales du raphé fortement coudées, dans la face valvaire . . . . ***Didymosphenia*** vol. 2/1 p. 380
1b Combinaison de caractéristiques différente, en particulier les fissures terminales en prolongement du raphé **2**
2a Chacune, une ligne apicale +/- éloignée du bord valvaire ou en vue connective apparaissant sur la face du manteau (vol. 2/1 Fig. 166: 12–14) . . . . ***Gomphoneis*** vol. 2/1 p. 579
2b Sans lignes longitudinales visibles sur la face valvaire ou celle du manteau **3**
3a Aire centrale sans stigma **39**
3b Aire centrale avec 1, 2 ou 4 stigma **4**
4a Quatre stigma entourant le nodule central de façon carrée (vol. 2/1 Fig. 165:14–18) . . . **24.** ***G. olivaceum*** var. ***minutissimum*** vol. 2/1 p. 375
4b Aire centrale avec 1 ou 2 stigma **5**
5a Deux stigma sur une face de l'aire centrale (vol. 2/1 Fig. 162: 8, 9) . . . . **14.** ***G. bipunctatum*** vol. 2/1 p. 368
5b Valves avec un stigma dans l'aire centrale, soit en prolongement de la strie moyenne, souvent raccourcie ou en position presque centrale (des points isolés supplémentaires sont rarement reconnaissables) **6**
6a Valves fortement étranglées transapicalement dans la partie supérieure **7**
6b Valves sans un tel étranglement, tout au plus au contour légèrement triondulé ou concave et plat. Dans la partie centrale enflées ou à pôle étiré en forme de tête, de bec ou en forme de tête supplémentaire **9**
7a Pôle de tête +/- cunéiforme ou étiré et arrondi-pointu (vol. 2/1 Fig. 160: 1–12) . . . . **9.** ***G. acuminatum*** vol. 2/1 p. 365
7b Pôle de tête plus tronqué à arrondi large **8**

8a Pôle de tête petit, étranglé, valves étroites sublinéaires (vol. 2/1 Fig. 162: 10–13) . . . . **15. *G. subtile*** vol. 2/1 p. 369
8b Pôle de tête étranglé plus grand, arrondi-large, stries moyennes alternativement plus courtes et plus longues (vol. 2/1 Fig. 159: 11–18) . . . . **16. *G. truncatum*** vol. 2/1 p. 369
9a Plus grande largeur de la valve très près, sous le pôle de tête **10**
9b Plus grande largeur près du milieu ou valves presque linéaires **13**
10a Pôle de tête étiré, en forme de bec ou avec tête supplémentaire **11**
10b Pôle de tête sans bec +/- cunéiforme (vol. 2/1 Fig. 160: 1–12) . . . . **9. *G. acuminatum*** vol. 2/1 p. 365
11a Pôle de tête visiblement en forme de bec ou avec tête supplémentaire (vol. 2/1 Fig. 157: 1–8) (cf. aussi *G. pseudoaugur*, Fig. 159: 1–4) . . . . **6. *G. augur*** vol. 2/1 p. 363
11b Pôle de tête arrondi ou avec légère pointe étirée **12**
12a Pôle de tête large-arrondi (vol. 2/1 Fig. 159: 11–18) . . . . **16. *G. truncatum*** vol. 2/1 p. 369
12b Pôle de tête étiré, +/- arrondi-pointu (vol. 2/1 Fig. 154: 21–22; aussi Fig. 156: 9) . . . . **1. *G. parvulum*** vol. 2/1 p. 358
(cf. aussi *G. augur*, Fig. 157: 3) . . . . **4. *G. gracile*** vol. 2/1 p. 361
13a Valves étirées au pôle de tête, s'étirant +/- longuement en pointe, arrondi-pointu ou avec tête **14**
13b Pôle de tête tronqué-cunéiforme ou arrondi-tronqué **25**
14a Pôles de tête ou les deux pôles avec tête étirée +/- étranglée **15**
14b Pôle de tête en forme de bec, s'étirant longuement en pointe, mais sans tête **17**
15a Seulement pôle de tête avec tête étirée, parfois aussi cunéiforme-arrondie (vol. 2/1 Fig. 162: 10–13) . . . . **15. *G. subtile*** vol. 2/1 p. 369
15b Aux deux pôles, étirement en forme de tête **16**
16a Aire centrale grande, élargie apicalement (vol. 2/1 Fig. 162: 14–18) . . . . **21. *G. helveticum*** vol. 2/1 p. 373
16b Aire centrale petite, non élargie apicalement (vol. 2/1 Fig. 157: 10) . . . . . ***G. sphaerophorum* sensu Mayer** vol. 2/1 p. 363 (voir sous **6. *G. augur***)
17a Valves longues et étroites, presque symétriques (naviculoïdes) avec bords +/- triondulés (vol. 2/1 Fig. 155: 22–24; aussi Fig. 156: 12–14) . . . . **3. *G. lagerheimii*** vol. 2/1 p. 361
. . . . **5. *G. hebridense*** vol. 2/1 p. 362
17b Valves non triondulées **18**
18a Largeur valvaire pas au-dessus de 4 µm (vol. 2/1 Fig. 164: 22–24) . . . . **20. *G. pseudotenellum*** vol. 2/1 p. 372
18b Largeur valvaire régulièrement au-dessus 4 µm **19**
19a Valves presque symétriques par rapport à l'axe transapical (naviculoïde), un ou les deux côtés arrondi-pointu jusqu'à très étiré-pointu, généralement formes +/- rhomboédriques-lancéolées (vol. 2/1 Fig. 156: 1–11; 154: 26–27). Si bord valvaire triondulé ou enflé au milieu, cf. *G. hebridense* et *G. lagerheimii* . . . . **4. *G. gracile*** vol. 2/1 p. 361
19b Forme valvaire autre **20**
20a Pôle de tête étiré jusqu'en forme de bec (semblable à *G. augur* var. *augur*), mais largeur plus grande vers le milieu ou soudainement fortement rétréci cunéiformément juste en-dessous de la pointe (vol. 2/1 Fig. 158: 1–6), voir aussi formes "*turris*" des variétés nominatives et autres taxons (vol. 2/1 Fig. 157: 7; 159: 4) . . . . **6. *G. augur*** var. ***turris*** vol. 2/1 p. 363
20b Pôle de tête pas nettement en forme de bec, tout au plus légèrement étiré **21**

21a Plus grande largeur au-dessus du milieu (vol. 2/1 Fig. 159: 1–3) (cf. aussi petites formes de *G. acutiusculum*, Fig. 162:1) . . . . **7. *G. pseudoaugur*** vol. 2/1 p. 364
21b Plus grande largeur au milieu ou valves presque linéaires **22**
22a Valves généralement +/- ovales-lancéolées, aire centrale à peine marquée, densité moyenne des stries dans une population généralement plus grande ou égale à 12/10 µm (vol. 2/1 Fig. 154: 1–25) . . . . **1. *G. parvulum*** vol. 2/1 p. 358
22b Valves avec autres combinaisons de caractéristiques **23**
23a Valves généralement linéaires-lancéolées, structure plus grossière, densité moyenne des stries dans une population inférieure à 12/10 µm, aire centrale unilatérale et plus marquée (vol. 2/1 Fig. 155: 1–21); cf. aussi *G. bohemicum* sensu Hustedt (vol. 2/1 Fig. 154: 29–32), si bord valvaire-triondulé cf. aussi *G. lagerheimii* (vol. 2/1 Fig. 155: 22–24) . . . . **2. *G. angustatum*** vol. 2/1 p. 360
23b Valves avec autres combinaisons de caractéristiques **24**
24a Valves lancéolées avec pôles arrondis-pointus (vol. 2/1 Fig. 156: 1–11; 154: 26–27) . . . . **4. *G. gracile*** vol. 2/1 p. 361
24b Valves en forme de cuisse avec partie de pied fortement rétrécie (vol. 2/1 Fig. 162: 1–3); cf. aussi formes de "*turris*" de *G. gracile*, Fig. 156: 5–10) . . . . **8. *G. acutiusculum*** vol. 2/1 p. 365
25a Aire centrale relativement large, unilatérale, élargie jusqu'au bord valvaire **26**
25b Aire centrale différente **27**
26a Stigma presqu'au centre des larges valves ovales en forme de cuisse (vol. 2/1 Fig. 162: 6, 7) (cf. aussi petites formes de *G. angustum*) . . . . **22. *G. tergestinum*** vol. 2/1 p. 373
26b Stigma plus éloigné du centre, valves étroites, linéaires en forme de cuisse (vol. 2/1 Fig. 164: 1–16); cf. aussi *G. angustatum* et *G. bohemicum* sensu Hustedt (vol. 2/1 Fig. 154: 29–32) . . . . **18. *G. angustum*** vol. 2/1 p. 370
27a Valves avec pôle de tête +/- cunéiforme **28**
27b Pôle de tête non cunéiforme **29**
28a Pôle de tête cunéiforme ou +/- coniquement pointu, partie la plus large généralement au-dessus du milieu valvaire. Stigma près des stries centrales courtes dans l'aire centrale (vol. 2/1 Fig. 160: 1–12); cf. aussi *G. augur* var. *turris* ainsi que les formes "*turris*" d'autres taxons) . . . . **9. *G. acuminatum*** vol. 2/1 p. 365
28b Partie de tête (dans les cas critiques) +/- cunéiforme, cependant plus largement arrondie à l'extrémité ou tuteurée, partie la plus large dans la partie centrale de la valve (vol. 2/1 Fig. 163: 1–12) . . . . **12. *G. clavatum*** vol. 2/1 p. 367
29a Plusieurs stries centrales alternativement courtes et longues (chez les formes plus petites, cette caractéristique peut disparaître) **30**
29b Stries différentes **31**
30a Valves s'étirant vers le pôle de tête remarquablement large-arrondi en se rétrécissant à peine (vol. 2/1 Fig. 159: 11–18) . . . . **16. *G. truncatum*** vol. 2/1 p. 369
30b Valves allant du pôle de tête au pôle de pied plus étroites, aire centrale ronde et agrandie par un nombre considérable de stries courtes, structure grossière (vol. 2/1 Fig. 162: 4, 5) . . . **23. *G. ventricosum*** vol. 2/1 p. 373
31a Stries en bordure de telle sorte que les aires axiale et centrale forment une face hyaline large (vol. 2/1 Fig. 164: 20, 21) . **19. *G. clevei*** vol. 2/1 p. 372
31b Stries moins fortement raccourcies **32**
32a Valves linéaires-lancéolées à elliptiques (sans tenir compte d'un renflement

central pour les valves plus grandes), aires axiale et centrale relativement éloignées. Les fissures internes du raphé s'écartent distalement et latéralement pendant que les fissures externes du raphé continuent leur course jusqu'aux pores centraux. Chez les petites formes, souvent à peine reconnaissable (vol. 2/1 Fig. 164: 1–16) . . . . **18. *G. angustum*** vol. 2/1 p. 370

32b Formes et structures combinées autrement **33**

33a Valves peu en forme de cuisse **34**

33b Valves plus en forme de cuisse, plus étirées vers le pôle du pied **36**

34a Stries pas nettement ponctuées en vues valvaire et connective (vol. 2/1 Fig. 161:9–11). Si présence de petites formes, cf. aussi *G. parvulum*, *G. angustatum*, *G. minutum* . . . . **13. *G. amoenum*** vol. 2/1 p. 368

34b Stries nettement ponctuées **35**

35a Aire centrale unilatérale, généralement petite (vol. 2/1 Fig. 161: 1–3) . . . . **10. *G. affine*** vol. 2/1 p. 366

35b Aire centrale plus grande, souvent développée des deux côtés (vol. 2/1 Fig. 161: 4, 5, 7, 8) . . . . **11. *G. insigne*** vol. 2/1 p. 366

36a Formes plus petites (longueur 10–35 µm), ponctuation des stries pratiquement pas reconnaissable (doubles points), en vue connective un point isolé devant chaque strie près du manteau valvaire (vol. 2/1 Fig. 159: 5–10) . . . . **17. *G. minutum*** vol. 2/1 p. 370

36b Généralement plus grandes formes, ponctuation sur la valve +/- reconnaissable, stries en vue connective allongées chacune par une ligne de points (vol. 2/1 Fig. 161: 8; 163: 12) **37**

37a Stries au manteau valvaire en vue connective allongées par quelques points plus grossiers, valves à tendance triondulaire (vol. 2/1 Fig. 163: 1–12) . . . . **12. *G. clavatum*** vol. 2/1 p. 367

37b Forme et/ou structures autres **38**

38a Raphé toujours nettement latéral, en vue connective fins points rapprochés en prolongement des stries des bords (vol. 2/1 Fig. 161: 4, 5, 7, 8; aussi Fig. 161: 1–3) . . . . **11. *G. insigne*** vol. 2/1 p. 366
. . . . **10. *G. affine*** vol. 2/1 p. 366

38b Raphé généralement faiblement latéral, il manque des points isolés en prolongement des stries au manteau valvaire en vue connective (vol. 2/1 Fig. 164: 1–16) . . . . **18. *G. angustum*** vol. 2/1 p. 370

39a Frustules en vue connective pliées, branches de raphé très raccourcies sur une des valves (convexes) (vol. 2/1 Fig. 91: 20–28) . . . . ***Rhoicosphenia abbreviata*** vol. 2/1 p. 381

39b Caractéristiques différentes **40**

40a Stries courtes, se situant en bordure, parfois prolongées transapicalement par des points uniques (vol. 2/1 Fig. 166: 1–11) . . . . **27. *G. grovei*** vol. 2/1 p. 377

40b Stries se prolongeant jusqu'au raphé **41**

41a Stries très délicates et rapprochées, > 25/10 µm (vol. 2/1 Fig. 165: 22–24) . . . . **26. *G. tackei*** vol. 2/1 p. 377

41b Stries < 25/10 µm **42**

42a Formes d'eaux douce et saumâtre, aire centrale +/- élargie transapicalement, *var. minutissimum* avec 4 stigma disposés en carré (vol. 2/1 Fig. 165: 1–18) . . . . **24. *G. olivaceum*** vol. 2/1 p. 374

42b Formes d'eau saline, sans aire centrale marquée ou aire centrale présente des deux côtés suite à la disparition des stries centrales, petites formes étroites-linéaires (vol. 2/1 Fig. 165: 19–21) . **25. *G. exiguum*** vol. 2/1 p. 376

17. Genre *Caloneis* Cleve 1894 (vol. 2/1 p. 382)

Typus generis: *Caloneis amphisbaena* (Bory) Cleve 1894; [*Navicula amphisbaena* Bory 1824] (Boyer 1927)

Clef pour les espèces

1a Grandes formes larges-lancéolées ou larges-elliptiques avec lignes longitudinales visibles dans la face valvaire **2**
1b Valves plus étroites, lignes longitudinales marginales, souvent non reconnaissables en vue valvaire **6**
2a Aires axiale et centrale formant une grande aire rhomboédrique ou lancéolée **3**
2b Aire centrale marquée d'un seul côté **4**
3a Extrémités arrondies-tronquées, aire centrale avec taches lunaires (vol. 2/1 Fig. 169: 5–7) . . . . **4. *C. obtusa*** vol. 2/1 p. 386
3b Extrémités arrondies, cunéiformes jusqu'à capitées, taches lunaires manquantes (vol. 2/1 Fig. 168: 4, 5) . . . **2. *C. amphisbaena*** vol. 2/1 p. 385
4a Valves lancéolées avec extrémités arrondies cunéiformes (vol. 2/1 Fig. 168: 1–3; 169: 4) . . . . **1. *C. permagna*** vol. 2/1 p. 384
4b Valves elliptiques-lancéolées, extrémités arrondies-tronquées
5a 12–14 stries/10 µm (vol. 2/1 Fig. 170: 1, 2) . . **5. *C. westii*** vol. 2/1 p. 386
5b 17–21 stries/10 µm (vol. 2/1 Fig. 169: 1–3) **3. *C. latiuscula*** vol. 2/1 p. 385
6a Stries > 30/10 µm **5**
6b Stries < 30/10 µm **8**
7a Rayures longitudinales non reconnaissables au microscope optique (vol. 2/1 Fig. 173: 6–8) . . . . **12. *C. hyalina*** vol. 2/1 p. 390
7b Rayures longitudinales submarginales, bien reconnaissables au microscope optique (vol. 2/1 Fig. 173: 22–24) . . . . . **9. *C. branderii*** vol. 2/1 p. 393
8a Taches lunaires dans l'aire centrale **9**
8b Aucune tache lunaire distincte dans l'aire centrale, tout au plus présence de structures tachetées très faiblement contrastantes **10**
9a Valves linéaires à linéaires-elliptiques, petite aire centrale (vol. 2/1 Fig. 170: 3–7) . . . . **6. *C. alpestris*** vol. 2/1 p. 386
9b Valves linéaires à lancéolées, bords souvent triondulés, grande aire centrale (vol. 2/1 Fig. 171: 1–11) . . . . **7. *C. schumanniana*** vol. 2/1 p. 387
10a Valves à côtés ondulés **11**
10b Valves à côtés différents **13**
11a Extrémités larges-capitées (vol. 2/1 Fig. 175: 1–6) . . . . **21. *C. undulata*** vol. 2/1 p. 394
11b Extrémités arrondies-tronquées ou étirées **12**
12a Formes plus grandes, plus larges au contour variable, extrémités arrondies-tronquées (vol. 2/1 Fig. 172: 1–13) . . . . . . **8. *C. silicula*** vol. 2/1 p. 388
12b Formes généralement plus petites, plus étroites, extrémités largement étirées (vol. 2/1 Fig. 174: 5–10) . . . . **17. *C. tenuis*** vol. 2/1 p. 392
13a Valves nettement renflées transapicalement au milieu **14**
13b Valves non renflées au milieu **15**
14a Extrémités arrondies-tronquées, aire centrale grande, ronde (vol. 2/1 Fig. 174: 1–4) . . . . **16. *C. pulchra*** vol. 2/1 p. 392
14b Extrémités cunéiformes pointues, aire centrale avec un large fascia (vol. 2/1 Fig. 174: 13–15) . . . . **24. *C. leptosoma*** vol. 2/1 p. 396
15a Côtés valvaires dans la partie centrale parallèles, faiblement concaves ou convexes **16**
15b Côtés valvaires relativement convexes (contour linéaire-lancéolé ou linéaire-elliptique) **20**

16a Stries nettement convergentes au milieu de la valve (vol. 2/1 Fig. 173: 1) . . . . **9.** ***C. clevei*** vol. 2/1 p. 389
16b Stries parallèles ou radiales au milieu de la valve **17**
17a Valves plus étroites que 6 µm **18**
17b Valves plus larges que 6 µm **19**
18a Stries radiales au milieu de la valve (vol. 2/1 Fig. 173: 21) . . . . **14.** ***C. lepidula*** vol. 2/1 p. 391
18b Stries parallèles au milieu de la valve (vol. 2/1 Fig. 174: 11, 12) . . . . **18.** ***C. sublinearis*** vol. 2/1 p. 393
19a Raphé fortement courbé, extrémités arrondies-cunéiformes (vol. 2/1 Fig. 173: 2, 3) . . . . **10.** ***C. lauta*** vol. 2/1 p. 389
19b Raphé faiblement courbé, extrémités largement arrondies (vol. 2/1 Fig. 173: 25, 26) . . . . **15.** ***C. thermalis*** vol. 2/1 p. 391
20a Contour linéaire-lancéolé **21**
20b Contour linéaire-elliptique **23**
21a Stries nettement radiales au milieu de la valve (vol. 2/1 Fig. 174: 16–21) . . . . **20.** ***C. molaris*** vol. 2/1 p. 394
21b Stries parallèles au milieu de la valve **22**
22a Extrémités arrondies-tronquées, raphé faiblement courbé (vol. 2/1 Fig. 173: 9–20) . . . . **13.** ***C. bacillum*** vol. 2/1 p. 390
22b Extrémités arrondies-cunéiformes, raphé fortement courbé (vol. 2/1 Fig. 173: 4) . . . . **11.** ***C. aerophila*** vol. 2/1 p. 389
23a Extrémités capitées à largement étirées et arrondies **24**
23b Extrémités tronquées à plates, cunéiformes et arrondies **25**
24a Fissures terminales grandes (vol. 2/1 Fig. 175: 10–13) . . . . **23.** ***C. budensis*** vol. 2/1 p. 395
24b Fissures terminales petites (vol. 2/1 Fig. 175: 7–9) . . . . **22.** ***C. macedonica*** vol. 2/1 p. 395
25a Extrémités largement arrondies (vol. 2/1 Fig. 172: 1–13) . . . . **8.** ***C. silicula*** vol. 2/1 p. 388
25b Extrémités plates, cunéiformes (vol. 2/1 Fig. 174: 16–21) . . . . **20.** ***C. molaris*** vol. 2/1 p. 394

18. Genre *Pinnularia* Ehrenberg 1843 nom. cons. (vol. 2/1 p. 397)

Typus generis: *Pinnularia viridis* (Nitzsch) Ehrenberg 1843
(ty vol. 2/1 p. cons.); [*Bacillaria viridis* Nitzsch 1817]

Clefs de détermination pour les espèces artificielles

1a Stries larges et espacées, ne se touchant pas . . . . **1. Groupe autour de *P. borealis*** (vol. 2/1 p. 398) (p. 277)
1b Stries et côtes transapicales plus étroites **2**
2a Stries très divergentes au milieu et aux extrémités valvaires, changeant soudainement de direction **3**
2b Stries relativement divergentes ou non **4**
3a Valves larges-lancéolées ou rhomboédriques-lancéolées . . . . **3. Groupe autour de *P. acoricola*** (vol. 2/1 p. 399) (p. 277)
3b Valves différentes . . . . **4. Groupe autour de *P. divergentissima*** (vol. 2/1 p. 400) (p. 278)
4a Valves linéaires avec aire plus large . . . . **2. Groupe autour de *P. brevistriata*** (vol. 2/1 p. 399) (p. 277)
4b Valves avec aire relativement large ou étroite **5**

5a Raphé fortement latéral, grandes formes . . . . **6. Groupe autour de *P. viridis*** (vol. 2/1 p. 402) (p. 280)
5b Raphé moins latéral, petites à grandes formes . . . . **5. Groupe autour de *P. microstauron*** (vol. 2/1 p. 400) (p. 278)

1. Groupe autour de *P. borealis* (Distantes) (vol. 2/1 p. 398)
1a Valves elliptiques-lancéolées, grandes formes (vol. 2/1 Fig. 176: 1, 2) . . . . **1. *P. alpina*** vol. 2/1 p. 403
1b Valves linéaires à linéaires-elliptiques, grandes et petites formes **2**
2a Valves plus courtes que 15 µm, pores centraux éloignés (vol. 2/1 Fig. 176: 11, 12) . . . . ***4. *P. balfouriana*** vol. 2/1 p. 404
2b Valves plus longues que 16 µm **3**
3a Formes plus petites avec 7–10 stries/10 µm **4**
3b Généralement formes plus grandes avec 6 ou moins de stries/10 µm . . **5**
4a Stries parallèles au milieu, souvent seulement aux bords, devenant marginalement plus larges (vol. 2/1 Fig. 176: 8–10) . . . . **3. *P. lagerstedtii*** vol. 2/1 p. 404
4b Stries radiales au milieu, devenant plus larges à l'aire, aire axiale étroite (vol. 2/1 Fig. 178: 1–6) . . . . ***6. *P. intermedia*** vol. 2/1 p. 406
5a Extrémités proximales du raphé recourbées, mais pas revers-latérales (vol. 2/1 Fig. 177: 1–12; 178: 7) . . . . ***5. *P. borealis*** vol. 2/1 p. 405
5b Extrémités proximales du raphé légèrement revers-latérales, fissure externe du raphé se dirige d'abord vers la médiane et se prolonge ensuite dans celle-ci (vol. 2/1 Fig. 176: 3–7) . . . . **2. *P. lata*** vol. 2/1 p. 403

2. Groupe autour de *P. brevistriata* (Brevistriatae) (vol. 2/1 p. 399)
1a Surface de l'aire axiale ornementée (particulièrement visible lors de l'observation dans le contraste différentiel-interférence ou en contraste de phase) **2**
1b Aire axiale non ornementée **4**
2a Raphé accompagné d'une côte axiale nette, bords triondulés (vol. 2/1 Fig. 181: 4–10) . . . . ***14. *P. nodosa* vol. 2/1 p. 409**
2b Côtes axiales manquantes, valves non triondulées, tout au plus renflées au milieu **3**
3a Valves plus larges que 8 µm, extrémités larges, en forme de tête, ornementation très visible (vol. 2/1 Fig. 181: 1–3) . . . . ***13. *P. acrosphaeria*** vol. 2/1 p. 409
3b Valves plus étroites que 6 µm, extrémités pas ou peu étirées, ornementation peu visible (vol. 2/1 Fig. 188: 4–8) . . . . ***40. *P. schwabei*** vol. 2/1 p. 423
4a Extrémités largement arrondies (vol. 2/1 Fig. 182: 1–3) . . . . ***16. *P. brevicostata*** vol. 2/1 p. 410
4b Extrémités cunéiformes arrondies **5**
5a 8–10 stries/10 µm, large bande longitudinale sur les stries, stries parallèles au milieu (vol. 2/1 Fig. 182: 1–3) . . . . ***15. *P. hemiptera*** vol. 2/1 p. 410
5b 12–15 stries/10 µm, étroite bande longitudinale sur les stries, stries convergentes au milieu (vol. 2/1 Fig. 175: 14–18) . **28. *P. schroederi*** vol. 2/1 p. 418

3. Groupe autour de *P. acoricola* (vol. 2/1 p. 399)
1a Formes délicatement rayées, 13–16 stries/10 µm (vol. 2/1 Fig. 183: 8–12) . . . . **18. *P. acoricola*** vol. 2/1 p. 411
1b Formes grossièrement rayées, < 13 stries/10 µm **2**
2a Largeur valvaire 9 µm et plus (vol. 2/1 Fig. 183: 1–3) . . . . **17. *P. suchlandtii*** vol. 2/1 p. 411

2b Largeur valvaire 8 µm et moins (vol. 2/1 Fig. 183: 4–7) . . . . **19. *P. cuneola*** vol. 2/1 p. 412

4. Groupe autour de *P. divergentissima* (vol. 2/1 p. 400)

1a Formes linéaires plus grandes aux extrémités arrondies tronquées, stries larges et épaisses, 7–10/10 µm (vol. 2/1 Fig. 186: 6–8) . . . . **34. *P. superdivergentissima*** vol. 2/1 p. 420
1b Formes plus petites et plus grandes avec stries plus fines, 9 et plus/10 µm **2**
2a Extrémités non marquées, largement arrondies **3**
2b Extrémités étirées ou capitées **5**
3a Extrémités plates courtes (vol. 2/1 Fig. 185: 18, 19) . . . . ***31. *P. subrostrata*** part. vol. 2/1 p. 419
3b Extrémités arrondies-tronquées ou cunéiformes **4**
4a Contour large, linéaire-elliptique (vol. 2/1 Fig. 185: 20–23) . . . . ***32. *P. obscura*** vol. 2/1 p. 420
4b Contour étroit, rhomboédrique-lancéolé (vol. 2/1 Fig. 185: 1, 2) . . . . ***29. *P. similis*** vol. 2/1 p. 418
5a 9–12 stries/10 µm, divergence des groupes de rayures relativement grande (vol. 2/1 Fig. 185: 11–17) . . . . ***31. *P. subrostrata*** vol. 2/1 p. 419
5b 12–14 stries/10 µm, divergence des groupes de rayures grande (vol. 2/1 Fig. 185: 3–10) . . . . ***30. *P. divergentissima*** vol. 2/1 p. 419

5. Groupe autour de *P. microstauron* (vol. 2/1 p. 400)

1a Aire centrale avec des marques distinctes **2**
1b Aire centrale sans marques, parfois présence de points ou autres structures faiblement visibles **4**
2a Marques situées au niveau du nodule central **3**
2b Marques situées au bord du manteau (vol. 2/1 Fig. 179: 3–8) . . . . ***10. *P. divergens*** vol. 2/1 p. 407
3a Marques en forme de petites taches lunaires (vol. 2/1 Fig. 178: 8–10; 179: 1) . . . . ***7. *P. stomatophora*** vol. 2/1 p. 406
3b Marques en forme de grandes taches lunaires constituées de stries articulées (vol. 2/1 Fig. 178: 11–14) . . . . . . ***8. *P. brandelii*** vol. 2/1 p. 407
4a Petites formes, généralement avec plus de 15 stries/10 µm **5**
4b Formes plus grandes avec moins de 15 stries/10 µm **8**
5a Extrémités largement arrondies avec formation visible de petites têtes **32**
5b Extrémités seulement étirées ou avec des petites têtes indistinctes . . . **6**
6a Stries faiblement radiales au milieu de la valve (vol. 2/1 Fig. 185: 24, 25) . . . . **33. *P. lapponica*** vol. 2/1 p. 420
6b Stries distinctement radiales au milieu de la valve **7**
7a Aire axiale s'élargissant continuellement de l'extrémité au centre (vol. 2/1 Fig. 185: 26) . . . . **27. *P. kneuckeri*** vol. 2/1 p. 418
7b Aire axiale généralement étroite aux extrémités valvaires et s'élargissant de façon lancéolée au milieu (vol. 2/1 Fig. 193: 19–29) . . . . ***45. *P. appendiculata*** vol. 2/1 p. 427
8a Les deux fissures terminales construites différemment, au bord de l'aire axiale se trouve une rangée de points se développant apicalement (vol. 2/1 Fig. 188: 1–3) . . . . **37. *P. platycephala*** vol. 2/1 p. 422
8b Les deux fissures terminales identiques **9**
9a Côtés valvaires parallèles, extrémités pointues, cunéiformes (vol. 2/1 Fig. 188: 13) . . . . **39. *P. balatonis*** vol. 2/1 p. 422
9b Côtés valvaires différents, extrémités formées autrement **10**
10a Côtés valvaires triondulés **11**
10b Côtés valvaires différents **20**

11a Ondulation du milieu des côtés valvaires plus large que les autres . . . **12**
11b Ondulation du milieu des côtés valvaires aussi large ou plus étroite que les autres **14**
12a Ondulation du milieu beaucoup plus large que les autres (vol. 2/1 Fig. 184: 5) . . . . ***26. *P. polyonca*** vol. 2/1 p. 417
12b Ondulation du milieu moins large que les autres **13**
13a Aire axiale relativement large (vol. 2/1 Fig. 184: 2) . . . . ***21. *P. legumen*** part.vol. 2/1 p. 413
13b Aire axiale étroite (vol. 2/1 Fig. 190: 1) . . . . ***42. *P. interrupta*** part. vol. 2/1 p. 424
14a Ondulation du milieu d'égale largeur que les autres **15**
14b Ondulation du milieu plus étroite que les autres **18**
15a Bandes longitudinales dans la face valvaire formées par des ouvertures alvéolaires internes **16**
15b Valves sans bandes longitudinales, parfois simulées par des angles de flexion dans la face valvaire **17**
16a Bandes longitudinales larges sur les stries (vol. 2/1 Fig. 179: 2) . . . . **9. *P. esox*** vol. 2/1 p. 407
16b Bandes longitudinales étroites, marginales sur les stries (vol. 2/1 Fig. 184: 22. 7–10) . . . . **22. *P. pulchra*** part. vol. 2/1 p. 414
17a Aire axiale relativement large (vol. 2/1 Fig. 184: 3) . . . . ***21. *P. legumen*** part.vol. 2/1 p.413
17b Aire axiale généralement étroite (vol. 2/1 Fig. 190: 2, 3, 5) . . . . ***42. *P. interrupta*** part. vol. 2/1 p. 424
18a Extrémités arrondies cunéiformes (vol. 2/1 Fig. 183: 17) . . . . **20. *P. infirma*** part.vol. 2/1 p. 412
18b Extrémités étirées et arrondies-tronquées **19**
19a Valves larges de 6–11 µm (vol. 2/1 Fig. 184: 6) . . . . **22. *P. pulchra*** part.vol. 2/1 p. 414
19b Valves larges de 15–23 µm (vol. 2/1 Fig. 184: 1) . . . . ***21. *P. legumen*** part. vol. 2/1 p. 413
20a Valves concaves au milieu **21**
20b Côtés valvaires parallèles ou convexes **23**
21a Extrémités arrondies, cunéiformes (vol. 2/1 Fig. 183: 14–16) . . . . **20. *P. infirma*** part. vol. 2/1 p. 412
21b Extrémités largement arrondies, capitées ou étirées **22**
22a Aire très large, lancéolée (vol. 2/1 Fig. 187: 4) . . . . **24. *P. braunii*** part. vol. 2/1 p. 416
22b Aire plus étroite (vol. 2/1 Fig. 190: 6, 10) . . . . ***42. *P. interrupta*** part. vol. 2/1 p. 424
. . . . (voir aussi Fig. 206: 3, *P. lundii* vol. 2/1 p. 415)
23a Extrémités arrondies-tronquées, stries du milieu légèrement radiales (vol. 2/1 Fig. 186: 9, 10) (cf. aussi 43. *P. microstauron* var. *brebissonii*) . . . . ***36. *P. rupestris*** vol. 2/1 p. 421
23b Extrémités légèrement marquées **24**
24a Extrémités rétrécies de façon tronquée, cunéiforme (vol. 2/1 Fig. 186: 4, 5) . . . . ***35. *P. sudetica*** vol. 2/1 p. 421
24b Extrémités bien marquées et étirées ou capitées **25**
25a Valves larges linéaires **26**
25b Valves plus étroites linéaires **28**
26a Bande longitudinale nette dans face valvaire, aire centrale généralement rhomboédrique, rarement un fascia, extrémités très peu marquées (vol. 2/1 Fig. 188: 9–12) . . . . **38. *P. karelica*** vol. 2/1 p. 422
26b Bandes longitudinales manquantes, aire centrale en forme de fascia, ex-

trémités toujours largement marquées (exception: 43. *P. microstauron* var. *brebissonii*, Fig. 191:7–9 avec extrémités arrondies à cunéiformes-arrondies, non marquées) **27**
27a Fissures terminales en forme de virgule **33**
27b Fissures terminales en forme de point d'interrogation (vol. 2/1 Fig. 191: 2–7; 192: 11–16) . . . . ***43. *P. microstauron*** part. vol. 2/1 p. 425
28a Aire centrale très large (vol. 2/1 Fig. 187: 1–3, 5) . . . . **24. *P. braunii*** vol. 2/1 p. 416
28b Aire centrale plus étroite **29**
29a Valves pas plus larges que 7 µm (vol. 2/1 Fig. 193: 1–18) . . . . ***44. *P. subcapitata*** vol. 2/1 p. 426
29b Valves plus larges **30**
30a Extrémités larges étirées (vol. 2/1 Fig. 191: 1, 9 et autres) . . . . ***43. *P. microstauron*** part. vol. 2/1 p. 425
30b Extrémités étroites étirées ou capitées **31**
31a Extrémités étroites étirées ou capitées, presqu'aussi large que le reste de la valve, aire axiale généralement large (vol. 2/1 Fig. 189: 1–9; 186: 1–3) . . . . ***41. *P. gibba*** vol. 2/1 p. 423
31b Extrémités étirées ou capitées, dans les deux cas relativement plus étroites que le reste de la valve, aire plus étroite (vol. 2/1 Fig. 190: 1–11) . . . . ***42. *P. interrupta*** vol. 2/1 p. 424
32a Contour elliptique, petites têtes relativement plus étroites que le milieu de la valve (vol. 2/1 Fig. 187: 9; 206: 8, 9, 11) . **25. *P. krookii*** vol. 2/1 p. 416
32b Contour linéaire, petites têtes seulement légèrement plus étroites que le milieu de la valve (vol. 2/1 Fig. 187: 6–8, 9'; 206: 12–19) . . . . **25a. *P. ignobilis*** vol. 2/1 p. 417
33a Petites têtes plus étroites que le milieu de la valve, côtés parallèles, concaves ou convexes (vol. 2/1 Fig. 187: 10–16; 206: 1–3) . . . . **23. *P. lundii*** vol. 2/1 p. 415
33b Petites têtes presqu'aussi large que le milieu de la valve, côtés renflés au milieu (vol. 2/1 Fig. 206: 4–7) . . . . . . **23a. *P. globiceps*** vol. 2/1 p. 415

Groupe autour de *P. viridis* (vol. 2/1 p. 402)
1a Extrémités se rétrécissant prématurément et se prolongeant largement, cunéiformément, milieu valvaire pas ou peu renflé **2**
1b Valves à peine rétrécies aux extrémités, généralement fortement renflées au milieu, extrémités arrondies tronquées à quelque peu élargies capitées **3**
2a Aire centrale généralement relativement petite, elliptique (vol. 2/1 Fig. 194: 1–4; 195: 1–6) . . . . ***46. *P. viridis*** vol. 2/1 p. 428
2b Aire centrale composé d'un demi fascia, large de 4–5 stries (vol. 2/1 Fig. 195: 7) . . . . **47. *P. semicruciata*** vol. 2/1 p. 438
3a Fissure externe du raphé peu ou pas ondulée **4**
3b Fissure externe du raphé fortement ondulée (raphé complexe) **8**
4a Valves sans bandes longitudinales **5**
4b Valves avec bandes longitudinales **6**
5a Stries fortement convergentes aux extrémités (vol. 2/1 Fig. 180: 1, 2) . . . . **11. *P. episcopalis*** vol. 2/1 p. 408
5b Stries peu convergentes ou parallèles aux extrémités (vol. 2/1 Fig. 180: 3–5) . . . . **12. *P. cardinaliculus*** vol. 2/1 p.408
6a Stries entièrement parallèles jusqu'à faiblement radiales, 7–10/10 µm (vol. 2/1 Fig. 197: 1, 2) . . . . ***49. *P. macilenta*** vol. 2/1 p. 429
6b Stries radiales au milieu de la valve, < 7/10 µm **7**
7a 4–5 stries/10 µm, bande longitudinale large (vol. 2/1 Fig. 198: 1) . . . . ***51. *P. dactylus*** vol. 2/1 p. 430

7b 5–7 stries/10 µm, bande longitudinale étroite (vol. 2/1 Fig. 196: 1–4) . . . . ***48. *P. maior*** vol. 2/1 p. 429
8a 6 stries et plus/10 µm **9**
8b Généralement, moins de 6 stries/10 µm **10**
9a Valves renflées au centre et aux extrémités (vol. 2/1 Fig. 198: 4) . . . . ***53. *P. gentilis*** vol. 2/1 p. 431
9b Valves non renflées au centre et aux extrémités (vol. 2/1 Fig. 197: 3, 4) . . . . **50. *P. aestuarii*** vol. 2/1 p. 430
10a Valves renflées au centre et aux extrémités (vol. 2/1 Fig. 198: 2, 3) . . . . ***52. *P. nobilis*** vol. 2/1 p. 430
10b Valves linéaires, centre et extrémités non élargis **11**
11a Fissure externe du raphé fortement ondulée, aire centrale peu développée (vol. 2/1 Fig. 199: 1–3) . . . . ***54. *P. streptoraphe*** vol. 2/1 p. 431
11b Fissure externe du raphé un peu moins ondulée, aire centrale formant un large fascia (vol. 2/1 Fig. 199: 4, 5) . . . . **55. *P. cardinalis*** vol. 2/1 p. 432

19. Genre *Mastogloia* Thwaites in W. Smith 1856 (vol. 2/1 p. 432)

Typus generis: *Mastogloia elliptica* var. *dansei* (Thwaites) Cleve; [*Dickieia dansei* Thwaites 1848]

Clef pour les espèces

1a Stries à doubles points (vol. 2/1 Fig. 202: 3–5) . . . . **6. *M. grevillei*** vol. 2/1 p. 435
1b Stries à simples rangées de points **2**
2a Aires latérales et centrale forment un dessin en forme de H **3**
2b Aires latérales manquantes **4**
3a Valves larges, 14–28 µm, raphé ondulé (vol. 2/1 Fig. 200: 1–5) . . . . **1. *M. braunii*** vol. 2/1 p. 433
3b Valves plus étroites, 6–10 µm, raphé filiforme (vol. 2/1 Fig. 200: 6, 7) . . . . **2. *M. pumila*** vol. 2/1 p. 433
4a Aire axiale accompagnée de fortes côtes siliceuses (vol. 2/1 Fig. 200: 8–10) . . . . **3. *M. baltica*** vol. 2/1 p. 433
4b Côtes siliceuses manquantes **5**
5a Raphé filiforme, tout au plus légèrement ondulé au milieu des branches de raphé (vol. 2/1 Fig. 201: 1–9) . . . . **4. *M. smithii*** vol. 2/1 p. 434
5b Fissure externe du raphé fortement sortie au milieu (vol. 2/1 Fig. 201: 10–14; 202: 1, 2) . . . . **5. *M. elliptica*** vol. 2/1 p. 434

20. Genre *Diatomella* Greville 1855 nom. cons. (vol. 2/1 p. 436)

Typus generis: *Diatomella balfouriana* Greville 1855
Une seule espèce, *Diatomella balfouriana*

21. Genre *Oestrupia* Heiden ex Hustedt 1935 (vol. 2/1 p. 436)

Typus generis: *Oestrupia powellii* (Lewis 1861) Heiden ex Hustedt (Patrick & Reimer 1966)

Clef pour les espèces

1a Valves linéaires à linéaires-elliptiques, côtés légèrement convexes (vol. 2/1 Fig. 202: 6–8) . . . . **1. *Oe. zachariasii*** vol. 2/1 p. 437
1b Valves linéaires avec côtés triondulés (vol. 2/1 Fig. 202: 9–12) . . . . **2. *Oe. bicontracta*** vol. 2/1 p. 437

22. Genre *Entomoneis* Ehrenberg 1845 (vol. 2/1 p. 438)

Typus generis: *Entomoneis alata* (Ehrenberg) Ehrenberg 1845; [*Navicula alata* Ehrenberg 1840]

Clef pour les espèces

1a Carène avec côtes solides (vol. 2/1 Fig. 203: 5; 204: 1) . . . . **2. *E. costata*** vol. 2/1 p. 439
1b Carène sans côtes **2**
2a Ligne de séparation très fortement ondulée (vol. 2/1 Fig. 205: 1–3) . . . . **4. *E. ornata*** vol. 2/1 p. 440
2b Ligne de séparation moins ondulée **3**
3a Carène particulièrement bien ponctuée à la ligne de séparation, sur la valve 15–17 stries/10 µm (vol. 2/1 Fig. 203: 1–4) . . **1. *E. alata*** vol. 2/1 p. 438
3b Carène pas ponctuée, stries très délicates sur la valve, +/- 24/10 µm, les valves sont généralement très délicatement siliceuses (vol. 2/1 Fig. 204: 2–4; 205: 9) . . . . **3. *E. paludosa*** vol. 2/1 p. 439

4. Famille Bacillariaceae Ehrenberg 1840 (vol. 2/2 p. 6)

Typus familiae: *Bacillaria* Gmelin 1791

Clef pour les genres:

1a Fibules absentes. Le canal raphéen repose en position distale sur une aile et est relié avec l'intérieur de la valve par des canaux alaires. Genre monotypique (vol. 2/2 Fig. 84: 13–19) . . . . **6. *Simonsenia*** (vol. 2/2 p. 135) (one species, vol. 212 p. 135)
1b Fibules toujours présentes, ailes et canaux alaires absents **2**
2a Valves 2 à 3 fois tordues autour de l'axe apical d'où un canal raphéen en hélice ((vol. 2/2 Fig. 87: 3) . . . . **5. *Cylindrotheca*** (vol. 2/2 p. 134) (one species, vol. 2/2 p. 134)
2b Caractéristique différente **3**
3a Axe apical hétéropolaire, semblable au *Gomphonema* (vol. 2/2 Fig. 92: 12–14, (sans diagnostic) . . . . ***Gomphonitzschia*** (vol. 2/2 p. 133)
3b Axe apical isopolaire **4**
4a Axe transapical des valves régulièrement hétéropolaire, donc valves dorsiventrales **5**
4b Axe transapical des valves régulièrement isopolaire ou très faiblement hétéropolaire (vol. 2/2 p. e. par une carène rentrante en son centre) . . **6**
5a Valves toujours petites, contour ressemblant au genre *Cymbella* (vol. 2/2 Fig. 92: 10, 11) . . . . **4. *Cymbellonitzschia*** (vol. 2/2 p. 133) (one species, vol. 2/2 p. 133)
5b Valves à contour différent, extrémités toujours rétrécies +/- cunéiformes et variables. Carène et canal raphéen toujours sur un côté des frustules . . . . **3. *Hantzschia*** (vol. 2/2 p. 126) (p. 302)

6a Raphé sur toutes les valves d'un population, toujours sur un côté d'un frustule . . . . **3.** ***Hantzschia*** (vol. 2/2 p. 126) (p. 302)
6b Raphé chez au moins la moitié de la population sur des côtés différents d'un frustule (ou toujours en position centrale) **7**
7a Frustules réunis en bandes tabloïdes; les individus peuvent glisser les uns le long des autres avec leurs carènes centrales. Cette formation peut donc s'associer et se dissocier variablement. Individus isolés avec caractéristiques selon. Vol. 2/2 Fig. 87; 4–7. . . . . **1.** ***Bacillaria*** (vol. 2/2 p. 8) (p. 283)
7b Caractéristiques différentes . . . . . . **2.** ***Nitzschia*** (vol. 2/2 p. 8) (p. 283)

1. Genre ***Bacillaria*** Gmelin 1781 (vol. 2/2 p. 8)

Typus generis: *Bacillaria paradoxa* Gmelin 1791
Une seule espèce, *Bacillaria paradoxa.*

2. Genre ***Nitzschia*** Hassall 1845 nom. cons. (vol. 2/2 p. 8)

Typus generis: *Bacillaria sigmoidea* Nitzsch 1817; (*Nitzschia sigmoidea* (Nitzsch) W. Smith 1853; *Nitzschia elongata* Hassal nom. illeg.)

Clef de détermination des sections (vol. 2/2 p. 9)

1a Carène raphéenne doublée d'un conopeum à l'extérieur de la valve, reconnaissable directement (microscope électronique, Fig. 1: 1–5) ou à une ligne parallèle (microscope optique, Fig. 6: 5 et 8: 6). Raphé jamais avec nodule central, en position centrale ou légèrement excentrée. Valves linéaires ou convexes au milieu, jamais concaves, le plus souvent courbées de façon sigmoïde: **(Sous-groupe *Nitzschia*)** **13**
1b Conopeum non reconnaissable à 100% **2**
2a Frustules +/- sigmoïdes en vues connective et/ou valvaire **14**
2b Frustules non sigmoïdes en vues connective et/ou valvaire **3**
3a Les côtes transapicales s'avancent loin dans la surface valvaire; ce sont les fibules allongées ou des éléments indépendants à côté des fibules courtes **10**
3b Pas de telles structures reconnaissables **4**
4a Valves à extrémités longues, étirées, rostrées . . . . **"Nitzschiellae"** (vol. 2/2 p. 121) (p. 301) . . . . **"Lanceolatae** part." (vol. 2/2 p. 76) (p. 292)
4b Extrémités toujours plus faiblement étirées et rostrées **5**
5a Valves avec des plis longitudinaux +/- marqués. Ponctuation des stries relativement grossière ou stries (côtes transapicales) correspondant en nombre aux fibules. Points parfois en rangées doubles. Espèces selon Fig. 35: 7 à 38: 11. . . . . **"Tryblionellae"** part. (vol. 2/2 p. 35)
5b Caractéristiques différentes **6**
6a Valves avec pli longitudinal +/- marqué provoquant soit une perturbation optique des stries ou apparaissant parfois comme une aire hyaline interrompant les stries. Stries proprement dites le plus souvent très rapprochées, délicates, difficiles à résoudre et doublées de côtes de soutien. Raphé toujours avec nodule central (vol. 2/2 Fig. 28: 1 à 35: 6 et 38: 12 à 39: 9) . . . . . . **"Tryblionellae"** part.,"**Apiculatae**","**Panduriformes** (vol. . . . . 2/2 p. 35) (p. 287)
6b Valves sans pli longitudinal marqué, en tout cas pas comme en 6a . . . **7**
7a Frustules en vue connective très larges à cause de nombreuses bandes. En état préparé, les valves se trouvent également dans cette position ; aussi la

carène raphéenne peu ou moyennement excentrée apparaît-elle comme marginale. Les valves dans cette position sont +/- en forme de 'parenthèse' ou de 'canoé'. Le bord avec carène raphéenne est plus renfoncé au milieu que le bord sans carène. Nodule central toujours présent . . . . **"Bilobatae", "Dubiae" part** (vol. 2/2 p. 53) (p. 289)

7b Caractéristiques différentes ou non reconnaissables comme tels à 100% **8**

8a Frustules et valves ayant les caractéristiques des taxons de Fig. 55: 1 à 58:15. Ici comme groupe-reliquat des "**Lineares**" historiques et d'autres formes encore difficilement définissables jusqu'à présent . . . . **"Lineares"** (vol. 2/2 p. 69)

8b Caractéristiques différentes **9**

9a Carène raphéenne en position valvaire peu ou moyennement excentrée. Jamais de nodule central. Fig. 11:1 à 14:17A . . . . **"Dissipatae"** (vol. 2/2 p. 18) (p. 285)

9b Carène raphéenne très excentrée, située au bord entre face et manteau valvaires . . . . **"Lanceolatae"** (vol. 2/2 p. 76) (p. 292) . . . . **"Dubiae"** part. (vol. 2/2 p. 53) (p. 289)

10a Raphé sans nodule central. Les côtes qui ne vont pas d'un bord à l'autre de la valve sont des prolongements des fibules. Stries relativement espacées et grossièrement ponctuées **11**

10b Raphé avec nodule central. Stries très rapprochées **12**

11a Valves très longues (vol. 2/2 Fig. 25: 1–4); cf. aussi (vol. 2/2 Fig. 26: 1–6) . . . . **"Insignes""Scalares"** (vol. 2/2 p. 34)

11b Valves beaucoup plus courtes (vol. 2/2 Fig. 39: 10 à 40: 8); cf. aussi *Denticula kuetzingii* . . . . **"Grunowia"** (vol. 2/2 p. 52)

12a Côtes traversant au moins en partie la face valvaire d'un bord à l'autre; ce sont des éléments qui se développent à côté des fibules. Stries très rapprochées >= 40/10 µm (vol. 2/2 Fig. 40: 9–17) . . . . **"Epithemiodeae"** (vol. 2/2 p. 50)

12b S'il y a des fibules, elles sont bien +/- prolongées en côtes mais sans s'avancer loin dans la face valvaire. Stries max. 30/10 µm. Cf. *N. bremensis, N. palustris, N. homburgiensis, Hantzschia spp.*

13a Frustules +/- courbés en vue connective ; plus rarement valves en cette position définie . . . . **"Sigmoideae"** (vol. 2/2 p. 11) (p. 285)

13b Frustules non régulièrement sigmoïdes ni en vue connective ni en vue valvaire (vol. 2/2 Fig. 11: 1 à 15: 17a). Cf. aussi "**Spathulatae**" qui rassemble des formes marines très proches des Dissipatae (vol. 2/2 Fig. 14: 6) . . . . **"Dissipatae"** (vol. 2/2 p. 18) (p. 285)

14a Raphé avec nodule central **15**

14b Raphé continu, pas de nodule central **16**

15a Formes correspondant à Fig. 70: 25–28 (cf. aussi fausses "Sigmata" Fig. 24: 3–7 et *N. lorenziana* Fig. 86: 6–10) **"Sigmata"** part. (vol. 2/2 p. 32) (p. 287)

15b Caractéristiques différentes, courbure sigmoïde en vue valvaire parfois seulement visible aux pôles . . . . . . **"Obtusae"** (vol. 2/2 p. 23) (p. 286)

16a Frustules nettement sigmoïdes en vue connective, extrémités +/- en forme de canoë; en vue valvaire non ou légèrement sigmoïdes seulement dans une certaine position. Carène raphéenne centrale à moyennement excentrée **17**

16b Valves en vue connective aux extrémités à peine en forme de canoë, en vue valvaire plus fortement sigmoïdes (vol. 2/2 Fig. 23: 1–9). Carène raphéenne plus excentrée, cf. aussi "**Insignes**" (vol. 2/2 Fig. 26: 1–6) et quelques "**Lanceolatae**" qui peuvent sporadiquement avoir des formes aussi régulièrement courbées (vol. 2/2 Fig. 70: 8, 28 ou 83: 2) . . . . **Sigmata"** part. (vol. 2/2 p. 32) (p. 287)

17a Stries très espacées, < 15/10 µm, grossièrement ponctuées. Les fibules se prolongent en partie nettement à l'intérieur de la face valvaire . . . . *"Hantzschia spectabilis"* vol. 2/2 p. 132

17b Stries beaucoup plus rapprochées > 20 et jusqu'à plus de 40/10 µm (vol. 2/2 Fig. 4: 1 à 10: 9) . . . . **"Sigmoideae"** (vol. 2/2 p. 11) (p. 285)

Sigmoideae (vol. 2/2 p. 12)

1a Formes exclusivement marines ou d'eau salée continentale (seules deux espèces ici, mais cf. aussi les vues valvaires plus courbées des formes autour de *N. sigma*, des autres **"Insignes"** et *Hantzschia spectabilis*) **8**

1b Formes prédominantes ou ne vivant qu'en eaux douces **2**

2a Stries +/- 25/10 µm **7**

2b Stries >= 30/10 µm **3**

3a Fibules larges. Espaces interfibulaires relativement plus étroits ou pas nettement plus larges. Stries +/- 30/10 µm (vol. 2/2 Fig. 4: 3; 6: 1–6) L.: 140–250 µm/l. 8–8,5 µm/F./ 5,5–6,5/10 µm/str. 30–31,5/10 µm . . . . **3.** ***N. wuellerstorffii*** vol. 2/2 p. 13

3b Espaces interfibulaires toujours plus larges que les fibules apparaissant +/- minces **4**

4a Stries >= 40/10 µm, pratiquement non-résolvables sans effet optique particulier **5**

4b Stries +/- 30/10 µm (vol. 2/2 Fig. 4: 4, 5; 7: 1–7) L.: 72–250 µm/l.: 3,5–7 µm/F. 5–12/10 µm/str. 30–40/µm . . . . **4.** ***N. vermicularis*** s. str. vol. 2/2 p. 14

5a Frustules fortement sigmoïdes en vue connective (vol. 2/2 Fig. 4: 6; 9: 1–4) L.: 60–160 µm/l.: 4–6 µm/F.: 7–12/10 µm/ str. n. r. . **6.** ***N. flexa*** vol. 2/2 p. 16

5b Frustules plus faiblement courbées **6**

6a Longueur max. 100 µm dans des cas extrêmes, largeur 3–3,5 µm, fibules > 13/10 µm (vol. 2/2 Fig. 10: 1–5) L.: 40–103 µm/l.: 3–3,5 µm/F.(13) 14–16(18)/10 µm/str. n. r. . . . . **7.** ***N. flexoides*** vol. 2/2 p. 17

6b Valves généralement plus longues, plus larges et en particulier fibules plus espacées, < 13/10 µm (vol. 2/2 Fig. 8: 1–8a) L.: 90–150 µm/F.: 5–9/10 µm/ str. 35–36(40)/10 µm . . . . **5.** ***N. acula*** vol. 2/2 p. 16 . . . . **4.** ***N. vermicularis*** s. lato vol. 2/2 p. 14

7a Largeur des valves < 7 µm, fibules très minces mais ne débouchant chacune que dans 1 côte transapicale (strie) (vol. 2/2 Fig. 10: 6–9) L.: 120–140 µm/l.: 6–6,5 µm/F. 11,5–12,5/10 µm/str. 24,5–25,5/10 µm . . . . **9.** ***N. eglei*** vol. 2/2 p. 18

7b Caractéristiques différentes (vol. 2/2 Fig. 4: 1, 2; 5: 1–5) (cf. aussi *N. speciosa* (vol. 2/2 Fig. 5: 6, 7) L.: (90)150–500 µm/l.: 8–15 µm/F.: 5–7/10 µm/str.(21)23–27/10 µm . . . . . . **1.** ***N. sigmoidea*** vol. 2/2 p. 12

8a Valves à peine plus étroites, extrémités très tronquées. Stries +/- 46/10 µm au microscope optique, à peine résolvables sans effet particulier (vol. 2/2 Fig. 9: 5, 6) L.: 250–300 µm/l.: 7–8,5 µm/F.: 5–7/10 µm/str. 46/10 µm . . . . **8.** ***N. geitleri*** vol. 2/2 p. 17

8b Extrémités +/- coniques, rétrécies; stries apparaissant en arêtes de poisson (vol. 2/2 Fig. 14: 1–5, sans diagnostic) . . . . ***N. macilenta*** vol. 2/2 p. 17

Dissipatae (vol. 2/2 p. 18)

1a Stries relativement espacées 20–25/10 µm (vol. 2/2 Fig. 13: 1–5) L.: 70–190 µm/l.: 4–7 µm/F.: 10–11(14)/10 µm/str. 20–24(26)/10 µm . . . . **13.** ***N. heufleriana*** vol. 2/2 p. 22

1b Stries beaucoup plus rapprochées **2**

2a Fibules exceptionnellement très petites, minces, rapprochées, 20–25/10 µm

(vol. 2/2 Fig. 13: 6, 7) L.: 90–140 µm/l.: 4–5,5 µm/F. 12 mais 20,5 (vis micro)-25/10 µm/vol. 2/2 p. 22str. n.r. . . . . **14. *N. fibulafissa***
2b Fibules différentes **3**
3a Formes de milieux à humidité variable, vol. 2/2 p.ex. sols temporairement trempés, zones supérieures des marées ou zones éclaboussées en général (vol. 2/2 Fig. 15: 1–17) L.: 30(21–43) µm/l.: 3–4,5 µm/F. 7–10(12) µm/str. +/– 45/10 µm . . . . **12. *N. harderi*** vol. 2/2 p. 23
. . . . ***N. incrustans*** vol. 2/2 p. 23
3b Formes vivant en eaux douces ou salées sous conditions “normales” . **4**
4a 2 lignes parallèles d’accompagnement à la carène raphéenne centrale +/- visibles. Formes marines (ici sans diagnostic, seulement présentée comme exemple sous Fig. 14: 6). Cf. aussi *N. acula* (vol. 2/2 Fig. 8: 6) . . . . . .
. . . . **“Spathulatae”**
4b Lignes d’accompagnement peu reconnaissables à cause de la faible résolution au microscope optique. Formes d’eaux douces **5**
5a Stries relativement espacées, 26–32/10 µm (vol. 2/2 Fig. 12: 1); cf. aussi *N. speciosa* (vol. 2/2 Fig. 5: 6, 7) L.: 35–100(130) µm/l.: 6–7 µm/F. 5–10/10 µm/ str. 26–32/10 µm . . **11. *N. recta*** var. ***robusta*** vol. 2/2 p. 21
5b Stries >= 40/10 µm **6**
6a Carène raphéenne plutôt centrale, reconnaissable au milieu de la valve. Rapport 50/50 dans les populations: Nitzschioïdes/Hantzschioïdes (vol. 2/2 Fig. 11: 1–7) L.: 12,5–85 µm/l.:(3)3,5–7(8) µm/F.: 5–11/10 µm/str. 39–50/10 µm . . . . **10. *N. dissipata*** var. ***dissipata*** vol. 2/2 p. 19
6b Carène raphéenne moins centrale, modérément excentrée. La tenue symétrique chez *N. recta* var. *recta* doit être exclusivement nitzschioïdes (vol. 2/2 Fig. 11: 8–14, voir aussi Fig. 12: 2–11) L.: 12,5–85 µm/l.: (3)3,5–7(8) µm/f.: 5–11/10 µm/str. 39–50 µm . . . .
. . . . **10. *N. dissipata*** var. ***media*** vol. 2/2 p. 19
L.: 35–100(130) µm/l.: 6–7 µm/f.: 5–10/10 µm/str. 26–32/10 µm . . . . . .
. . . . **11. *N. recta*** part. vol. 2/2 p. 20

Obtusae (vol. 2/2 p. 24)

1a Valves longues, étirées; à stries relativement espacées et grossièrement ponctuées (vol. 2/2 Fig. 21: 1–4a) L.: 60–165 µm/l.: 2,9–6(6,5) µm/f. 5–9/10 µm/str. 22–25,5/10 µm . . . . . . **20. *N. prolongata*** vol. 2/2 p. 29
L.: 130–165 µm/l.: 6–6,5 µm/f.: 7/10 µm/str. 22–24/10 µm
. . . . **21. *N. improvisa*** vol. 2/2 p. 30
1b Caractéristiques différentes **2**
2a Longueur valvaire régulièrement > 110 µm pour une largeur minimum de 7 µm (vol. 2/2 Fig. 17: 1–3). Si carène raphéenne en position nettement centrale cf. aussi *N. vidovichii* (vol. 2/2 Fig. 18: 6, sans descriptif) L.: 120–350 µm/l.: 7–13 µm/f. 5–6/10 µm/str. 22–30/10 µm . . . .
. . . . **15. *N. obtusa*** vol. 2/2 p. 25
2b Longueur et/ou largeur inférieures **3**
3a Courbure sigmoïde en vue valvaire uniquement reconnaissable à l’asymétrie du contour des pôles. Ceux-ci pourtant ne sont pas rostrées ni capitées
. . . . **4**
3b Caractéristiques différentes **5**
4a Extrémités plus étroites, plus asymétriques juste avant les pôles, de ce fait apparaissant en forme de lame de scalpel (vol. 2/2 Fig. 18: 2–5). Cf. aussi *N. vidovichii* (vol. 2/2 Fig. 18: 6, sans descriptif) L.: 20–110 µm/l.: 4,5–7,4 µm/ f.: 7–10/10 µm/str.(25) 27–38/10 µm . . . . . **16. *N. scalpelliformis*** p. 26
4b Extrémités s’amincissant plutôt graduellement et moins asymétriquement

(vol. 2/2 Fig. 19: 7 à 20:7) L.: 20–100 µm/l.: 4–6 µm/f.: 7–11/10 µm/str. 27–32/10 µm . . . . **18.** ***N. filiformis*** vol. 2/2 p. 27

5a Extrémités sigmoïdes, la plupart du temps déjà bien avant les pôles (vol. 2/2 Fig. 17: 4–8). Cf. aussi *N. subcohaerens* var. *scottica* (vol. 2/2 Fig. 20: 8–12) L.: 35–120 µm/l.: 3–4,5 µm/f. 7–11/10 µm/str. 30–36/10 µm . . . . **17.** ***N. nana*** vol. 2/2 p. 26

5b Caractéristiques différentes **6**

6a Stries comparativement espacées par rapport à la petite taille de la valve, 24–28/10 µm (vol. 2/2 Fig. 18: 8–10, sans descriptif) . . . . ***N. aremonica***

6b Caractéristiques différentes **7**

7a Extrémités valvaires plus ou moins rostrées et +/- "pliées" d'un côté. Str. 38–42/10 µm (vol. 2/2 Fig. 19: 1–7) L.: 20–55 µm/l.: 3–5 µm/f. 10–13/10 µm/str. (32)38–42/10 µm . . . . . . **24.** ***N. clausii*** vol. 2/2 p. 31

7b Caractéristiques différentes **8**

8a Les 2 fibules centrales remarquablement espacées l'une de l'autre, un peu plus longues que les autres et tenant ensemble dans un angle obtus (vol. 2/2 Fig. 21: 5–7). Cf. aussi *N. submarina* (vol. 2/2 Fig. 21: 8) L.: 17–38 µm/l.: 4–6 µm/f. 12–15/10 µm/str. 33–40/10 µm . . . . **25.** ***N. amplectens*** vol. 2/2 p. 31

8b Caractéristiques différentes **9**

9a Valves au milieu +/- concaves, extrémités +/- rostrées **10**

9b Caractéristiques différentes au moins de façon indicative chez quelques exemplaires (vol. 2/2 Fig. 20: 8–12) L.: +/- 20–45 µm/l.: 3–5 µm/f. 7–11/10 µm/str. 30–32/10 µm . . . . **19.** ***N. subcohaerens*** vol. 2/2 p. 28

10a (Apparemment) longueur valvaire plus limitée (vol. 2/2 Fig. 22: 1–6) L.: 18–54 µm/l.: 3,5–6,5 µm/f. 5–10/10 µm/str. 30–38/10 µm . . . . **22.** ***N. brevissima*** vol. 2/2 p. 30

10b Longueur moins fortement limitée (vol. 2/2 Fig. 22: 7–11) L.: 25–115 µm/l.: 3–5 µm/f. 5–8/10µm/str. 32–35/10 µm . **23.** ***N. terrestris*** vol. 2/2 p. 30

Sigmata (vol. 2/2 p. 32)

1a Les deux fibules du centre régulièrement plus espacées que les autres. Formes à petites valves (vol. 2/2 Fig. 70: 25–27). Cf. aussi *N. Lorenziana* (vol. 2/2 Fig. 86: 6–10), les "fausses *N. sigma*" (vol. 2/2 Fig. 24: 3–7) et des populations sporadiques sigmoïdes d'autres sections avec ou sans nodule central (vol. 2/2 p.e. Fig. 59: 5; 70: 8, 26)L.: 12–22 µm/l.: 1,8–2,4 µm/f. 12/10 µm/str. 25–26/10 µm . . . . **28.** ***N. austriaca*** vol. 2/2 p. 34

1b Fibules du centre non espacées **2**

2a Fibules de la carène raphéenne peu excentrée se prolongeant très fortement dans la face valvaire (vol. 2/2 Fig. 22: 12–14); cf. aussi **"Insignes"** (vol. 2/2 Fig. 26: 1–6) L.: 45–95 µm/l.: 3–7 µm/f. 4–7/10 µm/str. 27–30/10 µm vol. 2/2 p. 33 . . . . **27.** ***N. fasciculata***

2b Carène plus excentrée sans fibules fortement prolongées ; taille de la valve et densité des stries très variables selon les formes (vol. 2/2 Fig. 23: 1–9; 24: 1–2b) L.: 35–1000 µm/l.: 4–15(26) µm/f. (3)7–12(+?)/10 µm/vol. 2/2 p. 32 str. (15)19–8/10 µm . . . . **26.** ***N. sigma*** s. lato

Tryblionellae (vol. 2/2 p. 35)

1a Raphé ininterrompu ; pas de nodule central ; nombre des fibules toujours correspondant au nombre des côtes transapicales d'où difficulté à différencier les deux **16**

1b Nodule central présent, parfois reconnaissable à effets spéciaux au microscope **2**

2a Ponctuation des stries sur toute la face valvaire (ou partiellement dessus)

reconnaissable en quinquonce ; côtes apicales et transapicales apparaissant souvent par endroit en réseau réticulé (vol. 2/2 Fig. 39: 1–9), ici seulement une espèce avec descriptif (vol. 2/2 Fig. 38: 13–15a) . . . . **"Panduriformes"** *N. coarctata* vol. 2/2 p. 50

2b Ponctuation non en quinquonce **3**

3a Nombre de fibules correspondant au nombre des côtes transapicales d'où difficulté à les différencier **15**

3b Fibules faciles à différencier **4**

4a Stries (dans le sens le plus strict du terme) très délicates constituant des lignes ponctuées (parfois +/- lacuneuses), dans certains cas très difficiles à résoudre, doublées chez de nombreuses formes par des bandes beaucoup plus grossières (rangs secondaires) constituant plutôt des plis transversaux et désignés ci-après comme des côtes de soutien **8**

4b Toujours stries simples (sauf ondulation longitudinale), "normalement" marquées, n'apparaissant pas +/- lacuneuses ou diffuses, parfois sans ponctuation reconnaissable car constituées de 2 séries d'aréoles ou plus **5**

5a Points très grossiers et valves plus concaves au milieu sur les côtés (vol. 2/2 Fig. 38: 13–15a)vol. 2/2 p. 50 . . . . **48. *N. coarctata*** vol. 2/2 p. 50

5b Caractéristiques différentes **6**

6a Largeur valvaire >= 9 µm (vol. 2/2 Fig. 33: 6); si valves à ondulation longitudinale "diffuse" cf. aussi *N. plana* (vol. 2/2 Fig. 33: 1–3) L.: 43–110 µm/l.: 9–17 µm/f. 9–14/10 µm/str. 19–28/10 µm . . . . **42. *N. marginulata*** vol. 2/2 p. 45

6b Largeur valvaire atteignant toujours 9 µm **7**

7a Stries 16–20/10 µm (vol. 2/2 Fig. 34: 1–3) L.: (20)35–130 µm/l.: (4,5)5–9 µm/f. 7–10(12)/10 µm/ str. 19–28/10 µm . . . . **37. *N. hungarica*** vol. 2/2 p. 42

7b Stries 24–30/10 µm (vol. 2/2 Fig. 51: 7–15) L.: 23–31 µm/l.: 6–8 µm/f. 9–11/10 µm/str. 24–30/10 µm . . . . **41. *N. aerophila*** vol. 2/2 p. 44 . . . . **40. *N. parvula*** vol. 2/2 p. 44

8a Côtes de soutien très visibles sur toute la face valvaire, ininterrompues ou coupées par l'ondulation longitudinale dans la médiane, ou opposées **9**

8b Côtes de soutien non ou à peine reconnaissables, d'un seul côté ou seulement en bougeant la vis micrométrique **11**

9a Valves (8–11 µm) relativement minces par rapport à la longueur; côtes de soutien également relativement minces et très serrées (9–17/10 µm) (vol. 2/2 Fig. 30: 1–5) L.: 18–65(82) µm/l.: 8–23(26) µm/f. 6–12/10 µm/str. 36–36/10 µm . . . . **33. *N. calida*** vol. 2/2 p. 40

9b Caractéristiques différentes **10**

10a Les côtes de soutien dominent toujours nettement par rapport aux stries finement ponctuées; les formes à nombreuses variantes sont difficiles à délimiter par des caractéristiques communes (vol. 2/2 Fig. 28: 1 à 29: 5) L.: (50)60–80 µm/l.: 8–23(26) µm/f. 6–12/10 µm/str. 35–36/10 µm c. de soutien: 5–12(14?) µm . . . . **31. *N. levidensis*** vol. 2/2 p. 37

10b Côtes de soutien moins fortement dominantes. Valves toujours relativement grandes, longueur 50–180 µm, largeur 16–35 µm, aux extrémités moins abruptes et plutôt graduellement amincies, cunéiformes (vol. 2/2 Fig. 27: 1–4) L.: (50)60–180 µm/l.: 16–30(35) µm/f. 5–9/10 µm/str. 30–35/10 µm c. de soutien: 6–10/10 µm **30. *N. tryblionella*** vol. 2/2 p. 37

11a Valves relativement petites, longueur presque toujours < 30 µm, largeur < 7 µm (vol. 2/2 Fig. 27: 5–11) L.: 13–26(31) µm/l.: 7–10 µm/f. (6?)8–10/10 µm/str. 35–36/10 µm c. de soutien: 15/10 µm . . . . **32. *N. debilis*** vol. 2/2 p. 39

11b Valves plus grandes **12**

12a Largeur valvaire +/- 50–65 µm (vol. 2/2 Fig. 32: 1–4) L.: 130–300 µm/l.: 50–65 µm/f. 3–5/10 µm/str. 25–30/10 µm **36.** ***N. circumsuta*** vol. 2/2 p. 42
12b Largeur nettement moindre **13**
13a Stries 16–22/10 µm (vol. 2/2 Fig. 33: 1–3) L.: 50–290 µm/l.: 18–26 µm/f. 3,5–8/str. 16–22/10 µm . . . . **35.** ***N. plana*** vol. 2/2 p. 42
13b Stries nettement plus serrées **14**
14a Largeur des valves 12–30 µm, stries 30–38/10 µm (vol. 2/2 Fig. 30: 6 à 31: 5) . . . . **34.** ***N. littoralis*** vol. 2/2 p. 41
14b Valves plus larges: 35–40 µm (vol. 2/2 Fig. 33: 4, 5, sans descriptif) . . . . ***N.*** **species**
15a Largeur des valves > 12 µm (vol. 2/2 Fig. 34: 4–6) L.: 50–100(+?) µm/l.: 13–18 µm/f. 12–16/10 µm . . . . **39.** ***N. acuminata*** vol. 2/2 p. 44
15b Largeur des valves < 9 µm (vol. 2/2 Fig. 35: 1–6) L. 20–58 µm/l.: 4,5–8,5 µm/ f= c. tr.ap .= str.: 14(15)-20/10 µm . . . . **38.** ***N. constricta*** vol. 2/2 p. 43
16a Ponctuation marginale des stries, en rangées doubles (vol. 2/2 Fig. 35: 7–8) L.: 30–80 µm/l.: 15–25 µm/f. = c. tr.a 6–8/10 µm . . . . **43.** ***N. navicularis*** vol. 2/2 p. 45
16b Caractéristique différente **17**
17a Valves +/- elliptiques avec pôles très tronqués et arrondis. Stries au centre très grossièrement ponctuées, à ponctuation double au moins près de la carène raphéenne (vol. 2/2 Fig. 35: 9–13) L.: 26–44 µm/l.: 12–20 µm/f. = c. tr./str. 6–7/10 µm . . . . **44.** ***N. granulata*** vol. 2/2 p. 45
17b Caractéristiques différentes **18**
18a Stries grossièrement ponctuées (vol. 2/2 Fig. 37: 1–10) L.: 10–130 µm/l.: 3,5–26 µm/f. = c. tr. a./str. 5–21/10 µm **45.** ***N. compressa*** part vol. 2/2 p. 46
18b Stries plus finement ponctuées **19**
19a Valves comparativement plus longuement linéaires, largeur 4–12 µm (vol. 2/2 Fig. 36: 1–5) L.: (20)25–180 µm/l.: 4–12 µm/f. = c. tr.a vol. 2/2 p./str. 1–18/10 µm . . . . **46.** ***N. angustata*** vol. 2/2 p. 48
19b Valves linéaires-lancéolées, lancéolées ou elliptiques (si plutôt linéaires, largeur seulement +/- 4 µm) **20**
20a Stries non résolvables comme de simples rangées de points (vol. 2/2 Fig. 36: 6–10), cf. aussi *N. siliqua* (vol. 2/2 Fig. 36: 11–13) L.: 13–24 µm/l.: +/- 4 µm/f. 16–20/10 µm . . . . **47.** ***N. angustatula*** vol. 2/2 p. 48
20b Stries simplement ponctuées, extrémités très courtes à rostrées ou papilliformes (vol. 2/2 Fig. 38: 1–8) L.: 10–130 µm/l.: 3,5–26 µm/str. 5–21/10 µm . . . . **45.** ***N. compressa*** part. vol. 2/2 p. 46

Dubiae et Bilobatae (vol. 2/2 p. 53)

1a Ponctuation des stries en quinquonce (vol. 2/2 Fig. 54: 1–3a, voir 54: 4–5a) L.: (-?)60–100 µm/l.: 5–9 µm/f.:7–10(14)/10 µm/str. +/- 30/10 µm ponct.: > 20/10 µm . . . . **69.** ***N. littorea*** vol. 2/2 p. 67
L.: (35)50–75 µm/l.: 5–8 µm/f.: 7–10/10 µm str. (28)30–32/10 µm . . . . **70.** ***N. lacunarum*** /vol. 2/2 p. 68
1b Ponctuation non en quinquonce 2
2a Fibules étroites en largeur ou quelques-unes fortement amincies en racines de dents. Ces fibules s'étirent +/- loin à l'intérieur de la face valvaire . 17
2b Caractéristique différente, en tout cas fibules s'étirant moins 3
3a Carène raphéenne comparativement faible, le plus souvent ponctuellement renfoncée au niveau du nodule central. Valves le plus souvent linéaires. Fibules dans l'ensemble très minces, régulièrement en liaison avec une seule strie. Stries >= 30/10 µm **15**
3b Caractéristiques différentes 4

4a Toutes les fibules très larges, épaissies, chacune liée à 3 stries ou plus . **14**
4b Fibules liées à 1, 2 ou (irrégulièrement épaissie) max. 3 stries, en tout cas pas très larges **5**
5a Frustules très larges en vue connective par la présence de nombreuses ceintures connectives. Bord de la valve avec carène raphéenne très constricté, fortement concave, au niveau du nodule central. En vue connective, valves dans l'ensemble en forme de "canoë" ou de "parenthèse" avec extrémités asymétriques rostrées. En position appropriée (rare), la position peu excentrée de la carène raphéenne est reconnaissable. Fibules toujours minces et liées avec une seule strie **12**
5b Caractéristiques différentes ou difficiles à reconnaître **6**
6a Frustules et valves comme 5a, mais carène raphéenne plus fortement excentrée et moins renfoncée au niveau du nodule central (vol. 2/2 Fig. 41: 1, 2) L.: (-?)80–160 µm/l.: l.: (6?)12–16 µm/f. 9–10/10 µm/ str. 21–24/10 µm ponct.: +/– 30/10 µm . . . . **51** ***N. dubia*** vol. 2/2 p. 36
6b Caractéristiques différentes **7**
7a Valves toujours inhabituellement courtes par rapport à la largeur (vol. 2/2 Fig. 65: 1–2a) L.: (5)12–26,5 µm/l.: 2,5(4,4)-7 µm/f. 10–14/10 µm . . . . **73.** ***N. laevis*** vol. 2/2 p. 72
7b Valves différentes **8**
8a Valves linéaires ou linéaires-lancéolées plus ou moins larges. Chaque fibule est mince et liée à une seule strie. Stries comparativement denses, 28 à > 40/10 µm (vol. 2/2 Fig. 55: 1–10) L.: 34–228 µm/l.: 2,5–7,5 µm/f.: 8–17(+?)/10 µm/str. 28–41/10 µm . . . . . . **71.** ***N. linearis*** vol. 2/2 p. 72
8b Caractéristiques différentes **9**
9a Toutes les fibules sont courtes et liées avec plus d'une strie. Stries 24–30/10 µm (vol. 2/2 Fig. 51: 1–6a) L.: 22–125 µm/l.: (5)6–9 µm/f.: 7–10/10 µm/str.: 24–30/10 µm . . . . . . **65.** ***N. umbonata*** vol. 2/2 p. 65
9b Caractéristiques différentes **10**
10a Fibules proximales liées à 2, 3 ou plus de stries, distalement avec 1 seule strie. En eau saumâtre stries moins fines – 19–22, en eau douce très alcaline jusqu'à 24/10 µm (vol. 2/2 Fig. 43: 1–4) L.: 45–95 µm/l.: 7–12 µm/f.: 6–10/10 µm/str. 19–22(24)/10 µm . **54.** ***N. commutatoides*** vol. 2/2 p. 58
10b Caractéristiques différentes **11**
11a Valves légèrement renfoncées au niveau du nodule central mais difficile à observer. Les fibules du centre aussi sont minces et liées à une seule strie chacune. Str. +/- 20/10 µm (vol. 2/2 Fig. 42: 7, 8) L.: 45–100 µm/l.: 5–8 µm/ f. 9–12/10 µm/str. (18)19–23(24)/10 µm . . . . **53.** ***N. gisela*** vol. 2/2 p. 57
11b Valves moins minces-linéaires, plus fortement constrictées à la carène raphéenne. Fibules proximales souvent plus larges que les distales, les 2 du centre nettement plus écartées (vol. 2/2 Fig. 42: 1–6) L.: 46–100(et +) µm/l.: 5–8 µm/f.: 9–12/10 µm/ str. (18)19–23(24)/10 µm . . . . **52.** ***N. commutata*** vol. 2/2 p. 56
12a Stries espacées, 16–20/10 µm mais présentant des problèmes de délimitation (vol. 2/2 Fig. 46: 1, 2) L.: 50–150 µm/l. 5–14 µm/f.: 5–8/10 µm/str. 15–27/10 µm . . . . **59.** ***N. bilobata*** vol. 2/2 p. 61
12b Stries > 20/10 µm **13**
13a Frustules rappelant fortement *N. bilobata*, cependant régulièrement plus petits ou moins robustes et stries plus fines, 21–27/10 µm (vol. 2/2 Fig. 46: 3 à 47: 3) L.: 48–103 µm/l.: 8–14(5,5 rare) µm/f.: 8–10/10 µm/str. (21)24,5–27/10 µm . . . . **60.** ***N. hybrida*** vol. 2/2 p. 61
13b Frustules en général encore plus petits et/ou plus minces, apparaissant moins robustes. Stries +/- 30–40/10 µm (vol. 2/2 Fig. 47: 4 à 48: 9; 44: 8–10). Cf. également les "formes critiques hétérogènes autour de *N.*

thermaloides, N. normanii et *N. dubiiformis*" (vol. 2/2 Fig. 45: 1–16) L.: var./l: var./f.: 12–18/10 µm/str. 30–40/10 µm . . . . **61. *N. pellucida*** vol. 2/2 p. 63
L.: 40–50 µm/l.: 5–7 µm (milieu)/f.: 16–18/10 µm/str. " 40/10 µm . . . . **57. *N. dubiiformis*** vol. 2/2 p. 60

14a Frustules très grands. Largeur en vue connective +/- 40 µm. Stries 17–21/10 µm, L.: 100–230 µm/l.: 10–16 µm/f.: 2–3,5/10 µm/str. 17–21/10 µm (vol. 2/2 Fig. 52: 1, 2) . . . . . . **66. *N. kittlii*** vol. 2/2 p. 66

14b Frustules plus petits. Stries beaucoup plus fines (vol. 2/2 Fig. 52: 3 à 53: 8 et Fig. 56: 8, 8a). Cf. aussi *N. polaris* (vol. 2/2 Fig. 56: 9) L.: 30–10 µm/l.: 5–10 µm/f.: 4–6/10 µm/str.: 38–42/10 µm . **67. *N. dippelii*** vol. 2/2 p. 66
L.: 30 µm/l.: +/- 5 µm/stries +/- 27/10 µm . . **68. *N. vasta*** vol. 2/2 p. 66

15a Valves "trappues" par le rapport L/l., larges-linéaires avec extrémités cunéiformes. Stries interrompues par une mince ondulation longitudinale +/- au milieu de la valve (vol. 2/2 Fig. 43: 5–7) L.: 38–55 µm/l.: 7,5–8,5 µm (+?)/f.: 14–15/10 µm/str. 30–32(36)/10 µm . . . . **55. *N. normanii*** vol. 2/2 p. 58

15b Caractéristiques différentes **16**

16a Les 2 fibules du centre sont nettement plus espacées. Formes possédant de nombreuses variétés en eaux douces. Les formes pouvant éventuellement être confondues avec *N. thermaloides* ont toutefois à peine plus de 12 fibules/10 µm (vol. 2/2 Fig. 55: 1–10) L.: 38–224 µm/l.: 2,5–7,5 µm/f.: 8–17/10 µm/str.: 28–41/10 µm . . . . **71. *N. linearis*** vol. 2/2 p. 69

16b Fibules >= 16/10 µm, les 2 du centre régulièrement et comparativement peu espacées. Surtout formes marines côtières, apparaissant rarement dans les eaux salées intérieures (vol. 2/2 Fig. 44: 1–7). Cf. aussi les formes critiques hétérogènes autour de *N. thermaloides* (vol. 2/2 Fig. 45: 1–16) L.: 20–73 µm/l.: 4–6 µm/f.: 16–20/10 µm/str. 30–36/10 µm . . . . **56. *N. thermaloides*** vol. 2/2 p. 59

17a Valves comparativement grandes, longueur > 60 µm, largeur > 6 µm L.: 60–90 µm/l.: 6–9 µm/f. 5–9/10 µm/str. 26–32/10 µm (vol. 2/2 Fig. 49: 1–5) . . . . **62. *N. bremensis*** vol. 2/2 p. 63

17b Valves régulièrement plus petites **18**

18a Fibules en moyenne plus espacées, 6–10/10 µm et (avec focalisation conséquente) étirées jusqu'à plus de la moitié de la largeur valvaire (vol. 2/2 Fig. 50: 1–3) L.: 30–60 µm (+?)/l.: 4–7 µm/f.: 6–10/10 µm/str. 22–28/10 µm . . . . **63. *N. palustris*** vol. 2/2 p. 64

18b Fibules plus rapprochées, 9–15/10 µm, le plus souvent plus courtes et moins nettement étirées (vol. 2/2 Fig. 50: 4–9) L.: 32–52 µm/l.: 4,5 µm (c.) et 5–6 µm/f.: 9–15/10 µm/ str. 34–40/10 µm . . . . **64. *N. homburgiensis*** vol. 2/2 p. 64

Groupe hétérogène d'appartenance difficile à reconnaître ou appartenance de section jusqu'à présent non définie, e.a. groupe-reliquat des "Lineares" insuffisamment fondées (vol. 2/2 p. 69)

1a Raphé avec nodule central. Les deux fibules du centre régulièrement plus écartées que leurs voisines 7

1b Caractéristique différente 2

2a Fibules larges à très larges, reliées à 3 stries ou plus 3

2b Fibules plus minces, souvent très minces 5

3a Fibules 2–3/10 µm (vol. 2/2 Fig. 57: 1–4) L.: 200–300 µm/l.: 7–10 µm/f.: 2–3/10 µm/str. 18–21/10 µm . . . . **75. *N. peisonis*** vol. 2/2 p. 74

3b Fibules 4–8/10 µm 4

4a Valves relativement grandes. Stries grossièrement ponctuées, +/- 20/10 µm (vol. 2/2 Fig. 56: 1, 2) L.: 30–220 µm/l.: (4)5–14 µm/f.: 4–8/10 µm/str. 17–35/10 µm . . . . **74. *N. vitrea*** part. vol. 2/2 p. 72

4b Valves plus petites. Stries beaucoup plus rapprochées (vol. 2/2 Fig. 56: 3–7) . . . . **74. *N. vitrea*** part vol. 2/2 p. 74
. . . . **74A. *N. ebroicensis*** vol. 2/2 p.73

5a Valves de 100 µm ou plus longues. Fibules 5–7/10 µm (vol. 2/2 Fig. 57: 5–8) L.: 100–180 µm/l.: 4–6 µm/f.: 5–7/10 µm/str. 33–36/10 µm . . . . **78. *N. monachorum*** vol. 2/2 p. 75

5b Valves plus courtes rappelant plus les “Lanceolatae”. Fibules plus minces et beaucoup plus étroites **6**

6a Valves linéaires-lancéolées à linéaires. Stries 34–38/10 µm faciles à résoudre avec éclairage oblique (vol. 2/2 Fig. 58: 10–15) L.: (20)30–90 µm/l.: 4–6 µm/f.: 13–17/10 µm/str. 34–38/10 µm **76. *N. sublinearis*** vol. 2/2 p. 74

6b Valves en moyenne légèrement plus étroites, linéaires-lancéolées à lancéolées. Stries 40–50/10 µm, difficiles à résoudre au microscope optique (vol. 2/2 Fig. 58: 1–9) L.: 35–50 µm/l.: 4–5 µm/f.: (14)16–20(24)/10 µm/str. 40–50/10 µm . . . . **77. *N. pura*** vol. 2/2 p. 75

7a Formes marines appartenant éventuellement aux Lanceolatae mais les dimensions des valves sont inhabituelles, courtes et larges avec des fibules du centre très espacées comme Fig. 65: 1–2a L.: (5)12–26,5 µm/l.: (2,5)4,4–7 µm/f.: 10–14/10 µm/str. 32–36(50)/10 µm . . . . **3. *N. laevis*** vol. 2/2 p. 72

7b Caractéristiques différentes

8a Fibules larges, apparaissant épaisses, les 2 du centre très espacées (Fig. 56: 8, 8a). Cf. aussi *N. polaris* (vol. 2/2 Fig. 56: 9) L.: 30 µm/l.: 5 µm/stries 27(15)/10 µm . . . . **68. *N. vasta*** vol. 2/2 p. 67

8b Caractéristiques différentes, surtout fibules beaucoup plus étroites . . **9**

9a Forme marine, un seul individu connu selon Fig. 56: 10 . . . . **79. *N. pseudocommunis*** vol. 2/2 p. 75

9b Formes d’eaux douces riches en variétés. Les fibules étroites liées chacune à une seule strie sont caractéristiques pour l’ensemble (vol. 2/2 Fig. 55: 1–10) . . . . **71. *N. linearis*** vol. 2/2 p. 68
L.: 34–228 µm/l.: 2,5–7,5 µm/f.: 8–17/10 µm/str.: 28–41/10 µm . . . . **72. *N. subtilis*** vol. 2/2 p. 71

Lanceolatae (vol. 2/2 p. 76)

1a Fibules du centre régulièrement équidistantes dû au raphé ininterrompu de pôle à pôle (reconnaissable sous microscope électronique seulement) . . . . **groupe-clef A** (vol. 2/2 p. 78) (p. 292)

1b Course du raphé interrompu au milieu entre les deux pôles par un pont de silice, appelé ici toujours nodule central, apparaissant plus ou moins fortement siliceux sous microscope électronique. Les 2 fibules du centre, toujours voisines du nodule central, sont plus espacées que les autres . . . . **Groupe-clef B** (vol. 2/2 p. 82)) (p. 297)

Lanceolatae groupe A (vol. 2/2 p. 78)

1a Fibules très étroites, rétrécissement allongé ou cunéiforme ou en forme de petites dents (si les stries sont résolvables), elles sont chacune reliée à une seule strie (= côte transapicale), qui éventuellement n’est reconnaissable que grâce à un ajustement très précis du microscope **55**

1b Fibules plus larges. Même en cas de stries non-résolvables, grâce à leur forme épaissie et surtout large et non-rétrécie, on voit qu’elles sont reliées à plus d’une strie **2**

2a Stries relativement espacées, < 15–25/10 µm (toujours facilement résolvables même sans condensateur d'huile immergé), le plus souvent ponctuées **3**
2b Densité des stries > 25/10 µm **15**
3a Stries grossières, 17–20/10 µm ; néanmoins ponctuation non-résolvable car points situés en quinconce ou en rangées doubles très serrées **4**
3b Caractéristiques différentes **5**
4a Extrémités arrondies tronquées (vol. 2/2 Fig. 84: 1–8) . . . . **136.** ***N. valdecostata*** vol. 2/2 p. 121
4b Extrémités arrondies pointues (vol. 2/2 Fig. 84: 13–19) L.: 7–15 µm/l.: 1,5–2 µm/f.: 16–22/10/µm . . . . . ***Simonsenia delognei*** vol. 2/2 p. 135
5a Stries < 15/10 µm, grossièrement ponctuées. Fibules se prolongeant +/- dans la face valvaire (vol. 2/2 Fig. 78: 27–29) L.: 6–50 µm/l.: 4–6 µm/f.: 5–7/10 µm/str. 11–14/10 µm . . . . **115.** ***N. amphibioides*** vol. 2/2 p. 109
5b Stries plus rapprochées **6**
6a Fibules minces, chacune débouchant sur une seule côte transapicale. Stries <= 24/10 µm. Valves de 70 à 200 µm de long (vol. 2/2 Fig. 13: 1–5) L.: 70–200 µm/l.: 4–7 µm/f.: 10–11(14)/10 µm/str. 20–24(26)/10 µm . . . . **13.** ***N. heufleriana*** vol. 2/2 p. 22
6b Caractéristiques différentes **7**
7a Fibules (lors de la focalisation) taillées en pointes courtes (rappelant de petites dents) débouchant le plus souvent chacune sur une côte transapicale. Str. 24–28/10 µm (vol. 2/2 Fig. 71: 1–12) L.: 18–50 µm/l.: 4–6 µm/f.: 11–16/10 µm/str. 24–28/10 µm . . . . **96.** ***N. solita*** vol. 2/2 p. 99
7b Caractéristiques différentes **8**
8a Valves restent toujours comparativement courtes. Extrémités +/- arrondies capitées ou étirées **9**
8b Caractéristiques différentes **10**
9a Valves le plus souvent faiblement concaves au centre. Extrémités étroites et capitées (vol. 2/2 Fig. 83: 20–24) L.: 10–29 µm/l.: 2,5–4 µm/f.: 10–15/10 µm/str. 23–32/10 µmpart. . . **134.** ***N. elegantula*** vol. 2/2 p. 120
9b Valves linéaires ou faiblement convexes au centre. Extrémités larges, capitées (vol. 2/2 Fig. 74: 18–26)L.: 12–25 µm/l.: 3–4 µm/f. 11–12/10 µm/str. 23–24/10 µm . . . . **101.** ***N. bacilliformis*** vol. 2/2 p. 102
10a Stries < 19/10 µm. Formes marines (vol. 2/2 Fig. 69: 21) . . . . **92.** ***N. liebetruthii*** var. ***major*** vol. 2/2 p. 96
10b Stries plus rapprochées. Formes (aussi) en eaux continentales **11**
11a Stries 19–21/10 µm. Contour valvaire correspondant à Fig. 74: 11–17 . . . . **102.** ***N. modesta*** vol. 2/2 p. 102
11b Caractéristiques différentes **12**
12a Valves rarement > 40 µm de long. Stries 21–25/10 µm. Extrémités +/- tronquées. Formes d'eaux douces peu chargées en électrolytes (vol. 2/2 Fig. 74: 1–10) L.: 8–48 µm/l.: 3–5 µm/f.: (7)8–14/10 µm/str. 21–25/10 µm . . . . **100.** ***N. alpina*** vol. 2/2 p. 101
12b Caractéristiques différentes **13**
13a Valves toujours comparativement courtes selon Fig. 70: 1–9. Cf. aussi *N. liebetruthii* part. (vol. 2/2 Fig. 69: 14–20) L.: 8–24 µm/l.: 3–4 µm/f.: 10–12/10 µm/str.21–25/10 µm . **94.** ***N. angustiforaminata*** vol. 2/2 p. 98
13b Valves comparativement longues linéaires **14**
14a Largeur valvaire 4–7 µm (vol. 2/2 Fig. 61: 1–10) L.: [25(et -)] 40–150 µm (20?)/l.: 4–7 µm/f.: 7–13/10 µm/stries 20–33/10 µm . . . . **82.** ***N. intermedia*** vol. 2/2 p. 87
14b Largeur valvaire 2,5–3,5 µm (vol. 2/2 Fig. 77: 15–18) L.: 38–57 µm/l.:

2,5–3,5 µm/f.: 10–12/10 µm/str. 24–27/10 µm . . . . 110. *N. diversa* vol. 2/2 p. 107

15a Stries de +/- 25 à +/- 35/10 µm (encore relativement faciles à résoudre avec condensateur d'huile immergé, dans la partie supérieure éventuellement avec éclairage légèrement contrastant oblique) **16**

15b Stries plus rapprochées, > 35/10 µm (en microscopie optique seulement résolvables au contour ou à l'aide de moyens contrastants élevés ou pas du tout) **37**

16a Frustules formant des colonies étoilées en aigrette **17**

16b Caractéristique différente ou non-observable **18**

17a Largeur valvaire > 3,5 µm. Stries < 30/10 µm. Fibules < 13/10 µm (vol. 2/2 Fig. 61: 1–10). En cas de fibules nettement allongées et coupant une ligne longitudinale, cf. aussi *N. recta* var. *robusta* (vol. 2/2 Fig. 12: 1) L.: [25(et -)] 40–150 µm (200?)/l.: 4–7 µm/f.: 7–13/10 µm/part. str. 20–33/10 µm . . . . 82. ***N. intermedia*** vol. 2/2 p. 87 (forma. actinastroides)

17b Largeur valvaire < 3,5 µm. Stries 29–36/10 µm. Fibules 13–18/10 µm. (vol. 2/2 Fig. 60: 8–12) . . . . 81. ***N. fruticosa*** vol. 2/2 p. 86

18a Valves toujours comparativement courtes avec, pour la majorité de la population, des extrémités étirées en forme de petites têtes **19**

18b Caractéristiques différentes **20**

19a Bords valvaires légèrement concaves au milieu. Stries comparativement plus espacées, 23–32/10 µm. Formes d'eaux (très) alcalines (vol. 2/2 Fig. 88:20–24) L.: 10–29 µm/l.: 2,5–4 µm/f.: 10–15/10 µm/str. 23–32/10 µm . . . . 134. ***N. elegantula*** vol. 2/2 p. 120

19b Caractéristiques différentes, en particulier stries presque toujours beaucoup plus rapprochées (vol. 2/2 Fig. 83: 10–19) L.: 7–19 µm/l.: 2,3–4 µm/f.: 9–19/10 µm/str. 30–41/10 µm . . . 133. ***N. microcephala*** vol. 2/2 p. 120

20a Fibules (à la focalisation) courtes en pointe, débouchant le plus souvent dans une seule côte transapicale (vol. 2/2 Fig. 71: 1–12) L.: 18–50 µm/l.: 4–6 µm/f.: 11–16/10 µm/str. 24–28/10 µm . . . 96. ***N. solita*** vol. 2/2 p. 99

20b Caractéristiques différentes **21**

21a Extrémités largement arrondies, rarement peu étirées et alors fortement tronquées. Stries +/- 30/10 µm (vol. 2/2 Fig. 79: 1–6) L.: 6–40(60) µm/l.: 4–5,8 µm/f.: (8)10–14/10 µm/str. 28–38/10 µm . . . . 116. ***N. communis*** vol. 2/2 p. 110

21b Caractéristiques différentes **22**

22a Valves en moyenne comparativement grandes, tout au moins longues et (à tendance) linéaires **23**

22b Population avec en majorité des valves plus courtes, le plus souvent également plus étroites **26**

23a Largeur 4–7 µm (vol. 2/2 Fig. 61: 1–10; 59: 11, 12, 18) L.: [25(et -)] 40–150(200?) µm/l.: 4–7 µm/f.: 7–13/10 µm/str. 20–33/10 µm . . . . 82. ***N. intermedia*** vol. 2/2 p. 87
. . . . 80. ***N. palea*** part.

23b Valves plus étroites **24**

24a Extrémités valvaires étirées rostrées (vol. 2/2 Fig. 67: 11; 67: 4–10) L.: 20–80 µm/l.: (1,5)2–3 µm/f.: 12–16/10 µm/str. (26)27–33/10 µm . . . . 131. ***N. rostellata*** vol. 2/2 p.119
f.: 12/10 µm/stries +/- 31/10 µm 130. ***N. subacicularis*** part. vol. 2/2 p. 118

24b Caractéristique différente **25**

25a Stries 24–27/10 µm (vol. 2/2 Fig. 77: 15–18) L.: 38–57 µm/l.: 2,5–3,5 µm/f.: 10–12/10 µm/str. 24–27/10 µm . . . . 110. ***N. diversa*** vol. 2/2 p.107

25b Stries 29–36/10 µm (vol. 2/2 Fig. 60: 8–12). *N. subtilioides* (vol. 2/2 Fig. 60:

13, sans descriptif) L.: 20–83 µm/l.: < 3,5 µm/f. 13–18/10 µm/str. 29–36/10 µm . . . . **81.** ***N. fruticosa*** vol. 2/2 p. 86

26a Valves plus longues, linéaires et plus étroites **27**

26b Valves plus longues, +/- lancéolées, mais non linéaires-étroites **29**

27a Formes d'eaux alcalines. L.: < 40 µm/stries > 20/10 µm (vol. 2/2 Fig. 69: 14–32) . . . . **92.** ***N. liebetruthii*** vol. 2/2 p. 97

27b Formes d'eaux plutôt acides **28**

28a Stries 34–37/10 µm (vol. 2/2 Fig. 66: 12–16) L.: 28–50 µm/l.: +/- 3 µm/f.: 13–16/10 µm/str. 34–37/10 µm . . . . . . **88.** ***N. suchlandtii*** vol. 2/2 p. 93

28b Stries 26–32/10 µm (vol. 2/2 Fig. 72: 1–23a) L.: 8–45 µm/l.: 2,5–3 µm/f. 10–16/10 µm/str. 26–32(36)/10 µm . . . . **97.** ***N. perminuta*** vol. 2/2 p. 99

29a Formes à répartition nordique-alpine, vivant sur support bryophytique d'humidité changeante selon Fig. 74: 27–30 L.: 15–26,5 µm/l.: 4,5 µm/f.: 8(9)-10(12)/10 µm/str. 30–32/10 µm . . **103.** ***N. bryophila*** vol. 2/2 p. 103

29b Caractéristiques différentes **30**

30a Valves étroites et lancéolées **31**

30b Valves +/- plus larges et lancéolées **34**

31a Biotope marin ou eaux continentales très alcalines **32**

31b Formes d'eaux douces, d'alcalinité moyenne ou modérément plus élevée **33**

32a Stries 32–35/10 µm (vol. 2/2 Fig. 82: 1–5) L.: 18–40 µm/l.: 2,9–4,6 µm/f.: 13–20/10 µm/str. (30)32–35/10 µm . . . **127.** ***N. aequorea*** vol. 2/2 p. 116

32b Stries plus éloignées (vol. 2/2 Fig. 69: 14–32) . . . . **92.** ***N. liebetruthii*** vol. 2/2 p. 97

33a Longueur valvaire plus limitée étroitement, le plus souvent < 20 µm (vol. 2/2 Fig. 78: 7–12) L.: 12–20(24) µm/l.: 2–3,5(5) µm/f.: 12–16/10 µm/str. 27–32/10 µm . . . . **113.** ***N. bacillum*** vol. 2/2 p. 108

33b Valves de la population plus longues (vol. 2/2 Fig. 67: 4–10) L.: 20–80 µm/ l.: (1,5)2–3 µm/f.: 12–16/10 µm/str. (26)27–33/10 µm . . . . **130.** ***N. subacicularis*** vol. 2/2 p. 118

34a Stries (même 30–35/10 µm) difficiles à résoudre, dû au relief très plat des alvéoles et côtes transapicales (vol. 2/2 Fig. 59: 1–12, 19–23; 60: 1–7) L.: 15–70 µm/l.: 2,5–5 µm/f. 9–17/10 µm/str. 28–40/10 µm . . . . **80.** ***N. palea*** part. vol. 2/2 p. 85

34b Stries toujours faciles à résoudre, même sans éclairage oblique **35**

35a Largeur valvaire >= 5 µm, stries 28–32/10 µm (vol. 2/2 Fig. 59: 13–18) L.: 15–70 µm/l.: 2,5–5 µm/f.: 9–17/10 µm/str. 28–40/10 µm . . . . **80.** ***N. palea*** part. **(formes minuta)** vol. 2/2 p. 85

35b Valves plus étroites ou stries plus éloignées **36**

36a Largeur valvaire 4,5 µm. Stries 25–26/10 µm. Fibules du centre beaucoup plus écartées (+/- 10/10 µm) qu'aux extrémités (vol. 2/2 Fig. 70:10–13) L.: 17–20 µm/l.: 4–5 µm/f. 10–15/10 µm/str. 25–26/10 µm . . . . **95.** ***N. desertorum*** vol. 2/2 p. 98

36b Fibules moins espacées également au centre (vol. 2/2 Fig. 70: 14–21) L.: 10–25 µm/l.: 2,5–4 µm/f.: 14(18)-20/10 µm/str. 25–34/10 µm . . . . **93.** ***N. supralitorea*** vol. 2/2 p. 97

37a Valves toujours courtes (< 20 µm) avec extrémités microcéphales ou étirées rostrées (vol. 2/2 Fig. 83: 10–19) L.: 7–19 µm/l.: 2,3–4 µm/f.: 9–19/10 µm/ str. 30–41/10 µm . . . . **133.** ***N. microcephala*** vol. 2/2 p. 120

37b Caractéristiques différentes **38**

38a Extrémités tronquées à largement arrondies **39**

38b Extrémités arrondies pointues ou +/- capitées ou +/- rostrées, cf. indépendamment de cela aussi *N. recta* (vol. 2/2 Fig. 12: 1–11) **44**

39a Populations avec des valves toujours courtes et relativement larges, elliptiques **40**
39b Valves le plus souvent linéaires-elliptiques ou plus étroitement elliptiques **42**
40a Valves de 4,5–6,6 µm de large. Fibules 12–16/10 µm (vol. 2/2 Fig. 79: 7–11) L.: 13–22,5 µm/l.: 4,5–6,6 µm/f.: 12–16/10 µm/str.+/- 42/10 µm . . . . **117.** ***N. ovalis*** vol. 2/2 p. 110
40b Valves de 2,5–5 µm de large, fibules en moyenne plus rapprochées . . **1**
41a Valves délicatement siliceuses. Sous microscope électronique aréoles en quinquonce (vol. 2/2 Fig. 80: 16–21) L.: 6,5–18 µm/l.: 2,5–3,5(4) µm/f.: 15–18/10 (13) µm . . . . **21.** ***N. aurariae*** vol. 2/2 p. 113
41b Valves normalement siliceuses, comme chez les *"Lanceolatae"* comparables. Aréoles non en quinquonce. Vraisemblablement 2 formes hétérospécifiques d'eaux douces et salées (vol. 2/2 Fig. 79: 12–15) L.: 8–33 µm/l.: 2,5–5 µm/f. 14–20/10 µm/str. (40)43–55/10 µm . . . . **18.** ***N. pusilla*** part. vol. 2/2 p. 111
42a Valves comparativement un peu plus larges, 4–5 µm. Stries 35–40/10 µm (vol. 2/2 Fig. 80: 10–15) L.: 14 B 60 µm/l.: 4–5 µm/f.: 14–18(20)/10 µm/str. 35–40/10 µm . . . . **120.** ***N.*** **(?)** ***bergii*** vol. 2/2 p. 113
42b Stries toujours > 43/10 µm **43**
43a Populations ne montrant aucune tendance à des pôles étirés. Formes d'eaux saumâtres (vol. 2/2 Fig. 80: 1–9) L.: 17–25(36) µm/l.: 3–4 µm/f.: 15–17(20)/10 µm/str. 50/10 µm . . . . . **119.** ***N. perspicua*** vol. 2/2 p. 112
43b Les valves plus grandes montrent des tendances à l'étirement des pôles. Vraisemblablement 2 formes hétérospécifiques d'eaux douces et saumâtres (vol. 2/2 Fig. 79: 12–15 et 16–28) L.: 8–33 µm/l.: 2,5–5 µm/f. 14–20(24)/10 µm/ str. (40)43–55/10 µm . . **118.** ***N. pusilla*** vol. 2/2 p. 111
44a Valves comparativement courtes, +/- lancéolées, de largeur < 5 µm avec pôles arrondis-pointus à légèrement capités **45**
44b Caractéristiques différentes **51**
45a Stries 35–40/10 µm. Présence dans des lacs oligo- et mésotrophes moyennement alcalins (vol. 2/2 Fig. 78: 1–6) L: 10–20 µm/l.: 2–3 µm/f. 13–18/10 µm/str. 35–40/10 µm . . . . . . **112.** ***N. lacuum*** vol. 2/2 p. 107
45b Caractéristiques différentes, en particulier stries moins espacées **46**
46a Extrémités ne montrant aucune tendance à la microcéphalie, arrondies, pointues. Fibules plus larges à la base qu'au bord de la valve. Uniquement en eau douce (vol. 2/2 Fig. 83: 1–9) L.: 20–60 µm/l.: 3–5 µm/f.: 9–12/10 µm/str. +/- 50/10 µm . . . . . **132.** ***N. sociabilis*** vol. 2/2 p. 119
46b Caractéristiques différentes **47**
47a Formes d'eaux douces +/- moyennement alcalines **49**
47b Formes d'eaux saumâtres **48**
48a Valves plus courtes (jusqu'à +/- 20 µm) et de ce fait apparaissant plus larges-lancéolées (vol. 2/2 Fig. 81: 17–21) L.: 8–16 µm/l.: 3–4 µm/f. 17–20/10 µm . . . . **126.** ***N. rosenstockii*** vol. 2/2 p. 116
48b Valves plus longues > 30 µm (vol. 2/2 Fig. 82: 9–11a) . . . . **128.** ***N. agnita*** vol. 2/2 p. 117
49a Valves aux extrémités étirées et rostrées. Stries > 50/10 µm (vol. 2/2 Fig. 81: 14–16). Cf. aussi *N. palea* part. avec str. +/- 40/10 µm (vol. 2/2 Fig. 59: 23) L.: 30–37 µm/l.: 2,5–3 µm/f. 14–18/10 µm/str. > 50/10 µm . . . . **125.** ***N. pumila*** vol. 2/2 p. 115
49b Extrémités toujours courtes **50**
50a Fibules très minces en forme de traits (vol. 2/2 Fig. 58: 1–9) L.: 35–50 µm/ l.: 4–5 µm/f. (14)16–20(24)/10 µm . . . . **77.** ***N. pura*** vol. 2/2 p. 75
50b Fibules plutôt courtes en forme de points ; str. > 46/10 µm (vol. 2/2 Fig. 81:

10–12, ?13). Cf. aussi *N. palea* part. avec stries +/- 40/10 µm L.: 15–40 µm/l.: 2–3 µm/f. 14–19/10 µm/str. 46–55/10 µm . . . . **124. *N. archibaldii*** vol. 2/2 p. 115

51a Fibules en forme de traits (vol. 2/2 Fig. 58: 10–15). Cf. aussi *N. pura* (vol. 2/2 Fig. 58: 1–9) L.: (20)30–90 µm/l.: 4–6 µm/f. 13–17/10 µm/str. 34–38/10 µm . . . . **76. *N. sublinearis*** vol. 2/2 p. 74

51b Fibules en forme de points, courtes **52**

52a Valves lancéolées à linéaires-lancéolées, pas étroites lancéolées **53**

52b Valves généralement étroites dans la population, au moins partiellement linéaires **54**

53a Extrémités nettement capitées, fibules très épaisses (vol. 2/2 Fig. 65: 11–13) L.: 19–40 µm/l.: 4,5–5,5 µm/f. 7–12/10 µm/str. 42/10 µm . . . . **86. *N. pseudofonticola*** vol. 2/2 p. 92

53b Caractéristiques différentes n'apparaissant pas ou pas régulièrement (vol. 2/2 Fig. 59: 1 à 60: 7) L.: 15–70 µm/l.: 2,5–5 µm/f. 9–17/10 µm/str. 28–40/10 µm . . . . **80. *N. palea*** part. vol. 2/2 p. 85

54a Longueur toujours < 40 µm ; str. 34–37/10 µm. Formes nord-alpines. Vraisemblablement toujours en eaux acides (vol. 2/2 Fig. 66: 12–16) L.: 28–50 µm/l.: +/- 3 µm/f. 9–17/10 µm/str. 34–37/10 µm . . . . **88. *N. suchlandtii*** vol. 2/2 p. 93

54b Caractéristiques différentes (vol. 2/2 Fig. 66: 1–11; 67: 1–3) L.: 30 (et -)-110 µm/l.: 2,5–4 µm/f.: 12–18/10 µm/str. n. r. . . . . **87. *N. gracilis*** vol. 2/2 p. 93
. . . . **129. *N. acicularioides*** vol. 2/2 p. 118

55a Bien qu'apparaissant relativement longues et minces, fibules débouchant à la base dans plus d'une strie, et coupant une ligne longitudinale délicate (vol. 2/2 Fig. 12: 1–11) L.: 35–100(130 et +) µm/l.: 3,5–7(8) µm/f. 5–10/10 µm/str. 35–52/10 µm . . . . **11. *N. recta*** vol. 2/2 p. 20

55b Caractéristiques différentes **56**

56a Fibules en forme de traits ; stries < 25/10 µm (vol. 2/2 Fig. 13: 1–5) L.: 70–190 µm/l.: 4–7 µm/f. 10–11 à 14/10 µm/str. 20–24(26)/10 µm . . . . **13. *N. heufleriana*** vol. 2/2 p. 22

56b Caractéristiques différentes **57**

57a Fibules plus courtes et pointues ou avec une “petite dent”, chacune reliée à une strie (vol. 2/2 Fig. 71: 1–12) L.: 18–50 µm/l.: 4–6 µm/f. 11–16/10 µm/str. 24–28/10 µm . . . . **96 *N. solita*** vol. 2/2 p. 99

57b Fibules formées autrement, stries beaucoup plus proches **58**

58a Fibules exceptionnellement petites par rapport à la taille valvaire et en termes absolus, > 20/10 µm, apparaissant souvent groupées irrégulièrement (vol. 2/2 Fig. 13: 6, 7)L.: 90–140 µm/l.: 4–5,5 µm/f. +/- 12/10 µm (vis. micro. 20–25/10 µm) . . . . **14. *N. fibulafissa*** vol. 2/2 p. 22

58b Caractéristiques différentes **59**

59a Stries +/- 35/10 µm. Valves en moyenne un peu plus larges, avec tendance plus forte à un contour linéaire (vol. 2/2 Fig. 58: 10–15) L.: (20)30–90 µm/l.: 4–6 µm/f. 13–17/10 µm/str. 34–38/10 µm . . . . **76. *N. sublinearis*** (vol. 2/2 p. 74)

59b Stries +/- 40/10 µm. Valves comparativement un peu plus étroites avec plus forte tendance au contour lancéolé (vol. 2/2 Fig. 58: 1–9) L.: 35–50 µm/l.: 4–5 µm/f. (14)16–20(24) . . . . **77. *N. pura*** vol. 2/2 p. 75

Lanceolatae groupe B (vol. 2/2 p. 82)

1a Fibules très étroites, en général allongées ou cunéiformes, rétrécies en forme de racine dentaire, débouchant régulièrement sur une seule strie (côte transapicale) (seulement reconnaissable lors d'une focalisation minutieuse) **26**

1b Fibules plus larges. Même si stries non-résolvables, leur forme +/- épaisse et non-effilée montre qu'elles doivent être liées à plus d'1 strie **2**
2a Stries comparativement plus écartées, < 15–25/10 µm, le plus souvent reconnaissable quand elles sont ponctuées **3**
2b Stries > 25/10 µm **10**
3a Stries grossières, valves toujours petites, 16–19/10 µm. Ponctuation non-résolvable car en doubles rangées serrées (vol. 2/2 Fig. 84: 9–12) L.: 5–13 µm/l.: 2,5–3 µm/f. 6–14/10 µm/str. 16–19/10 µm . . . . **135.** ***N. valdestriata*** vol. 2/2 p. 121
3b Caractéristiques différentes **4**
4a Stries toujours < 20/10 µm. Fibules (sous microscope) débouchant en pointe dans les côtes transapicales (vol. 2/2 Fig. 78: 13–26). Cf. aussi *N. amphibioides* (vol. 2/2 Fig. 78: 27–29) et *N. gisela* (vol. 2/2 Fig. 42: 7, 8) L.: 6–50 µm/l.: 4–6 µm/f. 7–9/10 µm/str. 13–18/10 µm . . . . **114** ***N. amphibia*** vol. 2/2 p. 108
4b Caractéristiques différentes **5**
5a Valves comparativement longues, linéaires. Fibules 18–21/10 µm débouchant chacune dans une côte transapicale (vol. 2/2 Fig. 42: 7, 8) . . . . **53** ***N. gisela*** vol. 2/2 p. 57
5b Caractéristiques différentes **6**
6a Valves +/- lancéolées. Formes d'eaux douces moyennement alcalines Fig. 76: 8–16) L.: 30–85 µm/l.: 3,5–5 µm/f. 7–9/10 µm/str. 18–21/10 µm . . . . **107.** ***N. fossilis*** vol. 2/2 p. 105
6b Caractéristiques différentes **7**
7a Stries +/- 20/10 µm. Formes d'eaux très alcalines, en particulier salines. Sous microscope électronique, rangées d'aréoles dans la zone de la carène raphéenne non-doublées (vol. 2/2 Fig. 68: 11–17) L.: 5–60 µm/l.: 2–4,5 µm/f. 10–16/10 µm/str. 18–21/10 µm . . . . **89.** ***N. frustulum*** var. ***bulnheimiana*** vol. 2/2 p. 94
7b Caractéristiques différentes **8**
8a Valves plus grandes, régulièrement linéaires. Sous micr. électronique, rangées d'aréoles dans la zone de la carène raphéenne doublées. Formes plutôt d'eau acide (vol. 2/2 Fig. 73: 9–18) L.: 8–50 µm/l.: 3–4(5) µm/f. 7–12,5/10 µm/str. 20–26/10 µm . . . **99.** ***N. hantzschiana*** vol. 2/2 p. 101
8b Caractéristiques différentes, en particulier au niveau de l'écologie des eaux **9**
9a Valves comparativement plus longues, lancéolées ou linéaires-lancéolées à extrémités plutôt pointues à faiblement capitées (vol. 2/2 Fig. 75: 1 à 76: 7) L.: 10–65 µm/l.: 2,5–5 µm/f. 9–16/10 µm/str. 23–33/10 µm . . . . **104.** ***N. fonticola*** part. 103
L.: 10–65 µm/l.: 2,5–4 µm/f. 8–12(16)/10 µm/str. 23–25/10 µm . . . . **105.** ***N. tropica*** part. vol. 2/2 p. 104
9b Valves restant toujours courtes. Extrémités tronquées (vol. 2/2 Fig. 69: 1–13) L.: 3–22 µm/l.: 2,5–3,5 µm/f. 8–13/10 µm/str. 23–32/10 µm . . . . **90.** ***N. inconspicua*** vol. 2/2 p. 95
10a Frustules formant des colonies étoilées en aigrette (vol. 2/2 Fig. 77: 1–5, ?6) L.: 20–70 µm/l.: 2–3 µm/f. 10–15/10 µm/str. 28–30/10 µm . . . . **108.** ***N. incognita*** vol. 2/2 p. 106
10b Caractéristique différente ou n'étant plus définissable **11**
11a Valves relativement larges et proportionnellement restant exceptionnellement courtes. Les 2 fibules du milieu très espacées (vol. 2/2 Fig. 65: 1–2a). Cf. aussi *N. pseudocommunis* (vol. 2/2 Fig. 56: 10) L.: (5)12–26,5 µm/l.: 2,4–4,4–7 µm/f. 10–14/10 µm/ str. 32–36(50)/10 µm . . . . **73.** ***N. laevis*** vol. 2/2 p. 72

11b Caractéristiques différentes **12**
12a Fibules en forme de traits (vol. 2/2 Fig. 44: 1–7) L.: 20–73/l.: 4–6/f. 16–20/10 μm/str. 30–36/10 μm . . . . **56.** ***N. thermaloides*** vol. 2/2 p. 59
12b Fibules plutôt ponctiformes ou plus larges **13**
13a Les valves plus grandes d'une population sont linéaires, non lancéolées alors que les extrémités sont +/- courtes, cunéiformes et rétrécies . . . **14**
13b Les valves les plus longues ont tendance à être lancéolées ou linéaires-lancéolées alors que les extrémités sont plus longuement cunéiformes et rétrécies **16**
14a Largeur valvaire 3 à 6 μm ; formes d'eaux alcalines **15**
14b Largeur valvaire 2,5 à 3 μm. Le plus souvent nettement plus courtes que les précédentes. Formes dominant dans des eaux acides (vol. 2/2 Fig. 73: 1–8). Cf. aussi *N. incognita* très semblable, dans des eaux plus alcalines (vol. 2/2 Fig. 77: 1–5) . . . . **98.** ***N. acidoclinata*** vol. 2/2 p. 100
15a Str. 26–29/10 μm, facilement résolvables même sans éclairage oblique (relief obligatoire) (vol. 2/2 Fig. 77: 11–14) L.: 45–95 μm/l.: 3–4 μm/f. 10–13(16)/10μm/Str. 26–29/10 μm . . . . **109.** ***N. gessneri*** vol. 2/2 p. 106
15b Str. 29 jusqu'à >35/10 μm, plus difficiles à résoudre (vol. 2/2 Fig. 63: 8 à 64: 16) Cf. aussi *N. capitellata* (vol. 2/2 Fig. 62: 1–13a) L. 14–70 μm/l.: 3,5–6 μm/f. 7–13/10 μm/str. 29–35(40)/10 μm . . . . **84.** ***N. tubicola*** vol. 2/2 p. 90
16a Valves toujours courtes, elliptiques à linéaires-elliptiques selon Fig. 69: 1–13. L.: 3–22 μm/l.: 2,5–3,5 μm/f. 8–13/10 μm/str. 23–32/10 μm . . . . **90.** ***N. inconspicua*** vol. 2/2 p. 95
16b Caractéristiques différentes, en particulier pôles moins émoussés, plutôt pointus à capités-arrondis **17**
17a Valves nettement lancéolées, pôles pointus-arrondis L.: 10–65 μm/l.: 2,5–5 μm/f. 9–19/10 μm/str. 23–33/10 μm . . . . **104.** "***N. fonticola*** part." vol. 2/2 p. 103
17b Caractéristiques différentes **18**
18a Espèces avec valves strictement lancéolées. Longueur le plus souvent nettement < 40 μm. Formes d'eaux moyennement à plus alcalines. Sous ME doublure des rangées d'aréoles près de la carène raphéenne. (vol. 2/2 Fig. 75: 1–23). Cf. aussi les valves plus longues et étroites de *N. radicula* (vol. 2/2 Fig. 77: 7–10) L.: 10–65 μm/l.: 2,5–5 μm/f. 9–16/10 μm/str. 23–33/10 μm . . . . **104.** ***N. fonticola*** part. 103
18b Caractéristiques différentes, plus particulièrement les rangées d'aréoles qui sont toujours simples **19**
19a Valves lancéolées à linéaires-lancéolées, atteignant à peine 30 μm, le plus souvent plus courtes et dépassant rarement 3 μm de large. Extrémités non capitées (vol. 2/2 Fig. 68: 1–8; 68: 20–24a) L.: 5 à 60 μm/l.: 2–4,5 μm/f. 10–16/10 μm/str. 19–30/10 μm . . . **89.** ***N. frustulum*** part. vol. 2/2 p. 94 . . . . **91.** ***N. leistikowii*** vol. 2/2 p. 96
19b Caractéristiques différentes **20**
20a Valves étroites et lancéolées, jamais concaves des 2 côtés, largeur 2,5–3 μm; stries comparativement faciles à résoudre (vol. 2/2 Fig. 77: 7–10; 77: 1–5, ?6) L.: 33 à > 70 μm/l.: 2,5–3 μm/f. 10–13/10 μm/str. 28–30/10 μm . . . . **106.** ***N. radicula*** vol. 2/2 p. 105
L. 20–70 μm/l. 2–3 μm/f. 10–15/10 μm/ str. 28–30/10 μm . . . . **108.** ***N. incognita*** vol. 2/2 p. 106
20b Valves plus larges 3,5–6,5 μm, souvent +/- concaves au centre des 2 côtés. Stries plus difficiles à résoudre selon le relief **21**
21a Fibules plus larges. Ecarts entre elles irréguliers (vol. 2/2 Fig. 63: 8 à 64: 16)

L.: 14–70 µm/l.: 3,5–6 µm/f. 7–13/10 µm/str. 29–35/10 µm (genre espèces) . . . . **84.** ***N. tubicola*** (complexe) vol. 2/2 p. 90

21b Fibules plus étroites. Ecarts entre elles plus réguliers (vol. 2/2 Fig. 62: 1 à 63: 3). Cf. aussi *N. calcicola* (vol. 2/2 Fig. 63: 4–6) et sous *N. paleaeformis* "espèce d'Irlande" (vol. 2/2 Fig. 65: 9, 10) L.: 20–70 µm/l.: 3,5–6,5 µm/f. 10–18/10 µm/str. 23–40/10 µm . . . . **83.** ***N. capitellata*** (complexe incl. *N. subinvicta*) vol. 2/2 p. 88
L.: 22 µm/l.: 5 µm/f. 14/10 µm/str. 40/10 µm . . . . ***N. subinvicta*** vol. 2/2 p. 90

21c Caractères differents **22**

22a Valves étroites et lancéolées à étroites et linéaires-lancéolées, ou aux extrémités longuement rostrées. Largeur rarement > 3 µm. Stries toujours résolvables à l'aide de moyens contrastants très puissants du microscope optique **23**

22b Caractéristiques différentes. Stries comparativement faciles à résoudre avec éclairage oblique **24**

23a Extrémités longues, rostrées (vol. 2/2 Fig. 81: 8, 9) L.: 58–150 µm/l.: 2–2,5 µm/f. 16–21/10 µm/str. 45–60/10 µm . . . . **123.** ***N. graciliformis*** vol. 2/2 p. 115

23b Caractéristiques différentes. Contours variables (vol. 2/2 Fig. 81: 1–7) L.: 8–55(80) µm/l.: 1,5–4 µm/f.(12)14–19/10 µm/str. 44–55/10 µm . . . . **122.** ***N. paleacea*** vol. 2/2 p. 114

24a Fibules plus larges. Ecarts entre elles très irréguliers (vol. 2/2 Fig. 63: 8 à 64: 16) L.: 14–70 µm/l.:3,5–6 µm/f. 7–13/10 µm/str. 29–35/10 µm (genre espèces) . . . . **84.** ***N. tubicola*** (complexe) part. vol. 2/2 p. 90

24b Fibules plus étroites. Ecarts plus réguliers. **25**

25a Formes d'eaux moyennement à plus fortement alcalines (vol. 2/2 Fig. 62: 1 à63: 3) L.: 20–70 µm/l.: 3,5–6,5 µm/f. 10–18/10 µm/str. 23–40/10 µm . . . . **83.** ***N. capitellata*** (complexe) part. vol. 2/2 p. 88

25b Formes d'eaux plutôt moins alcalines (vol. 2/2 Fig. 65: 3–10) L.: 30–90 µm/l.: 3–5 µm/f. 10–13/10 µm/str. 35–40/10 µm . . . . **85.** ***N. paleaformis*** vol. 2/2 p. 92

26a Fibules n'apparaissant pas comme +/- des traits mais relativement épaisses et débouchant dans côtes transapicales, cunéiformes ou "en forme de racines étroites". Valves toujours comparativement petites avec des stries inhabituellement grossières, < 20/10 µm (vol. 2/2 Fig. 78: 13–26). Cf. aussi *N. amphibioides* (vol. 2/2 Fig. 78: 27–29) L.: 6–50 µm/l.: 4–6 µm/f. 7–9/10 µm/str. 13–18/10 µm . . . . . . **114.** ***N. amphibia*** vol. 2/2 p. 108

26b Caractéristiques différentes **27**

27a Stries +/- 20/10 µm (vol. 2/2 Fig. 42: 7, 8) . . **53.** ***N. gisela*** vol. 2/2 p. 57

27b Stries nettement plus rapprochées **28**

28a Largeur +/- 8 µm avec un pli longitudinal étroit (vol. 2/2 Fig. 43: 5, 7) L.: 38–55 µm/l.: 7,5–8,5 µm/f. 14–15/10 µm/str. 30–32/10 µm . . . . **55.** ***N. normanii*** vol. 2/2 p. 58

28b Caractéristiques différentes **29**

29a Largeur +/- 4 µm (vol. 2/2 Fig. 55: 1–10) . . **72.** ***N. subtilis*** vol. 2/2 p. 71
. . . . **71.** ***N. linearis*** part vol. 2/2 p. 69.

29b Valves plus larges **30**

30a Fibules en général très serrées, > 15/10 µm, les 2 du centre moins écartées. Formes marines (vol. 2/2 Fig. 44: 1–7) L.: 20–73 µm/l.: 4–6 µm/f. 16–20/10 µm/str. 30–36/10 µm . . . . . **56.** ***N. thermaloides*** vol. 2/2 p. 59

30b Fibules pas exceptionnellement rapprochées, (ici) < 15/10 µm, les 2 du centre plus visiblement écartées. Formes surtout d'eaux douces (vol. 2/2 Fig. 55: 1–10) L.: 34–228 µm/l.: 2,5–7,5 µm/f. 8–17(+?)/10 µm/str. 28–41/10 µm . . . . **71.** ***N. linearis*** part. vol. 2/2 p. 69

Nitzschiellae (vol. 2/2 p. 121)
(Y compris les formes de Lanceolatae qui portent à confusion)

1a Axe apical sigmoïde ou (plus rarement) en croissant 2
1b Axe apical droit 6
2a Stries grossières, le plus souvent < 20/10 µm (vol. 2/2 Fig. 86: 6–10) L.: 37(50)-190 µm/l.(3)4–7 µm/f. 6–10/10 µm/str. 13–19/10 µm . . . . **141.** ***N. lorenziana*** vol. 2/2 p. 125
2b Stries (beaucoup) plus rapprochées 3
3a Fibules de la carène raphéenne qui est modérément ou peu excentrée très larges; fibules du centre équidistantes (vol. 2/2 Fig. 86: 1–4) L.: 220–250 µm/l.: 5–6,5 µm/f.6 à 15/10 µm ext./str. 40–42/10 µm . . . . **142.** ***N. behrei*** vol. 2/2 p. 126
3b Caractéristiques différentes 4
4a Frustules très faiblement siliceuses. Extrémités apparaissant “fines comme des cheveux”, facilement déformables ou destructibles par le procédé de préparation. La plupart en forme de croissant, plus rarement sigmoïdes, souvent réunies en amas épais. Surtout dans le plancton marin, plus rarement dans les eaux intérieures salines (vol. 2/2 Fig. 87: 1, 2) L.: 30–400 µm/l.: (1)2–6 µm/f. 12–37/10 µm/str. 70–100 (100 non-résolvable) . . . . **140.** ***N. closterium*** vol. 2/2 p. 125
4b Caractéristiques différentes. Valves plus fortement siliceuses comme d’autres Nitzschiae de largeur valvaire comparable 5
5a Caractéristiques correspondant à la Fig. 85: 7–10 . . . . **139.** ***N. reversa*** vol. 2/2 p. 124
5b Si les valves sont en forme de croissant, il pourrait s’agir de *N. acicularis* var. *closterioides*. Si elles sont sigmoïdes, il pourrait s’agir d’autres taxons marins non présentés ici
6a Raphé avec nodule central. Les 2 fibules du centre écartées 7
6b Caractéristique différente, toutes les fibules du centre équidistantes . . 9
7a Partie centrale des valves fusiforme, 3–4,5 µm de large, avec des becs brusquement étirés. Fibules 18–21/10 µm, stries > 50/10 µm (vol. 2/2 Fig. 85: 5, 6) L.: 40–110 µm/l.: 3–4,5 µm/f. 18–21/10 µm/str. 55–64/10 µm . . . . **138.** ***N. draveillensis*** vol. 2/2 p. 123
7b Caractéristiques différentes 8
8a Extrémités de la partie centrale plus étroite (< 3 µm de large), moins brusquement étirées. Stries 45–60/10 µm, dans ou au-delà du champ de résolvabilité en microscopie optique (vol. 2/2 Fig. 81: 8, 9) L.: 58–150 µm/ l.: 2–2,5 µm/f. 16–21/10 µm/str. 45–60/10 µm . . . . **123.** ***N. graciliformis*** vol. 2/2 p. 115
8b Valves plus larges et stries plus écartées . . . . **83.** ***N. capitellata*** part. (formes tenuirostris) vol. 2/2 p. 88
9a Extrémités brusquement étirées, rostrées à partir de la partie centrale fusiforme. Stries non résolvables en microscopie optique (vol. 2/2 Fig. 85: 1–4) L.: 30–150 µm/l.: 2,2–5 µm/f. 15–22/10 µm/str. 60–72/10 µm . . . . **137.** ***N. acicularis*** vol. 2/2 p. 123
9b Caractéristiques différentes 10
10a Stries +/- 30/10 µm, relativement faciles à résoudre 11
10b Stries plus rapprochées, toujours résolvables avec éclairage oblique . . 12
11a Largeur des valves +/- 3 µm maximum (vol. 2/2 Fig. 67: 4–10) L.: +/- 20–80 µm/l.: (1,5)2–3 µm/f. 12–16/10 µm/ str. (26)27–33/10 µm . . . . **130.** ***N. subacicularis*** vol. 2/2 p. 118
11b Valves plus larges (vol. 2/2 Fig. 67: 11). Cf. aussi *N. longirostris* (LB-K, 1987) . . . . **131.** ***N. rostellata*** vol. 2/2 p. 119

12a Stries +/- 40/10 µm encore relativement faciles à résoudre par éclairage oblique L.: 15–70 µm/l.: 2,5–5 µm/f. 9–17/10 µm/str. 28–40/10 µm . . . . **80.** ***N. palea*** vol. 2/2 p. 85
L.: 40–60 µm/l.: 2,5–4 µm/f. 12–18/10 µm/str. 38–42/10 µm . . . . **129.** ***N. acicularioides*** vol. 2/2 p. 118
L.: 30(-)-110 µm/l.: 2,5–4 µm/f. 12–18/10 µm/str. 38–42(50)/10 µm . . . . **87.** ***N. gracilis*** vol. 2/2 p. 93

12b Stries encore plus rapprochées, plus difficiles à résoudre (vol. 2/2 Fig. 81: 14–16) (cf. *N. agnita* et autres formes (vol. 2/2 Fig. 82: 9–14) et autres espèces de LB-K, 1987) L.: 30–37 µm/l.: 2,5–3 µm/f. 14–18/10 µm/str. non résolvables . . . . **125.** ***N. pumila*** vol. 2/2 p. 115

3. Genre ***Hantzschia*** **Grunow 1877 nom. cons.** (vol. 2/2 p. 126)

Typus generis: *Eunotia amphioxys* Ehrenberg 1843 ty vol. 2/2 p. cons.; (*Hantzschia amphioxys* (Ehrenberg) Grunow in Cleve & Grunow 1880)

Clef pour les espèces

1a Nombre de fibules identique au nombre de côtes transapicales. Stries (+/– visibles) à double rangées de points en quinquonce (vol. 2/2 Fig. 93: 1–6) L.: 40–100 µm/l.: (4,5)14–27 µm/f. -/tr.c.: 6–13/10 µm . . . . **10.** ***H. marina*** vol. 2/2 p. 127

1b Caractéristiques différentes **2**

2a Raphé sans interruption d'un pôle à l'autre. Fibules centrales équidistantes **3**

2b Raphé interrompu au centre, les 2 fibules centrales nettement écartées **4**

3a Frustules faiblement sigmoïdes; valves dorsiventrales non typiquement hantzschioïdes (vol. 2/2 Fig. 91: 1–4) L.: 150–500 µm/l.: 10–15 µm/f. 4 à 6/10 µm/str. 9–12/10 µm . . . . **11.** ***H. spectabilis*** vol. 2/2 p. 132

3b Valves dorsiventrales typiquement hantzschioïdes, frustules non sigmoïdes L.: 65–260 µm/l.: 5–14 µm/f. 5–6/10 µm/str. 11–14/10 µm (vol. 2/2 Fig. 91: 5, 6) . . . . **4.** ***H. vivax*** vol. 2/2 p. 129

4a Stries comparativement très rapprochées, >= 30/10 µm. Formes marines (vol. 2/2 Fig. 92: 1–4, 5–7, 8, 9) L.: 55–76 µm/l.: 6–7 µm/f. 9–11/10 µm/str. 33–35/10 µm . . . . **5.** ***H. weyprechtii*** vol. 2/2 p. 129
. . . . **6.** ***H. petitiana*** vol. 2/2 p. 130
L.: 48–76 µm/l.: 7–10 µm/10 µm/str. 30–34/10 µm . . . . **7.** ***H. baltica*** vol. 2/2 p. 130

4b Caractéristiques différentes **5**

5a Fibules étroites d'un bout à l'autre, toujours reliées à une seule côte transapicale, principalement de forme marine, plus rarement en eaux continentales saumâtres **6**

5b Fibules au moins partiellement plus larges et reliées à plus d'une côte transapicale. Formes principalement d'eaux douces **7**

6a Stries très grossièrement ponctuées par rapport aux relatives grandes valves. Nombre de stries moins du double du nombre des fibules à peine étirées (vol. 2/2 Fig. 88: 8–10) L.: 40–85 µm/l.: 5–8,5 µm/f. -/str. 8,5–18/10 µm . . . . **9.** ***H. distinctepunctata*** vol. 2/2 p. 131

6b Nombre de stries plus du double du nombre des fibules très étirées (vol. 2/2 Fig. 90: 1–8) L.: 50–150 µm/l.: 5–12(18) µm/f. 3,5–7/10 µm/str. 7–15/10 µm . . . . **8.** ***H. virgata*** vol. 2/2 p. 130

7a Valves étroites (10–14 µm) par rapport à la longueur (230–430 µm) (vol.

2/2 Fig. 89: 1–2a) L.: 230–430 µm/l.: 10–14 µm/f. (7)8–10/10 µm/str. 18–19/10 µm . . . . **2. *H. elongata*** vol. 2/2 p. 128

7b Proportions moins accentuées par la longueur **8**

8a Valves toujours très grandes (vol. 2/2 Fig. 89: 3–5) L.: 170–265 µm/l.: 17–22 µm/f. (2)4–6/10 µm/str. 14–16/10 µm **3. *H. rhaetica*** vol. 2/2 p. 129

8b Proportions des valves extrêmement variables, mais n'atteignant presque jamais les dimensions de *H. rhaetica* (vol. 2/2 Fig. 88: 1–7) L.: 20–210(300) µm/l.: 5–15(25) µm/f. 4–11/10 µm/str. 11–28/10 µm . . . . **1. *H. amphioxys*** (groupe d'espèces) vol. 2/2 p. 128

5. Famille Epithemiaceae sensu Karsten in Engler & Prantl 1928 (vol. 2/2 p. 135)

Typus familiaris: *Epithemia* Brébisson ex Kützing 1844

Clef pour les genres

1a Axe transapical hétéropolaire. De ce fait, valves dorsiventrales **2**

1b Axe transapical isopolaire. Valves non dorsiventrales . . . . **1. *Denticula*** (vol. 2/2 p. 137) (p. 303)

2a Le raphé repose distalement plus nettement sur le côté ventral de la valve et monte plus ou moins proximalement dans la surface valvaire . . . . **2. *Epithemia*** (vol. 2/2 p. 145) (p. 304)

2b Valves avec côté dorsal étroit tombant à pic, et côté ventral large et plat tombant à pic. Raphé sur l'arête entre les côtés ventral et dorsal . . . . **3. *Rhopaladia*** (vol. 2/2 p. 157) (p. 305)

1. Genre ***Denticula*** Kützing 1844 (vol. 2/2 p. 137)

Typus generis: *Denticula elegans* Kützing 1844 (Boyer 1927)
(f = parois fibulaires)

Clef pour les espèces

1a Doubles rangées d'aréoles entre les côtes transapicales, bien visibles en microscopie optique à grandes ouvertures (vol. 2/2 Fig. 99: 1–10) L.: 14–42 µm/l.: 3–7 µm/f. 35–50/100 µm/str. 12–14/10 µm . . . . **7. *D. thermalis*** vol. 2/2 p. 144

1b Toujours une rangée simple d'aréoles entre les côtes transapicales . . . **2**

2a Septi formant sur les parois fibulaires des "petites têtes" relativement grosses, bien visibles surtout en vue connective **3**

2b Petites têtes absentes ou minuscules en vue connective **5**

3a Stries 15–20/10 µm **4**

3b Stries +/- 12/10 µm (vol. 2/2 Fig. 98: 8–18) L.: 12–78 µm/l.: 4–8 µm/f. 25–50/10 µm/str. 12–13/10 µm . . . . **5. *D. eximia*** vol. 2/2 p. 142

4a En moyenne plus de 40 fibules/100 µm (mesures faites sur plusieurs exemplaires pour exclure les irrégularités). Petites têtes des septi relativement grandes (vol. 2/2 Fig. 96: 10–33; 97: 1–5) L.: 15–45 µm/l.: 4–8 µm/f. 25–50/100 µm/str. 15–18/10 µm . . . . **3. *D. elegans*** vol. 2/2 p. 141

4b Moins de 40 fibules/100 µm. Petites têtes des septi très grandes. Valves de construction très robuste (vol. 2/2 Fig. 97: 9–17; 98: 1–7) L.: 28–65 µm/l.: 7–11 µm/f. 25–40(50)/100 µm/str. 16–20/10 µm . . . . **4. *D. valida*** vol. 2/2 p. 142

5a Parois fibulaires très basses et donc visibles seulement avec mise au point adéquate. Protubérances sur ces parois en position marginale sur le côté portant le raphé. Septi courts sur le bord, structure grossière (vol. 2/2

Fig. 99: 11–23; 100: 1–14, 18–22) L.: 10–120 µm/l.: 3–8 µm/f. 50–80/100 µm/str. (13)14–18(20)/10 µm . **6. *D. kuetzingii*** vol. 2/2 p. 143

5b Parois fibulaires aussi hautes que le manteau valvaire sur toute la largeur de la valve. Septi avec ponts transapicaux entiers. Structure relativement délicate (+ de 22 stries/10 µm) **6**

6a Canal raphéen marginal, le plus souvent difficilement visible (vol. 2/2 Fig. 96: 1–9) L.: 7–20 µm/l.: 2–3 µm/f. 60–100/100 µm/str. 28–30/10 µm . . . . **2. *D. subtilis*** vol. 2/2 p. 140

6b Canal raphéen légèrement à côté de la ligne médiane, toujours bien visible en microscopie optique (vol. 2/2 Fig. 95: 4–25) L.: 6–42(60) µm/l.: 3–7 µm/f. 50–70/100 µm/ str. (22)25–30(32)/10 µm . . **1. *D. tenuis*** vol. 2/2 p. 139

2. Genre *Epithemia* Brébisson ex Kützing 1844 (vol. 2/2 p. 145)

Epithema Brébisson 1838 (non *Epithema* Blume 1826)
Typus generis: *Eunotia turgida* Ehrenberg 1838 (Boyer 1927)
(f = parois fibulaires)

Clef pour les espèces

1a "Petites têtes" des septi aisément visibles sur les parois fibulaires (cf. aussi *E. cistula* (vol. 2/2 Fig. 105: 7–11)) en vue connective **2**

1b Petites têtes des septi difficiles à discerner sur les parois fibulaires en vue connective **3**

2a Raphé s'avançant le plus souvent à plus de la moitié de la partie dorsale valvaire. Les trous dans les sillons des septi en position ventrale (vol. 2/2 Fig. 102: 1–9; 103: 1–5) L.: 20–130 µm/l.: 4–18 µm/f. 10–20(30)/100µm/ . . . . **1. *E. argus*** vol. 2/2 p. 148

2b Raphé ne s'avançant pas à plus de la moitié de la partie dorsale de la valve. Trous dans les sillons des septi en position dorsale (vol. 2/2 Fig. 103: 6–9) L.: 40–120 µm/l.: 12–18 µm/f. 10–20/100 µm/str. 10–12/10 µm . . . . **2. *E. goeppertiana*** vol. 2/2 p. 150

3a Le raphé se situe en totalité sur le bord ventral de la valve. Seuls les pores centraux sont visibles. Parois fibulaires parallèles (cf. aussi *E. adnata* (vol. 2/2 Fig. 104: 1–7) L.: 20–78 µm/l.: 9–15 µm/f. 25–40/100 µm/str. 10–12/10 µm . . . . **3. *E. frickei*** vol. 2/2 p. 150

3b Raphé montant plus haut en direction dorsale ou situé dans la surface valvaire **4**

4a Branches du raphé presque droites, se développant sur la totalité de la face valvaire à proximité du côté dorsal de la valve (vol. 2/2 Fig. 105: 7–11) L.: 25–70 µm/l.: 7–12 µm/f. 20–40/100 µm/str. 11–14/10 µm . . . . **4. *E. cistula*** vol. 2/2 p. 151

4b Branches du raphé plus ou moins courbées, se développant au moins dans la partie centrale sur le côté ventral de la valve **5**

5a En moyenne (aussi lorsqu'il n'y a qu'une seule valve) 3 ou plus de stries par chambre interfibulaire **6**

5b 3 ou moins de stries par chambre interfibulaire **7**

6a Les branches du raphé se développent sur toute leur longueur dans la face valvaire et "montent" de façon courbe jusqu'au milieu du côté dorsal (vol. 2/2 Fig. 105: 1–6) L.: 20–80 µm/l.: 9–18 µm/f. 18–40/100 µm/str. 8–12/10 µm . . . . **5. *E. smithii*** vol. 2/2 p. 152

6b Les branches du raphé se développent dans presque toute leur longueur sur la partie ventrale de la valve et ne sont visibles qu'au milieu de la valve où elles atteignent à peine le centre valvaire (vol. 2/2 Fig. 107: 1–11; 108: 1–3) L.: 15–150 µm/l.: 7–14 µm/f. 20–80/100 µm/str. 11–14/10 µm . . . . **6.** ***E. adnata*** vol. 2/2 p. 152

7a Valves avec 5 fibules et plus / 10 µm, et plus de 10 stries/10 µm. Côté dorsal fortement convexe ; les branches du raphé atteignent le bord dorsal au centre de la valve (vol. 2/2 Fig. 106: 1–14) L.: 8–70 µm/l.: 6,5–16 µm/f. 50–75/100 µm/str. 10–15/10 µm . . . . **7.** ***E. sorex*** vol. 2/2 p. 154

7b Le plus souvent valves avec moins de 5 fibules/10 µm et 9 ou moins d'aréoles/10 µm. Bord dorsal moins convexe. Branches du raphé n'atteignant que le milieu de la valve dorsalement **8**

8a Extrémités larges et tronquées-arrondies. Côtés des valves presque parallèles. Valves fortement courbées (vol. 2/2 Fig. 104: 8–10) L.: 100–310 µm/l.: 20–26 µm/f. 25–40/100 µm/str. 6–8/10 µm . **8.** ***E. hyndmanii*** vol. 2/2 p. 154

8b Extrémités plus étroites, souvent étirées. Valves plus faiblement courbées (vol. 2/2 Fig. 108: 4–8; 109: 1–7) L.: 45–200 µm/l.: 13–35 µm/f. 30–500/100 µm/str. 7–9/10 µm . . . . **9.** ***E. turgida*** vol. 2/2 p. 155

3. Genre *Rhopalodia* O. Müller 1895 (vol. 2/2 p. 157)

Typus generis: *Navicula gibba* Ehrenberg 1830 (Boyer 1927)
(f = parois interfibulaires)

Clef pour les espèces

1a Valves en forme de parenthèses, aux extrémités pliées ventralement et arrondies en pointe (vol. 2/2 Fig. 111: 1–13; 111A: 1–7) L.: 22–300 µm/l.: 7–13 µm/f.50–80/100 µm/str. 30/10 µm . . . **1.** ***Rh. gibba*** vol. 2/2 p. 159

1b Contour valvaire différent **2**

2a Ponctuation reconnaissable en microscopie optique avec grande ouverture; moins de 30 points/10 µm sur les stries **3**

2b Ponctuation difficilement ou pas reconnaissable en microscopie optique; même à grande ouverture toujours plus de 30 points/10 µm **7**

3a Contour valvaire large en forme de croissant. Nombre de points < 16/10 µm sur les stries. Valves avec bord dorsal fortement convexe et contour trappu (vol. 2/2 Fig. 114: 1–11) L.: 12–80 µm/l.: 10–16 µm/f. 30–50/100 µm/str. 15–20/10 µm ponctuation 10–15/10 µm de stries . . . . **4.** ***Rh. musculus*** vol. 2/2 p. 163

3b Structure plus fine **4**

4a Bord ventral concave, contour valvaire en croissant **5**

4b Bord ventral droit ou faiblement convexe, contour valvaire en forme de demi-cercle ou lancéolé **6**

5a Carène peu développée ; raphé sur le bord dorsal à peine visible en vue valvaire (vol. 2/2 Fig. 112: 1–6; 113: 4–6) L.: 25–100 µm/l.: 5–12 µm/f. 30–100/100 µm/str. 15–19/10 µm . . . . **2.** ***Rh. gibberula*** vol. 2/2 p. 160

5b Carène et côté dorsal plus visibles, le canal raphéen est toujours bien visible en observant la face valvaire (vol. 2/2 Fig. 112: 7–10; 113: 1–3) L.: 22–112 µm/l.: 7,5–11 µm/f. 40–60/100 µm/str. 16–19/10 µm . . . . **3.** ***Rh. acuminata*** vol. 2/2 p. 162

6a Grandes formes avec bord ventral droit et d'un bout à l'autre des points doubles irrégulièrement ordonnés sur les stries, apparaissant le plus souvent comme points simples en microscopie optique (vol. 2/2 Fig. 113A: 1–6) L.: 24–75 µm/l.: 9–18 µm/f. 35–60/100 µm/str. 15–20/10 µm . . . . **5.** ***Rh. constricta*** vol. 2/2 p. 164

6b Formes plus petites, délicates, avec bord ventral droit ou concave. Stries constituées de points simples et donc fortement contrastées dans la partie ventrale des valves (vol. 2/2 Fig. 113: 7–13; 113A: 7–12) L.: 15–40 µm/l.: 5–8,5 µm/f. 35–60/100 µm/str. 17–22/10 µm . . . . **6. *Rh. brebissonii*** vol. 2/2 p. 164

7a Plus de 50 points/10 µm sur les stries. L/l des surfaces valvaires ventrales > 5,5 µm (vol. 2/2 Fig. 115: 1–8) L.: 20–74 µm/l.: 5–8 µm/f. 30–50/100 µm/str. 15–17/10 µm . . . . **7. *Rh. rupestris*** vol. 2/2 p. 165

7b Moins de 45 points/10 µm sur les stries. L/l des surfaces valvaires ventrales < 5 µm (vol. 2/2 Fig. 115: 9–12) L.: 18–52 µm/l.: 5–10 µm/f. 30–60/100 µm/str. 16–18/10 µm . . . . **8. *Rh. operculata*** vol. 2/2 p. 165

6. Famille Surirellaceae Kützing 1844 (vol. 2/2 p. 166)

Typus familiaris: *Surirella* Turpin

Clef pour les genres

1a Valves voûtées en forme de selle. Les axes apicaux des hypo- et épithèques se croisent en angle droit . . **4. *Campylodiscus*** (vol. 2/2 p. 211) (p. 310)

1b Structure des valves différente. Axes apicaux des hypo- et épithèques parallèles **2**

2a Valves étirées, linéaires. Axe apical courbé en "S" ou droit. Aire médiane étroite-linéaire, toujours nettement délimitée. Véritables côtes transapicales sur la partie extérieure de la valve, entre lesquelles se trouvent plusieurs rangées de petites aréoles . . . **3. *Stenopterobia*** (vol. 2/2 p. 207) (p. 310)

2b Caractéristiques différentes **3**

3a Ondulations transapicales et stries interrompues, au moins dans le secteur de la ligne médiane . . . . **2. *Surirella*** (vol. 2/2 p. 172)(p. 307)

3b Ondulations transapicales et stries non interrompues même au niveau de la ligne médiane . . . . **1. *Cymatopleura*** (vol. 2/2 p. 168) (p. 306)

1. Genre *Cymatopleura* W. Smith 1851 (vol. 2/2 p. 168)

Typus generis: *Surirella solea* Brébisson (Boyer 1927)

Clef pour les espèces

1a Les fibules continues transapicalement avec ondulations étroites atteignant l'aire axiale. Structure des creux et des crêtes des ondulations identiques aux ondulations apicales. Contour valvaire presque toujours panduriforme (vol. 2/2 Fig. 117: 1–5; 118: 1–8) L.: 30–300 µm/l_{max}.: 10–45 µm/f. 6–9/10 µm/str. 25–32/10 µm . . . . **1. *C. solea*** vol. 2/2 p. 168

1b Structure des valves différente. Structure des creux et des crêtes des ondulations différente **2**

2a Contour linéaire, lancéolé ou rhomboédrique (vol. 2/2 Fig. 119: 1–4; 120: 1–6; 121: 1–3) L.: 60–220(280) µm/l.: 30–90 µm/f. 2,5–6/10 µm/str. 15–20/10 µm . . . . **2. *C. elliptica*** vol. 2/2 p. 170

2b Contour largement elliptique à presque rond (vol. 2/2 Fig. 122: 1, 2) L.: 80–100 µm/l.: 70–100 µm . . . . **3. *C. brunii*** vol. 2/2 p. 171

2. Genre *Surirella* Turpin 1828 (vol. 2/2 p. 172)

Typus generis: *Surirella striatula* Turpin 1828

Clef pour les groupes

1a Valves avec ailes et canaux alaires. Formation de nœuds souvent visibles sur le bord des valves. Contour souvent linéaire-ovale, du moins pour les formes d'eaux douces . . . **2. Groupe Robustae** (vol. 2/2 p. 191) (p. 308)

1b Caractéristiques ci-dessus absentes. Contour souvent large-ovale, plus rarement linéaire. Des portules relient le canal raphéen à l'intérieur de la cellule. Formes d'eaux douces ou marines **2**

2a Genres caractérisés par des infondibules, des fibules en forme de fenêtre, et une zone centrale en forme de couronne. Formes exclusivement marines (non représentées dans ce volume) . . . . **3. Groupe Fastuosae**

2b De petites espèces pour la plupart, avec des pseudo-infondibules. Zone centrale absente ou seulement en forme de toit en bâtière ou d'aire +/- hyaline, ou ne consistant qu'en une carène apicale très étroite, ondulée. Fibules en forme de rigoles délimitant les chambres interfibulaires . . . . **1. Groupe Pinnatae** (vol. 2/2 p. 176) (p. 307)

1. Groupe des Pinnatae (vol. 2/2 p. 177)

(f = pseudo-infondibules)

Clef pour les espèces

1a Axe apical isopolaire **2**

1b Axe apical hétéropolaire **5**

2a Zone marginale avec pseudo-infondibules très nets jusqu'à la face valvaire et délimités (vol. 2/2 Fig. 138: 1–5; 139: 1–8) L.: 15–120 µm/l.: 12–20 µm/f. 30–40/100 µm/str. 14–16(18)/10 µm . . . **14. *S. amphioxys*** vol. 2/2 p. 189

2b Dans la zone marginale, seules des fibules étroites en forme de côtes sont visibles **3**

3a Formes grandes, robustes, linéaires avec moins de 17 stries/10 µm (vol. 2/2 Fig. 136: 1–4) L.: 40–190 µm/l.: 9–20 µm/f. 40–60/100µm/str. 11–16/10 µm . . . . **13. *S. gracilis*** vol. 2/2 p. 188

3b Formes linéaires fines et délicates avec plus de 22 stries/10 µm **4**

4a Formes plus courtes. L/l +/- 5/1 en vue connective. Les ondulations transapicales vont jusqu'à la ligne médiane (vol. 2/2 Fig. 133: 6–13; 134: 1, 6–10) L.: 18–70 µm/l.: 6–15 µm/f. 55–80/100 µm/str. (20)22–28/10 µm . . . . **11. *S. angusta*** vol. 2/2 p. 187

4b Formes avec des valves longues et étroites. L/l jusqu'à 15/1 en vue connective (vol. 2/2 Fig. 135: 15–17) L.: 50–110 µm/l.: 8–12 µm/f. 50–70/100 µm/str. 20–28/10 µm . . . . . **12. *S. lapponica*** vol. 2/2 p. 188

5a Ondulations transapicales parallèles jusqu'à la ligne médiane. L/l jusqu'à 5/1 ; 60 à 80 fibules/100 µm (vol. 2/2 Fig. 127: 14; 134: 2, 11, 12; 135: 1–14) L.: 9–47 µm/l.: 9–11 µm/f. 60–80/100 µm/str. 21–29/10 µm . . . . **10. *S. minuta*** vol. 2/2 p. 186

5b Caractéristiques différentes **6**

6a Valves larges en forme de poire **7**

6b Valves ovales à linéaires-ovales **9**

7a Valves faiblement on non concentriquement ondulées (vol. 2/2 Fig. 129: 1–5) L.: 30–65µm/l.: 27–31 µm/f. 35–80/100 µm/str. 17–25/10 µm . . . . **4. *S. crumena*** vol. 2/2 p. 182

7b Valves fortement concentriquement ondulées **8**

8a Largeur valvaire le plus souvent > 40 µm. Fibules et pseudoinfondibules formant un anneau marginal étroit (vol. 2/2 Fig. 131: 1–3) L.: 60–120 µm/l.: 40–70 µm/f. 30–60/100 µm/str. 15–18/10 µm . **6. *S. peisonis*** vol. 2/2 p. 183
8b Largeur valvaire généralement < 40 µm. Fibules et pseudo-infondibules formant un anneau marginal large (vol. 2/2 Fig. 132: 1–8; 133: 1–4) L.: 15–80 µm/l.: 10–45 µm/f. 40–45/100 µm/str. 14–19/10 µm . . . . **7. *S. brightwellii*** vol. 2/2 p. 183
9a Valves tordues autour de l'axe apical. Face valvaire ondulée de façon transapicale (vol. 2/2 Fig. 130: 1–8) L.: 15–40 µm/l.: 13–22 µm/f. 4,5–5/100 µm/str. 16–18/10 µm . . . . **5. *S. hoefleri*** vol. 2/2 p. 182
9b Valves non tordues **10**
10a Les 2 extrémités de la valve cunéiformes (visibles seulement chez les formes plus grandes) **11**
10b Chez les formes moyennes et plus grandes, un ou les deux pôles largement arrondis **13**
11a Formes linéaires-elliptiques avec le plus souvent moins de 35 fib./100 µm (vol. 2/2 Fig. 137: 1–9) L.: 40–90 µm/l.: 18–30 µm/f. 24–40/100 µm/str. 15–19/10 µm . . . . **9. *S. patella*** vol. 2/2 p. 185
11b Formes larges, lancéolées avec plus de 30 fib./100 µm **12**
12a Pseudo-infondibules pas clairement délimitées. Valves à ondulations faiblement concentriques (vol. 2/2 Fig. 125: 1–71; 126: 1) L.: 16–120 µm/l.: 12–45 µm/fib. 35–60/100 µm/str. (16)17–19/10 µm . . . . **1. *S. ovalis*** vol. 2/2 p. 178
12b Pseudo-infondibules clairement délimitées. Valves à ondulations fortement concentriques (vol. 2/2 Fig. 132: 1–8; 133: 1–4) L.: 15–80 µm/l.: 10–45 µm/côtes 30–45/100 µm/str. 14–19/10 µm . . **7. *S. brightwellii*** vol. 2/2 p. 183
13a Ondulations transapicales ou stries nettes **14**
13b Ondulations transapicales ou stries non nettes **15**
14a Longueur des valves toujours < 50 µm (vol. 2/2 Fig. 128: 1–10) L.: 15–48 µm/l.: 8–16 µm/f. 30–45/100 µm/str. 10–13/10 µm . . . . **3. *S. subsalsa*** vol. 2/2 p. 181
14b Longueur des valves toujours > 70 µm **16**
15a Contour valvaire linéaire-elliptique (vol. 2/2 Fig. 130: 9, 10; 134: 3–5) L.: 20–46 µm/l.: 11–13 µm/f. 30–40/100 µm/str. 15–17/10 µm . . . . **8. *S. visurgis*** vol. 2/2 p. 184
15b Contour valvaire ovale (vol. 2/2 Fig. 123: 4, 5; 126: 2–11; 127: 1–13) L.: 8–70 µm/l.: 8–30 µm/f. 35–60(70)/100 µm/ str. (16)17–19(20)/10 µm) . . . . **2. *S. brebissonii*** vol. 2/2 p. 179
16a 20 à 30/100 µm de côtes étroites s'avancent en prolongement des fibules jusqu'à une côte médiane étroite (vol. 2/2 Fig. 140: 1–3) L.: 70–140 µm/l.: 30–55 µm/f. 20–30/100 µm/str. 18–25/10 µm **16. *S. gemma*** vol. 2/2 p. 191
16b 6 à 15 ondulations transapicales/100 µm. Crêtes des ondulations presque aussi larges que les creux (vol. 2/2 Fig. 140: 4, 5) L.: 100–240 µm/l.: 50–160 µm/f. 6–15/100 µm/str. 14–20/10 µm **15. *S. striatula*** vol. 2/2 p. 190

2. Groupe des Robustae (vol. 2/2 p. 191)

Clef pour les espèces

1a Frustules fortement tordus autour de l'axe apical (vol. 2/2 Fig. 168: 1–7) L.: 40–220 µm/l.: 50–80 µm/str. env. 25/10 µm . **35. *S. spiralis*** vol. 2/2 p. 207
1b Frustules peu ou pas tordus autour de l'axe apical **2**
2a Axe apical isopolaire **3**
2b Axe apical hétéropolaire **13**

3a Valves délicates à structure fine, 40–75 canaux alaires/100 µm, extrémités nettes et le plus souvent étirées (vol. 2/2 Fig. 173: 1–8). Cf. . . . . ***Stenopterobia delicatissima*** vol. 2/2 p. 210
3b Extrémités non étirées **4**
4a Contour lancéolé **5**
4b Contour lancéolé à linéaire-lancéolé **8**
5a > 35 canaux alaires/100 µm (vol. 2/2 Fig. 154: 1–5; 155: 1) L.: 60–140 µm/l.: 20–32 µm/can.al. 25–50/100 µm . . . . . **21. *S. constricta*** vol. 2/2 p. 198
5b < 30 canaux alaires/100 µm **6**
6a Les ondulations partant des canaux alaires forment des cercles marginaux (vol. 2/2 Fig. 148: 1–4) L.: 32–64 µm/l.:9–15 µm/str. 24–30/10 µm . . . . **19. *S. birostrata*** vol. 2/2 p. 197
6b Valves autrement structurées, nettement ponctuées **7**
7a Aire médiane étroite-lancéolée, bords des crêtes des ondulations nettement pourvus de spinules (vol. 2/2 Fig. 145: 2–4; 146: 1–4; 147: 1–5; 150: 4–6) L.: 76–150 µm/l.: 30–60 µm/can. al. 12–22/10 µm **18. *S. bifrons*** vol. 2/2 p. 196
7b Aire médiane large et lancéolée. Ondulations peu visibles, spinules très nettes et irrégulièrement disposées dans l'aire (vol. 2/2 Fig. 152: 1–5) L.: 50–125 µm/l.: 33–50 µm/can. al. 15–30/100 µm **20. *S. turgida*** vol. 2/2 p. 97
8a 35–50 canaux alaires/100 µm **9**
8b < 32 canaux alaires/100 µm **10**
9a Espèce d'eau douce (vol. 2/2 Fig. 148: 5–9) L.: 22–61 µm/l.: 8–11 µm/str. 25–30/10 µm . . . . **23. *S. roba*** vol. 2/2 p. 200
9b Espèce d'eau saumâtre (vol. 2/2 Fig. 154: 1–5; 155: 1) L.: 60–140 µm/l.: 20–32 µm/can.al. 25–50/100 µm . . . . . **21. *S. constricta*** vol. 2/2 p. 198
10a Ondulations transapicales visibles jusqu'à l'aire médiane au microscope optique **11**
10b Ondulations transapicales marginalement indiquées au microscope optique **12**
11a Petites formes délicates à > 20 canaux alaires/100 µm. (vol. 2/2 Fig. 149: 1–9; 150: 1; 151: 1–4) L.: 20–120 µm/l.: 9–25 µm/str. 20–22/10 µm . . . . **22. *S. linearis*** vol. 2/2 p. 198
11b Grandes formes robustes à < 20 canaux alaires/100 µm. Souvent de nombreuses et très délicates spinules sur les crêtes des ondulations (vol. 2/2 Fig. 141: 1–3; 142: 1–5; 143: 1–9; 144: 1–3; 145: 1) L.: 80–400 µm/l.: 30–90 µm/str. 10–20/10 µm . . . . **17. *S. biseriata*** vol. 2/2 p. 195
12a En vue valvaire, pôles échancrés (vol. 2/2 Fig. 153: 5–8) L.: 60–135 µm/l.: 23–40 µm/can.al. 20–25/100 µm . . . **25. *S. barrowcliffia*** vol. 2/2 p. 200
12b Pôles sans échancrure (vol. 2/2 Fig. 153: 1–4) L.: 60–95 µm/l.: 16–20 µm/can.al. 28–35/100 µm . . . . **24. *S. didyma*** vol. 2/2 p. 200
13a Valves avec de véritables épines robustes avant le pôle large et parfois également avant le pôle étroit (vol. 2/2 Fig. 166: 1–4; 167: 1–4) L.: 120–350 µm/l.: 60–125 µm/can.al. 7–15/10 µm . . . . **32. *S. capronii*** vol. 2/2 p. 205
13b Epines typiques absentes **14**
14a > 35 canaux alaires/100 µm, ondulations transpicales peu ou marginalement visibles **15**
14b < 35 canaux alaires/100 µm, nettes ondulations transapicales également visibles au milieu de la valve **16**
15a 85–100 canaux alaires/100 µm, face valvaire presqu'hyaline (vol. 2/2 Fig. 151: 5–7) L.: 2–38 µm/l.: 8–12 µm/can.al. 85–100/100 µm/str. 32–37/10 µm . . . . **33. *S. suecica*** vol. 2/2 p. 206
15b 35–50 canaux alaires/100 µm, ondulations transapicales faibles (vol. 2/2

Fig. 151: 8, 9) L.: 22–46 µm/l.: 6,5–10 µm/can. al. 30–50/100 µm/str. insolv. . . . . **34. *S. tenuis*** vol. 2/2 p. 206

16a Nette projection des ailes **17**

16b Projection moins visible à invisible, grandes formes (vol. 2/2 Fig. 160: 5; 161: 1, 2; 162: 1–7; 163: 1–4) L.: 110–400 µm/l.:35–90 µm/can.al. 12–21/100 µm . . . . **31. *S. elegans*** vol. 2/2 p. 204

17a > 20 canaux alaires/100 µm (cf. aussi *S. splendida*) **18**

17b < 20 canaux alaires/100 µm **19**

18a Formes plus grandes avec côte médiane nette (vol. 2/2 Fig. 164: 1–4; 165: 1–3) L.: 40–185 µm/l.: 13–45 µm/can. al. 20–30/100 µm . . . . **29. *S. tenera*** vol. 2/2 p. 203

18b Formes plus petites, côte médiane étroite ou absente (vol. 2/2 Fig. 155: 2–9) . . . . **30. *S. bohemica*** vol. 2/2 p. 204

19a Face valvaire légèrement tordue (vol. 2/2 Fig. 160: 1, 2) L.: 70–165 µm/l.: 30–55 µm/can. al. 9–15/100 µm . . . . . . **27. *S. astridae*** vol. 2/2 p. 202

19b Face valvaire non tordue **20**

20a 9–15 canaux alaires/100 µm (le plus souvent 10 ou moins) ondulations transapicales très nettes (vol. 2/2 Fig. 156: 1–5; 157: 1–4) L.: 150–400 µm/l.: 50–150 µm/can. al. 7–12/100 µm . . . . **26. *S. robusta*** vol. 2/2 p. 201

20b Ondulations transapicales plus délicates **21**

21a Projection des ailes étroite et peu nette, le plus souvent 15 canaux alaires/ 100 µm (vol. 2/2 Fig. 158: 1–3; 159: 1–6; 160: 3, 4) L.: 75–250 µm/l.: 40–70 µm/can. al. 12–25/100 µm . . . . . **28. *S. splendida*** vol. 2/2 p. 202

21b Projection des ailes toujours nette (cf. **11b**) L.: 80–400 µm/l.: 30–90 µm/str. 10–20/10 µm . . . . **17. *S. biseriata*** vol. 2/2 p. 195

3. Genre *Stenopterobia* Brébisson in litt. ex Habirshaw *et al.* 1878 (vol.- 2/2 vol. 2/2 p. 207)

Typus generis: *Surirella anceps* Lewis 1863

Clef pour les espèces

1a Axe apical sigmoïde **2**

1b Axe apical droit (vol. 2/2 Fig. 173: 1–8; 174: 1–12) L.: 63–110 µm/l.: 4,5–7 µm/str. 26–30/10 µm . . . . **4. *St. delicatissima*** vol. 2/2 p. 210

2a Côtes de soutien marginales, aire médiane au moins 1/3 de la largeur valvaire (espèce nord-américaine) (vol. 2/2 Fig. 171: 1–4) L.: 140–320 µm/ l.: 9–12 µm/str. 14–16/10 µm . . . . **1. *St. anceps*** vol. 2/2 p. 208

2b Aire médiane très étroite mais nette **3**

3a Stries < 24/10 µm (vol. 2/2 Fig. 171: 5–9; 172: 1–3) L.: 70–280 µm/l.: 6–9 µm/fib. env. 30–60/100 µm . . . . **2. *St. curvula*** vol. 2/2 p. 209

3b Stries > 26/10 µm (vol. 2/2 Fig. 172: 4,6) L.: 63–110 µm/l.: 4,5–7 µm/str. 26–30/10 µm . . . . **3. St. *densestriata*** vol. 2/2 p. 210

4. Genre *Campylodiscus* Ehrenberg 1840 (vol.- 2/2 vol. 2/2 p. 211)

Lectotype: *Cocconeis* (?) *clypeus* Ehrenberg 1938 (Boyer 1927)

Clef pour les espèces

1a Face valvaire avec percées en forme de petites touches marginales, en rangées radiales plus ou moins irrégulières et qui sont +/- désordonnées au centre de la valve (vol. 2/2 Fig. 175: 1, 2; 176: 1–3) . . . . **1. *C. echeneis*** vol. 2/2 p. 213

1b Structure différente **2**
2a Creux des ondulations très étroits apparaissant comme des côtes étroites, radiales mais à contour très net **3**
2b Ondulations radiales irrégulières à creux plus larges **5**
3a Valves nettement en forme de selle et enroulées autour de l'axe apical, ondulations radiales 13–18/100 µm (vol. 2/2 Fig. 181: 4–6) L.: 100–125 µm/ l.: env. 70 µm/fib. 13–18/100 µm . . . . . . **6. *C. levanderi*** vol. 2/2 p. 215
3b Valves en forme de selle mais non enroulées autour de l'axe apical . . **4**
4a Ondulations radiales 20–30/100 µm, atteignant souvent le centre de la valve (vol. 2/2 Fig. 182: 1–5) . . . . **2. *C. noricus*** vol. 2/2 p. 213
4b Ondulations radiales 10–20/100 µm, laissant toujours un grand champ libre au centre (vol. 2/2 Fig. 175: 5; 179: 1–4; 180: 1–7; 181: 1–3) . . . . **3. *C. hibernicus*** vol. 2/2 p. 214
5a Ondulations radiales marginales interrompues par un anneau hyalin, concentrique (vol. 2/2 Fig. 175: 3, 4; 177: 1–5) . **4. *C. clypeus*** vol. 2/2 p. 214
5b Ondulations radiales marginales non-interrompues. Champ central structuré comme la zone du bord (vol. 2/2 Fig. 178: 1–6) . . . . **5. *C. bicostatus*** vol. 2/2 p. 215